T0295291

Advanced Carpentry: Frame and Finish

SIXTH EDITION

Pearson **NCCER** | National Center for Construction Education and Research

NCCER

President and Chief Executive Officer: Boyd Worsham
Vice President of Innovation and Advancement: Jennifer Wilkerson
Chief Learning Officer: Lisa Strite
Senior Manager, Curriculum Development: Chris Wilson
Production Manager: Graham Hack
Technical Writing Manager: Gary Ferguson
Technical Writers: Jo Ann Bartusik, Mitch Ryan
Art Manager: Bree Rodriguez
Technical Illustrators: Judd Ivines, Liza Wailes
Production Artist: Chris Kersten
Permissions Specialists: Adam Black, Sherry Davis

Pearson

Commercial Product Manager – Associations: Andy Dunaway
Senior Digital Content Producer: Shannon Stanton
Associate Project Manager: Monica Perez-Kim
Content Producer: Alexandrina Wolf
Development Editor: Nancy Lamm
Executive Marketing Manager: Mark Marsden
Designer: Mary Siener
Rights and Permissions: Jenell Forschler
Composition: Integra Software Services
Printer/Binder: Lakeside Book Company
Cover Printer: Lakeside Book Company
Text Typefaces: Palatino LT Pro and Helvetica Neue

Cover Image

The home pictured was framed and built in High Springs, FL, in 2020 by Amira Custom Homes. It was built for architect Jim West as a retirement home. As a personal home for an architect, it incorporates many fun features and design details, including a 14/12 roof pitch, many vaulted ceilings throughout the home, and three separate structures connected by covered walkways.

10 9 8 7 6 5 4 3 2 1

1 2024

Paper
ISBN-10: 0-13-817241-2
ISBN-13: 978-0-13-817241-1

Hardcover
ISBN-10: 0-13-817243-9
ISBN-13: 978-0-13-817243-5

National Center for
Construction Education
and Research

PREFACE

To the Trainee

Carpentry is an ancient trade. Discoveries from the earliest periods of civilization confirm that carpentry is one of the world's oldest trades. Although today's carpentry tools are different, many of the same building concepts, construction methods, and mathematical principles remain unchanged. The industry, however, has changed dramatically. What started out as a necessity for shelter has grown into one of the world's largest and most rewarding industries.

In a growing industry, regional variations in materials, building methods, and hazards made it necessary to protect the health and safety of construction professionals and the public. In addition to local codes and ordinances, the *International Building Code®* (*IBC®*) and *International Residential Code®* (*IRC®*) were created and are widely adopted to ensure safe and efficient building practices. These codes play an important role in all construction projects, and every carpenter must carefully consider how they affect each job.

Today's carpenters construct, repair, and install building frameworks and structures made from wood and a variety of other materials. Carpenters earn extremely competitive wages, and the demand for carpenters is expected to remain strong through the coming decade. This development is highlighted by recent reports from the US Department of Labor's Bureau of Labor Statistics that estimates many of the more experienced construction professionals will retire within the next 10 years. This trend, along with continued strong growth in many building sectors, is one of the reasons carpentry is an excellent choice for anyone pursuing a career in construction.

NCCER wishes you success as you embark on your advanced training in the carpentry craft, and we hope that you will continue your training outside of this series. By taking advantage of training opportunities as they arise, you will develop the confidence and creativity needed to succeed in the construction industry.

New with *Advanced Carpentry: Frame and Finish*

NCCER is proud to present *Advanced Carpentry: Frame and Finish*. This edition contains the following new modules: *The Building Process*, *Mass Timber Construction*, and *Interior Finish, Trim, and Cabinets*, which covers finish trim for doors, windows, walls, and ceilings. Additional information was added about structural insulated panels, fire-resistance-rated construction, and shear walls. The Suspended Ceilings module contains new information on acoustical tile and ceilings, and new content about exterior siding, flashing, soffits, and trim materials has been added to the Exterior Finish and Trim module.

We wish you success as you progress through this training program. If you have any comments on how NCCER might improve upon this textbook, please complete the User Update form using the QR code on this page. NCCER appreciates and welcomes its customers' feedback. You may submit yours by emailing support@nccer.org. When doing so, please identify feedback on this title by listing #Carpentry in the subject line.

Our website, www.nccer.org, has information on the latest product releases and training.

SCAN ME

NCCER Standardized Curricula

NCCER is a not-for-profit 501(c)(3) education foundation established in 1996 by the world's largest and most progressive construction companies and national construction associations. It was founded to address the severe workforce shortage facing the industry and to develop standardized training processes and curricula. Today, NCCER is supported by hundreds of leading construction and maintenance companies, manufacturers, and national associations. The NCCER Standardized Curricula were developed by NCCER in partnership with Pearson, the world's largest educational publisher.

Some features of the NCCER Standardized Curricula are as follows:

- An industry-proven record of success
- Curricula developed by the industry, for the industry
- National standardization providing portability of learned job skills and educational credits
- Compliance with the Office of Apprenticeship requirements for related classroom training (*CFR 29:29*)
- Well-illustrated, up-to-date, and practical information

NCCER maintains a secure online database that provides certificates, digital credentials, transcripts, and wallet cards to individuals who successfully complete programs under an NCCER-accredited organization or through one of NCCER's self-paced, online programs. This system also allows individuals and employers to track and verify industry-recognized credentials and certifications in real time.

For information on NCCER's credentials, contact NCCER Customer Service at 1-888-622-3720 or visit **https://www.nccer.org**.

Digital Credentials

Show off your industry-recognized credentials online with NCCER's digital credentials!

NCCER is now providing online credentials. Transform your knowledge, skills, and achievements into digital credentials that you can share across social media platforms, send to your network, and add to your resume. For more information, visit **www.nccer.org**.

Cover Image

Amira Custom Homes, located in Alachua, FL, was founded by Corey Amira in 2015. After spending his adult life working in residential construction, it was Corey's goal to build a custom home-building company dedicated to high-quality construction and helping clients build their dream homes. Corey enjoys being very active on the jobsites, ensuring every step goes to plan and the highest standards are adhered to throughout the build.

DESIGN FEATURES

Content is organized and presented in a functional structure that allows trainees to access the information where they need it.

Trainees can navigate *Advanced Carpentry* using color coded tabs on the upper right hand corner.

The Objectives list the skills and knowledge trainees need in order to complete the module successfully.

The Performance Tasks give you an opportunity to apply your knowledge to real-world tasks.

Section Openers provide a visual organizational structure for the information. Objectives and Performance Tasks are broken out for each section.

Trade Terms appear on the page adjacent to the text where they are first presented.

Step-by-step presentations and math equations help make the concepts clear and easy to grasp.

QR codes link trainees directly to digital resources that highlight current content.

MODULE 27215

Interior Finish, Trim, and Cabinets

Source: Hendrickson Photography/Shutterstock

Objectives

Successful completion of this module prepares you to do the following:

1. Describe the job preparation and safety planning required to complete a finish carpentry project.
   a. Identify tool and equipment safety guidelines for finish carpentry tools and materials.
   b. Explain how to perform a material takeoff for window, door, and ceiling trim.
2. Identify the different types of standard moldings and materials.
   a. Identify the different types of base, wall, and ceiling moldings.
   b. Identify the different types of window and door trim.
3. Explain how to install different types of molding and trim.
   a. Describe common techniques for cutting and fastening trim.
   b. Explain the process for installing door and window trim.
   c. Describe the steps for installing base and ceiling molding.
4. Identify common types of cabinets and countertops.
   a. Identify common types of wall, base, and island cabinets.
   b. Describe common types of countertops.
5. Identify common cabinet components and hardware.
   a. Describe various cabinet components, such as doors, drawers, and shelves.
   b. Identify common fasteners and hardware used in cabinet assembly.
6. Explain how to lay out and install a basic set of cabinets.
   a. Describe common surface preparation techniques required before cabinet installation.
   b. Explain how to install wall cabinets.
   c. Describe the process for installing base cabinets and countertops.

CODE NOTE

Codes vary among jurisdictions. Because of the variations in code, consult the applicable code whenever regulations are in question. Referring to an incorrect set of codes can cause as much trouble as failing to reference codes altogether. Obtain, review, and familiarize yourself with your local adopted code.

Performance Tasks

Under supervision, you should be able to do the following:

1. Make square and miter cuts to selected moldings using a hand miter box and a compound miter saw.
2. Make a coped joint using a jigsaw or coping saw.
3. Install interior door, window, base, and ceiling trim using a finish nailer.
4. Identify and install selected hardware on cabinet doors.
5. Identify and lay out various types of base and wall units following a specified layout scheme.
6. Install a selected type of cabinet and countertop.

Digital Resources for Carpentry

Scan this code using the camera on your phone or mobile device to view the digital resources related to this craft.

SCAN ME

	2.0.0 Types of Advanced Wall Systems
Performance Tasks	**Objective**
1. Construct a shear wall for light-frame construction.	Describe the different types of advanced wall systems.
2. Construct firewalls in accordance with specifications.	a. Describe structural insulated panels (SIPs) and explain the carpenter's role in their installation.
	b. Identify the carpenter's role in the installation of masonry veneer walls.
	c. Define fire-resistance-rated construction and explain methods used to develop wall systems that protect occupants from smoke and fire.
	d. Describe how shear walls reinforce buildings against lateral forces.

Trim: Millwork placed around wall openings and at wall intersections at floors and ceilings.

Moldings: A defined, decorative element that outlines the edges of a structure and covers gaps at intersections of walls, ceilings, floors, and openings.

Interior finish and trim installation are a few of the finishing touches that complete the building process. Because trim and cabinets are some of the most visually striking interior components, a highly trained, detail-oriented craft professional must install every trim piece with straight lines and tight-fitting joints to create a clean and professional final product.

This module covers the materials, equipment, and procedures used to install interior trim, cabinets, and countertops.

Trim is installed to create an attractive finished look and hide joints at the intersection of walls, floors, ceilings, and various wall coverings. Trim also protects vulnerable edges. Trim should blend with the overall design of the building or room. For modern residential and commercial buildings, the interior trim typically uses simple, modern moldings to create a clean aesthetic, while traditional architectural styles tend to use moldings with elegant and complex shapes and grooved profiles (*Figure 1*).

For this example, convert 45'-4$\frac{3}{8}$" to decimal feet.

Step 1 Convert the inch-fraction $\frac{3}{8}$" to a decimal. This is done by dividing the numerator of the fraction (top number) by the denominator of the fraction (bottom number). For this example, $\frac{3}{8}$" = 0.375".

Step 2 Add the 0.375" to 4" to obtain 4.375".

Step 3 Divide 4.375" by 12 to obtain

$$0.3646' = 0.36' \text{ (rounded off).}$$

Step 4 Add 0.36' to 45' to obtain 45.36';

thus, 45'-4$\frac{3}{8}$" equals 45.36'.

Converting Decimal Feet to Feet and Inches

To convert values given in decimal feet into equivalent decimal feet and inches, use the following procedure. For this example, convert 45.3646' to feet and inches (convert to the nearest $\frac{1}{8}$").

Step 1 Subtract 45' from 45.3646' = 0.3646'.

Step 2 Convert 0.3646' to inches by multiplying 0.3646' by 12 = 4.3752".

Step 3 Subtract 4" from 4.3752" = 0.3752".

Step 4 Convert 0.3752" into eighths of an inch by multiplying 0.3752" by 8 = 3.0016 eighths or, when rounded off, $\frac{3}{8}$". Therefore, 45.3646' = 45'-4$\frac{3}{8}$".

Important information is highlighted, illustrated, and presented to facilitate learning.

Placement of images near the text description and details such as callouts and labels help trainees absorb information.

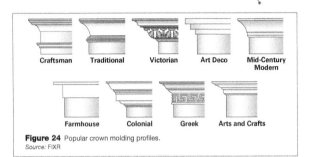

Figure 24 Popular crown molding profiles.
Source: FIXR

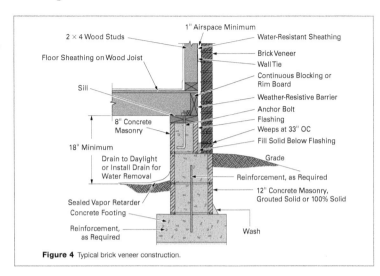

Figure 4 Typical brick veneer construction.

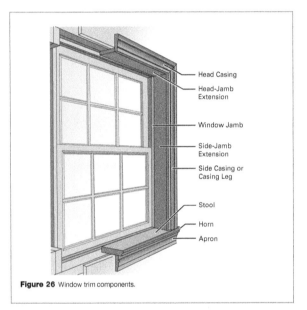

Figure 26 Window trim components.

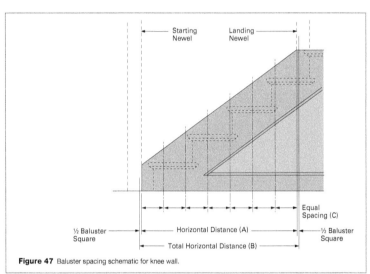

Figure 47 Baluster spacing schematic for knee wall.

New boxes highlight safety and other important information for trainees. Warning boxes stress potentially dangerous situations, while Caution boxes alert trainees to dangers that may cause damage to equipment. Notes boxes provide additional information on a topic.

WARNING!

Only trained and authorized personnel can operate a laser instrument. Inform all people within the operating range of the instrument that a laser instrument will be in use. Never look directly into a laser beam or point the beam into the eyes of others. Set up the laser transmitter so that the height of the emitted beam is above or below normal eye level.

Going Green looks at ways to preserve the environment, save energy, and make good choices regarding the health of the planet.

CAUTION

If the screwgun is running too fast, the tip of the screw may burn out before it penetrates the steel. Once the screw is properly seated, the screwgun automatically stops spinning the screw, preventing the screw from stripping.

NOTE

Fireproofing is not a code-defined term. However, it is used in the trade to explain the practice of applying fire protection material to extend the time something can be exposed to fire before it is compromised. In construction, fireproofing is referred to as *passive fire protection*, as discussed in Section 2.3.8.

Going Green

LEED

The US Green Building Council Leadership in Energy and Environmental Design (LEED) Green Building Rating System promotes a whole-building approach with criteria in five areas: sustainable site development, water savings, energy efficiency, material selection, and indoor environmental quality. Building systems can contribute toward points used to achieve the various LEED levels through their energy efficiency, materials credits, reduced building footprint, indoor air quality, and durability.

These boxed features provide additional information that enhances the text.

Guidelines for Fastening Trim

- Remove any pencil marks left along the edge of a cutline before fastening the trim. Pencil marks in interior trim corners are difficult to remove after the pieces are fastened into position. Pencil marks can also show through a stained or clear finish, making the joint appear open. When marking interior trim, make light, fine pencil marks to reduce your cleanup time.
- Ensure that all joints fit tightly by measuring, marking, and cutting carefully. Take pride in your work and never leave an interior trim project with misaligned trim and ill-fitting corners. If you make a mistake, take the time to replace trim pieces until you are satisfied with the final product.
- Repair any dings or dents left by finish nailers or hammers.
- Set finish nails below the surface of the trim and fill nail holes with putty matching the paint or stain finish. The set depth for finish nails should equal the nail head width. Prefinished molding often requires color-matching nails that are driven flush.

Did You Know?

Fire Protection Materials

The *International Building Code®* (*IBC®*) 2021, Section 202, defines types of fire protection materials, such as the following:

- *Intumescent Fire-Resistant Coatings:* Thin film liquid mixture applied to substrates by brush, roller, spray, or trowel that expands into a protective foamed layer to provide fire-resistant protection of the substrates when exposed to flame or intense heat.
- *Sprayed Fire-Resistant Materials:* Cementitious or fibrous materials that are sprayed to provide fire-resistance protection of the substrates.

Review questions at the end of each section and module allow trainees to measure their progress

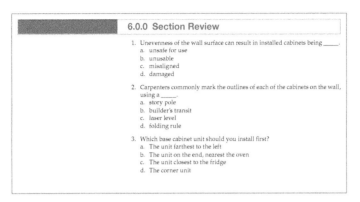

6.0.0 Section Review

1. Unevenness of the wall surface can result in installed cabinets being _____.
   a. unsafe for use
   b. unusable
   c. misaligned
   d. damaged

2. Carpenters commonly mark the outlines of each of the cabinets on the wall, using a _____.
   a. story pole
   b. builder's transit
   c. laser level
   d. folding rule

3. Which base cabinet unit should you install first?
   a. The unit farthest to the left
   b. The unit on the end, nearest the oven
   c. The unit closest to the fridge
   d. The corner unit

Module 27215 Review Questions

1. Maintaining an organized work area can reduce _____.
   a. trip hazards
   b. productivity
   c. waste
   d. safety planning

2. When changing or tightening router bits, you should always _____.
   a. fill out a hot work permit
   b. wear safety goggles
   c. disconnect the router from the power source
   d. notate the activity in a maintenance log

3. A key difference between a power miter saw and a compound miter saw is _____.
   a. the power miter saw uses smaller blades
   b. the compound miter allows you to tilt the blade at a vertical angle
   c. the compound miter saw has a rotating miter table for angled cuts
   d. a power miter saw only uses fine-toothed blades

NCCERconnect

This interactive online course is a unique web-based supplement in the form of an electronic book that provides a range of visual, auditory, and interactive elements to enhance training. Visit **www.nccerconnect.com** for more information!

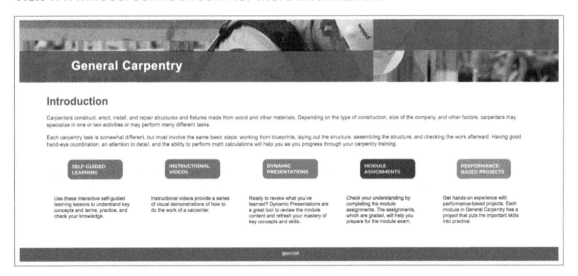

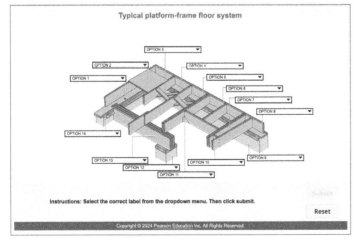

Typical platform-frame floor system

Instructions: Select the correct label from the dropdown menu. Then click submit.

1.1.3 Apprenticeship Standards

All apprenticeship standards prescribe certain work-related or on-the-job learning (OJL). This OJL is broken down into specific tasks in which the apprentice receives hands-on training while on the job. A specified number of hours are required in each task. The total amount of OJL (*Figure 3*) for a carpentry apprenticeship program is traditionally 8,000 hours, which amounts to four years of training and qualifies an individual to become a journeyman carpenter. In a competency-based program, it may also be possible to shorten this time by testing out of specific tasks through a series of performance exams.

Figure 3 On-the-job learning is an important part of an apprenticeship.

Source: Tony Vazquez, The Hubbard Construction Company

In a traditional program, the required OJL may be acquired in increments of 2,000 hours per year. Layoffs or illness may affect the duration. The apprentice must log all work time and turn it in to

ACKNOWLEDGMENTS

This curriculum was revised as a result of the vision and leadership of the following sponsors:

ClarkDietrich
The Haskell Company
International Code Council
Price Daniel Unit – Windham School District
Wayne Brothers

This curriculum would not exist without the dedication and unselfish energy of those volunteers who served on the Authoring Team. A sincere thanks is extended to the following:

Buddy Showalter
Don Allen
Jake Reece
Kendall Purvis
Michael Osborne
Tim Mosely
Tim Thompson

A final note: This book is the result of a collaborative effort involving the production, editorial, and development staff at Pearson Education, Inc., and NCCER. Thanks to all of the dedicated people involved in the many stages of this project.

NCCER PARTNERS

To see a full list of NCCER Partners, please visit
www.nccer.org/about-us/partners.

CONTENTS

Module 27404 Wall Systems and Installations

Module 27205 Steel Framing Systems

Module 27203 Thermal and Moisture Protection

Module 45104 Drywall Installation

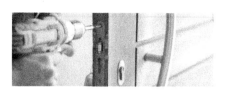

1.0.0   Door Installation Safety..364

    1.1.0   **General Safety Guidelines**... 364

    1.0.0   **Section Review**.. 365

2.0.0   Common Door Frames..365

    2.1.0   **Wood Door Frames**... 366

        2.1.1 Double Wood Door Frames 370

        2.1.2 Door-Stop Strips...371

    2.2.0   **Metal Door Frames**... 371

        2.2.1 Installing Welded Metal Door Frames in Wood-Framed
             Construction..373

        2.2.2 Sound Attenuation and Grouting for Metal Frames 374

        2.2.3 Installing Welded Metal Door Frames in Steel-Framed
             Construction..374

        2.2.4 Installing Unassembled Metal Door Frames in Drywall
             Construction..375

        2.2.5 Installing Welded Metal Door Frames in Existing Masonry
             Construction..376

        2.2.6 Installing Welded Metal Door Frames in New Masonry
             Construction..378

    2.3.0   **Door Swing** 379

    2.0.0   **Section Review**.. 381

3.0.0   Residential Door and Hardware Installation 381

    3.1.0   **Residential Doors** 381

        3.1.1 Bypass Doors 381

        3.1.2 Bifold Doors..382

        3.1.3 Pocket Doors...383

        3.1.4 Folding Doors 384

        3.1.5 Metal Doors..385

        3.1.6 Doors between Dwellings and Garages............................386

        3.1.7 Garage Doors 387

    3.2.0   **Residential Door Hardware**....................................... 388

        3.2.1 Door Hinges 388

        3.2.2 Locksets...390

    3.3.0   **Residential Door and Door Hardware Installation** 391

        3.3.1 Door Jack...391

        3.3.2 Manufactured Prehung Door-Unit Installation.................391

        3.3.3 Door-Hinge Installation 394

Module 27209 Suspended and Acoustical Ceilings

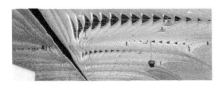

Module 27405 Finished Stairs

Module 27204 Exterior Finish and Trim

Module 46100 Introduction to Leadership

Appendices

The Building Process

Source: unkas_photo/Getty Images

Objectives

Successful completion of this module prepares you to do the following:

1. Understand the primary codes and standards used in the construction industry.
   a. Identify the primary codes used in the construction industry.
   b. Identify the standards that apply to the construction industry.
2. Explain the primary functions within the construction industry.
   a. Describe the construction sectors.
   b. Describe the roles of individuals in the construction industry.
3. Describe the construction process.
   a. Explain the stages of the construction process.

Performance Tasks

This is a knowledge-based module. There are no Performance Tasks.

Overview

Construction is about much more than just building something. It includes all the steps from idea through finished product. Every construction project requires planning, budgeting, code compliance, building, and oversight to ensure its successful completion. This module examines the codes and standards that apply to construction projects, the various types of construction projects, the people involved in construction projects, and the stages of the building process for construction projects.

Digital Resources for Carpentry

Scan this code using the camera on your phone or mobile device to view the digital resources related to this craft.

Performance Tasks	# 1.0.0 Codes and Standards

There are no Performance Tasks in this section.	**Objective**
	
	Understand the primary codes and standards used in the construction industry.
	  a. Identify the primary codes used in the construction industry.
	  b. Identify the standards that apply to the construction industry.

Building codes and standards guide the design of construction projects as well as the work involved during construction. Construction specifications are written to conform to building codes and the best practices of the construction trade. Model building codes provide minimum standards to guard the life, health, and safety of the public by regulating and controlling the design, construction, and quality of materials used in construction. Model building codes have also come to govern the use and occupancy, location of a type of building, and the ongoing maintenance of buildings and facilities. Once adopted by the local jurisdiction, building codes become the legal instruments that enforce public safety in construction of any building or structure where a person or group of people might enter. They are used not only in the construction industry but also by the insurance industry for compensation appraisals and claims adjustments, and by the legal industry for court litigation.

Did You Know?

Codes Versus Standards

Codes generally are rules, suggested practices, or recommendations. They clarify what needs to be done and can become law when they are adopted by a government agency.

    Standards typically set testing methodology, material specifications, guidance documents, and practices. They clarify how something should be done. Standards can become law if they are referenced by the code.

1.1.0 Building Codes

Until 2000, there were three model building codes. The three code-writing groups—Building Officials and Code Administrators® (BOCA®), International Conference of Building Officials® (ICBO®), and Southern Building Code Congress International® (SBCCI®) combined their efforts under the umbrella of the International Code Council® (ICC®) with the purpose of writing one nationally accepted family of building and fire codes. The first editions of the *International Residential Code®* and *International Building Code®* were published in 2000 and continue to be updated on a three-year cycle.

- The *International Residential Code®* (IRC®) addresses the design and construction of one- and two-family dwellings and townhouses (*Figure 1*).
- The *International Building Code®* (IBC®) pertains to all other building construction not covered by the IRC (*Figure 2*).

Model building codes are written to be just that: models. Local jurisdictions, such as states, counties, cities, boroughs, or townships, model their building codes after the *IRC®/IBC®* and adopt them as their own. Once adopted by a local jurisdiction, the model building codes then become law. It is common for local jurisdictions to change and add new requirements to adopted model code requirements to provide more stringent requirements and/or meet local needs.

Figure 1 *IRC®* topics.

The provisions of the building codes apply to the construction, alteration, movement, demolition, repair, structural maintenance, and use of any building or structure within the local jurisdiction. An important general rule about codes is that in almost every case the most stringent local code will apply.

The *ICC®* also publishes the *International Energy Conservation Code®* (*IECC®*) on a three-year update cycle. The *IECC®* covers energy efficiency, which includes cost, energy usage, use of natural resources, and the impact on the environment. The *IECC®* provides separate provisions for Commercial and Residential construction.

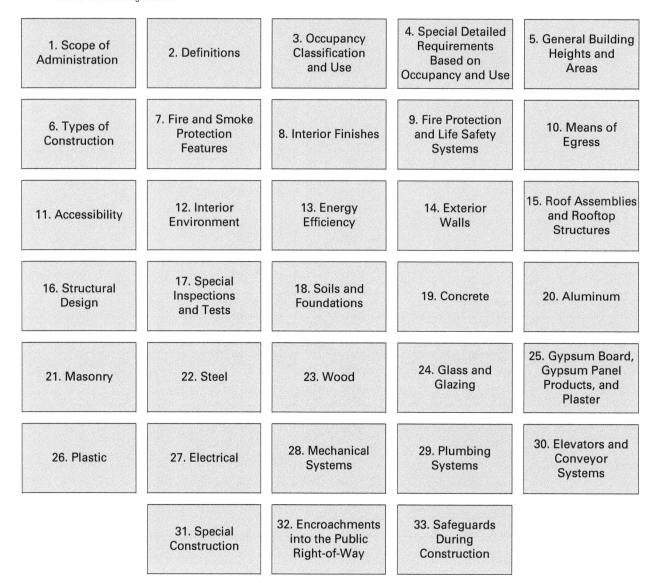

1. Scope of Administration	2. Definitions	3. Occupancy Classification and Use	4. Special Detailed Requirements Based on Occupancy and Use	5. General Building Heights and Areas
6. Types of Construction	7. Fire and Smoke Protection Features	8. Interior Finishes	9. Fire Protection and Life Safety Systems	10. Means of Egress
11. Accessibility	12. Interior Environment	13. Energy Efficiency	14. Exterior Walls	15. Roof Assemblies and Rooftop Structures
16. Structural Design	17. Special Inspections and Tests	18. Soils and Foundations	19. Concrete	20. Aluminum
21. Masonry	22. Steel	23. Wood	24. Glass and Glazing	25. Gypsum Board, Gypsum Panel Products, and Plaster
26. Plastic	27. Electrical	28. Mechanical Systems	29. Plumbing Systems	30. Elevators and Conveyor Systems
	31. Special Construction	32. Encroachments into the Public Right-of-Way	33. Safeguards During Construction	

Figure 2 *IBC*® topics.

The *National Fire Protection Association*® (*NFPA*®) also publishes codes that apply to the construction industry. *NFPA*® offers a variety of services to the public (*Figure 3*) to promote its primary mission of providing information and training to minimize the risks associated with fire, electrical, and other hazards. *NFPA 70*®, also known as the *National Electrical Code*® (*NEC*®), sets minimum standards for the safe installation of electrical wiring and equipment and is one of the most important tools for an electrician. The *NEC*® is also updated every three years. The primary purpose of the *NEC*® is to keep people and property safe from the dangers of electricity.

NOTE

Keep in mind that the *IBC*®, *IRC*®, *IECC*®, and *NEC*® specify only minimum requirements. Always check the local codes or job requirements, which may be more stringent.

Did You Know?

State and Local Codes

Many state and local governments also have their own codes that apply to the construction industry. These can be found on the *ICC*® (**https://codes.iccsafe.org**) by searching Locations or typing in the location you are looking for.

Some locations adopted the codes as they are written, while others adopt amended versions. For example, Pennsylvania's adopted codes are called the Uniform Construction Code.

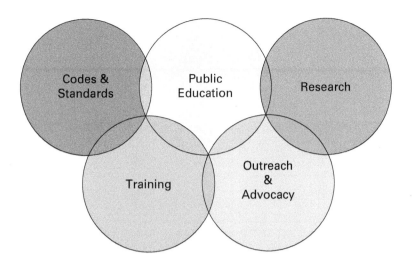

Figure 3 *NFPA®* services.

Here's the good news: The codes that exist when project plans are submitted are the ones that are enforced once a project is completed. That means that if the codes change during construction, the contractor won't be required to alter the project to meet the new codes. That said, if new plans get submitted, or if additions or alterations happen after the job is done, the contractor will have to follow the new code.

1.2.0   Standards

Standards tend to be more technical than building codes. Standards are often referenced within building codes because they generally provide more details about systems or products, how to perform the work, and how the final system or product must work.

There are various organizations that develop standards for the construction industry to establish testing methodologies and methods, as well as specifications for materials used. Standards also provide guidance to define quality and establish safety and performance criteria.

The *IBC®* and *IRC®* each provide a list of standards organizations that are referenced throughout the code. *Figure 4* shows a list of all the standards that are used in either or both codes. The full list of organizations and the standards referenced can be found in *Chapter 35* of the *IBC®* and *Chapter 44* of the *IRC®*.

The following are just a few examples of the organizations that provide standards specific to the construction industry and are referenced by the code:

- *ASHRAE* – ASHRAE serves as a source of technical guidance and standards for heating, ventilation, and air conditioning (HVAC). Some of the most popular ASHRAE industry standards cover ventilation and indoor air quality, energy standards, design and construction of green buildings, and energy efficiency. ASHRAE also offers the following professional certifications:
  - Building Commission Professional
  - Building Energy Assessment Professional
  - Building Energy Modeling Professional
  - High-Performance Building Design Professional
  - Healthcare Facility Design Professional
  - Operations and Performance Management Professional

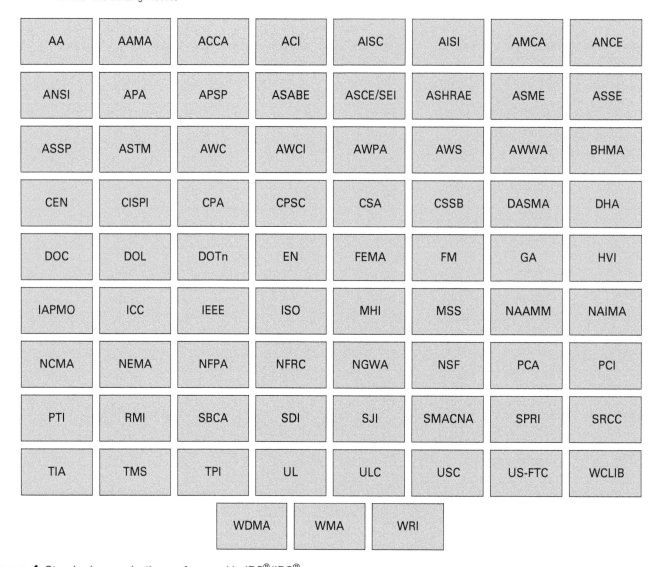

Figure 4 Standards organizations referenced in *IBC®/IRC®*.

- *ASTM International* – ASTM International is the largest voluntary standards development organization in the world. It develops technical standards for materials, products, systems, and services. Along with standards development, ASTM offers the following:
  - Product Certification with the Safety Equipment Institute
  - Personnel Certifications
  - Environmental Product Declarations
  - ASTM Training and eLearning
- *UL* – UL is a global leader in applied safety science and has testing, inspection, and compliance services geared towards the building and construction industry. The UL enterprise has three organizations that share its mission: UL Research Institute, UL Standards & Engagement, and UL Solutions. UL Solutions is the commercial business and what most people are familiar with. Its focus is to help solve safety, security, and sustainability challenges. UL Solutions services include testing and certification to meet applicable standards and requirements in the following areas:
  - Personnel
  - Products
  - Facilities
  - Processes
  - Systems

There are also organizations that promote green building standards for the construction industry. While these organizations are not referenced by the *IBC®/IRC®*, many construction companies choose to participate in their standards programs. Here are two examples of such organizations:

- *Green Building Initiative (GBI)* – As an international non-profit organization and Accredited Standards Developer through ANSI, Green Building Initiative (GBI) works to reduce climate impacts by improving the building environment (*Figure 5*). GBI provides the Green Globes® certification program and the federal Guiding Principles Compliance (GPC) building certification and assessment program. Green Globes® is a science-based certification system to evaluate environmental sustainability, health and wellness, and resilience in commercial real estate. The system allows building owners to determine which sustainability features best fit their building project. Guiding Principles Compliance (GPC) is a third-party assessment program for federal agencies to assess their building projects' compliance with the federal Guiding Principles. GBI offers certification in the following areas:

  ○ New Construction
  ○ Core & Shell
  ○ Sustainable Interiors
  ○ Existing Buildings
  ○ Multi-Family New Construction
  ○ Multi-Family Existing Construction
  ○ Global Real Estate Sustainability Benchmark

Figure 5 Improving the building environment.
Source: Jacob_09/Shutterstock

- *LEED* – The purpose of LEED is to encourage the adoption of sustainable construction and building management standards as established by the US Green Building Council (USGBC). The basic concept of LEED is to provide a template for building design and construction that is environmentally and socially responsible. There are eight different LEED rating systems, based on the type of construction project. While each rating system has different criteria, each is based on a point or credit structure. The project goes through a verification and review process by Green Business Certification, Inc.™ to earn points/credits to achieve the LEED certifications: Certified, Silver, Gold, or Platinum (*Figure 6*). Building owners can work toward certification in many ways, including the following:

- Using recyclable material during construction
- Minimizing construction waste
- Creating an energy-efficient building envelope that minimizes heating and cooling losses while maximizing occupant health and comfort
- Ensuring excellent indoor air quality
- Using energy-efficient equipment and appliances
- Installing automated plumbing controls for fixtures that reduce water usage
- Capturing and using rainwater for horticultural use
- Ensuring that all building construction characteristics and operations are sustainable without significant environment impact

Figure 6 LEED certification.
Source: Tada Images/Shutterstock

1.0.0 Section Review

1. How often are the *IBC*® and *IRC*® updated?
   a. Every 2 years
   b. Every 3 years
   c. Every 5 years
   d. Every 6 years

2. What do building standards explain?
   a. What needs to be done
   b. Electrical safety
   c. How to perform the work
   d. Federal regulations

2.0.0 Organization within the Construction Industry

Objective

Explain the primary functions within the construction industry.

  a. Describe the construction sectors.
  b. Describe the roles of individuals in the construction industry.

Performance Tasks

There are no Performance Tasks in this section.

While we often think of a construction project as building something, it is actually a much larger process. There are many different types of construction projects, and each one requires an entire team to ensure its successful completion. The type and size of the project will determine the team that is needed and the role those individuals will play in its completion.

2.1.0 Sectors of the Construction Industry

Four primary sectors make up the construction industry (*Figure 7*). Each sector includes specific types of construction, materials, equipment, and skills. The four sectors are:

(A) Residential

(B) Commercial

(C) Industrial

(D) Heavy Civil/Infrastructure

Figure 7 Construction sectors.
Sources: RichLegg/Getty Images (7A), Starcevic/Getty Images (7B), hdere/Getty Images (7C); RZUS_Images/Alamy Stock Photo (7D)

- *Residential* – Includes design, construction, and maintenance of single-family homes and multi-family homes, such as apartment buildings, public housing developments, and even separate garages and sheds
- *Commercial* – Includes design, construction, and maintenance of schools, government buildings, medical facilities, hotels, sports arenas and stadiums, shopping centers, and large office buildings
- *Industrial* – Includes construction of businesses that produce something for distribution, such as manufacturing plants, oil refineries, electrical generating plants, chemical processing plants, and large mills
- *Heavy civil/Infrastructure* – Often referred to as horizontal construction; includes bridges, roadways, airports, tunnels, and dams

Each construction sector has its own approach to initiating and paying for projects, selecting the required equipment, and ensuring its workforce is adequately trained and prepared for the job.

2.2.0 Roles in the Construction Industry

The construction industry employs more people and contributes more to the nation's economy than any other industry. Our society will always need new homes, highways, bridges and infrastructure, airports, hospitals, schools, factories, and office buildings. It takes an entire team (*Figure 8*) to complete a successful construction project.

Figure 8 Construction team.
Source: Robert Kneschke/Shutterstock

The size and makeup of this team depends on the size of the project, the type of project, and the size of the construction company working on the project. Larger companies and projects may require separate people to fulfill each role, while smaller companies and projects may combine some roles. The primary roles of the team for a construction project include:

- *Project owner* – The project owner initiates the project and usually finances the building endeavor. This can be an individual, a company, an organization, or a government agency.
- *Architect* – The architect meets with the project owner to transform the owner's thoughts and ideas into construction plans and specifications. The objective is to design a project that meets the owner's needs (*Figure 9*). The architect may also visit the construction site during the planning stage and throughout the project.

Figure 9 Architect.
Source: Iain Masterton/Alamy Stock Photo

- *Engineer* – The engineer designs, plans, and builds the construction project (*Figure 10*). The engineer may also assist with developing the budget for the project. Some projects may also require specialized engineers, such as civil, electrical, structural, or mechanical.

Figure 10 Engineering design.
Source: KRAUCHANKA HENADZ/Shutterstock

- *Building official/Plan examiner* – Construction plans and specifications are submitted to the building official/plan examiner for approval before a construction project can begin. The building official/plan examiner will review the plans to make sure they comply with all codes and regulations before issuing the necessary permits for the project.

- *Fire official* – The fire official inspects building plans to ensure they meet all fire codes. Fire officials may also inspect the building during and/or after construction to make sure fire codes are being met.

- *Cost estimator* – The cost estimator will review the approved construction documents to identify the factors that affect cost, such as time, materials, and labor. Based on this information, the cost estimator will prepare an estimate of the cost for the project, recommend ways to reduce costs, and prepare a budget. Throughout the project, cost estimators will also analyze the costs to keep the project on budget. Estimators have a great deal of responsibility, because errors in estimates can cost the contractor significant amounts of money. Depending on the size and type of the business, the estimating job may also be done by the contractor or the project manager.

- *Scheduler/Planner* – The scheduler/planner is responsible for creating a timetable for the construction project to keep it on schedule (*Figure 11*). The scheduler/planner also assesses the progress of the work and addresses any issues that may arise that would prevent the project from being completed on schedule.

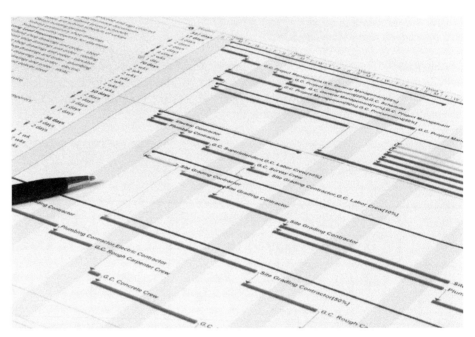

Figure 11 Planning and Scheduling.
Source: Paul Barnwell/Alamy Stock Photo

- *Contractor* – The contractor oversees many parts of a construction project, including resource management, budgeting, code adherence, quality assurance, and construction materials. Depending on the size of their business, contractors may work with the crew or may manage the business full time.

- *Subcontractor/Specialty contractor* – The subcontractor/specialty contractor employs craft professionals with specialized skills to complete specific tasks on a construction project, such as electrical, HVAC, plumbing, masonry, and site layout.

- *Project manager* – The project manager works with the rest of the construction management team (*Figure 12*) to control the scope and direction of a construction project and oversee the planning and delivery of the project on time and within budget. Large contracting companies and large construction projects may have several project managers.

Figure 12 Construction management team.
Source: Tashi-Delek/Getty Images

- *Safety manager* – The safety manager is another key onsite person. Safety managers are responsible for the safety and health of the workers on a project (*Figure 13*). They are responsible for developing safety plans and procedures, training workers in safety procedures, and ensuring that the project complies with applicable health and safety regulations.

Figure 13 Construction safety.
Source: John Williams RUS/Shutterstock

- *Site superintendent* – The site superintendent manages the day-to-day activities on a construction site and supervises work performed by the subcontractors. Depending on the size of the company and project, the site superintendent may also be involved in estimating and scheduling.

- *Foreman/Crew leader* – The foreman/crew leader is a front-line leader who directs the work of a crew of craftworkers (*Figure 14*). It is the foreman/crew leader's job to make sure that the work is completed correctly and on time. Each construction crew will have a separate foreman/crew leader. Foremen/crew leaders are also responsible for the safety of those under them.

Figure 14 Construction crew.
Source: Bannafarsai_Stock/Shutterstock

- *Master craftsman* – A master craftworker is one who has achieved and continuously demonstrates the highest skill levels in the trade. Master craftworkers can act as mentors and teachers to less experienced members of the craft. Check your location for Master requirements.

- *Journeyman* – A craftworker can become a journeyman after successfully completing an apprenticeship program or by having documented sufficient skills and knowledge in the craft. Check your location for journeyman requirements. A person can remain a journeyman or advance in the trade. Journeymen may have additional duties, such as supervisor or estimator. With larger companies and larger jobs, journeymen often become specialists.

- *Craftworker* – The craftworker serves as a physical laborer on a construction project. Craftworkers work in skilled occupations, such as electrician, carpenter, plumber, HVAC technician, painter, welder, heavy equipment operator, or roofer. In these occupations, craftworkers perform a wide variety of duties and operate all kinds of hand and power tools.

- *Inspector* – The inspector may work for a contractor or for the authority having jurisdiction (AHJ), such as a local or state government. Inspectors ensure that all construction follows applicable codes and quality requirements. Inspectors review and approve work at various stages of construction, depending on the project (*Figure 15*). A project may also require specialty inspectors for certain areas, such as electrical, mechanical, plumbing, elevators, and public works.

Figure 15 Building inspection.
Source: KomootP/Shutterstock

2.0.0 Section Review

1. Which construction sector includes apartment buildings, garages, and sheds?
   a. Commercial
   b. Heavy civil/infrastructure
   c. Industrial
   d. Residential

2. On a construction team, who is the front-line leader who directs the work of the craftworkers?
   a. Journeyman
   b. Foreman/crew leader
   c. Site superintendent
   d. Master craftworker

3.0.0 The Construction Process

Objective	Performance Tasks
Describe the construction process. a. Explain the stages of the construction process.	There are no Performance Tasks in this section.

The construction process includes all the steps surrounding building something. Every construction project starts with an idea and evolves through a series of stages until it is a completed project. Some parts of the construction process will need to be completed in a specific sequence. However, there will often be many tasks going on at the same time, and often overlapping with each other. Depending on the size of the project, the construction process can take weeks, months, or even years.

3.1.0 Stages of the Construction Process

Understanding the stages of the construction process is a crucial part of having a successful construction project. The construction process can be broken down into three main stages (*Figure 16*), with each having multiple steps. Different companies may approach the process differently or even have different names for the steps in the process, but they will all work through the same stages.

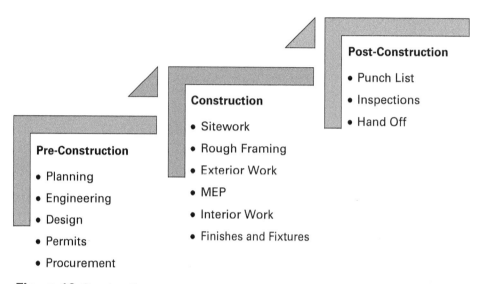

Pre-Construction

- Planning
- Engineering
- Design
- Permits
- Procurement

Construction

- Sitework
- Rough Framing
- Exterior Work
- MEP
- Interior Work
- Finishes and Fixtures

Post-Construction

- Punch List
- Inspections
- Hand Off

Figure 16 Construction process.

3.1.1 Pre-Construction

Every construction project begins with the owner having an idea of something to build (*Figure 17*). The owner needs to hire someone to plan, design, and build this project. Typically, this will be a general contractor, an architect, and/or an engineer. These individuals can work for the same company, or they can work for separate companies.

Figure 17 Project conception.
Source: Vladimir Vladimirov/Getty Images

There are several tasks that need to be accomplished during the pre-construction phase. The type and size of the project will determine which tasks apply to the project. Typical pre-construction tasks include:

- *Planning and budgeting* – Planning and budgeting are ongoing tasks in the construction process. Planning involves selecting the site and determining its feasibility for the project. It also sets a schedule for the project, as well as work and completion schedules for the individual tasks throughout the project. Budgeting includes the initial steps of acquiring the land and designing the structure. It should also determine actual costs associated with site preparation, building permissions, project management, materials, and labor.

- *Surveying* – Surveying looks at the condition of the land for the project to mark the boundaries of the plot and examine the layout of buildings, roads, electricity lines, gas lines, water lines, and any other infrastructure near the construction site (*Figure 18*).

Figure 18 Surveying.
Source: Sorn340 Studio Images/Shutterstock

- *Designing* – Designing is when an architect creates the plans and drawings for a project (*Figure 19*). This process can go through several phases before resulting in a finished design.

- *Engineering* – Engineers focus on the infrastructure and systems that make up the building plans. The scope of the project will determine what types of engineers are needed: civil, structural, electrical, plumbing, mechanical, and/or energy.

- *Approvals* – A variety of approvals, permits, and inspections will be needed for a construction project. The plans and specifications get submitted to a building official/plan examiner for approval before they can become the approved construction documents. At this point, the contractor will also need to apply for permits and schedule any necessary pre-construction inspections.

- *Procurement* – Procurement involves acquiring all the materials, services, and labor needed to complete the construction project on time and within budget. This includes hiring subcontractors; securing necessary services, such as water and electricity; and ensuring materials will be available when they are needed.

Did You Know?

Lean Construction

Lean construction focuses on ways to design and build projects in a manner that will maximize value and minimize waste of materials, effort, or time. Practicing lean construction is a company-wide effort. It requires the company to examine each step of the construction process to look for ways that it can be more efficient and that any waste of resources can be minimized. The benefits to the company include increased efficiency, enhanced safety, reduced costs, and quicker turnaround times.

There are several organizations that have lean construction training and certification programs, such as:

- Associated General Contractors of America (AGC) – Lean Construction Educating Program
- Lean Construction Institute (LCI)

Figure 19 Project design.
Source: gualtiero boffi/Shutterstock

3.1.2 Construction

The actual building occurs during the construction stage of the project. There will be many different tasks going on during this phase, many of them overlapping with each other. The primary tasks of the construction phase include:

- *Inspections* – Inspectors working for the company and the authority having jurisdiction will conduct inspections throughout the construction project. During the construction process, this may include inspections for site conditions and code compliance as well as to make sure the approved construction documents are adhered to.

- *Site setup* – Setting up the construction site often begins with demolition of any existing structures and site cleaning and leveling. Once the site is prepared, it needs to be set up for the current construction project. This often includes:
  - Installing a fence as a boundary around the plot
  - Setting up an office for the team
  - Setting up a storage area for tools and materials
  - Establishing power and water supplies for use during construction

- *Sitework and foundation* – This is the point where the project finally breaks ground. Sitework needs to be completed before the foundation can be built. The sitework can include grading the land; developing driveways for deliveries and workers; and digging or drilling for foundations, footings, wells, and underground utilities. Once the sitework is complete, construction of the foundation can begin (*Figure 20*). The type of foundation required for the project is a major consideration for how much sitework is needed.

- *Rough framing* – Rough framing is when a building starts to go vertical (*Figure 21*). In most buildings, the rough framing is constructed with either wood or steel. The walls are erected first and then attached to the floors and ceilings. This is a crucial part of the construction process because it sets the foundation for the entire building.

Figure 20 Building the foundation.
Source: georgeclerk/Getty Images

Figure 21 Rough framing.
Source: photovs/123RF

Reduce, Reuse, Recycle, and Rebuy

Construction and demolition (C&D) creates a significant amount of waste in the United States. The Environmental Protection Agency (EPA) promotes Sustainable Materials Management (SSM) in an effort to reduce the amount of waste generated through C&D. There are several ways companies can help reduce C&D waste:

- Reducing materials
- Reusing materials
- Recycling materials
- Rebuying materials

For more information on Sustainable Materials Management, visit the EPA **https://www.epa.gov** or your local government agency.

- *Exterior work* – The primary focus of exterior work is to "dry in" the building, or seal it off from the elements (*Figure 22*). During this phase, there will be several tasks going on at the same time:
  - Installing OSB on the exterior
  - Installing windows and doors
  - Installing roofing
  - Installing exterior insulation and siding
  - Completing any masonry, brick, or plasterwork

Figure 22 Exterior work.
Source: Cynthia Farmer/Alamy Stock Photo

- *MEP (Mechanical, Electrical, Plumbing)* – Once the building envelope is sealed, specialty subcontractors can begin working on the mechanical, electrical, and plumbing requirements for the building (*Figure 23*). These specialty subs will each focus on a different aspect of the building:
  - Mechanical contractors will install boilers, air handlers, ductwork, and other HVAC-related components.
  - Electricians will install panels, generators, conduit, and various other electrical devices as well as pull wire through the building.
  - Plumbers will run water supply pipes as well as waste, drain, and vent pipes.

(A) Mechanical **(B) Electrical** **(C) Plumbing**

Figure 23 MEP.
Sources: photovs/Getty Images (23A); photovs/Getty Images (23B); photovs/Getty Images (23C)

- *Interior work* – The primary focus of interior construction is to finish off the interior components of the structure (*Figure 24*). Interior work includes:
  - Installing insulation
  - Installing and finishing drywall
  - Hanging ceilings
  - Laying floors
  - Hanging interior doors

Figure 24 Interior work.
Source: photovs/123RF

- *Finishes and fixtures* – At this point, much of the work is complete, but many parts of the building will need finishing touches to complete the project (*Figure 25*). Some of the tasks in this phase include:

Figure 25 Finishing work.
Source: Feverpitched/Getty Images

- ○ Painting walls
- ○ Installing floor coverings
- ○ Installing light switches, outlets, and fixtures
- ○ Installing sinks, toilets, and faucets
- ○ Installing countertops and cabinets
- ○ Completing and testing the HVAC setup

3.1.3 Post-Construction

After the major stages of the construction process are complete, the contractor needs to wrap up the project. The primary tasks for post-construction include:

- *Punch list* – The contractor creates a punch list of any final work items remaining as well as any work that does not conform to the approved construction documents. This work needs to be completed or fixed before the final handoff.

- *Inspections* – Inspectors working for the company and the authority having jurisdiction will conduct a final inspection of the construction project. For some projects, this may include specialty inspectors for certain areas, such as electrical, mechanical, plumbing, elevators, fire protection systems, and public works. The construction project must pass these final inspections and the *Certificate of Occupancy* issued for the project to be considered complete.

- *Handoff* – The last step of the construction project is the handoff (*Figure 26*). During this step, the contractor should walk the owner through the entire building and go over how to work every system in the building. The contractor should also make sure the owner has copies of the as-built drawings; operations manuals for all technical systems; and all applicable permits, approvals, and licenses. Finally, all final paperwork will be signed, and payments made, bringing the construction project to an end.

Certificate of Occupancy

The required information for a *Certificate of Occupancy* is clearly defined in *Section 111.2* of the *IBC*® and *Section 110.3* of the *IRC*®.

Figure 26 Handoff.
Source: Ridofranz/Getty Images

3.0.0 Section Review

1. During which stage of a construction project does surveying take place?
   a. Pre-construction
   b. Conception
   c. Construction
   d. Post-construction

2. Which step in the construction process includes installing a boundary fence and preparing an onsite office?
   a. Sitework
   b. Procurement
   c. Site setup
   d. Exterior work

Module 27213 Review Questions

1. What industry uses model codes for compensation appraisals and claims adjustments?
   a. Construction
   b. Legal
   c. Government
   d. Insurance

2. Which code addresses the design and construction of one- and two-family dwellings and townhouses?
   a. $IBC^{®}$
   b. $IRC^{®}$
   c. $NEC^{®}$
   d. $IECC^{®}$

3. Which code sets the minimum standard for the safe installation of electrical wiring and equipment?
   a. $IECC^{®}$
   b. $NEC^{®}$
   c. $IRC^{®}$
   d. $IBC^{®}$

4. Which organization serves as a source of technical guidance and standards for HVAC?
   a. ASTM International
   b. LEED
   c. UL
   d. ASHRAE

5. Which organization works to reduce climate impacts by improving the building environment?
   a. LEED
   b. GBI
   c. UL
   d. ASHRAE

6. What determines the team that is needed for a construction project?
   a. Construction company
   b. Cost of the project
   c. Length of the project
   d. Type and size of the project

7. Which construction sector includes schools, stadiums, and shopping centers?
   a. Residential
   b. Heavy civil/infrastructure
   c. Commercial
   d. Industrial

8. Who transforms the owner's thoughts and ideas into construction plans and specifications?
   a. Engineer
   b. Architect
   c. Foreman/crew leader
   d. Scheduler/planner

9. Who inspects the building plans to ensure they meet all fire codes?
   a. Building official/plan examiner
   b. Safety manager
   c. Inspector
   d. Fire official

10. Who oversees the planning and delivery of the project on time and within budget?
    a. Project manager
    b. Scheduler/planner
    c. Cost estimator
    d. Foreman/crew leader

11. Who manages the day-to-day activities on a construction site and supervises work performed by the subcontractors?
    a. Project manager
    b. Foreman/crew leader
    c. General contractor
    d. Site superintendent

12. Which task examines the layout of buildings, roads, electricity lines, gas lines, water lines, and other infrastructure near the construction site?
    a. Surveying
    b. Engineering
    c. Planning and budgeting
    d. Designing

13. Which task involves erecting walls and attaching them to the floors and ceilings?
    a. Dry-in
    b. Rough framing
    c. Exterior work
    d. Interior work

14. At what point can MEP tasks begin?
    a. As soon as the rough framing is complete
    b. After drywall is installed
    c. After the ceilings are hung
    d. Once the building envelope is sealed

15. What is the last step of a construction project?
    a. Punch list
    b. Finishing and fixtures
    c. Handoff
    d. Inspection

Answers to Odd-Numbered Module Review Questions are found in *Appendix A*.

Answers to Section Review Questions

Answer	Section	Objective
Section One		
1. b	1.1.0	1a
2. c	1.2.0	1b
Section Two		
1. d	2.1.0	2a
2. b	2.2.0	2b
Section Three		
1. a	3.1.1	3a
2. c	3.1.2	3a

Site Preparation

Source: Sorn340 Studio Images/Shutterstock

Objectives

Successful completion of this module prepares you to do the following:

1. Describe how to set up and check instruments associated with differential leveling.
   a. List safety hazards associated with site layout.
   b. Summarize how to set up and check leveling instruments.
2. Explain how to determine site and building elevations.
   a. Describe how to use a leveling instrument to calculate a site's elevations.
3. Describe applications involving differential leveling.
   a. Explain how to transfer an elevation up a structure.
   b. Summarize applications for profile, cross-section, and grid leveling.
4. Lay out building lines using radial layout techniques.
   a. Explain how to measure distances by taping.
   b. Describe how to measure horizontal and vertical angles.
   c. Explain how to lay out building lines.

Performance Tasks

Under supervision, you should be able to do the following:

1. Set up, adjust, and field check leveling instruments.
2. Read transit level/theodolite scales and verniers.
3. Use a leveling instrument to calculate a site's elevations.
4. Record differential leveling data in field notes in accordance with accepted procedures.
5. Use differential leveling procedures to transfer elevations up a structure.
6. Use a transit level to lay out building lines using radial layout techniques.
7. Perform calculations pertaining to angular measurements.

Overview

While Module 27114, *Principles of Site and Building Layout*, described the basic concepts of site layout, this module covers more advanced site preparation topics, such as setting up and checking leveling instruments, and performing specific steps associated with differential leveling. Additionally, detailed steps are provided for measuring horizontal, vertical, and traverse angles. This information, along with hands-on experience you will gain in lab and jobsite settings, will help you accurately perform these advanced site layout activities.

CODE NOTE

Codes vary among jurisdictions. Because of the variations in code, consult the applicable code whenever regulations are in question. Referring to an incorrect set of codes can cause as much trouble as failing to reference codes altogether. Obtain, review, and familiarize yourself with your local adopted code.

Digital Resources for Carpentry

Scan this code using the camera on your phone or mobile device to view the digital resources related to this craft.

1.0.0 Setting Up and Checking Site Layout Instruments

Performance Tasks

1. Set up, adjust, and field check leveling instruments.
2. Read transit level/theodolite scales and verniers.

Objective

Describe how to set up and check instruments associated with differential leveling.

a. List safety hazards associated with site layout.

b. Summarize how to set up and check leveling instruments.

Differential leveling: A method of leveling used to determine the difference in elevation between two points.

Site layout instruments used for **differential leveling** were covered in detail in Module 27114, *Principles of Site and Building Layout*. Leveling instruments such as the builder's level and auto level can be used to make simple horizontal angular measurements only (for example, 45°, 20°30'). Transit levels and other similar levels can be used to measure angles in both the horizontal and vertical planes, or when making more precise horizontal measurements (for example, 26°15'35"). This module focuses on advanced site layout topics, including instrument setup and calibration, field checks, differential leveling applications, and detailed steps for measuring distances and angles.

1.1.0 Site Layout Safety

While performing site layout, it may be necessary to access and set up instruments in traffic areas to establish and transfer control. If you are working in traffic areas, whether on-site or off-site, implement the following recommendations along with any other contractor or OSHA requirements:

- Complete a job hazard analysis (JHA) to identify potential hazards that may be encountered when laying out the site.
- Familiarize yourself with your site-specific safety plan.
- Ensure craftworkers are properly trained before working in the area.
- Wear high-visibility clothing and other personal protective equipment required for the tasks at hand.
- Ensure that the working area is secure and well-marked using cones.
- Ensure that all vehicular traffic is properly redirected using proper warning and directional signage.
- Work with a partner who assists in setup and monitoring.
- Maintain and inspect all signage and barriers.

NOTE

Be careful and plan ahead. Always think safety. Create a job hazard analysis before undertaking any task. Accidents do not just happen; they are generally caused by carelessness and unsafe practices.

1.2.0   Field Checking Site Layout Instruments

Field testing transit levels, builder's levels, auto levels, or theodolites and other surveying equipment for correct calibration and adjustment should be done if the instrument has never been used, or if the instrument may be out of adjustment. To provide an understanding of the principles upon which instrument calibration is based, it helps to understand the basic geometry of angle-measuring instruments. Tests commonly performed to check instrument calibration are also described.

1.2.1 Geometry of Angle-Measurement Instruments

Transit levels, theodolites, and other instruments used to make angular measurements consist essentially of an optical line of sight, which is perpendicular (at a right angle, or 90 degrees) to, and supported on, a horizontal **axis**. As shown in *Figure 1*, the line of sight of the instrument is perpendicular to the horizontal axis, and the horizontal axis of the instrument is perpendicular to a vertical axis, about which it can rotate. The line of sight is perpendicular to the horizontal axis when the telescope level bubble is centered and when the vertical circle/arc is set at 90/270 degrees, or 0 degrees for **vernier** transit levels. Spirit levels mounted on the base of the instrument **alidade** are used to make the vertical axis coincide with the direction of gravity. These geometric relationships must be maintained in the instrument; otherwise, the instrument will be out of calibration and any angles measured or laid out with it will be incorrect.

Axis: An imaginary line or plane.

Vernier: A short auxiliary scale set parallel to a primary scale.

Alidade: The upper part of a transit level or theodolite that contains the telescope and related components.

1.2.2 Initial Setup and Adjustment

The initial setup of a leveling instrument, such as a builder's level or auto level, typically requires the use of a leveling rod. Several accessories may be used with leveling rods. A movable red-and-white metal disk called a target (*Figure 2*) is used to help make more precise rod readings. The target's vernier scale is set parallel to and beside the primary scale of the leveling rod. Its use enables readings to the nearest eighth of an inch (architect's rod) or nearest hundredths of a foot (engineer's rod). The target is moved up or down on the rod until the 0 on the vernier scale is lined up with the **crosshairs** of the leveling instrument. The target is then clamped in place. To read the vernier scale, count the number of vernier divisions up from the 0 (index mark) until one of the vernier divisions lines up exactly with a division on the rod scale itself. This number is added to the last division on the rod, just below the vernier's index mark.

Crosshairs: A set of lines, typically horizontal and vertical, placed in a telescope used for sighting purposes.

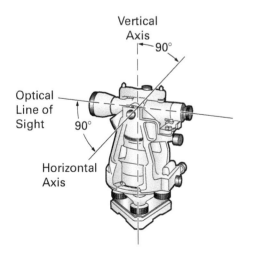

Figure 1 Geometry of an angle-measurement instrument.

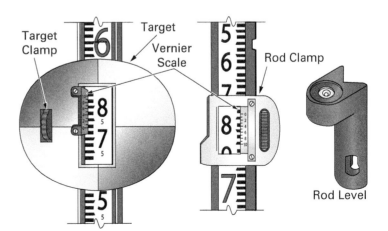

Figure 2 Rod accessories.

NOTE

When using an auto level, skip Steps 4 and 5.

Slope: A measurement of how much the ground varies from horizontal.

Removing a Leveling Instrument from Its Case

When removing a leveling instrument from its case, pay close attention to how it sits in the case. The more moving parts an instrument has, the harder it will be to fit it back into the case if all the parts are not aligned properly.

Parallax: The apparent movement of the crosshairs in a surveying instrument caused by movement of the eyes.

Use the following steps to set up a leveling instrument:

Step 1 Select a location to set up the instrument so that its horizontal line of sight will be at a correct height to intercept the leveling rod, as shown in *Figure 3*.

Step 2 Set up a tripod, making sure to spread its legs wide enough (at least 3' between the legs) to provide a firm foundation for the instrument. Push the legs firmly into the ground and fasten them securely. If setting up on sloping ground, make sure to place one leg of the tripod into the **slope**. Also, make sure the head of the tripod is horizontal. If the tripod head is too far out of level, there is little chance of correctly leveling the instrument on top of it.

Step 3 Carefully remove the leveling instrument (*Figure 4*) from its case and attach it to the horizontal head of the tripod.

Step 4 Level the instrument according to the manufacturer's instructions. Avoid overtightening the leveling screws.

Step 5 Sight along the top of the telescope tube to aim the telescope in the direction of a distant leveling rod or other target, then look through the telescope and adjust the focus. When the crosshairs (*Figure 5*) are positioned on or near the target, make the final settings with the horizontal tangent drive to bring the crosshairs exactly on point. Focus the crosshairs by turning the eyepiece one way or another until the crosshairs are as dark and crisp as they can possibly be. Then, adjust the telescope's focusing knob until the graduations on the rod are legible, sharp, and crisp. Keep both eyes open. This eliminates squinting, does not tire the eyes, and gives the best view through the telescope. Failure to focus the telescope crosshairs properly will cause **parallax**.

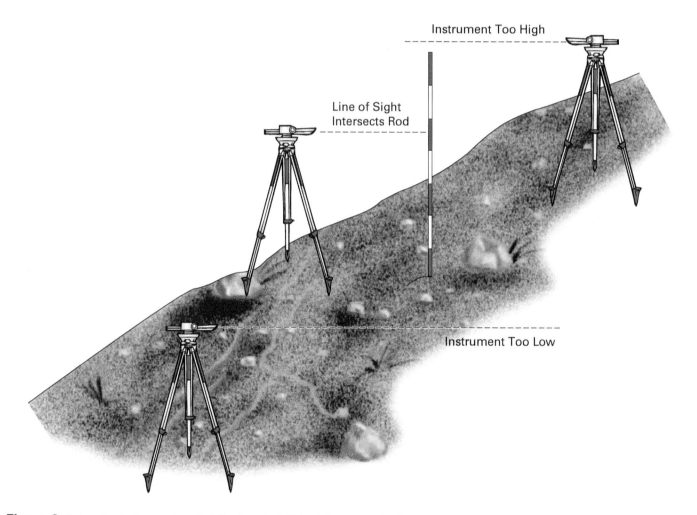

Figure 3 Set up the instrument so that the line of sight is at the proper level.

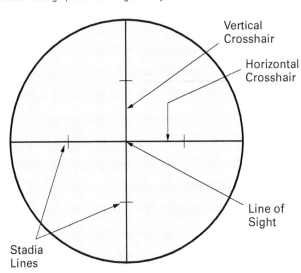

Figure 5 Crosshairs when looking through a telescope.

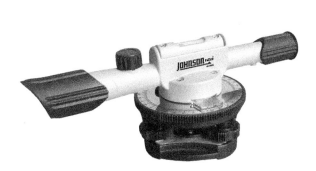

Figure 4 Typical builder's level operator controls.
Source: Johnson Level & Tool

NOTE

When turning two leveling screws simultaneously (as is required when leveling four-screw and three-screw instruments), always rotate them in opposite directions (turn one in a counter-clockwise direction and the other in a clockwise direction, when viewed from above) and turn them at the same rate. When rotating the two leveling screws, the spirit-level bubble will always follow the direction of the left thumb. That is, if the left thumb is turning the leveling screw in a counter-clockwise direction, the bubble will move towards the right; if turning it in a clockwise direction, the bubble will move towards the left. Note that the left-thumb rule also applies if the left hand is used to adjust a single leveling screw, such as is necessary when leveling a three-screw system.

1.2.3 Instrument Checks

This section describes field checking that applies to most site layout instruments, including levels, transit levels, theodolites, and total-station instruments. The field checks described here are generic in nature. All checks for a specific make and model of instrument must be made as directed in the manufacturer's instructions for that instrument. In all cases, the instrument must be sent to a calibration/repair facility for adjustment. Checks described here include:

- Horizontal-Crosshair Checks
- Plate-Level Checks
- Circular-Level Checks
- Optical-Plummet Checks
- Line-of-Sight Checks

Horizontal-Crosshair Test

The object of the horizontal-crosshair test is to ensure that the instrument's horizontal crosshair is in a plane that is perpendicular to the vertical axis of the instrument. With a properly adjusted instrument, you should be able to place any part of the horizontal crosshair on the object or point being viewed with the telescope and still get an accurate reading. The horizontal-crosshair test is simple to perform. First, level the instrument, then sight the horizontal-crosshair reticule on a distant nail head or other well-defined point (*Figure 6*). Once the crosshair is placed on the point, turn the instrument's horizontal tangent screw so that the instrument slowly rotates about its vertical axis. The crosshair should stay fixed on the point as the instrument

NOTE

Parallax occurs when there is an apparent movement of the crosshairs on the rod or object being viewed as the eye moves. If this occurs when reading a leveling rod, major errors can occur. You can easily check for parallax by looking at the rod or object being viewed and moving your head slightly while looking at the crosshairs. If they stay on the same spot, no parallax exists.

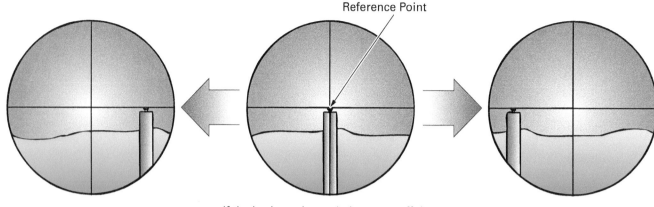

If the horizontal crosshair moves off the
reference point, the reticule needs adjustment.

Figure 6 Horizontal-crosshair test.

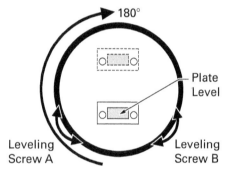

Figure 7 Plate-level test positions.

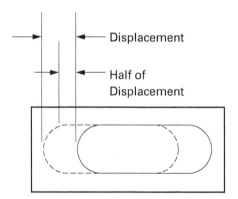

Figure 8 Plate-level adjustment.

Optical plummet: A device incorporated
into a transit level, theodolite, or similar
instrument that allows the operator to
sight a point below that is exactly plumb
with the center of the instrument. This
enables quick and accurate setup of the
instrument over a point.

is rotated. If any part of the crosshair moves above or below the reference point, the
instrument needs adjustment and should be returned to a repair facility.

Plate-Level Check

The purpose of the plate-level check is to make sure that the vertical axis of the
instrument is precisely vertical. Adjustment is required if the axis of the plate
level is not perpendicular to the vertical axis. To determine whether the instrument requires adjustment, level the instrument carefully as follows:

Step 1 Rotate the instrument alidade so that the plate-level vial is parallel with the
centers of leveling screws A and B as shown in *Figure 7*. Adjust leveling
screws A and B only to center the bubble in the plate level.

Step 2 Rotate the instrument alidade 180 degrees around its vertical axis, and check
the position of the plate-level bubble. If the bubble remains centered, the plate
level is adjusted correctly. If the bubble is displaced from the centered position,
proceed with Steps 3 and 4.

Step 3 Adjust the plate level as directed in the manufacturer's instructions to move the
position of the bubble back toward the center of the vial for a distance equal to
one-half of the displacement (*Figure 8*). This adjustment is typically done using
a capstan screw provided for this purpose. Next, center the bubble again using
leveling screws A and B.

Step 4 Rotate the instrument alidade 180 degrees around its vertical axis again and
check for bubble movement. If the bubble is displaced, repeat the adjustment
again as necessary until the position of the plate-level bubble does not change.
This indicates the plate level is adjusted properly.

Circular-Level Check

The only purpose served by the circular-level check is to aid in rough leveling
of the instrument during its initial setup. After the instrument has been properly leveled, a perfectly centered bubble in the circular level indicates that it
is in adjustment. If the bubble is displaced, this indicates that an adjustment is
required. Adjustment of the circular level should be done only after the instrument has been carefully leveled using the plate level only. Next, the circular-level
adjustment screws on the bottom of the circular level are adjusted as needed to
center the bubble in the circular level.

Optical-Plummet Check

The alignment of the optical plummet's axis relative to the vertical axis of the
instrument should be checked periodically for accuracy. An optical plummet can
get out of adjustment, causing the instrument to be set over erroneous points.

The alignment of an optical plummet can be checked by placing its crosshairs
over a reference point on the ground with the instrument at 0 degrees (*Figure 9*).

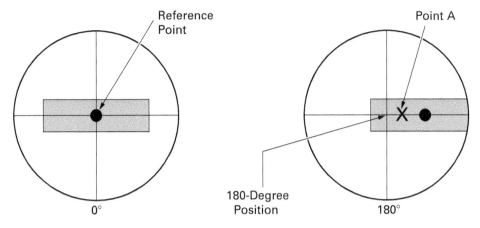

Figure 9 Optical-plummet check.

Next, the instrument is rotated 180 degrees, and the position of the crosshairs is checked again. If the crosshairs are superimposed on the reference point at both positions, the plummet is aligned. If not, the plummet should be adjusted following the instrument manufacturer's instructions. Typically, this is done by turning the adjustment screws on the optical plummet so that its crosshairs are positioned over a point (point A) midway between the original points sighted for the 0-degree and 180-degree positions. This procedure should be repeated as necessary until the crosshairs remain superimposed on the reference point when the instrument is at the 0-degree and 180-degree positions.

Line-of-Sight Check (Levels and Transit Levels)

The line-of-sight check, commonly called a **peg test**, determines if the instrument's telescope line of sight is horizontal. This means that the line of sight is parallel to the barrel of the telescope and the axis of the telescope bubble tube. The line of sight is checked as follows:

Step 1   In a fairly level and clear area, place two stakes at a distance of about 200' apart.

Step 2   Set up and level the instrument at a point exactly midway between the two stakes (*Figure 10A*). While working with a rod person, take several elevation readings at both stake locations (points A and B). Average the set of **backsight (BS)** readings and set of **foresight (FS)** readings, then subtract the averages to determine the actual difference in elevation between the two stake points. Record this difference.

Step 3   Move and set up the instrument as close as possible (within a foot) of the stake at point A (*Figure 10B*).

Step 4   While the rod person holds a rod plumb on stake A, sight backwards through the objective lens of the telescope at a pencil point that is being held and moved slowly up and down the rod by the rod person. When the pencil point is exactly centered in your view, read the rod and record the backsight elevation value.

Step 5   Rotate the telescope and take a rod reading on stake B, 200 feet away, in a normal manner. Record the elevation as a foresight and subtract from the elevation obtained in Step 4. Do this several times and record the average as the difference in elevation readings.

Step 6   Compare the difference in the elevation readings obtained in Steps 2 and 5. Ideally, the difference in **elevations** should be the same. This indicates no adjustment is required. Any difference greater than 2 or 3 thousandths of a foot should be considered significant enough to require adjustment of the instrument. This is done by adjusting the telescope level vial per the manufacturer's instructions.

Peg test: A procedure used to check for an out-of-adjustment bubble vial on levels and other instruments.

NOTE

With the instrument placed midway between the two points, any error that may be caused by a line-of-sight problem will result in an identical amount of error in both the rod readings. Because the errors are identical, the calculated difference in elevation between the two points is the true difference in elevation.

Backsight (BS): A reading taken on a leveling rod held on a point of known elevation to determine the height of the leveling instrument.

Foresight (FS): A reading taken on a leveling rod held on a point in order to determine the elevation.

Elevations: Vertical distances above a datum point. For leveling purposes, a datum is normally based on the ocean's mean sea level (MSL).

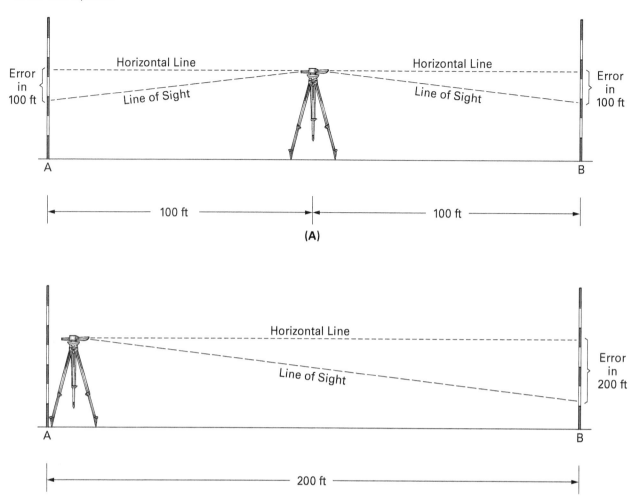

Figure 10 Peg test.

Line-of-Sight Check (Theodolites and Total-Station Instruments)

Horizontal and vertical collimation checks make sure the line of a total-station telescope, as represented by the vertical and horizontal crosshairs and the barrel of the telescope, are aligned. These checks can be done in several different ways. One simple way is described here:

Step 1 Set up and initialize the instrument about 200 feet away from a clearly defined target point.

Step 2 With the telescope in the normal position, record the horizontal and vertical angles measured from the instrument to the target point.

Step 3 **Invert the telescope**, release the upper motion clamp if so equipped, and view the target point again. Measure and record the horizontal and vertical angles.

Step 4 Repeat Steps 2 and 3 several times. Separately average the values for the horizontal angles and the values for the vertical angles measured in Step 2. Do the same for the horizontal- and vertical-angle values measured in Step 3.

Step 5 Compare the average values for the horizontal and vertical angles measured in Step 2 with those measured in Step 3. With proper collimation, the difference between the horizontal angles should be exactly 180 degrees, with any difference being twice the horizontal collimation error. The sum of the vertical-angle readings should be exactly 360 degrees, with any difference being twice the vertical collimation error. Small differences of one or two seconds are generally considered inconclusive as to whether the instrument is in or out of calibration. Larger differences indicate that the instrument should be calibrated before using it. The definition or magnitude of larger differences depends on the instrument and the type of measurements for which it will be used.

Invert the telescope: To reverse the direction of an instrument telescope around its horizontal axis.

1.2.4 Setting Up and Adjusting Transit Levels

This section outlines the procedures for setting up a transit level directly over an established (fixed) point (this point being the vertex for a subsequent horizontal-angle measurement). Procedures are given both for setting up an older-style instrument over a point using a plumb bob and for setting up newer-style instruments over a point using a built-in optical plummet.

Setting Up Over a Point Using an Instrument and Plumb Bob

To set up over a point using an instrument and plumb bob, proceed as follows:

Step 1   Set up a tripod and closely center it over the established point. Spread its legs wide enough (at least 3' between the legs) to provide a firm foundation for the instrument and also a comfortable working height. If setting up on sloping ground, make sure to place one leg of the tripod into the slope. The tripod legs should be positioned and/or adjusted so that the tripod's head plate is as level as possible.

Step 2   Carefully remove the transit level from its case and attach it securely to the tripod.

Step 3   Suspend a plumb bob from the hook provided on the instrument and allow it to swing freely. Adjust the tripod legs as necessary so that the suspended plumb bob hangs very close (1/4" to 1/2") to the established point without touching it.

Step 4   Level the instrument in the horizontal plane by adjusting the leveling screws per the manufacturer's instructions. When properly leveled, the bubble in the leveling vial should remain exactly centered as the telescope is rotated in a complete circle around its base.

Step 5   Loosen one pair of the instrument's leveling screws, then carefully shift the instrument on the tripod head in the required direction until the plumb bob is directly over the established point. Be careful not to rotate the instrument. If the instrument is inadvertently rotated, you must repeat the entire leveling procedure.

Step 6   Repeat Steps 4 and 5 as necessary until the instrument is level, the plumb bob is centered directly over the established point, and the tripod legs are firmly pushed into the ground so that the tripod is stable.

Step 7   Sight along the top of the telescope tube to aim the telescope in the direction of a distant rod or other target, then look through the telescope and adjust the focus. When the crosshairs are positioned on or near the target, tighten the horizontal clamp screw and make the final settings with the horizontal tangent screw to bring the crosshairs exactly on point. Focus the crosshairs by turning the eyepiece one way or another until the crosshairs are as dark and crisp as they can possibly be. Then, adjust the telescope's focusing knob until the graduations on the rod or details of the object are legible, sharp, and crisp. Keep both eyes open. This eliminates squinting, does not tire the eyes, and gives the best view through the telescope.

Leveling Transit Levels

When leveling a transit level, never overtighten the adjustment screws. The screws should be equally tensioned but not too tight. Overtightening can make it difficult to maintain level. It is a good practice to recheck the level before making each angle measurement.

Setting Up Over a Point Using an Instrument with an Optical Plummet

As noted earlier, newer transit levels are equipped with a device called an *optical plummet*. This allows the instrument to be optically aligned over a reference point by looking through an eyepiece and aligning the crosshairs over the point. An optical plummet consists of a set of lenses and mirrors that enable the user of the instrument to look into a viewing port on the side of the instrument (*Figure 11*). The optics and mirrors are in the lower part of the instrument so that when the base of the instrument is perfectly level, the crosshairs of the optical plummet will fall on a point exactly under the center of the instrument. The procedure for setting up and positioning the instrument when using an optical plummet is basically the same as described when using an instrument with a

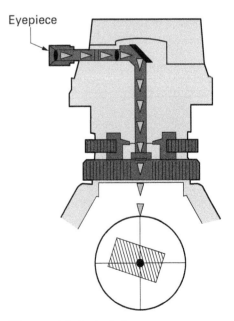

Eyepiece

Figure 11 Optical plummet.

plumb bob. The exception is that the crosshairs of the optical plummet are used to position the instrument exactly over the point, rather than the plumb bob. In many instances, a plumb bob can be used for rough leveling and instrument placement, then moved aside and the optical plummet used to verify the exact placement of the instrument over a point.

1.2.5 Setting Up and Checking Laser Levels

The accuracy and calibration of a laser level can be altered because of severe shock and vibration. Laser levels should therefore be checked for proper calibration at regular intervals, particularly before starting on a new construction site. Calibration of the instrument should be performed by a qualified person and in accordance with the manufacturer's instructions. When an instrument is calibrated in the field, errors can be introduced because of temperature, humidity, and wind conditions. For this reason, it is recommended that the instrument be checked and calibrated in a controlled environment.

Although it is typically rugged enough for construction work, a laser level is a precision instrument requiring the same care as any high-quality leveling device. Observing the following maintenance guidelines will help ensure trouble-free laser operation:

- Always follow the manufacturer's recommended maintenance procedures as directed in the operator's manual for the instrument.
- To prevent moisture and dirt from settling inside the unit, always make sure the laser level, accessories, and carrying case are clean and dry before storage. To clean a laser level, electronic distance meters, and similar instruments, wipe them off with a soft, nonabrasive cloth and clear, solvent-free water.
- When not in use, store the laser level and its accessories in the original case(s).
- Make sure the batteries in the unit and any spare batteries are fully charged to prevent work delays resulting from discharged batteries. If using rechargeable nickel-cadmium batteries, also make sure that a battery charging unit is available to recharge the batteries when they become discharged.
- During periods of prolonged storage, it is advisable to set up and operate the laser for a minimum of eight hours per month.
- Do not attempt internal repairs to a laser instrument. Return it to a dealer, manufacturer, or other qualified organization for repair.

Laser-Level Beam Check

A laser level's beam should be checked for proper calibration at regular intervals, particularly if using the instrument for the first time or if the unit has been handled roughly. Severe shock or vibration may have caused the instrument to be out of calibration. The laser transmitter needs calibration when the laser beam emitted from one side of the unit is above true level, and the beam emitted from the opposite side is below true level (*Figure 12*). When the instrument is correctly calibrated, it emits a 360-degree horizontal level-plane beam. If the unit is turned 180 degrees or 90 degrees from its original position, the reading is within the manufacturer's specifications, typically within $\pm^3/_{32}$" per 100 feet of the original position. The laser-transmitter calibration should be checked as described in the manufacturer's instructions. Manufacturers of different makes and models of instruments recommend checking their instruments in different ways. A general procedure for one method is given here:

WARNING!

Only trained and authorized personnel can operate a laser instrument. Inform all people within the operating range of the instrument that a laser instrument will be in use. Never look directly into a laser beam or point the beam into the eyes of others. Set up the laser transmitter so that the height of the emitted beam is above or below normal eye level.

Figure 12 Laser-transmitter emitted signals.

Step 1 Set up and level the laser transmitter at a location that has a clear line of sight to an object at least 100 feet away. Attach a laser detector (receiver) unit on a leveling rod (*Figure 13*) or other calibrated rod, and hold it plumb at the 100-foot location.

Step 2 Turn on the laser transmitter and rotate it so that the +X side is aimed at the laser detector.

Step 3 Move the receiver into the beam to get an on-grade reading. Mark and record the elevation, noting that it pertains to the +X axis.

Step 4 Rotate the laser transmitter 180 degrees so that the −X side is aimed at the laser detector. Mark and record the elevation, noting that it pertains to the −X axis.

Step 5 Compare the +X and −X elevation readings. If the difference between the readings is less than $3/32$", the X axis is within calibration. If within calibration, go to Step 9. If the difference in readings is more than $3/32$", X-axis calibration is required, as described in Steps 6 through 8.

Step 6 To correct for a calibration error, locate a new mark midway between the +X and −X marks. Move the receiver on the rod until its center-marking notch is aligned with the new midpoint mark.

Step 7 Adjust the X-axis calibration screw to obtain an on-grade laser-beam reading at the midpoint line. Most detectors have move-up and move-down indicators that show the direction the beams need to be moved relative to the on-grade point.

Step 8 Rotate the laser transmitter 180 degrees back to the original +X face. The on-grade reading should be on or within $3/32$" of the midpoint line. If not, repeat Steps 3 through 8.

Step 9 Rotate the laser 90 degrees. Repeat Steps 3 through 8 as required to check and adjust the beam emitted from the +Y and −Y sides of the unit.

Figure 13 Laser detector attached to a rod.

1.0.0 Section Review

1. Potential safety problems that might be encountered while laying out the site are identified by completing a(n) _____.
   a. job hazard analysis
   b. site safety checklist
   c. hazard warning report
   d. OSHA specifications survey

2. True or False: Instruments should always be checked and calibrated in the field.
   a. True
   b. False

3. Newer transit levels can be precisely aligned over a point by using an optical _____.
   a. plumb bob
   b. plummet
   c. angle
   d. placement aid

2.0.0 Differential Leveling

Performance Tasks

3. Use a leveling instrument to calculate a site's elevations.

4. Record differential leveling data in field notes in accordance with accepted procedures.

Objective

Explain how to determine site and building elevations.

a. Describe how to use a leveling instrument to calculate a site's elevations.

Differential leveling on a jobsite requires precise use of the leveling instrument, accurate conversion of dimensions, and exact math calculations. The information covered in this section will provide you the information you need to make accurate differential leveling calculations.

2.1.0 Determining Site and Building Elevations

When preparing to perform differential leveling, one of the first steps is to review the construction documents. These documents may show dimensions both in feet and inches, in feet and fractions of a foot expressed as a decimal (decimal feet), or both. The dimensions of structures are usually shown in feet and inches. Land measurements and ground elevations are typically shown in decimal feet.

2.1.1 Converting between Measurement Systems

Because different units of measure are provided, it is often necessary to convert between the measurement systems. Conversion tables are available in many trade-related reference books that can be used for this purpose. However, you should be familiar with the methods used to make the conversions mathematically in case conversion tables are not readily available.

Converting Feet and Inches to Decimal Feet

To convert values given in feet and inches (and inch-fractions) into equivalent decimal feet values, use the following procedure.

For this example, convert 45'-4⅜" to decimal feet.

Step 1 Convert the inch-fraction ⅜" to a decimal. This is done by dividing the numerator of the fraction (top number) by the denominator of the fraction (bottom number). For this example, ⅜" = 0.375".

Step 2 Add the 0.375" to 4" to obtain 4.375".

Step 3 Divide 4.375" by 12 to obtain

0.3646' = 0.36' (rounded off).

Step 4 Add 0.36' to 45' to obtain 45.36';

thus, 45'-4⅜" equals 45.36'.

Converting Decimal Feet to Feet and Inches

To convert values given in decimal feet into equivalent decimal feet and inches, use the following procedure. For this example, convert 45.3646' to feet and inches (convert to the nearest ⅛").

Step 1 Subtract 45' from 45.3646' = 0.3646'.

Step 2 Convert 0.3646' to inches by multiplying 0.3646' by 12 = 4.3752".

Step 3 Subtract 4" from 4.3752" = 0.3752".

Step 4 Convert 0.3752" into eighths of an inch by multiplying 0.3752" by 8 = 3.0016 eighths or, when rounded off, ⅜". Therefore, 45.3646' = 45'-4⅜".

Study Problems: Converting Different Values

Convert these decimals to percentages.

1. 0.62 = _____
2. 0.475 = _____
3. 0.7 = _____

Convert the following fractions to their decimal equivalents without using a calculator.

4. $\frac{1}{4}$ = _____
5. $\frac{3}{4}$ = _____
6. $\frac{1}{8}$ = _____
7. $\frac{5}{16}$ = _____
8. $\frac{20}{64}$ = _____

Convert the following decimals to fractions without using a calculator. Reduce them to their lowest terms.

9. 0.5 = _____
10. 0.12 = _____
11. 0.125 = _____
12. 0.8 = _____
13. 0.45 = _____

NOTE

When making measurement conversions, it is a good practice to use at least four or five decimal places during your calculations; then round off at the end to avoid compounding rounding errors. Also, using an appropriate calculator makes the conversion process easier with less chance of making mistakes.

2.1.2 Differential Leveling Process

The process of differential leveling is based on the measurement of vertical distances from a level line. Elevations are transferred from one point to another by using a leveled leveling instrument to first read a leveling rod held vertically on a point of known elevation, then to read a rod held on a point of unknown elevation (*Figure 14*). Following this, the unknown elevation is calculated by adding or subtracting the readings.

Benchmark (BM): A relatively permanent object with a known elevation located near or on a site. It can be iron stakes driven into the ground, a concrete monument with a brass disk in the middle, a chiseled mark at the top of a concrete curb, or similar items.

Turning point (TP): A temporary point whose elevation is determined by differential leveling. The turning-point elevation is determined by subtracting the foresight reading from the height-of-instrument elevation.

To determine elevations between two or more widely separated points, or points on a sloping terrain, the same basic differential leveling process is performed several times. For example, if the benchmark (BM) elevation in *Figure 14* is 112.8' and the first rod reading is 1.2', then the instrument height is 114.0' (112.8' + 1.2' = 114.0'). If the first turning point (TP) reading is 5.0', then the first turning point elevation is 109.0' (114.0' − 5.0' = 109.0').

When using any instrument that does not have an automatic leveling feature, such as a builder's level, first confirm that it is level by measuring the level of two points of known elevation before measuring a point of unknown elevation. To do this, set up the instrument between two benchmarks in the area in which you need to work (*Figure 15*). Read the rod set over the first benchmark and then do the same over the second benchmark. Perform the same calculation as above to confirm that the instrument is level. For example, if the first BM elevation is 108.2' and the first rod reading is 3.2', then the instrument height is 111.4' (108.2' + 3.2' = 111.4'). If the second BM reading is 5.0' and the second BM elevation is 106.4' (111.4' − 5.0' = 106.4'), then the instrument is level and may be used to measure an unknown elevation.

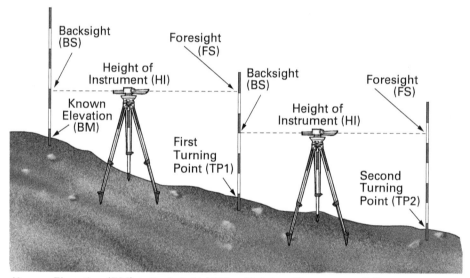

Known Elevation (BM) + Backsight (BS) = Height of Instrument (HI)
Height of Instrument (HI) − Foresight (FS) = Turning Point (TP) Elevation

Figure 14 Differential leveling relationships.

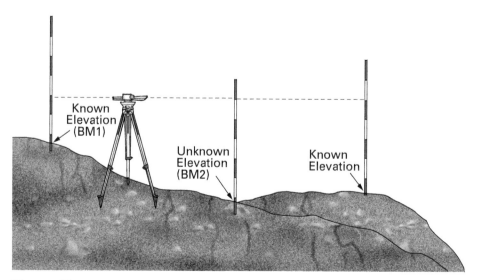

Figure 15 Measuring an unknown elevation.

2.1.3 Differential Leveling Steps

Before beginning the leveling process, select a benchmark that is closest to your work. If you do not know the exact location of the closest benchmark, refer to the site plan. Also determine the longest reasonable distance between your measurement points to minimize instrument setups. Note that some sites are relatively flat, while others have steep slopes. Regardless of the slope involved, the procedure for leveling is the same. When leveling at a site with a very steep slope, the procedure becomes more time consuming. This is because the line of sight of the instrument relative to intercepting a leveling rod is shorter, requiring more setups to cover the distance involved.

The differential leveling procedure generally involves two people working together and communicating with each other. As noted previously, one person is designated as the rod person and the other the instrument person. Depending on the complexity of the task, recording collected measurement data in the **field notes** may be done by either person, both, and sometimes by a third person. An example of a typical differential leveling procedure is described as follows; its path (traverse) is shown in *Figure 16*.

Field notes: A permanent record of field measurement data and related information.

Step 1 The differential leveling procedure begins by recording the starting point (BM) and its known elevation in the **station** and elevation columns of the field notes. For the example shown, the entries are BM (station) and 1,000.00' (elevation).

Station: Instrument-setting location in differential leveling.

Step 2 The instrument person sets up the leveling instrument at Station 1 (STA 1) in preparation for the first measurement. It should be located so that a level rod placed on the BM is in the line of sight of the level, and the rod can be clearly read. Note that this same point should also allow the line of sight of the level to intercept a level rod held on the proposed location of the first turning point (TP1). Set this point at an equal distance between the two points and no farther away than 150' to 200' from either point of measurement. This reduces the possibility of error if the instrument is out of calibration.

Step 3 While the rod person holds the leveling rod plumb on the BM, take a backsight rod reading, then record it in the field notes. For our example, the backsight (BS) reading of 7.77' is recorded in the BS (+) column of the notes. Following this, the **height of instrument** (HI) is calculated and recorded in the HI column of the field notes. For our example, the HI is 1,007.77' (HI = BM + BS = 1,000.00 + 7.77').

Height of Instrument: Elevation of the line of sight of the telescope relative to a known elevation. Determined by adding the backsight elevation to the known elevation.

Step 4 The rod person paces or otherwise measures the approximate distance between the BM and the leveling instrument and then advances an equal distance beyond the level in the desired direction of the first turning point (TP1). This point must be located so that when the leveling rod is placed on it, the line of sight of the leveling instrument will intercept the rod. The rod person selects an appropriate solid surface, such as a sidewalk or large rock, for the turning point. An unmarked point on grass or soil should never be used as a turning point. If no natural solid object is available, a metal turning pin, railroad spike, or wood stake driven into the ground can serve as a turning point. When a turning point on a solid surface such as a sidewalk or pavement is used, the point should be marked and identified by the turning point number.

Step 5 While the rod person holds the level rod plumb on TP1, take a foresight (FS) rod reading, then record it in the field notes. For our example, the FS reading of 5.23' is recorded in the FS (−) column. Following this, the elevation of TP1 is calculated and recorded in the elevation column of the field notes. For our example, the elevation is recorded as 1,002.54' (turning point elevation = HI − FS = 1,007.77' − 5.23').

Side Shots

Some surveyors/carpenters take intermediate readings to points that are not part of the main differential leveling loop. These readings are called side shots. It is important to make sure that your differential leveling loop is properly closed before making any side shots. After closing the loop, side shots can be taken from established turning points.

Step 6 In preparation for the next set of backsight and foresight readings, the instrument person moves the leveling instrument to a point beyond TP1 and sets up the instrument at Station 2, which is approximately midway between TP1 and TP2.

Step 7 Once the leveling instrument is set up, backsight and foresight readings are taken between points TP1 and TP2 in the same way as previously described in Steps 3 through 5, with the following exceptions. The known elevation of TP1 is used instead of the BM to calculate the instrument height (HI) at Station 2. Then, calculate the elevation of TP2 using the new HI and the foresight reading on TP2.

Step 8 Repeat Steps 3 through 7 as necessary to complete the differential measurement loop from the TP2 to the **temporary benchmark** TBM1, then back via TP3 to the starting point at BM.

Temporary benchmark: A point of known (reference) elevation determined from benchmarks through leveling that lasts for the duration of the project.

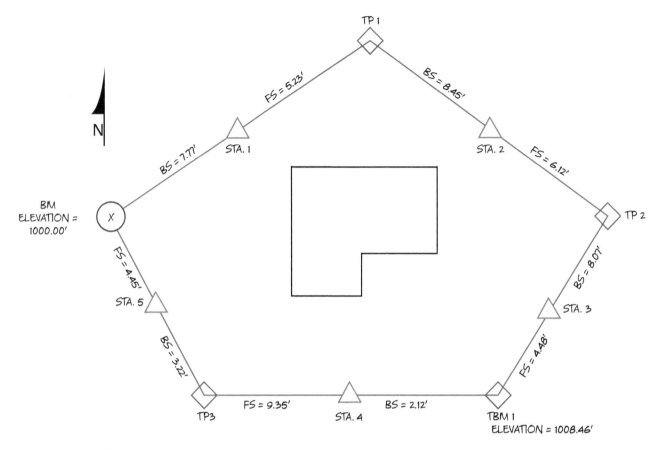

STATION (STA)	BS (+)	HI	FS (−)	ELEVATION
BENCHMARK (BM) TO TEMPORARY BENCHMARK 1 (TBM 1)				
BM	7.77'			1000.00'
STA. 1		1007.77'		
TP 1	8.45'		5.23'	1002.54'
STA. 2		1010.99'		
TP 2	8.07'		6.12'	1004.87'
STA. 3		1012.94'		
TBM 1	2.12'		4.48'	1008.46'
STA. 4		1010.58'		
TP 3	3.22'		9.35'	1001.23'
STA. 5		1004.45'		
BM			4.45'	1000.00'
Σ CHECK	29.63'		29.63'	

DIFFERENCE = 0.00'

DIFFERENCE = 0.00'

Figure 16 Differential leveling traverse and related field notes data.

In the example shown, the leveling traverse is run back to the starting point at BM. This is called closing the loop, or a closed level loop. Any leveling survey should close back either on the starting benchmark or on some other point of known elevation to provide a check of the measurements taken.

Always confirm that field notes are free of mathematical or calculator input errors. You can do this by summing the backsight (BS) and foresight (FS) columns and comparing the difference between them with the starting and ending elevations. As shown in *Figure 16*, the difference between the BS sum and the FS sum is 0.00'. Also, the difference between the starting elevation of 1,000.00' and the ending elevation of 1,000.00' is 0.00'. Because the differences are equal, the calculations are confirmed, and the loop is properly closed. An error would exist if the differences were not equal or were not within the established accuracy standard or tolerances specified for the project. Using the same example, the calculations for the traverse between the BM and TBM1 can be checked in the same manner. This is shown in *Figure 17*.

When performing differential leveling, it is easy to make mistakes. However, mistakes can be eliminated by constantly checking and rechecking your work. Some common mistakes to avoid when performing differential leveling are as follows:

- Using the incorrect crosshair when sighting
- Backsight and foresight distances not equal
- Instrument not leveled

STATION (STA)	BS (+)	HI	FS (-)	ELEVATION
BENCHMARK (BM) TO TEMPORARY BENCHMARK 1 (TBM 1)				
BM	7.77'			1000.00'
STA. 1		1007.77'		
TP 1	8.45'		5.23'	1002.54'
STA. 2		1010.99'		
TP 2	8.07'		6.12'	1004.87'
STA. 3		1012.94'		
TBM 1			4.48'	1008.46'
	24.29'		15.83'	

MATH CHECK: 1000.00
          +   24.29
          ─────────
              1024.29
          −   15.83
          ─────────
              1008.46

Figure 17 Example of a math check.

- Leveling rod not plumb (if not using a level, the rod should be rocked forward and backward, then the smallest reading recorded)
- Sections of an extended leveling rod not adjusted properly
- Dirt, ice, etc., accumulated on the base of the rod
- Misreading the rod
- Recording incorrect values in the field notes
- Moving the position of a turning point between backsight and foresight readings

2.1.4 Field Notes

In the differential leveling procedures described earlier, reference was made to recording measurement information in field notes. Writing a legible and accurate set of notes in a field book (*Figure 18*) is just as important as doing the

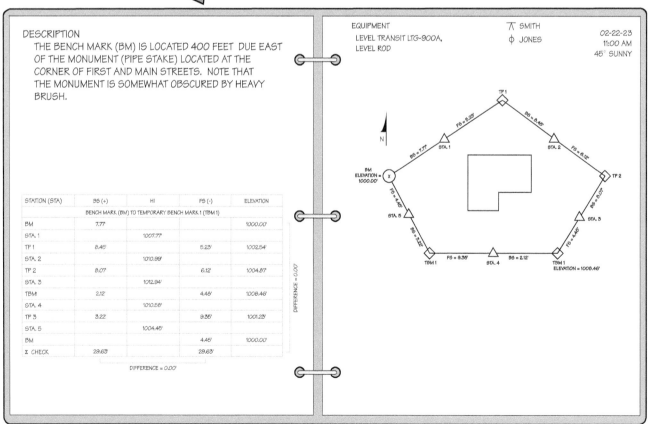

DESCRIPTION
THE BENCH MARK (BM) IS LOCATED 400 FEET DUE EAST OF THE MONUMENT (PIPE STAKE) LOCATED AT THE CORNER OF FIRST AND MAIN STREETS. NOTE THAT THE MONUMENT IS SOMEWHAT OBSCURED BY HEAVY BRUSH.

EQUIPMENT
LEVEL TRANSIT LTG-900A,
LEVEL ROD

SMITH
JONES

02-22-23
11:00 AM
45° SUNNY

STATION (STA)	BS (+)	HI	FS (-)	ELEVATION
BENCH MARK (BM) TO TEMPORARY BENCH MARK 1 (TBM 1)				
BM	7.77			1000.00'
STA. 1		1007.77		
TP 1	8.45'		5.23'	1002.54'
STA. 2		1010.99'		
TP 2	8.07		6.12'	1004.87'
STA. 3		1012.94'		
TBM1	2.12'		4.48'	1008.46'
STA. 4		1010.58'		
TP 3	3.22		9.35'	1001.23'
STA. 5		1004.45'		
BM			4.45'	1000.00'
Σ CHECK	29.63'		29.63'	

DIFFERENCE = 0.00'

DIFFERENCE = 0.00'

Figure 18 Example of field notes.

leveling or layout work itself, since the field notes provide a historical record of the work performed. They serve as a reference should there ever be a question about the correctness or integrity of the work performed, especially in a court of law. Field notes should leave no room for misinterpretation; notes should be written so that others can understand your work. General guidelines for writing and keeping field notes are as follows:

- All field notebooks should contain the name, address, and phone number of the owner.
- All pages should be numbered, and there should be a table of contents page.
- Make neatly printed entries in the book using a suitable sharp pencil with hard lead (3H or 4H). Never use cursive writing in a field book.
- Begin each new task on a new page. The left-hand pages generally are used for entering numerical data and the right-hand pages are for making sketches and notes.
- Always record the date, time, weather conditions, names of crew members and their assignments, and a list of the equipment used.
- Record each measurement in the field book immediately after it is taken. Do not trust it to memory.
- Record data exactly. Ideally, the data should be checked by two crew members at the time it is recorded.
- Make liberal use of sketches if needed for clarity. They should be neat and clearly labeled, including the approximate north direction. Do not crowd the sketches.
- Never erase. If you make a mistake, draw a single line through the incorrect entry and write the correct data above it.
- Draw a diagonal across the page and mark the word VOID on the tops of pages that, for one reason or another, are invalid. When marking the page, be careful not to make the voided information unreadable. The date and name of the person voiding the page should also be recorded.
- Mark the word COPY on the top of copied pages. Refer to the name and page number of the original document.
- Always keep the field book in a safe place on the jobsite. At night, lock it up in a fireproof safe. Original field books should never be destroyed, even if copied for one reason or another.

2.0.0 Section Review

1. The decimal feet value of 46'-5¾" is _____.
   a. 46.27
   b. 46.34
   c. 46.48
   d. 46.75

2. When performing site layout activities, a historical record of the work performed is recorded in _____.
   a. field notes
   b. floor plans
   c. legal forms
   d. site safety plans

3.0.0 Differential Leveling Applications

Performance Task	Objective
5. Use differential leveling procedures to transfer elevations up a structure.	Describe applications involving differential leveling. a. Explain how to transfer an elevation up a structure. b. Summarize applications for profile, cross-section, and grid leveling.

In addition to setting benchmarks, grade stakes, and similar site layout work, there are many applications involving leveling.

Some of the most common applications of differential leveling include the following:

- Transferring elevations up a structure
- Profile leveling
- Cross-section leveling
- Grid leveling

3.1.0 Transferring Elevations Up a Structure

When constructing multistory buildings and other tall structures, ground elevations frequently need to be transferred vertically up the structure as it is being built to maintain the design grades. One method for accomplishing this involves the use of both differential leveling and taping skills. The process begins by first establishing a temporary benchmark (see TBM 1 in *Figure 19*) with a known elevation at the base of the structure by using differential leveling methods. Following this, a tape is used to measure up from TBM 1 the vertical

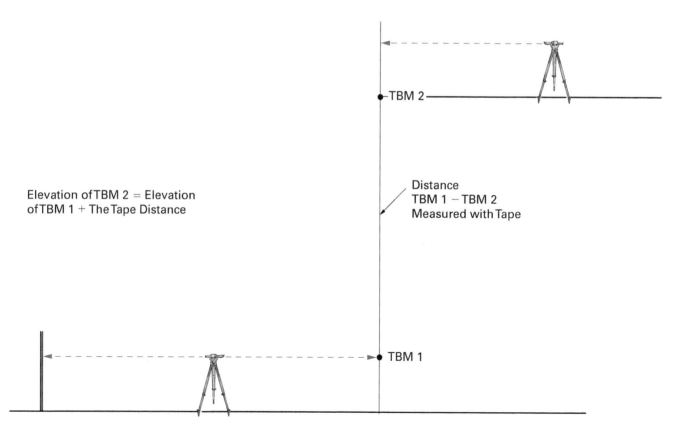

Elevation of TBM 2 = Elevation of TBM 1 + The Tape Distance

Distance TBM 1 − TBM 2 Measured with Tape

Figure 19 Transferring elevations up a structure.

distance needed to establish a second temporary benchmark (TBM 2) on the floor or level of the structure on which the elevation(s) are needed. Once TBM 2 has been established on the upper level, a leveling instrument can be set up and a height-of-instrument (HI) calculated in the normal way, by backsighting on and reading a rod held on TBM 2 (HI = BS + TBM 2 elevation). Note that the elevation of TBM 2 is equal to the elevation of TBM 1 plus the tape distance. Following this, any subsequent leveling tasks are performed on the upper floor or level just as if the instrument were placed on the ground.

Points that need elevations may occasionally be above the line of sight of the instrument, such as with elevations for the bottom of a beam and ceiling levels. Taking the elevation in these instances requires that the leveling rod be held upside down and placed against the beam, ceiling, etc. This results in a positive foresight reading that must be added to the HI rather than subtracted.

3.2.0 Profile, Cross-Section, and Grid Leveling

Profile, cross-section, and grid leveling are methods used to determine the profile of a terrain or surface. These methods are briefly described here. However, procedures for performing these leveling methods are beyond the scope of this curriculum. These types of procedures can be found in most surveying or field-engineering texts, or in reference books, some of which are listed as *References* at the back of this book.

3.2.1 Profile Leveling

Profile leveling is the process of determining the elevation of a series of points along the ground at approximately uniform intervals along a continuous center line, such as when determining the profile of the ground along the center line of a highway. The method and calculations used to perform profile leveling are the same as those used for differential leveling. Profile leveling consists of making a series of differential leveling measurements in the usual manner while traversing the project center line. However, from the instrument's HI position at each station, a series of additional intermediate foresight readings are taken on several points (profile points) along the center line to determine their elevations. These readings are taken at regular intervals or where the terrain changes abruptly, causing sudden changes in elevations to occur. After the field work has been completed, this data can be used to plot the profile of the land along the center line.

3.2.2 Cross-Section Leveling

Cross-section leveling is basically the same as profile leveling. The difference is that rather than determining intermediate elevations of several profile points along a center line, cross-section leveling determines elevations for several profile points that are perpendicular to the center line. Note that for a specific project, there is only one center line profile but there can be numerous cross sections. Cross-section profile plots derived from cross-section leveling data are used for estimating quantities of **earthwork** to be performed.

Earthwork: All construction operations connected with excavating (cutting) or filling earth.

3.2.3 Grid Leveling

Grid leveling is a process used to determine the existing topography of a construction site or other land area. It is also used when required to determine earthwork quantities related to an excavation (pit) or a mound. This is normally done when it is necessary to calculate the volume of material that has been cut or filled. Basically, this method requires that a rectangular profile grid be laid out on the building lot with grid intersections occurring at regular intervals spaced about 50' or 100' apart. Following this, differential leveling is done in a similar manner as for profile leveling, except that more intermediate elevation readings can be taken from one instrument position.

When performed in conjunction with earthwork, grid leveling is normally done both before and after the earthwork is accomplished. The difference between the original and final elevations is then used in a volume formula to calculate the volume of material excavated or filled. This is commonly known as cut and fill.

3.0.0 Section Review

1. True or False: Transferring elevations up a structure requires only taping skills.
   a. True
   b. False

2. Determining the elevation of a series of points along a continuous center line is called _____.
   a. profile leveling
   b. grid leveling
   c. centerline leveling
   d. cross-section leveling

4.0.0 Laying Out Building Lines

Performance Tasks

6. Use a transit level to lay out building lines using radial layout techniques.
7. Perform calculations pertaining to angular measurements.

Objective

Lay out building lines using radial layout techniques.
   a. Explain how to measure distances by taping.
   b. Describe how to measure horizontal and vertical angles.
   c. Explain how to lay out building lines.

With a basic understanding of the tools, instruments, and equipment, site layout can begin. Before using any instruments for layout tasks, they must be field checked to ensure precision measurements. Distances can then be determined by taping or by using one of the site layout instruments.

4.1.0 Taping Distances

To achieve both accuracy and precision when taping, a good understanding of the measurement process and its principles is required.

4.1.1 Accuracy and Tolerances

Accuracy in measurement is determined by how close the measured distance is to the actual distance. For example, when a crew measures as 202' a property boundary line that is known to be 212', that crew has a poor accuracy. When another crew measures the same boundary line as 212.01', that crew has a high degree of accuracy. In construction, it is important to maintain a high degree of accuracy to avoid costly errors. Errors exist in every measurement that is made because of both human and instrumentation limits, and while errors can never be eliminated, they can be reduced by using proper methods and techniques.

Human errors typically result from differences in eyesight, sense of touch and feel, physical strength, and other factors. Human errors can be reduced by ensuring that all members of the carpentry crew approach the task in an organized manner and practice proper measurement procedures. Mistakes can be detected by checking and rechecking your work.

One easy way to decrease errors is to perform the measurements twice, comparing the second set of measurements to the first. A rule of thumb is that for every 100' measured, the difference should be no more than $^1/_{100}$ (0.01') of a foot. When the entire distance has been measured twice, the two sets of measurements are averaged to decrease any errors.

Instrumentation errors are usually a result of instruments or equipment being damaged, out of calibration, or incapable of measuring within tolerances. These types of errors are compensated for by making sure that you always use well-maintained and calibrated measuring instruments. Other examples of instrumentation errors are a tape's length being different from its marked length because of stretching or temperature variations, or the angle of the tape not being horizontal during the measurement. These types of errors are compensated for by making corrections to the measured values via a mathematical formula.

Tolerances are used in construction layout to define how far off the exact design location something can be and still be acceptable. You must always meet or exceed the specified tolerances. Several factors determine the degree of accuracy required for a specific job or different **control points** within a jobsite. There is no single standard that defines required tolerances, so the specifications for each job must always be consulted. Once the specific tolerances are known, you must use measurement equipment that is capable of measuring to these tolerances.

Control points: Specific points created on the site, where they are used as elevation and dimension reference points during site and building layout.

4.1.2 Taping Guidelines

Guidelines that must be followed to achieve accuracy when taping are summarized in *Figure 20* and briefly explained below:

- *Use a calibrated tape* – Manufacturers furnish a data sheet with each precision tape they make. This sheet will tell you exactly how accurate the tape is and to which standard it is compared.

- *Know where zero (0) is on the tape* – Determine where the exact 0 point of the tape is and use it. It may be at the end of a loop or other fitting at the end of the tape, or it can be offset from the end of the tape. Some carpenters measure from the 1" mark and then subtract 1" when making a measurement.

- *Maintain good alignment* – Distances must be measured as a straight line. When measuring distances that are longer than the tape, intermediate measurement points must be used. Any such intermediate points should be directly in line between the beginning and end points being measured.

- *Apply correct tape tension* – Apply the correct tension to the tape during measurements. This is necessary to overcome any sag in the tape. Use the tension value specified by the manufacturer in the product literature supplied with the tape. As mentioned earlier, a tension spring can be attached to the end of the tape to aid you in applying the right tension.

- *Measure horizontally* – The tape must be read when it is in a horizontal position. On flat ground, this is not much of a problem. On a slope or incline, tape readings can be taken in smaller increments. This allows the downhill person to comfortably hold the tape in a horizontal position while still applying the right tension to the tape. Typically, this puts the end of the tape about chest high.

- Use a calibrated tape
- Know where zero(0) is on the tape
- Maintain good alignment
- Apply correct tape tension
- Measure horizontally
- Repeat measurements
- Make mathematical corrections as necessary
- Record readings immediately
- Use well-maintained equipment

Figure 20 Guidelines for distance measurements by taping.

- *Repeat measurements* – To avoid mistakes that can easily occur when taping, make all measurements at least twice. Reversing the direction of the measurements greatly reduces the chance of repeating a mistake. Greater accuracy can be achieved by averaging the two sets of readings.

- *Make mathematical corrections as necessary* – Mathematical corrections must be made to measurement data to compensate for conditions such as tape length differences resulting from calibration, the expansion or contraction of the tape's length due to temperature variations, the tape being positioned on a slope rather than held horizontal during a measurement, etc. All these conditions can affect the accuracy of your measurements. These conditions are described in more detail in a later section.

- *Record readings immediately* – The reading for each measurement should be recorded in the field notes immediately after it is taken to avoid omissions that can contribute to errors. Also, make sure to check that you have recorded each entry correctly. It is easy to transpose numbers or mis-apply a decimal point.

- *Use well-maintained equipment* – Accurate measurements can only be made with well-maintained taping equipment.

4.1.3 Taping a Distance

The task of taping involves two people working together and communicating with each other. The following procedure outlines one method for measuring a distance between two existing points, such as two control monuments. The procedure for measuring a distance between known and unknown points, such as when laying out a building, would be performed in basically the same manner.

In this procedure, the two people involved are designated as the rear tape person and the head tape person (*Figure 21*). It is assumed that a 100' tape is being used, the overall distance to be measured is greater than 100', and the terrain is relatively flat, allowing for horizontal measurements to be made with the tape on the ground.

Step 1   Determine the straight path for the overall measurement between the start and end points of the line to be measured. This path should be cleared of any brush, rocks, or other obstacles that will hamper making the measurements.

Step 2   Once the measurement path has been determined, the location of the start and end points of the overall measurement path are marked with range poles or laths with flagging.

Step 3   While the rear tape person holds onto the tape reel at the starting point, the head tape person takes the 0' end of the tape and advances along the measurement line to the location of the first intermediate measuring point, presumably at 100' from the beginning point. Some companies prefer that the head tape person advance with the reel while the rear tape person holds onto the 0' end of the tape. Either way is acceptable.

Step 4   The head tape person applies the proper tension to the tape as the rear tape person aligns the 100' mark on the tape exactly on the starting point. Once positioned over the starting point, the rear tape person signals the head tape person, who then marks the position of the 0' point of the tape with a chaining pin or other marker. If a chaining pin is used, the pin should be slanted slightly in a direction away from the tape.

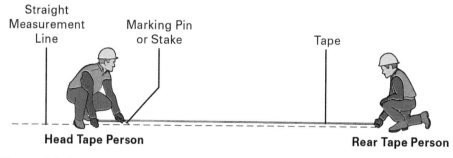

Figure 21 Taping procedure.

Step 5 Following this, both workers advance along the measurement line in preparation for the next measurement. The starting point for this measurement is at the location of the chaining pin established in Step 4.

Step 6 Steps 3 and 4 are repeated as required until the line has been measured from the original starting point to the end.

Step 7 The overall distance measured is equal to the total number of full tape lengths measured plus the reading of the last measurement. For example, if three 100' measurements were made and the last measurement recorded by the rear tape person is 30.25', then the total distance measured is 330.25'.

Step 8 Repeat Steps 1 through 7 and record the second set of measurements.

Step 9 Add the total distances obtained in Steps 7 and 8, then divide the sum by two, using the result as the measurement. For example, in Step 7 the total measured distance was 330.25'. Assume the distance measured in Step 8 was 330.27'. The sum of the two measurements is 660.52' (330.25' + 330.27' = 660.52'), so the average is 330.26' (660.52' ÷ 2 = 330.26').

The general procedure for measuring a distance over terrain with an excessive slope or incline is basically the same as described above.

However, to aid in maintaining the tape in the horizontal (level) position when making measurements, a method called **breaking the tape** is used. This means making measurements using a portion of the full tape's length in a series of steps until the full tape length has been traversed.

For example, if moving down the slope of the hill, the head tape person advances with the end of the tape along the line for a distance equal to the tape length. Then, leaving the tape on the ground, the head tape person returns as far along the tape as necessary to reach a point that will allow him or her to comfortably hold the tape in the horizontal position and still be able to apply the proper tension during the measurement (*Figure 22*). Typically, this is about chest or waist high.

The head tape person holds a plumb bob string over a convenient whole-foot mark, and when the tape is properly tensioned and determined to be horizontal, and both the head tape person and rear tape person agree that a measurement should be taken, the head tape person then marks the location of the whole-foot mark onto the ground below.

Because the tape is elevated above ground level during the measurement, this is accomplished by lowering a plumb bob suspended from the whole-foot mark on the tape to a point just above the ground. When the plumb bob is stationary, the pressure on the plumb bob string is released, allowing the tip of the plumb bob to contact the ground. The point where the plumb bob tip strikes the ground is then marked with a chaining pin.

The head tape person continues to hold the intermediate foot mark on the tape until the rear tape person arrives, at which time the head tape person hands the tape to the rear tape person while still maintaining the foot-mark position on the tape that he or she has been holding. This procedure is repeated as often as necessary while moving down the hill, until the end of the tape is reached, and the required total distance is measured or laid out.

When taping, it is easy to make mistakes. However, mistakes can be eliminated or minimized by constantly checking and rechecking your work. Avoid the following common mistakes when taping:

- Using incorrect measuring tools for the job being performed
- Having too much slack in the tape during measurement
- Not holding the tape horizontal
- Allowing the tape to be twisted or kinked
- Not aligning the tape correctly on measurement reference points
- Making measurements using a wrong reference point or points
- Reading tape graduations incorrectly
- Recording incorrect numbers
- Miscounting the number of full tape lengths measured
- Making errors in mathematical computations

Breaking the tape: Making measurements using a portion of the full tape's length in a series of steps until the full tape length has been traversed.

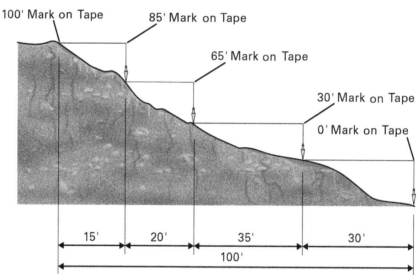

Figure 22 Breaking the tape to measure a distance on a steep slope.

It is sometimes desirable to express measured distances in terms of tape lengths. This is done by dividing the distance measured by the length of the tape used to make the measurement. For example, when using a 100' tape, a distance that measured 825.45' would be expressed as 8.2545 tape lengths. If using a 50' tape, the same distance would be expressed as 16.5090 tape lengths.

Estimating Distances by Pacing

The ability to pace a distance with reasonable accuracy can be very helpful. Pacing can be used to check measurements that have been made by others or to estimate an unknown distance. You can determine your average pace length by walking a known distance that has been previously measured with a steel tape and dividing that length by the number of paces taken. When pacing the distance, you should walk naturally with a consistent pace length. Some craftworkers count each step as a pace. Others only count full strides (two paces) instead of paces when stepping with their right or left foot. It does not matter how you count—just be consistent. Also, for the last pace in the measurement, which is normally less than a full pace, record to the nearest $\frac{1}{2}$ or even $\frac{1}{4}$ pace, if possible.

The following steps outline the procedure for using pacing to find your average pace length and determining an unknown distance.

Step 1 Use a tape to lay out a level distance of 100'.

Step 2 Starting at the beginning point, walk naturally to pace the 100' distance. Record the number of paces required to travel the distance.

Step 3 Repeat Step 2 a minimum of four more times and record the number of paces required for each time.

Step 4 Calculate your average number of paces per 100'. For example, assume the total number of paces for the five trips equals 199 paces (40.5 + 39 + 40 + 39.5 + 40 = 199). In this case, the average number of paces per 100' equals 39.8 paces (199 total paces ÷ 5 trips).

Step 5 Calculate your length of pace in feet by dividing the average number of paces into the distance traveled. For our example, the average pace length is 2.51' (100' ÷ 39.8 paces).

Once the average length of your pace is known, other distances can be determined by pacing the distance and calculating its length by multiplying your pace length by the number of paces needed to travel the distance. For example, if it takes 60 paces to travel a distance and your average pace length is 2.51', then the distance is approximately 151' (60 paces × 2.51' = 150.6' = 151' rounded off).

Measurement Errors

Errors resulting from tape length, tape temperature, or both have a cumulative effect. The more measurements you make, the larger the error can be at the completion of your measuring. Good practice is to take your time and repeat a measurement one or more times to make sure you are correct before moving on to the next measurement.

By using trigonometric relationships, the slope correction can be calculated when either the angle of the slope or the difference in elevations between the bottom and top of the slope are known.

4.2.0 Measuring Horizontal, Vertical, and Traverse Angles

Several methods commonly used to measure horizontal and vertical angles are described here. Depending on the type of instrument used, the procedure will vary. However, the general method described here applies to most instruments.

4.2.1 Measuring Horizontal Angles

In the following example, a repeating theodolite is used. The instrument is located at point A (*Figure 23*). You will be measuring the angle between lines AB and AD.

Reversing the Telescope

Reversing or inverting the telescope means to turn it 180 degrees on its horizontal axis. Older surveyors may call this plunging the scope.

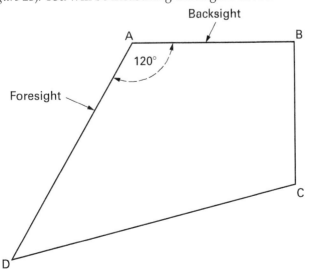

Figure 23 Measuring an angle.

NOTE

Measuring angles by reversing or doubling is one method commonly used to check for mistakes in readings. This increases the precision of the readings. It is done by performing Steps 1 through 4, and then performing Steps 5 through 8.

Step 1   Set up and level the instrument over the established reference point (point A) as described earlier in this module.

Step 2   Point the instrument at the target located at point B. This is called the backsight. It is the reference point from which all measurements are made. Carefully sight the target through the telescope, and then zero the instrument.

Step 3   Turn the instrument to the right to sight the target at point D. This is called the foresight.

Step 4   Read and record the angle.

Step 5   Reverse, or invert, the telescope and tighten the upper motion clamp to hold 90 degrees.

Step 6   Turn the instrument to the right and sight the target located at point B. The angle reading should still be 120 degrees.

Step 7   Turn the instrument and sight the target located at point D.

Step 8   Read and record the angle. The reading should be 240 degrees.

The angle was measured twice so the reading should be two times the angle (in this case 240 degrees). To obtain an average of the two measurements, divide the end result by two (240 ÷ 2 = 120).

Measuring Angles by Repetition

Measurement of angles by repetition is used when increased accuracy is required. Typically, this degree of accuracy pertains more to work performed by surveyors and field engineers than those performed by carpenters. It is used when it is desired to gain accuracy beyond the **least count** of the instrument being used. The least count is the finest reading that can be made directly on a vernier of a transit level or micrometer of a theodolite.

Least count: The finest reading that can be made directly on a vernier of a transit level or micrometer of a theodolite.

Measurement of an angle by repetition is identical to the procedure described previously for measurement by doubling except there are from four to eight repetitions made instead of only two. When recording the values measured for each angle, normal practice is to record only the first and last readings. Following this, the value for the accumulated (summed) angular measurements is divided by the number of repetitions to derive the average value for the angle.

For example, assume that after six repetitions (three direct and three reversed), the summed angular value is 240°00'. The average value for the angle is then equal to 40°00' (240°00' ÷ 6). Note that it is often necessary to add 360 degrees, or multiples of 360 degrees, to the final instrument reading in order to account for the number of complete 360-degree revolutions the telescope has been turned horizontally while making the repeated measurements. For example, a 60°00' angle measured eight times causes the instrument to be turned through 480°00'; however, the instrument's scale would read only 120°00'. Therefore, to calculate the average value for the angle being measured, it is necessary to add 360 degrees to 120°00' before dividing by 8 (480°00' ÷ 8 = 60°00').

Closing the Horizon

A technique called closing the horizon can be used to check the accuracy of angular measurements. Closing the horizon means that the unused angle is measured to complete the circle (*Figure 24*). When the horizon is closed and all angles at the station are added together, the sum should be exactly 360 degrees. Normally there will be some small error. Should the error be large (more than 1 minute), a mistake has been made in the measurements, and they should be redone.

4.2.2 Measuring Vertical Angles

Vertical angles are measured in a similar way as horizontal angles. As described previously, vertical angles are measured with reference to the horizon when

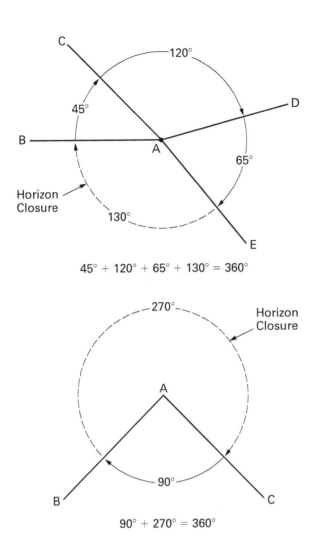

45° + 120° + 65° + 130° = 360°

90° + 270° = 360°

Figure 24 Simplified examples of horizon closures.

using a transit level. A measurement from the horizon to a high point is a positive (+) vertical angle, and from the horizon to a low point is a negative (−) vertical angle. Obviously, it is important to record whether a vertical angle is positive or negative.

A transit level must be carefully set up and leveled to measure a vertical angle. The telescope bubble should be centered and the vernier on the vertical vernier scale should read 0°00'. The procedure for measuring a vertical angle requires that the horizontal and vertical motion clamps be loosened so that the telescope can be rotated and vertically positioned so that the horizontal crosshair rests approximately on the point to which the vertical angle is to be measured. With the vertical motion clamp or clamps tightened, the vertical tangent screw is adjusted to set the horizontal crosshair exactly on the point. Following this, the value of the angle is read from the vertical circle and vernier.

Note that if the transit level has a full vertical circle and the telescope can be plunged, a more accurate reading can be obtained by measuring the vertical angle twice, once with the telescope direct (upright position) and once reversed (inverted position), then averaging the two readings.

With some instrument models, the vertical angular reading with the telescope level is 90 degrees instead of 0 degrees. If this is the case, when measuring a positive vertical angle, it is necessary to subtract the angle reading on the instrument from 90 degrees to obtain the actual angle. For example, if the instrument angular reading is 70°30', the actual angle being measured is 19°30' (90°00' − 70°30').

Similarly, when measuring a negative vertical angle, it is necessary to subtract 90 degrees from the instrument reading to obtain the actual angle. For example, if the instrument angular reading is 118°30', the actual angle being measured is 28°30' (118°30' − 90°00').

4.2.3 Measuring Traverse Angles

A traverse is a continuous series of points or stations that are tied together by angle and distance. The angles are measured using transit levels and theodolites, and the distances are measured with steel tapes or electronic distance measurement instruments (EDMIs). The procedure for measuring distances in a traverse was covered in Module 27114, *Principles of Site and Building Layout*.

Traverses can be open or closed. An open traverse is a series of measured straight lines and angles that do not geometrically close, such as encountered with a highway or pipeline. A closed traverse is one that begins and ends at the same point, such as is encountered with property boundary lines or building lines. The measurement of complex traverses is normally performed by surveyors and field engineers; however, carpenters do measure simpler closed traverses such as those encountered when laying out or measuring building lines. *Figure 25* shows an example of a four-sided traverse with four different interior angles.

When performing traverse angular measurements, each of the interior angles would be measured. For accuracy, each of the four angles would be measured by repetition, typically making two direct and two reversed measurements. When recording the measurement values for each angle, record only the first and last readings. If two direct and two reverse measurements were made for each angle, the final (accumulated) reverse measurement value for each angle would be divided by four to obtain the average angle value.

The main point to remember about measuring any closed traverses is that the sum of the interior angles is equal to (n − 2) × 180 degrees, where n is the number of sides. Therefore, for any four-sided traverse, the sum of the four interior angles should always equal 360 degrees [(4 − 2) × 180° = 360°]. As shown in *Figure 25*, the sum of the four interior angles equals 360 degrees. Another example is in a square or rectangular traverse, where the sum of the four interior

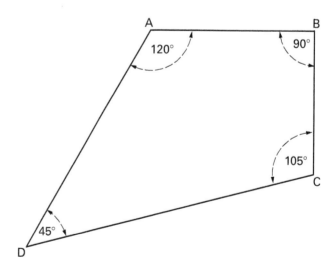

Sum of the Interior Angles = (n − 2) 180

Where n = Number of Sides

For Example Shown: (4 − 2) 180 = 360°

Check: 120° + 90° + 105° + 45° = 360°

Figure 25 Measuring traverse angles.

angles, each angle being 90 degrees, must always equal 360 degrees. Because of instrumentation tolerances, etc., it is expected that very small errors will occur in the measurement of each angle when measuring angles in a traverse. Thus, if all the interior angles of a traverse are measured and their total is very close to (n − 2) × 180 degrees, you can be sure that there is a minimal error in the reading. It is common practice to adjust these small errors out of the angles by distributing them around the traverse so that the sum of the interior angles will equal (n − 2) × 180 degrees. Any such adjustment should only be made to compensate for small errors resulting from instrumentation limitations, not for mistakes made in the measurements.

Measuring Angles

Some common mistakes made when making angular measurements include:

- Poor setup and leveling of instruments
- Misreading the instrument scale indications
- Transposing and/or recording the wrong angle values in field notes
- Sighting on the wrong targets, marks, or lines when measuring horizontal or vertical angles
- Using the wrong instrument tangent screw
- Failure to center the telescope bubble before measuring a vertical angle
- Failure to consider the algebraic sign for the values of vertical angles measured with a transit level

4.3.0　Laying Out Building Lines

Two common methods are used to lay out building lines. The traditional method was reviewed in Module 27114, *Principles of Site and Building Layout*. The radial layout method is described in this section.

4.3.1 Laying Out Building Lines Using the Radial Method

The layout of building foundation lines using the radial method involves making several distance and angle measurements from a single control point on site in order to locate the building foundation stakes. Layout using the radial method is typically done with a total station or with a theodolite and companion EDMI. It may also be done with a transit level and steel tape.

Determining Distances with Trigonometry

To do a radial layout, you need to know the distances and angles from the main control point to each of the individual building points. Normally, this information has been calculated and provided by others for your use in the field. There may be instances, however, when the **Pythagorean theorem** and right-angle trigonometry must be used to calculate the distance and angle to each point. Trigonometry is used when only the length of one side of the triangle and only one angle (other than the right angle) are known. Trigonometry recognizes that there is a relationship between the size of an angle and the length of the sides in a right triangle.

Recall that the sides of a triangle are called side opposite, side adjacent, and hypotenuse with respect to either of the acute angles. As shown for the angle A relationships in *Figure 26*, the side opposite angle A is labeled (a) and the side adjacent to angle A is labeled (b). A construction calculator with trigonometric functions (*Figure 27*) contains the keys for sine [SINE or SIN], cosine [COS], and tangent [TAN] functions, all of which represent relationships between the sides of a right triangle.

NOTE

Some instruments may not be accurate to the second. Therefore, choose the count nearest to the desired angle, which will result in the least amount of error. Because of this, it is important to check and recheck all distances and diagonals. A tiny angle error can result in a large error over a long distance.

Pythagorean theorem: A geometric theorem for right triangles stating that the sum of the squares of the legs of a right triangle is equal to the square of the hypotenuse. It is expressed mathematically as $a^2 + b^2 = c^2$.

Hypotenuse: The longest side of a right triangle. It is always opposite from the right angle.

NOTE

When making a series of calculations for angle measurements, a more accurate result is obtained if you make your calculations using all the decimal places displayed by your calculator, then round off your answer at the end.

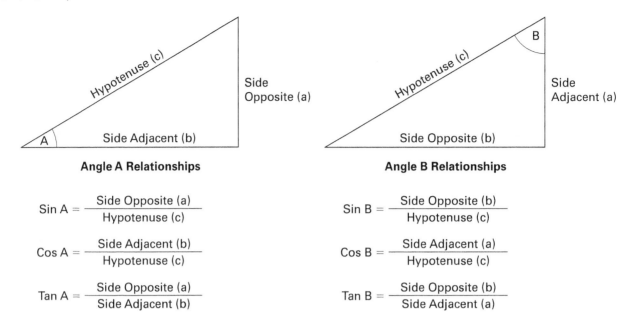

Figure 26 Angle-side relationships in a right triangle.

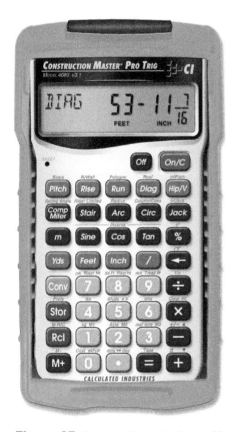

Figure 27 Construction calculator with trigonometric functions.
Source: Courtesy of Calculated Industries

These relationships for angle A are expressed as follows:

$$\sin A = \frac{a}{c} = \frac{\text{side opposite angle A}}{\text{hypotenuse}} = \frac{\text{opposite}}{\text{hypotenuse}}$$

$$\cos A = \frac{b}{c} = \frac{\text{side adjacent angle A}}{\text{hypotenuse}} = \frac{\text{adjacent}}{\text{hypotenuse}}$$

$$\tan A = \frac{a}{b} = \frac{\text{side opposite angle A}}{\text{side adjacent angle A}} = \frac{\text{opposite}}{\text{adjacent}}$$

The same relationships exist for angle B as shown for angle A. However, you must substitute B for A in the sin, cos, and tan formulas and make sure to substitute the correct labeling for the sides in the formulas. For example, the side opposite angle B is b and the side adjacent to angle B is a.

Here is an easy way to remember the sin, cos, and tan functions for both angles A and B:

$$SIN \textbf{ S}ome \textbf{ O}ld \textbf{ H}orse \frac{O}{H}$$

$$COS \textbf{ C}aught \textbf{ A}nother \textbf{ H}orse \frac{A}{H}$$

$$TAN \textbf{ T}aking \textbf{ O}ats \textbf{ A}way \frac{O}{A}$$

The specific formula that is used when making calculations involving right triangles is determined by what is known about the triangle. If you know the lengths of any two sides, you can calculate the angle using the basic formulas given in this section. If you know the angle and the length of one side, you can calculate the lengths of the other sides using the appropriate sin, cos, or tan formulas after they have been re-arranged to solve for the unknown. For angle A, these relationships are:

$$\text{Length of side a} = c \sin A \text{ or } b \tan A \text{ or } c \cos B$$

$$\text{Length of side b} = c \cos A \text{ or } c \sin B \text{ or } a \tan B$$

$$\text{Length of side c} = a/\sin A \text{ or } b/\cos A$$

$$\text{or } a/\cos B \text{ or } b/\sin B$$

Figure 28A, *Figure 28B*, and *Figure 28C* show examples for a radial layout, as well as the calculations used to determine the angle and distance for each of the points shown in the examples. Using this method, a carpenter can set up in one location, and with the use of a digital transit and taping, can successfully lay out a site.

Once the distances and angles have been calculated, the instrument is set up and leveled directly over the control point (point A) and site point D. For accuracy, always use a backsight (AD) that is equal to or greater than the foresight (A1, A2, A3, A4). Ideally, the backsight should be twice the foresight.

Following this, the various angles are turned relative to a known baseline and the distances measured from the control point to each of the building points. After all points have been located, it is a good practice to check all distances and diagonals with a steel tape.

4.3.2 Trigonometric Leveling

Trigonometric leveling can be used to determine elevations such as a point on a tall building or structure. As shown in *Figure 29*, it is done using a transit level to measure the angle, then calculating the unknown height using right angle trigonometry.

The trigonometric leveling procedure begins by first setting up and properly leveling the transit level over an established point at a known distance from the structure. The transit level must also be set horizontal to use it as a level. Then, the height of the instrument (HI) is determined in the normal way by backsighting (BS) on a leveling rod held on a benchmark (BM) of known elevation. The HI is then calculated using the formula HI = BM + BS.

NOTE

The method described here is used to determine the approximate elevation of a building. It assumes the building is plumb. In theory, this is a logical assumption, but in practicality, it is not always true.

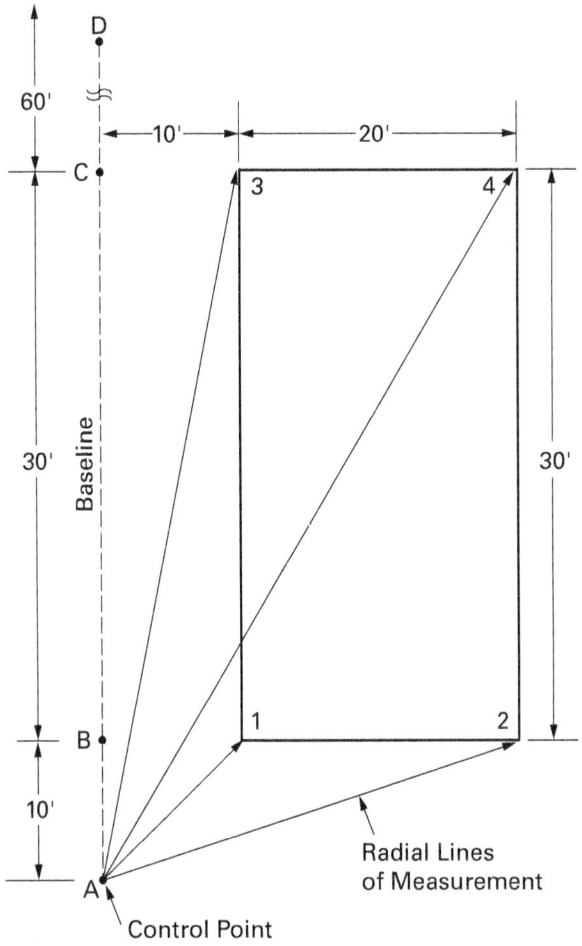

Figure 28A Site layout using the radial method (1 of 3).

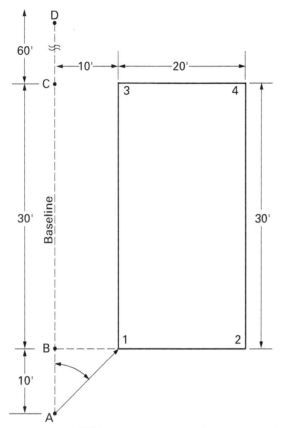

Scientific Calculator

Dist A-1 = $\sqrt{(10')^2 + (10')^2}$ = 14.14'

Tan $\angle$ = $\frac{10}{10}$ = 1

Inv Tan = 45.000000°
           = 45°00'00"

Construction Calculator

Enter: 10' Run, 10' Rise, Diag. = 14' 1-$^{11}/_{16}$",
Pitch = 45.00 degrees, convert to degrees
minutes seconds = **45.00.00 DMS**

(A) Distance and Angle from Control Point A to Building Point 1

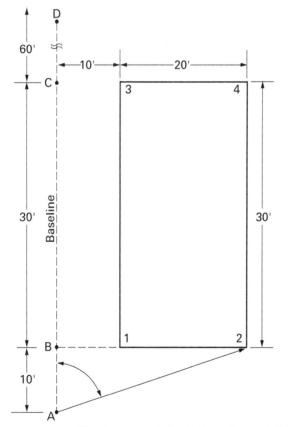

Scientific Calculator

Dist A-2 = $\sqrt{(10')^2 + (30')^2}$ = 31.62'

Tan $\angle$ = $\frac{30}{10}$ = 3

Inv Tan = 71.56505118°
           = 71°33'54"

Construction Calculator

Enter: 10' Run, 30' Rise, Diag. = 31' 7-$^1/_2$",
Pitch = 71.57 degrees, convert to degrees
minutes seconds = **71.33.54 DMS**

(B) Distance and Angle from Control Point A to Building Point 2

Figure 28B Site layout using the radial method (2 of 3).

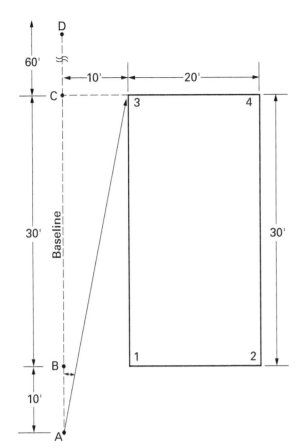

Scientific Calculator

Dist A-3 = $\sqrt{(10')^2 + (40')^2}$ = 41.23105626'

Tan $\angle$ = $\dfrac{10}{40}$ = 0.25

Inv Tan = 14.03624347°
= 14°02'10"

Construction Calculator

Enter: 40' Run, 10' Rise, Diag. = 41' 2–3/4",
Pitch = 14.04 degrees, convert to degrees
minutes seconds = **14.02.10 DMS**

(C) Distance and Angle from Control Point A to Building Point 3

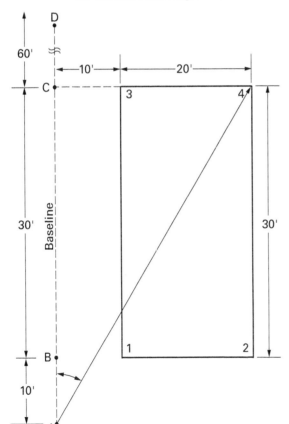

Scientific Calculator

Dist A-4 = $\sqrt{(30')^2 + (40')^2}$ = 50.00'

Tan $\angle$ = $\dfrac{30}{40}$ = 0.75

Inv Tan = 36.86989765°
= 36°52'11"

Construction Calculator

Enter: 40' Run, 30' Rise, Diag. = 50' 0",
Pitch = 36.87 degrees, convert to degrees
minutes seconds = **36.52.11 DMS**

(D) Distance and Angle from Control Point A to Building Point 4

Figure 28C Site layout using the radial method (3 of 3).

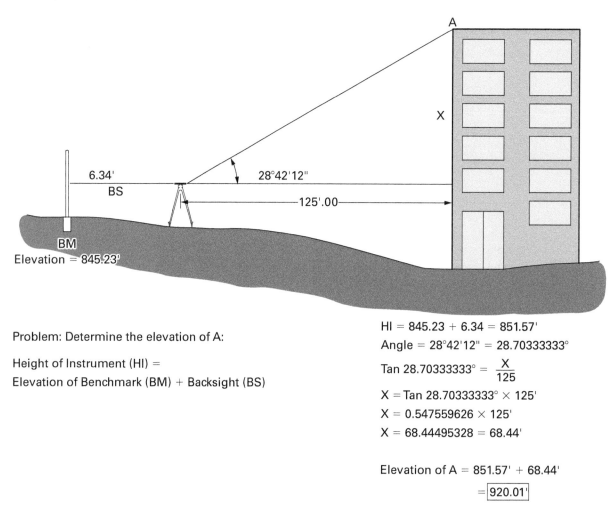

Problem: Determine the elevation of A:

Height of Instrument (HI) =
Elevation of Benchmark (BM) + Backsight (BS)

HI = 845.23 + 6.34 = 851.57'
Angle = 28°42'12" = 28.70333333°

Tan 28.70333333° = $\frac{X}{125}$

X = Tan 28.70333333° × 125'
X = 0.547559626 × 125'
X = 68.44495328 = 68.44'

Elevation of A = 851.57' + 68.44'
= 920.01'

Figure 29 Example of trigonometric leveling.

For the example shown in the figure, the HI is 851.57' (845.23' + 6.34'). Following this, the vertical angle to the point on the structure is measured with the transit level, and the angle is read on the vernier. For this example, the angle is 28°42'12". Using the values for the angle and the distance from the instrument to the building (125.00'), the height of the point on the building relative to the instrument (height X) can be calculated using the tangent function. For this example, height X is 68.44'. This height is then added to the instrument HI to yield the actual elevation. For this example, the elevation is 920.01' (851.57' + 68.44').

4.0.0 Section Review

1. When measuring a distance over an incline, you will mark the ground where a measurement was taken with a _____.
   a. plumb bob and chaining pin
   b. grade stake and chaining pin
   c. chaining pin and tensioning spring
   d. tensioning spring and plumb bob

2. When measuring a closed traverse shape with five sides, the sum of the interior angles should be _____.
   a. 360 degrees
   b. 540 degrees
   c. 720 degrees
   d. 900 degrees

3. True or False: The radial method of laying out building lines typically involves use of a hand sight level.
   a. True
   b. False

Module 27402 Review Questions

1. When working in high-traffic areas, it is important to _____.
   a. move quickly between other workers
   b. wear high-visibility clothing
   c. keep an eye on your supervisor
   d. carry a copy of the site-specific safety plan

2. When sighting through the telescope of a transit level, you should _____.
   a. squint to focus on the crosshairs
   b. use your left eye and close the right
   c. use your right eye and close the left
   d. keep both eyes open

3. To help make more precise readings, leveling rods can be equipped with an accessory called a _____.
   a. target
   b. vernier scale
   c. receiver
   d. bull's-eye level

4. To ensure that the vertical axis of an instrument is precisely vertical, you should perform the field check known as a(n) _____.
   a. plate-level check
   b. alidade-adjustment check
   c. circular-level check
   d. vertical-arc check

5. Calibration tests for a leveling instrument include a line-of-sight test and a _____.
   a. vertical-crosshair test
   b. parallax test
   c. convergence test
   d. horizontal-crosshair test

6. If a field check indicates an instrument is not calibrated correctly, then it should be _____.
   a. recalibrated quickly in the field
   b. sent to a qualified repair facility
   c. returned to the manufacturer for a refund
   d. repaired on-site only by the engineer

7. When a tripod is set up, the legs should have a spread of about _____.
   a. 18 inches
   b. 2 feet
   c. 3 feet
   d. 42 inches

8. If a benchmark elevation is 114.6' and the first rod reading is 2.2', then the instrument height is _____.
   a. 100.0'
   b. 112.2'
   c. 112.4'
   d. 116.8'

9. A common mistake to avoid when performing differential leveling is _____.
   a. using two people instead of one to take measurements
   b. being too careful to record accurate data in the field notes
   c. using the incorrect crosshairs when sighting
   d. selecting a benchmark prior to starting the differential leveling process

10. At night, the field book should be _____.
    a. placed in your toolbox or locker
    b. taken home with you
    c. locked in a fireproof safe
    d. kept with the leveling instrument

11. To estimate the amount of earthwork to be performed at a construction site, use the profiling method called _____.
    a. cross-section leveling
    b. true-line leveling
    c. grid leveling
    d. line-of-sight leveling

12. When performed in conjunction with earthwork, the _____ process is done both before and after earthwork is completed.
    a. cross-section leveling
    b. grid leveling
    c. center-line leveling
    d. line-of-sight leveling

13. The process of measuring a distance using a portion of a full tape's length in a series of steps is called _____.
    a. iterative measuring
    b. gradient taping
    c. step measuring
    d. breaking the tape

14. When measuring a negative vertical angle, how many degrees must be subtracted from the instrument reading to obtain the actual angle?
    a. 30 degrees
    b. 45 degrees
    c. 90 degrees
    d. 180 degrees

15. An example of a closed traverse is a _____.
    a. property boundary
    b. pipeline
    c. street
    d. railway

Answers to Odd-Numbered Module Review Questions are found in *Appendix A*.

Answers to Section Review Questions

Answer	Section	Objective
Section One		
1. a	1.1.0	1a
2. b	1.2.3	1b
3. b	1.2.4	1b
Section Two		
1. c	2.1.1	2a
2. a	2.1.4	2a
Section Three		
1. b	3.1.0	3a
2. a	3.2.1	3b
Section Four		
1. a	4.1.3	4a
2. b	4.2.3	4b
3. b	4.3.1	4c

Wall Systems and Installations

Objectives

Successful completion of this module prepares you to do the following:

1. Identify safety hazards to consider when installing an advanced wall system.
   a. Describe hazards that may be present when installing wall systems.
2. Describe the different types of advanced wall systems.
   a. Describe structural insulated panels (SIPs) and explain the carpenter's role in their installation.
   b. Identify the carpenter's role in the installation of masonry veneer walls.
   c. Define fire-resistance-rated construction and explain methods used to develop wall systems that protect occupants from smoke and fire.
   d. Describe how shear walls reinforce buildings against lateral forces.

Performance Tasks

Under supervision, you should be able to do the following:

1. Construct a shear wall for light-frame construction.
2. Construct firewalls in accordance with specifications.

Overview

Advanced wall systems include structural insulated panels, masonry veneer systems, fire-resistance-rated (FRR) construction, and shear walls. Although many basic construction concepts apply to advanced wall systems, additional knowledge is required to understand insulative characteristics, details about masonry veneers, wall fire protection, and the effects of lateral forces on buildings.

CODE NOTE

Codes vary among jurisdictions. Because of the variations in code, consult the applicable code whenever regulations are in question. Referring to an incorrect set of codes can cause as much trouble as failing to reference codes altogether. Obtain, review, and familiarize yourself with your local adopted code.

Digital Resources for Carpentry

Scan this code using the camera on your phone or mobile device to view the digital resources related to this craft.

1.0.0 Wall System Safety	
Performance Tasks There are no Performance Tasks in this section.	**Objective** Identify safety hazards to consider when installing an advanced wall system. a. Describe hazards that may be present when installing wall systems.

A good carpenter is always aware of the safety rules in every area of the construction trades and always wears appropriate PPE. Constructing advanced wall systems often requires craftworkers to work at elevated locations. Whether you are a few steps up a ladder or working from an aerial lift, precautions must be taken to keep workers safe (*Figure 1*). Core Module 00101, *Basic Safety (Construction Site Safety Orientation)*, provides in-depth coverage of construction site safety.

Figure 1 Work Safety Plan.
Source: Edgars Sermulis/Alamy Images

1.1.0 Potential Wall Installation Hazards

Overhead power lines must be identified (*Figure 2*) and carefully considered when raising walls, not only for the workers, but also for equipment operators moving materials to various parts of the building. Based on OSHA requirements, a safe minimum distance of 10' must be maintained between a power line and a piece of equipment for line voltages of 50kV (kilovolts) or less. For voltages above 50kV, the distance must be increased by 0.4' for each 1kV of voltage. Contact the utility company to see if the power can be shut down on the lines, when necessary. If this is not possible, ask if they can place insulation over the lines during the time you will be working in the location. If overhead power lines are present, ensure they are included in the JHA. Also, use non-conductive ladders when working around power lines.

Always complete a JHA before beginning a wall project to identify the potential hazards and recommend action that can be taken to eliminate or minimize each hazard. In addition to working at elevations, the following potential hazards may be encountered:

- Respiratory hazards when applying certain types of sealant
- Flying debris as a result of cutting or handling wall panels
- Tools falling from a roof, ladder, or other elevated surface onto workers below

NOTE

Projects may have more stringent height guidelines than those published by OSHA. Consult with your supervisor or proper authorities to determine the guidelines that are in effect.

Figure 2 Identify overhead power lines.
Source: A Sharma/Shutterstock

Falls from high places can cause serious injury or death when the wrong type of fall protection equipment is used, or when the right equipment is used improperly. A fall protection plan must be prepared for any project where workers will be more than 6' off the ground or on an elevated working surface. The fall protection plan should incorporate the use of guardrails, Personal Fall Arrest Systems (PFASs), and safety nets. The use of these devices is governed by OSHA's *Safety and Health Standards for the Construction Industry, Part 1926, Subpart M*. Rules covering guardrails on scaffolds are contained in *Subpart L*.

Advanced wall systems provide different, but important, safety issues to consider before undertaking the process. Keep the following in mind:

- Heavy and awkward objects must be lifted correctly, or strains and back problems may result.

- Large panels must be lifted in the correct manner, whether using a crane or forklift, or manually lifting wall panels.

- All scaffolding, platforms, and ladders must be in good condition and set up correctly.

- Adhesives and their solvents used for wall panels can be flammable and their vapors may be toxic. Always ensure the area you are working in is well-ventilated. Skin irritation may result from using these materials, so protective gloves may be required.

- Hand tools may be sharp and should be used with a great deal of care. Remember, always cut away from yourself.

- Power tools and their cords must be kept in a good state of repair and always grounded when required. Electrical cords should be checked at least once a week to see if they are in good condition. Never use a power tool in a wet area. A ground fault circuit interrupter (GFCI) is required by OSHA (Occupational Safety and Health Administration) at all times. A portable pigtail GFCI is required if the device is plugged into a standard wall outlet.

- Be careful and plan ahead. Always think safety. Create a job hazard analysis (JHA) before undertaking any task. Accidents do not just happen; they are generally caused by carelessness and unsafe practices.

Smaller wall panels are commonly lifted manually, resulting in twisting or lifting injuries. Many major contractors have instituted stretching programs to minimize these types of injuries. The programs involve stretching activities, similar to an athlete's, prior to and/or during the workday.

1.0.0 Section Review

1. A fall protection plan must be prepared for any project where workers will be more than _____ off the ground or on an elevated work surface.
   a. 2'
   b. 4'
   c. 6'
   d. 8'

2. The purpose of completing a JHA before beginning a wall project is to identify the potential hazards and _____.
   a. fulfill code requirements of the *IBC*®
   b. eliminate them before the project begins
   c. recommend action to eliminate or minimize each hazard
   d. test workers on their knowledge of job hazards

2.0.0 Types of Advanced Wall Systems

Performance Tasks

1. Construct a shear wall for light-frame construction.
2. Construct firewalls in accordance with specifications.

Objective

Describe the different types of advanced wall systems.
   a. Describe structural insulated panels (SIPs) and explain the carpenter's role in their installation.
   b. Identify the carpenter's role in the installation of masonry veneer walls.
   c. Define fire-resistance-rated construction and explain methods used to develop wall systems that protect occupants from smoke and fire.
   d. Describe how shear walls reinforce buildings against lateral forces.

Some of the most common types of advanced wall systems include the following:

- Structural insulated panels
- Masonry veneer systems
- Fire-resistance-rated assemblies
- Shear walls

Curtain wall: A nonbearing exterior wall that is set into, and attached to, the steel or concrete structure of a building.

Another common advanced wall system is a **curtain wall**. Curtain walls are constructed primarily from glass. As a result, they are installed by glaziers rather than carpenters. Curtain walls are nonbearing and are typically set into, and attached to, the structure of a building. One key advantage of a curtain wall is that it is thin, which allows it to add more floor space inside the building. Curtain walls support their own dead load and are designed to resist wind loads rather than support the weight of the building.

Curtain Wall

The invention of the curtain wall concept made it possible to create glass walls that could be directly attached to the building structure. Although curtain walls are fairly lightweight assemblies, they are capable of withstanding significant stresses. Before the invention of the curtain wall, individual bricks and stones were commonly used to create the exterior walls of a structure.

As shown in *Table 1*, other types of wall panel construction are available.

TABLE 1 Other Wall Panel Construction Options

Wall Panel	Composition
Insulation Concrete Forms (ICF)	Cast-in-place concrete sandwiched between two insulating foam panels.
Insulated Metal Panels (IMPs)	Rigid foam insulation sandwiched between two sheets of coated metal.
Single Skin Panels	Single interlocking layer of prefinished or natural metal that is pre-formed, or roll formed, into a specific shape or profile.
Aluminum Composite Material (ACM)	Three-layer panel made up of two pre-painted aluminum sheets bonded to a polyethylene core.
Metal Composite Material (MCM)	Factory-manufactured panel consisting of metal skins bonded to both faces of a solid plastic core (*IBC*® 2021 Section 202).

2.1.0 Structural Insulated Panels

Insulated wall systems use modern technology to construct wood, metal, or concrete wall systems that offer excellent insulative characteristics. A common characteristic of these systems is that they are created by layering different types of insulation materials.

Going Green

LEED

The US Green Building Council Leadership in Energy and Environmental Design (LEED) Green Building Rating System promotes a whole-building approach with criteria in five areas: sustainable site development, water savings, energy efficiency, material selection, and indoor environmental quality. Building systems can contribute toward points used to achieve the various LEED levels through their energy efficiency, materials credits, reduced building footprint, indoor air quality, and durability.

There are two types of **structural insulated panels (SIPs)**: panelized and modular. SIPs consist of oriented strand board (OSB) panels sandwiched around a foam core made of expanded polystyrene (EPS), **extruded** polystyrene (XPS) or rigid polyurethane foam. Other materials, such as plywood, pressure-treated plywood, or cement board, may also be used as face panels for specialized applications. The OSB panels are typically $\frac{7}{16}$" or $\frac{5}{8}$" thick with sizes ranging from 4' × 6' to 8' × 24'. Conduit is installed in the panels, or holes are drilled through the foam to provide channels for electrical wiring. Window and door openings may be cut at the manufacturing facility or may be cut on site.

SIPs are manufactured under controlled conditions in a manufacturing facility using computer-controlled equipment. Information from the electronic construction drawings is transferred to the panel-cutting equipment to provide precise quality control with little waste. When delivered to the construction site, the panels assemble quickly, thus reducing labor costs.

When constructing a SIP wall, the wall panels are lifted into position using a crane (*Figure 3*) or by hand, depending on the size and weight of the panel. SIP panels must be anchored to the foundation using a preservative-treated sill plate. The sill plate is equal in width to the space between the panels. The SIPs are connected to the sill plate using fasteners specified by the manufacturer.

Structural insulated panels (SIPs): A type of wall system that consists of an insulated foam core sandwiched between two wood structural panels (WSPs), such as oriented strand board (OSB) or plywood.

Extruded: The forming of desired shapes by forming or pressing a material through a shaped opening.

Figure 3 Structural insulated panels.
Source: brizmaker/Shutterstock

Splines or proprietary connectors are used to connect SIPs to one another. Dimensional lumber may be used, but this technique allows for thermal bridging and lowers insulation values. Depending on the manufacturer, panel connections include insulated lumber, composite splines, mechanical locks, or overlapping OSB panels.

Did You Know?

Insulating Concrete Forms

Insulating concrete forms (IFCs) are a type of concrete forming system that uses cast-in-place concrete sandwiched between two insulating foam panels. Unlike conventional concrete forming systems, the foam panels remain in place after the concrete has set. The forms consist of interlocking modular units that are stacked without mortar. ICF construction is commonly used for low-rise commercial buildings, and in some cases, residential structures.

Source: Radovan1/Shutterstock

Did You Know?

Insulated Metal Panels

Insulated metal panels (IMPs) consist of rigid foam insulation sandwiched between two sheets of coated metal. The panels are available in a variety of styles and sizes. Steel or aluminum outside panels create a vapor, air, and moisture barrier and provide long-term thermal stability. The metal panels provide great durability and are available in a wide variety of colors and finishes. Unlike brick, precast or tilt-up concrete, and other porous wall surfaces, IMPs do not absorb water, thus minimizing mold and mildew issues.

Source: Myfotoprom/Shutterstock

2.2.0 Masonry Veneer Walls

A masonry veneer wall system consists of a single **wythe** of masonry as a facing attached to a wood frame (*Figure 4*), metal frame, or other structural system. The masonry veneer acts as a protective **exterior wall envelope** for the structure—as a barrier to wind and rain, and as an insulator of sound and heat or cold. The veneer may be brick, stone, glass, tile, or solid concrete masonry units. The masonry veneer is not a load-bearing wythe but is designed to carry only its own weight. Although these masonry units are fastened to the framework, they are not bonded to that backing with masonry or mortar.

Wythe: Single thickness of a masonry wall.

Exterior wall envelope: A system or assembly of exterior wall components, including exterior wall covering materials, that provides protection of the building structural members, including framing and sheathing materials, and conditions interior spaces from the detrimental effects of the environment (*IBC®* 2021, Section 202).

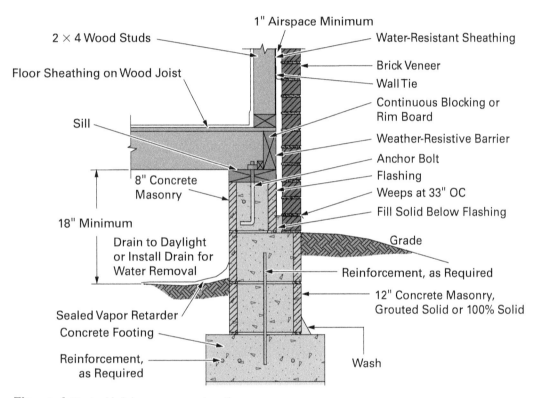

Figure 4 Typical brick veneer construction.

The masonry veneer is attached to the structural frame by noncorroding masonry veneer ties. Generally, building codes or construction drawings specify the gauge and spacing of metal anchors. These specifications vary based on the backing and airspace between the veneer and framing and whether the building is in a high seismic area. Energy conservation techniques using exterior foam sheathing require larger airspaces constructed between anchored masonry veneer and its backing in order to accommodate thicker continuous insulation used in colder climate zones. Masonry veneer with airspaces up to a maximum of $4^5/_8$" may be constructed using traditional tie configurations. Airspaces up to $6^5/_8$" must be constructed using stiffer tie configurations.

A minimum airspace of at least 1" is provided between the veneer and backing. The backing is usually covered with a water-resistive barrier and flashing. A **weephole** is included at the bottom of the airspace to eliminate water that may pass through the masonry veneer. Because the veneer wythe is exposed to changing weather conditions, the masonry units and mortar must be capable of withstanding the effects of the weather.

NOTE

In certain cases, plain masonry foundation walls are not required to use reinforcement. Refer to Table R404.1(1) in the *International Residential Code (IRC®)* for additional information.

Weephole: Channel that allows water that may enter the wall system to drain directly to the exterior.

2.3.0 Fire-Resistance-Rated Construction

Interior and exterior walls can be designed to deter the spread of fire and to reduce sound transmission from one occupancy to another. Firestopping and **fire blocking** is required in most commercial buildings and in some residential structures to prevent the spread of fire and/or smoke from one area to another. Firestopping and fire blocking are life-safety issues and should be taken seriously.

Did You Know?

Firestop Systems

Section 202 of the *International Building Code*® (*IBC*®) 2021, defines firestops and firestop systems:

- *Membrane-penetration firestop* — A material, device, or construction installed to resist for a prescribed time period the passage of flame and heat through openings in a protective membrane in order to accommodate cables, cable trays, conduit, tubing, pipes, or similar items.
- *Membrane-penetration firestop system* — An assemblage consisting of fire-resistance-rated floor-ceiling, roof-ceiling, or wall assembly, one or more penetrating items installed into or passing through the breach in one side of the assembly and the materials or devices, or both, installed to resist the spread of fire into the assembly for a prescribed period of time.
- *Penetration firestop* — A through-penetration firestop or a membrane-penetration firestop.
- *Through-penetration firestop system* — An assemblage consisting of a fire-resistance-rated floor, floor-ceiling, or wall assembly, one or more penetrating items passing through breaches in both sides of the assembly and the materials or devices, or both, installed to resist the spread of fire through the assembly for a prescribed period of time.

2.3.1 Fire-Resistance-Ratings

A **fire-resistance rating (FRR)** represents the ability of a wall, floor, roof, or structural member to withstand fire for a period of time, ranging from one to four hours. A rating of two hours, for example, means that a structural member such as a wall, floor, or roof would prevent flames or hot gases from passing through it for two hours. The FRR also refers to the ability of a structure or material to withstand the force of water sprayed from a hose.

The FRR of a frame wall is based on its construction materials. Fire-resistance-rated walls are built in different fashions, depending on their application. The FRR of masonry and concrete walls depends on thickness. *Table 2* shows examples of FRR based on minimum thickness for different types of concrete and masonry walls.

TABLE 2 Examples of Fire-Resistance Ratings for Various Concrete and Masonry Materials

Material	Minimum Thickness Required	
	One-Hour Rating	Four-Hour Rating
Lightweight concrete masonry unit (expanded slag or pumice)	2.1"	4.7"
Solid lightweight concrete	2.5"	5.1"
Solid brick of clay or shale	2.7"	6"
Siliceous aggregate concrete	3.5"	7"

Source: IBC® 2021, Table 721.1(2)

Although there are various design options for wood-framed, load-bearing partition walls, *Figure 5* shows specifications for a typical fire-resistance-rated wall assembly.

Quick Selector for Fire-Rated Assemblies
Partitions/Wood Framing (Load Bearing)

Single Layer	Ref.	Design No.	Description	STC	Test No.
45 MIN 	UL FM	U317 WI-45 MIN	½" Fire-shield gypsum board applied vertically to each side of 2 × 4 wood studs 16" OC. With 5D coated nails, 1⅝" long, 0.086" shank, 15/64" heads, 7" OC at edges and intermediate studs. Joints of square edge, bevel edge, or predecorated gypsum board may be left exposed.	34	NGC 2161
1 HR 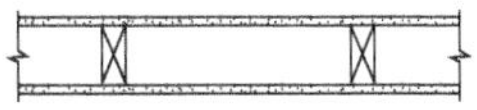	UL FM	U305 WI6A-1HR WP 3605	⅝" Fire-shield gypsum board or ⅝" XP fire-shield gypsum board applied horizontally or vertically to each side of 2 × 4 wood studs 16" OC. With 6D coated nails,1⅞" long, 0.0915" shank, ¼" heads, 7" OC at edges. Joints of square edge, bevel edge or predecorated gypsum board may be left exposed. Joints staggered 16" on opposite sides.	35	NGC 2403
1 HR 	UL FM	U309 WI6B-1HR WP 3510	⅝" Fire-shield gypsum board or ⅝" XP fire-shield gypsum board applied horizontally or vertically to each side of 2 × 4 wood studs 24" OC. With 6D coated nails, 1⅞" long, 0.0915" shank, ¼" heads, 7" OC at edges. Joints of square edge, bevel edge or predecorated gypsum board may be left exposed. Joints staggered 24" on opposite sides.	38	NGC 2404
1 HR 	FM GA	WIA-1HR (WP) 45 WP-1200	⅝" Fire-shield gypsum wallboard or ⅝" fire-shield MR board screw attached horizontally to both sides 3⅝" screw studs, 24" OC. All wallboard joints staggered.	42	NGC 2385
	OSU	T-1770	⅝" Fire-shield gypsum wallboard screw attached vertically to both sides 3⅝" screw studs, 24" OC. All wallboard joints staggered.		
Double Layer					
2 HR 	UL FM	U301 Based on WP 4135	⅝" Fire-shield gypsum board, two layers applied either horizontally or vertically to each side of 2 × 4 wood studs 16" OC. Base layer attached with 6D coated nails, 1⅞" long, 0.0915" shank, ¼" heads, 6" OC face layer ⅝" fire-shield gypsum board attached with 8d coated nails 2⅜" long, 0.113 shank, 9/32" heads, 8" OC vertical joints located over studs. Joints staggered 16" each layer and side.	40	NGC 2363
2 HR 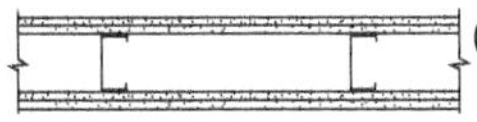	OSU GA	T-1771 Based on WP 1711	First layer ⅝" fire-shield gypsum wallboard screw attached vertically both sides 3⅝" steel studs, spaced 24" OC second layer laminated vertically both sides. Vertical joints staggered.	48	NGC 2282

Figure 5 Specifications for typical fire-resistance-rated walls.

Organizations such as UL Solutions test various construction assemblies and issue FRRs. There are separate sections of UL263/ASTM E119 that address various design types, including floor-ceilings, roof-ceilings, beams, columns, walls, and partitions. Different elements are assigned letters; for example, walls and partitions are U, V, and W. Each element is divided by type of construction; for example, partition walls made of wood studs with gypsum board are covered by design standard numbers U300-U399, V300-V399, and W300-W399. UL fire-resistance ratings are expressed in hours. For example, the specifications for a one-hour-rated bearing wall made from wood studs and gypsum board are set forth in design detail U305.

Wall assemblies are tested in a furnace to simulate fire conditions. They are subjected to specific temperatures for a set period of time. Loads are placed on bearing walls. Assemblies may also be subjected to fire-hose stream conditions.

In addition to UL Solutions listings, criteria for fire-resistance-rated wood-frame assemblies can also be found in a number of other sources, including the *International Building Code® (IBC®)*, Intertek Testing Services' *Directory of Listed Products*, and the Gypsum Association's *Fire Resistance Design Manual* (GA 600). These criteria provide detailed descriptions of wall-assembly construction, including specific products tested, specific construction methods, and alternative construction methods. For example, the design criteria for a steel stud wall with gypsum board includes specifications for screws, joints, insulation, steel framing members, caulking and sealants, and steel corner fasteners.

2.3.2 Fire-Resistance-Rated Wall Types

The *International Building Code (IBC®)* designates four types of fire-resistance-rated wall assemblies:

- *Firewalls* — Two- to four-hour fire-resistance-rated assemblies with protected openings. Firewalls extend continuously from the foundation to or through the roof, and have robust structural integrity.
- *Fire barriers* — Assemblies that are one- to four-hour fire-resistance rated but do not require the structural integrity of firewalls.
- *Fire partitions* — One-hour fire-resistance-rated vertical assemblies with protected openings.
- *Smoke barriers* — Assemblies that restrict the movement of smoke in a building and are also one-hour fire-resistance-rated.

Each of these assemblies requires a specific combination of materials in a certain configuration, but there are many different combinations of materials that can achieve a fire-resistance rating. If you are building fire-resistance-rated assemblies for a specific project, the approved construction documents should provide a detailed description of which materials to use and how the materials are to be configured to meet the FRR requirement.

2.3.3 Gypsum Board Used for Fire Protection

Wall and ceiling assemblies built with standard gypsum board cannot have an FRR because fire-resistance-rated assemblies must be able to withstand fire for at least one hour, as determined by tests conducted under laboratory conditions. Gypsum board, when installed in combination with certain materials, can protect building elements from fire, minimizing structural damage, slowing the spread of a fire, and allowing more time for people to escape.

Certain types of gypsum board provide greater protection times in fire-resistance-rated wall and ceiling assemblies. Type X gypsum board contains noncombustible glass fibers in its gypsum core, which keep the board from shrinking during the calcination that happens during a fire. Because Type X gypsum board retains its shape, it continues to act as a fire barrier even after calcination. This characteristic is vital, because any gap in a fire-resistance-rated assembly can allow a fire to spread. For example, one layer of $\frac{5}{8}$" Type X gypsum board, used with wood studs at 16" on-center, can be a part of a one-hour fire-resistance-rated assembly. Type X gypsum can also be used to protect mass timber elements in buildings up to 18 stories tall based on provisions adopted in the 2021 *IBC*® (*Figure 6*).

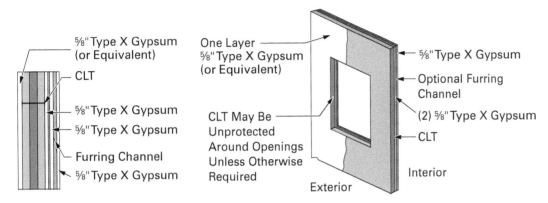

Figure 6 Type X gypsum board used to protect mass timber.

Type C gypsum board is an enhanced and more expensive version of gypsum board that can also be used in fire-resistance-rated assemblies. In addition to glass fibers, it contains unexpanded vermiculite in its gypsum core, which expands when heated to keep the board from shrinking and causing a gap in the assembly.

These two types of fire-rated gypsum board are not interchangeable, because assemblies that use Type C board typically have a higher FRR than assemblies that use Type X board. Make sure that materials used in a project meet all local codes and conform to the approved construction documents. Any substitutions need to be approved by the registered design professional or the building official.

2.3.4 Firewalls

One option available to the designer to increase the size of the structure is to construct a *firewall*. When constructed in accordance with the *IBC*®, each side of the structure on each side of the wall is considered a separate building. Therefore, the size of the building can be increased. A firewall can also serve to separate different types of construction. A firewall is the most highly regulated fire assembly in a structure. Firewalls that are located on lot lines are called *party walls* and are not permitted to have any openings. These walls are designed to prevent fire from passing through them into the adjoining space. Proper construction can provide fire protection for up to four hours.

Firewalls are assigned a FRR prior to construction in accordance with the local code. There are different methods available to meet these requirements. The FRR required, and the local codes, determine which method is used.

This section covers several systems designed around the use of gypsum board and steel framing to meet FRR requirements (*Figure 7*). The first system (top of *Figure 7*) requires a void space between the finished walls for the adjoining spaces. The firewall is constructed as a nonbearing wall in a void space between

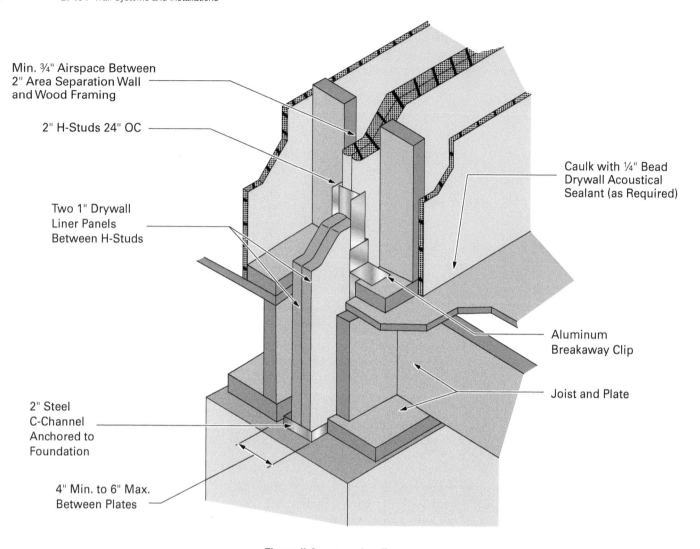

Min. ¾" Airspace Between
2" Area Separation Wall
and Wood Framing

2" H-Studs 24" OC

Two 1" Drywall
Liner Panels
Between H-Studs

2" Steel
C-Channel
Anchored to
Foundation

4" Min. to 6" Max.
Between Plates

Caulk with ¼" Bead
Drywall Acoustical
Sealant (as Required)

Aluminum
Breakaway Clip

Joist and Plate

Firewall Construction Features

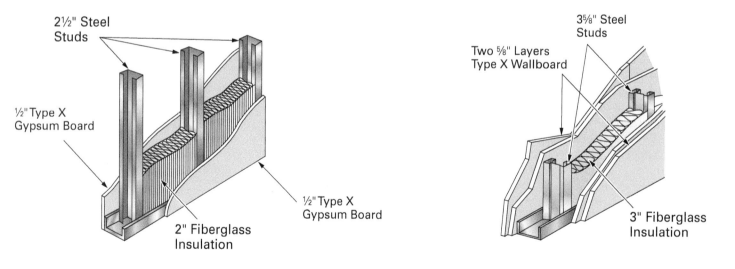

2½" Steel
Studs

½" Type X
Gypsum Board

2" Fiberglass
Insulation

½" Type X
Gypsum Board

3⅝" Steel
Studs

Two ⅝" Layers
Type X Wallboard

3" Fiberglass
Insulation

One-Hour, Fire-Resistance-Rated Wall

Two-Hour, Fire-Resistance-Rated Wall

Figure 7 Typical firewall construction.

the two building areas. Special steel components, called C-channels, are provided to mount two 1"-thick pieces of gypsum board placed face-to-face. They are supported along the edges by H-style supports. A single H-style support is placed at the top of the first two pieces of gypsum board. The lower channels provide a mounting frame for the lower two pieces of gypsum board. The upper H-style support provides a mounting frame for the next two pieces of gypsum board.

Special clips are used to attach the firewall to the interior walls on each side. These clips are designed to fail on the side exposed to the fire. The failed clips allow the loadbearing finish wall on the fire side to collapse. This design leaves the erect firewall—still attached to the other finish wall—in place.

The minimum FRR is based on the occupancy classifications in the adjoining buildings and in some cases the construction types. The bottom two details in *Figure 7* provide examples of one-hour and two-hour fire-resistance-rated assemblies. Firewalls must be constructed of noncombustible materials unless the building is of Type V construction. The openings in a firewall are limited and depend on whether a fire sprinkler system is present. Openings must be protected with a fire-protected door and must latch when closed. Ducts and air transfer openings must be protected with a fire damper.

In the construction of these walls, a sealant can be used to limit the passage of sound and smoke through the structure. Sound-attenuation materials can also be used in the construction. Be sure to follow the manufacturer's instructions when installing a firewall. Proper construction of these walls is critical in preventing the spread of fire between spaces in the building.

2.3.5 Fire Barriers

Fire barriers are similar to firewalls. They are designed to form a required separation to limit the spread of a fire within a building, but they are not required to extend from the foundation to the roof—just from the top of the floor/ceiling assembly to the underside of the floor or roof sheathing above. Fire barriers are intended to divide areas inside the building. For example, they are used to enclose stairwells and exit corridors. These walls are typically constructed using gypsum board for fire protection. Openings in these walls are rated based on the FRR of the wall. Utilities passing through these walls must be firestopped at the same rating as the wall. The firestopping process for utilities is described later in this module.

Shaft walls are used to enclose any multistory shaft space in a building and are constructed as fire barriers. Examples of these openings include elevator shafts (*Figure 8*), laundry chutes, shafts for ductwork, electrical conduits, and pipes. Walls for elevator shafts must be strong enough to withstand the air pressure and suction created by the elevator moving up and down in the shaft. They should also be soundproof to eliminate the noise caused by elevator operation. Shaft walls can be constructed using multiple layers of gypsum board and shaft liner products to enclose the shaft. Their purpose is to prevent fire from penetrating and spreading throughout the building inside these multistory shafts.

2.3.6 Fire Partitions

Fire partitions are typically used to separate dwelling units in an apartment or condominium building, as well as guest rooms in a hotel or motel. Fire partitions are also used to separate corridors from adjacent areas of buildings. These corridors provide a smoke-free environment for people when evacuating during a fire.

Fire partitions must be at least one-hour fire-resistance-rated construction. Doors must have a minimum 20-minute fire-protection rating, be self-closing, and latch. Fire partitions must extend from the floor assembly to either the floor or roof sheathing above, or to the ceiling of a fire-resistance-rated floor/ceiling or roof/ceiling assembly. Ducts and air transfer openings often must be protected with a fire damper, and corridors may require a smoke damper.

2.3.7 Openings and Penetrations

Fire-resistant-rated wall assemblies must be penetrated by doors to allow the flow of traffic within the building. These doors must provide a secure seal in case of fire. Special doors are required to provide these sealed passages. These doors typically are required to have an FRR equal to that of the wall in which they are

Labels for Opening Protectives in Fire-Resistance Rated (FRR) Assemblies

Codes generally require that only labeled windows, doors, and door frames be installed in fire-resistance-rated assemblies. Labels provide information on the FRR of the window or door assembly. The labels should never be painted over or covered in any fashion.

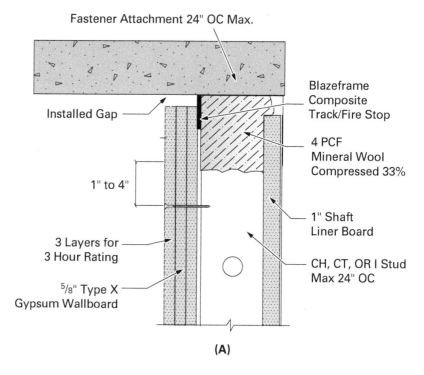

Fastener Attachment 24" OC Max.

Installed Gap

Blazeframe
Composite
Track/Fire Stop

4 PCF
Mineral Wool
Compressed 33%

1" to 4"

1" Shaft
Liner Board

3 Layers for
3 Hour Rating

CH, CT, OR I Stud
Max 24" OC

⁵/₈" Type X
Gypsum Wallboard

(A)

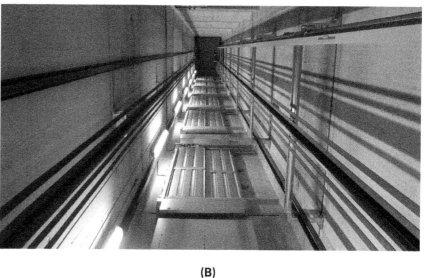

(B)

Figure 8 Shaft-wall construction examples.
Source: Wolfgang Beck/Shutterstock (9B)

installed. Windows can be mounted in the door to provide a view of the other side so the door can be opened safely. The thickness of the glass is determined by the required FRR. Wire reinforcing may be incorporated into these windows.

Fire-door assemblies consist of a fire door, frame, hardware, and accessories such as closers. Each of these components is required to be designed and tested to provide smoke and draft control. Fire-door assemblies are required to latch so the pressure of a fire does not push the door open. They are also typically required to be installed so they close automatically or on their own, so they are not left open under normal conditions. NFPA 80 is the standard for fire doors and fire windows. This standard establishes minimum criteria for installing and maintaining assemblies and devices used to protect openings in walls, floors, and ceilings from penetration by fire and smoke. FRRs are given in terms of the time that the door, window, or device will prevent the fire from passing through to the adjoining space.

Openings

Opening ratings typically correspond to the assembly FRR. For walls requiring 3- or 4-hour FRR, the openings require 3-hour ratings. Walls with 2-hour FRR require 90-minute doors. One-hour fire barriers require 45-minute fire doors and shutters. As previously noted, doors in one-hour fire partitions used for corridors must have a fire-protection rating of 20 minutes.

Fire-resistance-rated door frames are either pre-assembled at a manufacturing facility or assembled on site. They must be securely anchored at the jamb and at the floor in accordance with the manufacturer's instructions.

Door hardware is either provided by the manufacturer or by the contractor performing the installation. In either case, the manufacturer must have prepared the door and frame to receive the hardware to ensure the FRR is maintained. Fire doors are mounted using steel ball-bearing hinges and are required to be self-closing, either with self-operated closers or a fail-safe closing system. The fail-safe system automatically closes the door when a fire is detected by the system. Pairs of doors must be equipped with coordinators to make sure both doors close properly. Both the door head and jambs must be sealed with gaskets when smoke control is also a requirement.

Penetrations

Penetrations through fire-resistance-rated assemblies require protection. Modern building design and the use of fire-resistant insulation have effectively limited the spread of fire. However, smoke has proven to be more deadly than the fire itself because people can be easily overcome by smoke. New methods have been developed to limit the passage of smoke through structures.

Services such as plumbing, telecommunication lines, and electrical cabling are required to pass through fire-resistance-rated assemblies (*Figure 9*). These openings could provide a path for fire to spread vertically or horizontally through the building. Firestopping materials are available for sealing these penetrations and maintaining the FRR of the wall or floor. They also limit the ability of smoke to penetrate other areas in the structure. The materials are designed to expand when exposed to fire. In some cases, the materials are capable of expanding up to three

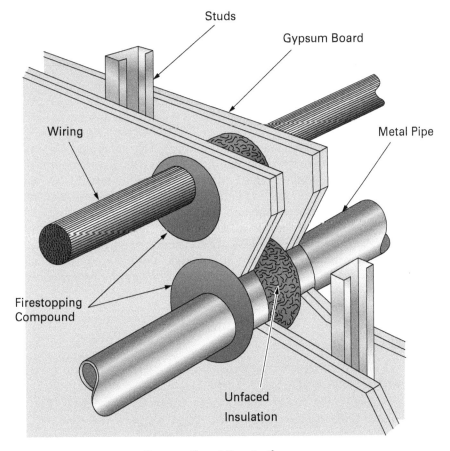

Gypsum Board Penetration

Figure 9 Firestopping openings.

NOTE

Fireproofing is not a code-defined term. However, it is used in the trade to explain the practice of applying fire protection material to extend the time something can be exposed to fire before it is compromised. In construction, fireproofing is referred to as *passive fire protection*, as discussed in Section 2.3.8.

or four times their original size. This allows the material to block the opening even if the fire burns through the wire, cable, or pipe passing through the wall or floor.

When any such material is used on the jobsite, special care must be taken to follow the manufacturer's instructions exactly. Only by proper installation of the material will it function as designed to stop the fire and smoke from penetrating the opening.

2.3.8 Fire Protection Materials

Among the most important concerns in any building is fire protection. Building codes mandate either active or passive fire protection, or both, for a structure. Passive fire protection materials can dramatically reduce fire damage to a structure. These materials work by slowing the transfer of heat from a fire to the structure by providing thermal separation between the fire and the structural elements. Structural frame members constructed of concrete and masonry often provide the required fire protection by insulating the steel reinforcement within the concrete or masonry (*Table 3* and *Table 4*). The thickness of the protection is listed in the building code or designed by an engineer and is called *cover*, which is the thickness of concrete or grout over the reinforcing steel.

TABLE 3   Minimum Dimensions of Concrete Columns (Inches)

Types of Concrete	Fire-Resistance Rating (Hours)				
	1	$1\frac{1}{2}$	2	3	4
Siliceous		8	9	10	12	14
Carbonate		8	9	10	11	12
Sand-lightweight		8	$8\frac{1}{2}$	9	$10\frac{1}{2}$	12

Source: 2021 *IBC®* Table 722.2.1 where minimum concrete cover is 1 inch times the number of hours of required fire-resistance rating or 2 inches, whichever is less.

TABLE 4   Minimum Dimensions of Reinforced Clay Masonry Columns for Fire-Resistance Rating (inches)

Column Size	Fire-Resistance Rating (Hours)			
	1	2	3	4
Minimum column dimension (inches)	8	10	12	14

Source: 2021 *IBC®* Table 722.4.1(6) where minimum cover for longitudinal reinforcement is 2 inches.

Steel structural members are typically protected by spraying a fire-resistant material to insulate the steel from the heat of the fire. These materials often require special inspection. As already discussed, gypsum board is also used to protect structural members such as light-frame wood and cold-formed steel, but it can also be used to protect concrete, masonry, steel, and mass timber. When used with active systems such as water sprinklers, passive fire-protection elements significantly slow the progress of a fire, thereby reducing fire damage.

Fire protection requirements are specified in the code and by local laws. Materials are usually rated by the time they can delay the progress of a fire, usually in 15-minute increments.

Did You Know?

Fire Protection Materials

The *International Building Code®* (*IBC®*) 2021, Section 202, defines types of fire protection materials, such as the following:

- *Intumescent Fire-Resistant Coatings:* Thin film liquid mixture applied to substrates by brush, roller, spray, or trowel that expands into a protective foamed layer to provide fire-resistant protection of the substrates when exposed to flame or intense heat.
- *Sprayed Fire-Resistant Materials:* Cementitious or fibrous materials that are sprayed to provide fire-resistance protection of the substrates.

Many different types of fire protection materials are available, including the following:

- Cement products, which work by absorbing heat from the fire as water evaporates from the coating.
- Subliming products, which when heated change from a solid to a gas without going through the liquid stage. They absorb heat as they change.
- Intumescent materials turn into foam and swell when exposed to heat, forming an insulative barrier between the fire and the structure.
- Fibrous materials made into boards and blankets are often used for fire protection because some do not burn at lower temperatures. Ceramic fibers are a good choice because they do not burn in the first stages of a fire, and some can withstand up to 2,100°F (1,150°C).

The method of fire protection of a wall depends on the materials used and the base to which the fire protection materials are applied. Concrete structures are protected somewhat differently than wood structures. Some of the methods for protecting walls are described as follows. The construction drawings will contain details on fire protection required for the specific project.

In many commercial buildings, the structural steel used for girders and columns must be encased in fire-retardant material in order for the structural element to meet the required FRR. The amount of fire protection material used depends on the FRR required. Two types of fire-retardant material are used: noncombustible, and combustible materials that are treated with a fire retardant.

Combustible materials include wood, fiberboard, paper, felt, plastics, asphalt, and pitch. Noncombustible materials are mineral based, and include stone, marble, concrete, gypsum plaster, and glass. As previously discussed, gypsum board is often used as a fire protection material. When concrete is used as a fire protection material, forms and shoring similar to those used in forming concrete beams and girders are used. *Figure 10* shows methods used to protect steel columns.

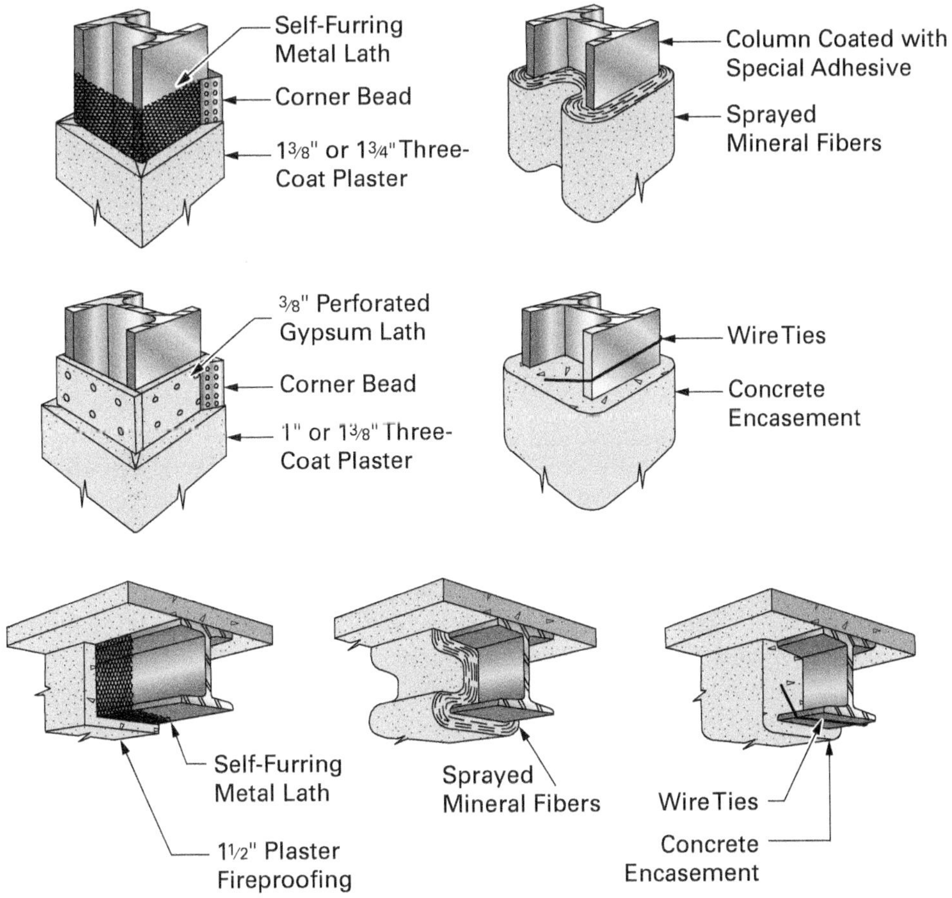

NOTE

Most fire protection material must be installed by a certified contractor.

Figure 10 Fire protection of steel columns.

One method involves the installation of a protective layer of plaster over the steel support column. Metal or perforated gypsum lath is wrapped around the column so the plaster can adhere. Another method involves spraying the support column with a layer of fire-resistant mineral fibers (*Figure 11*). The third method is to cover the column with concrete. The FRR for any of these methods depends on the thickness of the protective layer applied to the column.

Figure 11 Column sprayed with fire-resistant material.

2.4.0   Shear Walls

Shear wall: A wall that strengthens buildings by resisting lateral forces caused by soil loads, high wind, and/or seismic activity.

Braced wall: A nonbearing exterior wall that is set into, and attached to, the structure of a building. This type of wall is designed to strengthen walls and resist lateral forces.

Lateral forces: Forces exerted on the sides of a building due to soil loads, wind, or seismic activity.

Natural forces from high winds, earthquakes, and even soil loads can exert strong lateral forces on buildings. A **shear wall** is a type of advanced wall that strengthens buildings by resisting these types of forces. The *International Building Code* (*IBC*®) uses the term *shear* wall, while the *International Residential Code* (*IRC*®) refers to this type of wall as a **braced wall**. Both terms refer to walls that resist lateral forces exerted on buildings because of wind and seismic activity.

Lateral forces exert force or pressure on the sides of a building. In simple terms, shear walls prevent buildings from twisting, sliding, or racking. Unlike load-bearing walls designed to support the weight of the structure above it, shear walls use a vertical cantilevered diaphragm to resist lateral forces. Shear walls and braced wall panels serve the same purpose: to transfer the lateral load, or shear, through the sides of a building to the foundation.

Shear walls are used in both residential and commercial buildings. Types of shear wall construction include the following:

- *Vertical studs and sheathing* — This type of wall is common in residential buildings. Vertical wood or cold-formed steel studs and blocking are strengthened with sheathing made from Oriented Strand Board (OSB) or plywood. Interior gypsum board can also provide additional wall capacity.

- *Midply* — This type of wall uses one ply of sheathing material located at the wall's center between pairs of studs and plates positioned 90-degrees relative to those in standard shear walls. Steel rods are typically used at the end of a midply wall.
- *Reinforced concrete* — This type of wall is commonly used in larger commercial buildings. These walls use steel-reinforced concrete. Reinforcement can be provided in both vertical and horizontal directions.
- *Steel* — This type of wall uses large steel sheets, boundary columns, and horizontal beams. Steel braced frames and moment frames are examples of steel shear walls created with bolted or welded connections.
- *Concrete masonry units* — This type of wall is constructed of concrete masonry units (CMUs), steel reinforcement, and grout that fills voids in the CMU block.

2.4.1 Residential Shear Walls

Shear walls are common in light-frame construction such as one- and two-family dwellings, townhouses, apartment buildings, and offices or business occupancies up to 5 or 6 stories. They typically are built with vertical studs, horizontal blocking, and wood structural panels (WSPs). Additionally, they often use braced frames and/or diagonal rods for added strength. Anchor bolts and hold-downs (*Figure 12*) are typically used to secure the shear wall to the foundation.

Figure 12 Hold-downs and anchor bolts used with a shear wall.

When installing shear walls composed of vertical studs and WSP sheathing, observe the following rules:

- Check the stamp on each WSP to ensure that the sheathing is rated and complies with local building codes.
- WSP sheathing is commonly installed vertically with the long dimension parallel to the studs. It can also be installed horizontally with the long dimension perpendicular to the studs. For horizontally applied WSP, intermediate blocking to support panel edges might be specified on the approved construction documents. However, blocking may not be required. Always carefully check the construction documents for proper shear wall fabrication.
- Select panels long enough for your particular shear wall. Walls with longer panels typically require less blocking between the vertical studs.

Uplift Forces

Uplift forces due to wind loads exert force or pressure on a building in an upward vertical direction. In some cases, the building elements are heavy enough to offset this vertical upward force. In other cases, particularly with light-frame steel or wood construction, uplift hardware is used to counteract these forces. This is often an integral part of shear wall design and detailing.

- Make sure nail size and spacing requirements are in accordance with construction documents. Nail size and spacing is one of the most important aspects of shear wall strength.
- Ensure anchor bolts, uplift straps, hold-down devices, and any other hardware specified in the construction documents are properly located and installed.

2.4.2 Commercial Shear Walls

Shear walls play a critical role in larger commercial buildings, especially taller, multistory buildings subjected to strong winds or seismic forces. Commercial shear walls are typically constructed of reinforced concrete and/or steel. Shear walls may be located on the perimeter of a building, or they may exist as core wall systems within the interior (*Figure 13*). In some cases, structural engineers may even design interior elevator shafts to serve as core shear walls.

Figure 13 Interior shear walls during construction of a commercial building.
Source: RuzainiRx/Shutterstock

2.0.0 Section Review

1. A common characteristic of insulated wall systems is that they are created by _____.
   a. computer-controlled technology
   b. layering different types of insulation materials
   c. layering polyurethane foam
   d. combining concrete and brick

2. True or False: Masonry veneers are typically designed to be load-bearing.
   a. True
   b. False

3. To prevent smoke and heat from spreading, penetrations for utilities that pass through fire-resistance-rated walls must be sealed with _____.
    a. firestopping material
    b. plastic
    c. fireproofing materials
    d. acoustic sealant

4. The force or pressure exerted on the side of a building is known as _____
    a. shear wall
    b. load bearing
    c. lateral force
    d. horizontal blocking

Module 27404 Review Questions

1. When working around a power line carrying more than 50kV, the safe minimum distance of 10' must be increased for each additional 1kV by _____.
    a. 0.4'   c. 0.8'
    b. 0.5'   d. 1'

2. When using power tools to work on advanced wall systems, electrical cords should be checked _____ to see if they are in good condition.
    a. by a supervisor
    b. at least once a week
    c. at the end of every shift
    d. before they are used in a wet area

3. Workers should always use _____ ladders when working around power lines.
    a. conductive
    b. steel
    c. non-conductive
    d. aluminum

4. Curtain walls are constructed primarily from _____.
    a. oriented strand board
    b. plywood
    c. glass
    d. gypsum board

5. For structural insulated panels (SIPs), the width of the sill plate is _____ the space between the panels.
    a. half the size of
    b. equal to the size of
    c. twice the size of
    d. three-quarters the size of

6. Insulated wall systems consisting of foam material sandwiched between sheets of OSB or plywood are referred to as _____.
    a. IMPs   c. SIPs
    b. SFPs   d. ICFs

7. Masonry veneer is attached to the structural frame using _____.
    a. noncorroding metal anchors
    b. grout
    c. lag screws
    d. weepholes

8. A fire-resistance rating represents the ability of a wall, floor, roof, or structural member to withstand fire for a period of time as well as to _____.
   a. withstand the penetration of smoke
   b. provide load-bearing support for the building
   c. withstand the force of water sprayed from a hose
   d. create a decorative veneer for the structure

9. _____, which are two- to four-hour fire-resistance rated with protected openings, extend continuously from the foundation to or through the roof.
   a. Fire partitions
   b. Fire barriers
   c. Smoke barriers
   d. Firewalls

10. Type X gypsum board contains _____ in its core, which keeps the board from shrinking during the calcination that happens during a fire.
    a. expanded polystyrene
    b. rigid polyurethane foam
    c. vermiculite
    d. noncombustible glass fibers

11. _____ are used to seal penetrations in fire-resistance-rated assemblies, maintaining the fire-resistance rating of the wall or floor, and limit the ability of smoke to penetrate to other areas in the structure.
    a. Fire partitions
    b. Firestopping materials
    c. Smoke barriers
    d. Firewalls

12. To achieve a two-hour fire-resistance rating, a reinforced concrete column must have a minimum cover thickness of _____.
    a. 1"   c. 3"
    b. 2"   d. 4"

13. Shafts are constructed as _____.
    a. firewalls
    b. fire barriers
    c. fire partitions
    d. smoke barriers

14. Advanced walls that strengthen building by resisting lateral forces caused by high winds, earthquakes, soil load, and/or seismic activity are called _____.
    a. shear walls
    b. shaft walls
    c. load-bearing walls
    d. curtain walls

15. _____ is one of the most important aspects of shear wall strength.
    a. Nail size and anchor bolts
    b. Ratings and building codes
    c. Screw length and diameter
    d. Nail size and spacing

Answers to Odd-Numbered Module Review Questions are found in *Appendix A*.

Answers to Section Review Questions

Answer	Section	Objective
Section One		
1. c	1.1.0	1a
2. c	1.1.0	1a
Section Two		
1. b	2.1.0	2a
2. b	2.2.0	2b
3. a	2.3.5	2c
4. c	2.4.0	2d

Roofing Applications

Source: Felix Mizioznikov/Shutterstock

Objectives

Successful completion of this module prepares you to do the following:

1. Explain the safety requirements for roofing projects.
   a. Identify potential hazards when working on roofs.
   b. Identify proper personal protective equipment (PPE) and hazard control devices used when working on roofs.
2. Identify the tools and fasteners used in roofing.
   a. Identify the hand tools used when working on roofing projects.
   b. Identify the power tools used when working on roofing projects.
   c. Identify fasteners used on roofing projects.
3. Identify the different roofing systems and their associated materials.
   a. Identify composition shingles and their applications.
   b. Identify tile/slate roofing materials and their applications.
   c. Identify metal roofing and its applications.
   d. Identify built-up roofing and its applications.
   e. Identify single-ply roofing and its applications.
   f. Explain the purpose of underlayment and waterproof membrane.
   g. Discuss the purpose of drip edge, flashing, and roof ventilation.
4. Describe the installation techniques for common roofing systems.
   a. Describe how to properly prepare a roof deck.
   b. Explain how to install composition shingles.
   c. Explain how to install metal roofing.
   d. Discuss roof projections, flashing, and ventilation.
5. Describe the estimating procedure for roofing projects.

Performance Tasks

Under supervision, you should be able to do the following:

1. Demonstrate how to install composition shingles on a specified roof and valley.
2. Demonstrate the method to properly cut and install the ridge cap using composition singles.
3. Lay out, cut, and install a cricket or saddle.
4. Demonstrate the techniques for installing other selected types of roofing materials.

NOTE

Codes vary among jurisdictions. Because of the variations in code, consult the applicable code whenever regulations are in question. Referring to an incorrect set of codes can cause as much trouble as failing to reference codes altogether. Obtain, review, and familiarize yourself with your local adopted code.

Digital Resources for Carpentry

Scan this code using the camera on your phone or mobile device to view the digital resources related to this craft.

Overview

As you travel, take note of various types of residential and commercial structures. Note the kinds of roof construction, and how many different types of roofing materials are used. Part of your work as a carpenter will involve preparing roof decks to receive the finish roofing material, and you may even install roofing material. The roof is the most vulnerable part of a building. If not properly installed, it will leak. In some situations, it could even collapse. Safety is always a major consideration when working on a roof.

NCC:R   Industry-Recognized Credentials

If you are training through an NCCER-accredited sponsor, you may be eligible for credentials from NCCER. The ID number for this module is 27202. Note that this module may have been used in other NCCER curricula and may apply to other level completions. Contact NCCER at 1.888.622.3720 or go to **www.nccer.org** for more information.

You can also show off your industry-recognized credentials online with NCCER's digital credentials. Transform your knowledge, skills, and achievements into credentials that you can share across social media platforms, send to your network, and add to your resume. For more information, visit **www.nccer.org**.

1.0.0 Roofing Safety

Performance Tasks

There are no Performance Tasks in this section.

Objective

Explain the safety requirements for roofing projects.

a. Identify potential hazards when working on roofs.

b. Identify proper personal protective equipment (PPE) and hazard control devices used when working on roofs.

Pitch: The ratio of the rise to the span, indicated as a fraction. For example, a roof with a 6' rise and a 24' span will have a $^1/_4$ pitch.

Roofing materials protect a structure and its contents from the elements. In addition to providing rain protection, some materials are especially suitable for use in areas where fire, high wind, or extreme heat problems exist, or in areas where cold weather, snow, and ice are common. Materials can contribute to the attractiveness of the structure with the careful selection of texture, color, and pattern. However, the design of the structure as well as local building codes may limit the choice of materials because of the pitch of the roof or because of other considerations at a particular location. In every case, the project specifications must be checked to determine the type of roofing materials to be used.

1.1.0 Roofing Hazards

Roofing projects have many inherent hazards, such as working at elevated locations. The roof load capacity must be known so that the weight of roofing materials does not exceed the anticipated load. When possible, place loads directly over rafters or trusses.

Overhead power lines must be identified and carefully considered when working on roofing projects, not only for the workers, but also for equipment operators moving materials to various parts of the roof. Per OSHA (Occupational Safety and Health Administration), a safe distance of at least 10 feet must be maintained between a power line and a piece of equipment for line voltages of 50kV or less. (For voltages above 50kV, the distance must be increased by 0.4 feet for each 1kV of voltage.) Contact the power company to see if the power can be shut down on the lines. If this is not possible, ask if they can place insulation over the lines during the time you will be working in the location. If overhead power lines are present, ensure they are included in the job hazard analysis (JHA). Use nonconductive ladders when working around power lines.

Always complete a JHA before beginning a roofing project to identify the potential hazards and recommend action that can eliminate or minimize each hazard. In addition to working at elevations, the following other potential hazards may be encountered on a roofing project:

- Injuries that may occur from manually lifting roofing materials
- Injuries that may occur when cutting roofing materials
- Respiratory hazards when applying certain types of roofing materials such as asphalt-based roofing materials
- Flying debris caused by cutting or handling roofing materials
- Tools falling from a roof, ladder, or other elevated surface onto workers below

Review the NCCER Module 00101, *Basic Safety* and Module 00109, *Introduction to Materials Handling* to refresh your knowledge on fall protection, working from heights, proper lifting techniques, and handling of materials.

Skylights or roof windows are commonly installed in both residential and commercial structures. Although they cover the open space below, skylights and roof windows are not structural elements and should be treated with care when installing roofing materials. Do not sit or stand on skylights and roof windows, and do not place tools or supplies on them.

Roofers spend a major part of their time working on sloped roofs. Most construction injuries and deaths result from falls. Falls from high places can cause serious injury or death when using the wrong type of fall protection equipment, or when improperly using the right equipment. Any project where workers will be more than 6' off the ground or on an elevated working surface requires a fall protection plan. Appropriate signage also needs to be visible at the worksite (*Figure 1*).

Figure 1 Working at heights.
Source: Phojai Phanpanya/Shutterstock

WORKING AT HEIGHT SIGNS

1.2.0 PPE and Hazard Control

Observing the following guidelines when working on roofs will help ensure your safety and the safety of others:

- Wear boots or shoes with rubber or crepe soles that are in good condition.
- Always wear fall protection devices, even on shallow-pitch roofs.
- Rain, frost, and snow are all dangerous because they make a roof slippery. If possible, wait until the roof is dry; otherwise, wear special roofing footwear with skid-resistant cleats in addition to fall protection.
- Brush or sweep the roof periodically to remove any accumulated dirt or debris.
- Install any required **underlayment** as soon as possible. Underlayment usually reduces the danger of slipping. On sloped roofs, properly fasten underlayment before stepping on it.
- On pitched roofs, install necessary roof brackets as soon as possible. They can be removed and repositioned as shingle-type roofing is installed.
- Remove any unused tools, cords, and other loose items from the roof. They can be a serious hazard.
- Check and comply with any federal, local, and state code requirements when working on roofs.
- Be alert to any other potential hazards such as live power lines.
- Use common sense. Taking chances can lead to injury or death.

Underlayment: Asphalt-saturated felt protection for sheathing; 15 lb roofer's felt is commonly used. The roll size is 3' × 144', or a little over four squares.

When working outdoors or in high-heat conditions for extended periods of time, take precautions to avoid heat exhaustion and exposure to the sun's ultraviolet rays. Preventive measures include the following:

- Wear a hard hat.
- Wear light clothing made of natural fibers.
- If possible, wear tinted glasses or goggles.
- Use a sun protection factor (SPF) 30 or higher sunblock on exposed skin.
- Drink adequate amounts of water to prevent dehydration, especially in arid parts of the country.

1.2.1 Scaffolding and Staging

Scaffolding and staging have been the causes of many minor and serious accidents due to faulty or incomplete construction or inexperience on the part of the designer or craftworker constructing them. Therefore, to avoid hazards caused by faulty or incompetent construction, competent, certified persons should design and construct all scaffolding and staging. A competent, certified person must also inspect and tag the scaffolding/staging every morning or at the change of shift.

Familiarize yourself with safety rules and regulations for scaffolding construction even if you are not the one building or using it. Everyone on the jobsite should thoroughly understand and adhere to the following safety factors:

- Any type of scaffold used should have a minimum safety factor ratio of four to one; that is, it should carry at least four times its intended load. Roofing material weight adds up quickly when placed in one location.
- All staging or platform planks must have end bearings on scaffold edges with adequate support throughout their lengths to ensure the minimum safety factor ratio of four to one.
- Scaffolding timbers, if used, must be carefully selected and maximum nailing used for added strength.
- Because scaffolds are built for work that cannot be done safely from the ground, makeshift scaffolds using unstable objects for support, such as boxes, barrels, or piles of bricks, are prohibited.

When the scaffolding is placed on a solid, firm base and erected correctly, the roofer should be able to work with confidence.

Any scaffolding assembled for use should be tagged with one of the following three colors:

- *Green*—A green tag identifies a scaffold that is safe for use. It meets all OSHA standards.
- *Yellow*—A yellow tag means the scaffolding does not meet all applicable standards. An example is a scaffold where there is no railing because of equipment

Scaffold/Staging Platforms

All scaffold/staging walk boards or platform planks should be laminated wood or aluminum staging planks designed and rated specifically for use as scaffold/staging walk boards or platforms.

Aluminum Extension Plank   **Aluminum Decked Plank**   **Plywood Decked Plank**

Extended Platforms

Aluminum-pole pump jack systems may be extended across the length of a wall using appropriately placed poles supporting properly rated platforms.

interference. Workers can use a yellow-tagged scaffold; however, they must use a safety harness and lanyard. Other precautions may also apply.

- *Red*—A red tag means a scaffold is being erected or taken down. You should never use a red-tagged scaffold.

Other more common types of scaffolding include ladder jacks and pump jacks (*Figure 2*). Pump jacks and, to a lesser extent, ladder jacks are useful for applying the starter strip and lower courses of roofing. Ladder jacks can usually support a 2' wide, adjustable-length platform up to 10' or 18' long. Ladder jacks are attachable to either side of a ladder but cannot be used at heights above 20' or when separated by more than 8' intervals along the length of the platform. Pump jacks using aluminum posts for heights under 50' are movable platform supports that are raised or lowered vertically. They are operated by a foot lever and can raise a person plus a rated load.

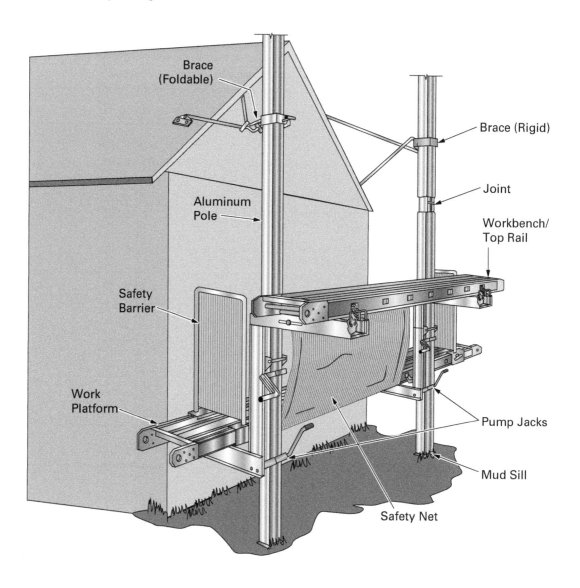

Figure 2 Pump jacks.

Ladder Placement

To check for proper ladder placement, stand straight with your toes touching the base of the ladder and with your arms extended straight out toward the ladder. If you can just touch a rung of the ladder, it is probably positioned properly for climbing.

1.2.2 Ladders

Ladders are useful and necessary pieces of equipment for roofing application and do not cause accidents if properly used and maintained. Causes of ladder accidents include the following:

- Improper use of ladders
- Ladders not secured properly
- Structural failure of ladders

- Improper handling of objects while on a ladder
- Lack of training on proper ladder use

All workers who will be using a ladder must be trained on the proper use of the ladder prior to its use.

Figure 3 shows a ladder erected correctly in relation to the roof eaves. If the ladder is to be left standing for a long period of time, it should be securely fastened at both the top and bottom.

The normal purpose of a ladder used in a roofing application is solely to gain access to the roof itself. For ladder safety, follow these precautions:

- Always use an OSHA-rated ladder sized appropriately for the job.
- Always inspect a ladder for defects or damage before use.
- Always use fiberglass ladders to reduce the possibility of accidental electrocution resulting from contact with power lines. Avoid using aluminum ladders. Aluminum ladders, if used, should never be raised or placed in situations where they can fall or accidentally come into contact with power lines.
- For longer ladders, two people should carry, position, and erect the ladder.
- Always face a ladder and grasp the side rails or rungs with both hands when going up or down.
- Take one step at a time. Always maintain three points of contact with the ladder.
- Remember that an ordinary straight ladder supports only one person at a time.
- Before using a ladder, be sure there is no oil, grease, or sand on the soles of your shoes. Due to the tread composition, some shoe types easily attract foreign objects that can cause you to slip on a ladder.
- Never carry tools or materials up or down a ladder. Use a rope or other device to raise or lower everything so that you can always have both hands free when climbing the ladder.
- Make sure that the base of the ladder is level and has adequate support. Shim the legs or use levelers, if necessary.

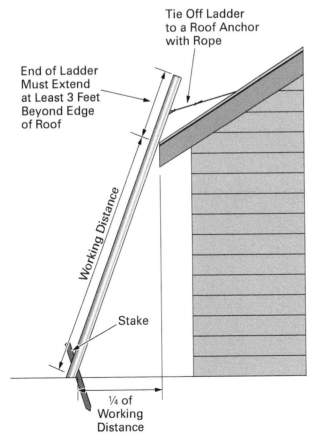

Figure 3 Safe ladder placement.

NOTE

A protective board should be placed along the outside edges of an elevated workbench to prevent tools, equipment, and materials from falling to a lower level and possibly injuring someone.

Bundle: A package containing a specified number of shingles. The number is related to square-foot coverage and varies with the product.

Slope: The ratio of rise to run. The rise in inches is indicated for every foot of run.

- Make sure the ladder is at the correct angle. Always tie off the ladder prior to use and secure it at its top to a rigid support.
- Never overreach.
- Fixed ladders must be provided with cages, wells, ladder safety devices, or self-retracting lifelines where the length of climb is less than 24' but the top of the ladder is at a distance greater than 24' above lower levels.

1.2.3 Material Movement

Construction workers often use mechanical devices to move materials on a jobsite. These devices include conveyor belts, power ladder conveyors (attached to a ladder), forklifts, and truck-mounted hydraulic lifts. Make sure all safety devices are in place before starting.

Exercise extreme caution to ensure that the lift does not make contact with the roof surface or the person unloading each **bundle** of shingles. Immediately distribute the bundles around the roof area. Do not pile them in any one spot. Roof structures have been designed to carry a specific dead load. Placing unnecessary strain on the roof structure by concentrating the shingles in one place on the roof can jeopardize the safety of the workers. The end result might prove to be disastrous, with the collapse of the roof itself. An immediate dispersal of the bundles will prevent any problems from occurring and also make the installation more efficient because you only want to move and place the load once.

1.2.4 Roofing Brackets

Roofing brackets provide firm footing and material storage points on steep-slope roofs. The **slope** of the roof determines what type of roofing bracket, if any, to use. Most roofers feel comfortable on a roof with a 4 in 12 or a 5 in 12 slope. An increased slope of 6 in 12 places more strain on the feet and the body. Therefore, the roofer must be very conscious of the height off the ground and be careful with each and every movement. *Figure 4* shows two types of roofing brackets that can be used with a 10' to 14', defect-free, 2" thick plank.

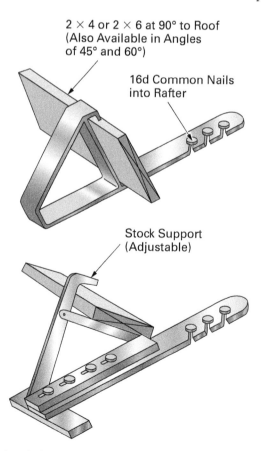

2 × 4 or 2 × 6 at 90° to Roof
(Also Available in Angles
of 45° and 60°)

16d Common Nails
into Rafter

Stock Support
(Adjustable)

Figure 4 Roofing brackets.

Both types of brackets can be nailed firmly to the roof, but the adjustable bracket, which is installed to correspond to the slope of the roof, makes standing and moving around more comfortable. Never get overconfident due to bracket usage. Be aware of the height at which you are working and be cautious.

When installing roof brackets, nail them to the rafters, not just to the **roof sheathing**.

Roof brackets and toeboards alone are not sufficient to meet OSHA fall standards. Any work above 6' requires a proper rail or safety harness.

Roof sheathing: Usually 4 × 8 sheets of plywood, but can also be 1 × 8 or 1 × 12 roof boards, or other new products approved by local building codes. Also referred to as *decking*.

1.0.0 Section Review

1. When doing roofing work, the minimum distance between any piece of equipment and a live power line of 50kV or less is at least _____.
   a. 3 feet
   b. 10 feet
   c. 4 yards
   d. 20 feet

2. Scaffolding with a ____ may be used if applicable precautions are observed.
   a. yellow tag
   b. green tag
   c. blue tag
   d. red tag

2.0.0 Tools and Fasteners

Objective

Identify the tools and fasteners used in roofing.
   a. Identify the hand tools used when working on roofing projects.
   b. Identify the power tools used when working on roofing projects.
   c. Identify fasteners used on roofing projects.

Performance Tasks

There are no Performance Tasks in this section.

Installing roofing requires a variety of hand and power tools. While some of these tools are unique to roofing, carpenters use most of these for other tasks as well. The type of roof being installed determines the fasteners required.

2.1.0   Hand Tools

Many of the tools used for roofing are common to other trades. These common tools include the following:

- Backsaw
- Crowbar
- Handsaw
- Carpenter's level
- Nail apron
- Sliding T-bevel
- Keyhole saw
- Pop riveter
- Chalk line
- Tape measure
- Angle square

- Caulking gun
- Tin snips
- Pry bar
- Scribing compass
- Utility knife
- Framing square
- Claw hammer
- Nail ripper
- Shingle shovel (for roofing material removal)

Other tools are specific to the installation of certain types of roofing. Some of these tools are shown in *Figure 5*.

The roofing hammer, also referred to as a *shingle hatchet*, is used primarily for shingle installation. The hatchet end can split shingles, and the top edge has a sliding gauge to set a dimension for the amount of weather exposure for the shingle.

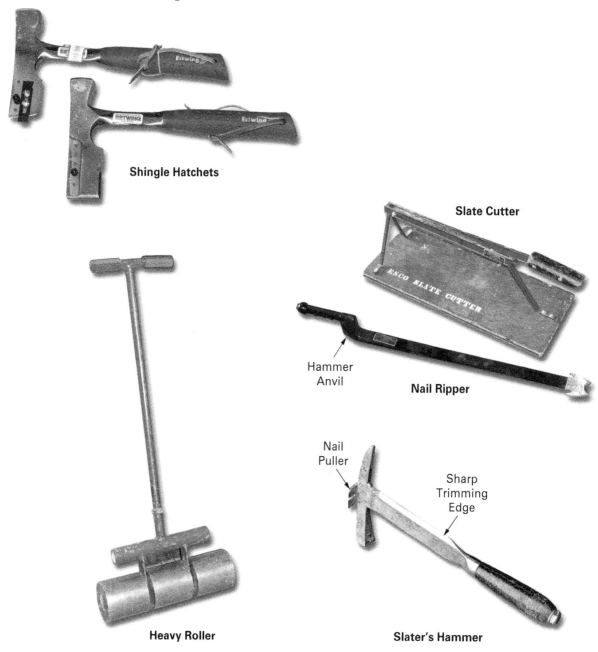

Shingle Hatchets

Slate Cutter

Hammer Anvil

Nail Ripper

Nail Puller

Sharp Trimming Edge

Heavy Roller

Slater's Hammer

Figure 5 Roofing hand tools.

A utility knife can trim or cut all types of composition shingles, including architectural shingles, using either a straight or hooked blade. It can also cut underlayment, cap shingles, roll roofing, and membrane roofing.

Slate-roofing installation usually requires three specialized tools: a slater's hammer, a nail ripper, and a slate cutter. The slater's hammer has a sharp edge for cutting slate and a point for poking nail holes through the slate. The nail ripper has sharp-edged barbs on one end used to shear off nails under a piece of slate. To shear nails, strike the ripper on the face of the anvil with a hammer. The slate cutter aids in the trimming of slate and the punching of nail holes.

Various types of tile cutters and nibblers are used in the installation of tile roofs for splitting or trimming tiles (*Figure 6*). Hand grinders with diamond wheels can cut slots in masonry for flashing installation. A wet saw with a diamond wheel can perform flat or shaped tile cutting on large projects.

Portable brakes can custom bend flashing material for any type of roof installation. Some roofers use heavy rollers to flatten underlayment to eliminate buckling under the finish roof and for the application of cold-cement, fully adhered roll roofs, built-up roofing (BUR), or single-ply membrane roofing. These rollers are sold as vinyl flooring rollers in weights ranging from 75 lb to 150 lb.

Worker safety is important on a construction site. Always use the correct tool for the job and use it safely.

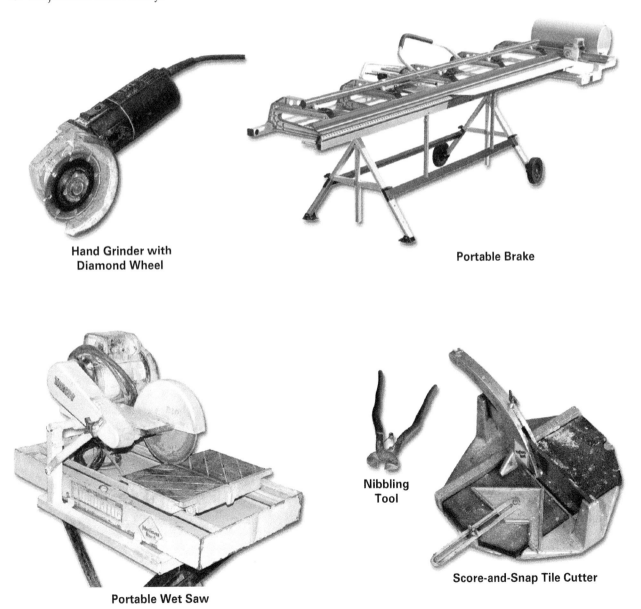

Hand Grinder with Diamond Wheel

Portable Brake

Portable Wet Saw

Nibbling Tool

Score-and-Snap Tile Cutter

Figure 6 Roofing equipment.

2.2.0 Power Tools

Just like the hand tools, many power tools used in roofing are common in other trades, including:

- Power circular saw
- Power saber saw
- Power drill and drill bit sets (regular and masonry)
- Pneumatic nailers

    Rules for the safe use of all power tools include the following:

- Keep all tools in good condition with regular maintenance.
- Do not attempt to operate any power tool before being trained by the instructor or a competent person on that particular tool.
- Use only equipment that is approved to meet Occupational Safety and Health Administration (OSHA) standards.
- Examine each tool for damage before use and do not use damaged tools.
- Always wear eye protection and other appropriate personal protective equipment (PPE) when operating power tools.
- Wear face and hearing protection when required.
- Wear proper respiratory equipment when necessary.
- Wear the appropriate clothing for the job. Always wear tight-fitting clothing that cannot become caught in the moving tool. Roll up or button long sleeves, tuck in shirttails, and tie back long hair. Do not wear any jewelry or watches.
- Do not distract others or let anyone be a distraction while operating a power tool.
- Do not engage in horseplay.
- Do not throw objects or point tools at others.
- Consider the safety of others, as well as yourself.
- Do not leave a power tool unattended while it is running.
- Do not carry, raise, or lower a tool by its electrical cord.
- Assume a safe and comfortable position before using a power tool.
- Do not remove ground plugs from electrical equipment or extension cords.
- Be sure that a power tool is properly grounded and connected to a ground fault circuit interrupter (GFCI) circuit before using it.
- Be sure to unplug portable or stationary power tools at the power source or disable the power source before performing maintenance or changing accessories.
- Do not use a dull or broken tool or accessory.
- Use a power tool only for its intended purpose.
- Keep your feet, fingers, and hair away from the blade and/or other moving parts of a power tool.
- Do not use a power tool with guards or safety devices removed or disabled.
- Do not operate a power tool if your hands or feet are wet.
- Keep the work area clean at all times.
- Become familiar with the correct operation and adjustments of a power tool before attempting to use it. Always follow the manufacturer's instructions pertaining to its intended use.
- Keep a firm grip on the power tool at all times.
- Use electric extension cords of sufficient rating to energize the particular power tool being used.

- Do not use worn or frayed extension cords.
- If a tool or extension cord is defective, bring it to the attention of the supervisor who can tag it and immediately remove it from service.
- Extension cords should not be hung by nails or wire or fastened with staples.
- Report unsafe conditions to your instructor or supervisor.

2.2.1 Power Nailers

There are many types of power nailers. Using one of these tools to apply asphalt or fiberglass shingles can cut the labor time by 50%. Carpenters often use portable electric or pneumatic (powered by air or carbon dioxide gas) units on a jobsite.

WARNING!

Never disable or modify the safety devices on power or pneumatic nailers.

Some of the listed safety practices for power nailers are as follows:

- Always wear safety glasses of an OSHA-approved and ANSI-designated type. The manufacturer of the nailing unit often provides recommendations.
- Before using a power nailer, operators require training.
- Because operating principles vary, study the manufacturer's operating manual.
- Be certain to use the type of fastener required by the manufacturer.
- If pneumatic, make sure to adjust the pressure at the nailer.
- Treat the machine as you would a gun. Do not point it at yourself or others.
- Always keep the unit tight against the surface to drive the fastener correctly.
- When not in use, disconnect the unit from the power source to prevent accidental release of fasteners.
- Keep air lines untangled on the roof to prevent tripping.

Power Nailers

Make sure that the nailer is equipped with a flush-mount attachment or that the impact pressure of the tool can be regulated at the tool to prevent overdriving the nails and cutting through the roofing material.

Pneumatic Roofing Nailer with Plastic Washer Attachment for Underlayment Application

Typical Pneumatic Roofing Nailer

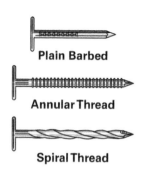

Plain Barbed

Annular Thread

Spiral Thread

Figure 7 Typical composition shingle nails.

Square: The quantity of shingles needed to cover 100 ft² of roof surface. For example, square means 10' square, or 10' × 10'.

Asphalt roofing cement: An adhesive that is used to seal down the free tabs of strip shingles. This plastic asphalt cement is mainly used in open valley construction and other flashing areas where necessary for protection against the weather.

Valley: The internal part of the angle formed by the meeting of two roofs.

Saddle: An auxiliary roof deck that is built above the chimney to divert water to either side. It is a structure with a ridge sloping in two directions that is placed between the back side of a chimney and the roof sloping toward it. Also referred to as a *cricket*.

2.3.0 Fasteners

Roofing nails or roofing cement typically fasten roofing materials to the underlayment. Always refer to the roofing manufacturer to determine the appropriate size and type of fasteners or the proper roofing cement.

2.3.1 Roofing Nails

Carpenters need to use nails of the proper length and made of a material that is compatible with or the same as the drip edges and flashing to fasten common roofing materials. *Figure 7* shows the most common nails used for composition shingles, and *Table 1* shows the typical nail length required for different applications. These nails are usually available in galvanized steel; some other types are aluminum, stainless steel, or copper. Normally, slate roofs require copper slater's nails, and tile roofs require stainless-steel nails. Fastening composition shingles requires about 1 pound of nails per **square** (100 ft²). For other types of nailed roofing, follow the manufacturer's recommendations. Check your local code for nail penetration requirements.

Using pneumatic-powered or electric-powered nailing guns to install composition roofs and underlayment makes the process much faster.

TABLE 1 Typical Nail Lengths

Nailing Application	$^3/_8$" Plywood or Waferboard	1" Sheathing
Strip or single (new construction)	$^7/_8$"	$1^1/_4$"
Over old asphalt layer	1"	$1^1/_2$"
Re-roofing over wood shingles	—	$1^3/_4$"

2.3.2 Cold Asphalt Roofing Cement

Carpenters can use cold **asphalt roofing cement**, consisting of modified asphalt and/or coal tar products, for the installation of composition shingles, underlayment, roll roofing, and cold asphalt built-up roofing (BUR). It is available in liquid non-fibered form or in plastic-fibered form. The non-fibered form is usually used in the lapped installation of underlayment and spread over large areas with a spreader or mop. The plastic-fibered cement is used for spot repairs or cementing nail heads, shingle tabs, **valley** overlaps, and **saddle** materials, and as an exposed sealer for gaps or flashing. Asphalt cement is generally available in 1-gallon and 5-gallon pails.

2.0.0 Section Review

1. When working with slate roofing, a nail ripper is used to _____.
   a. cut or trim slate tiles
   b. shear off nails under slate tiles
   c. poke nail holes through slate tiles
   d. hammer nails through slate tiles

2. Using a power nailer to install asphalt or fiberglass shingles can reduce labor time by _____.
   a. 10%
   b. 25%
   c. 35%
   d. 50%

3. Tile roofs are usually installed using nails made from _____.
   a. copper
   b. galvanized steel
   c. stainless steel
   d. aluminum

4. True or False: Asphalt roofing cement must be heated before application.
   a. True
   b. False

3.0.0 Roofing Systems and Materials

Objective

Identify the different roofing systems and their associated materials.
   a. Identify composition shingles and their applications.
   b. Identify tile/slate roofing materials and their applications.
   c. Identify metal roofing and its applications.
   d. Identify built-up roofing and its applications.
   e. Identify single-ply roofing and its applications.
   f. Explain the purpose of underlayment and waterproof membrane.
   g. Discuss the purpose of drip edge, flashing, and roof ventilation.

Performance Tasks

There are no Performance Tasks in this section.

Many types of roofing materials, as well as roofing systems, are available today. This section will cover the most common materials used on residential and commercial structures.

3.1.0 Composition Shingles

Composition shingles (*Figure 8*) are the most common roofing material in North America. They are available in a wide variety of colors, textures, types, and weights (thicknesses). Three-tab shingles are made of a fiber or fiber-mat material, coated or impregnated with asphalt, and then coated with various mineral granules to provide color, fire resistance, and ultraviolet protection.

In the past, composition shingles were made using asbestos fiber or organic fiber and were commonly referred to as *asphalt shingles*. The manufacture of asbestos-fiber shingles has been prohibited because asbestos poses a cancer risk and environmental disposal hazard.

Organic fiber shingles, which have a life span of 15 to 20 years, have been largely replaced by asphalt-coated, fiberglass mat shingles, simply called fiberglass shingles, which have a life span of 20 to 25 years.

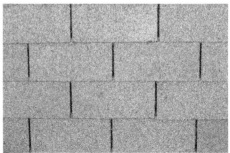

(A) Composite Shingle   **(B) Architectural Shingle**   **(C) Three-Tab Shingle**

Figure 8 Typical composition shingles.
Sources: Richard Goldberg/Shutterstock (8A); Oleksandr Rado/Alamy Images (8B); Phil Berry/Shutterstock (8C)

More expensive architectural shingles with life spans of 25 to 40 years or more are also available. Architectural shingles have multiple layers of fiberglass laminated together when manufactured or job-applied in layers to create a heavy shadow effect (*Figure 9*). Manufacturers may also provide their shingles in fungus- and/or algae-resistant versions for use in damp locations where unsightly fungus growth or black streaks caused by algae tend to be a problem. Copper granules incorporated with the mineral aggregates bonded to the surface of the shingle provide the fungus and algae resistance.

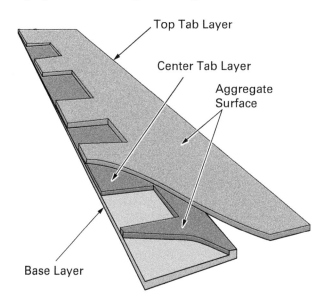

Figure 9 Typical architectural shingle.

Composition shingles generally suit every climate in North America and can normally be applied to any roof with a slope of 4 in 12 (*Figure 10*) up to 21 in 12, provided they have factory-applied, seal-down adhesive strips. With the

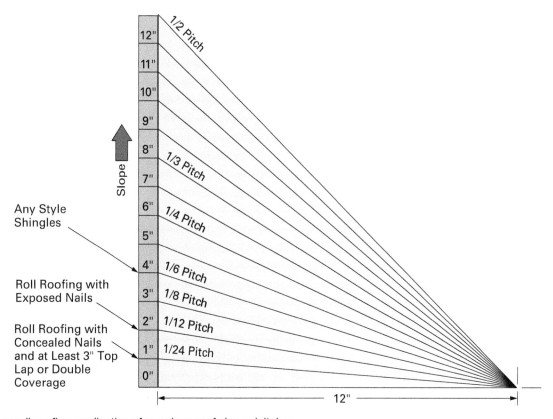

Figure 10 Typical shingle or roll-roofing applications for various roof slopes/pitches.

application of double-lap underlayment, they can be used on roofs with a slope as low as 2 in 12. However, unless appearance is a problem, roll roofing or membrane roofing is usually recommended on slopes lower than 4 in 12.

Steep Slope Applications

On steep-slope roofs greater than 21 in 12, such as a mansard or simulated mansard roof, the built-in sealant strips on shingles do not generally adhere to the underside of the overlaying shingles. Because of this, each shingle must be sealed down with spots of quick-setting asphalt cement at installation. Standard three-tab shingles need two spots of cement under each tab. Strip shingles need three or four spots under the strip. Laminated architectural shingles usually get extra nails and four spots of cement.

Standard three-tab composition shingles are supplied in squares (100 ft^2) with three bundles per square. They are also labeled by their weight per square (210 lb, 230 lb, 240 lb, or 250 lb shingles). The heavier the shingle, the longer its life.

Special Architectural Shingles

These shingles are very expensive and are used primarily in elegant residential applications. Various styles of special architectural shingles that provide different pattern effects on a roof are available.

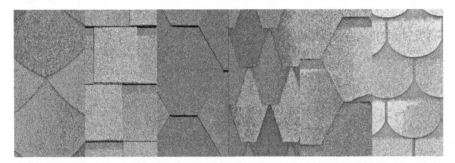

Source: Ivan Kovbasniuk/Shutterstock

Roll Roofing

Roll roofing is available in 50 lb to 90 lb weights with the same materials and colors as composition shingles. While inexpensive and quick to apply, it typically only lasts 5 to 12 years. Roll roofing works well on shallow-slope roofs where appearance is not a problem and as valley flashing to match the color of the roof shingles.

Selvage or Double-Coverage Roll   **Standard Roll**

Metric Shingles

Most shingles manufactured today are available in foot-pound (English) and metric dimensions. The standard foot-pound, three-tab shingle or laminated architectural shingle is 36" long. The metric versions are 39$\frac{3}{8}$" long.

3.2.0   Tile/Slate Roofing

Tile and slate are expensive as roofing materials and are heavier than composition shingles. Slate is not very popular today because it is so expensive. Although tile and slate are not common roofing materials, it is important to be able to recognize the materials and understand the proper methods of installation.

3.2.1   Tile Roofing

Like slate, glazed and unglazed clay and ceramic tile roofing products (*Figure 11*) are fire resistant and rot resistant, and last from 50 to 100 years. Tile roofing is expensive and heavy (7 to 10 pounds per square foot) and requires appropriate roof framing to support the tile weight and any other anticipated loads. In the South, and to a great extent in the Southwest areas of the country, tile roofing is very popular because it is fireproof and impervious to damage caused by intense sunlight. It is available in Spanish, Mission (barrel or S-style), and other styles known by various names such as French, English, Roma, and Villa. Other styles include flat tiles that may resemble slate or wood shakes. Matching hip and ridge tiles, rake/barge tiles, and hip starter tiles are also available. Every tile for a particular style furnished by a manufacturer is a uniform size in width, thickness, and length.

Tiles are fastened with noncorrosive nails (copper, galvanized, or stainless steel). Copper nails have the advantage of being soft enough to allow expansion and contraction without causing the tiles to crack if they are fastened too tight. They also allow easier replacement of any broken tiles. Because of its durability and resistance to any alkaline corrosion, copper is generally used as flashing.

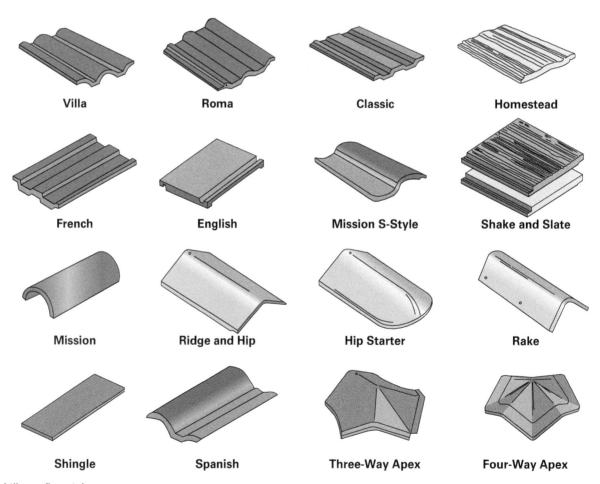

Figure 11 Typical tile roofing styles.

3.2.2 Slate Roofing

Today, real slate roofing (*Figure 12*) is probably the most expensive roofing option in terms of roof framing materials, roofing materials, and increased installation time. Even though it has a very long service life of 60 to 100 years or more, it is usually not financially practical to use slate unless it is required or desired strictly for architectural purposes. No other material can match the high-quality look of slate. Unfortunately, besides being expensive, slate is heavy—about 7 to 10 pounds per square foot. As a result, the roof framing must be engineered to be substantially stronger to support the slate load, as well as any anticipated snow and ice loads.

Figure 12 Slate roofing.
Source: jamesforddesign/Getty Images

Slate is completely fireproof and rot proof. It is available in a wide variety of grades, thicknesses, and colors ranging from gray or black to shades of green, purple, and red. The colors are qualified as unfading or, if subject to some change, weathering. The industry often uses the old federal grading system of A (best) through C to indicate the quality of slate; however, architectural specifications normally use the American Society for Testing and Materials (ASTM) International testing numbers to specify the desired slate quality. New York and Vermont slate types are very durable and have a uniform color and a straight, smooth grain running lengthwise. Slate types with streaks of color usually contain intrusions of sand or other impurities and are not as durable. Premium-quality slate in various colors can be obtained from quarries from Maine to Georgia. Copper slater's nails and flashing are normally used with slate roofs.

The Hazards of Working on Slate Roofs

Take extreme care when walking on slate roof surfaces and always wear fall protection and soft-soled shoes. Avoid walking on a damp or wet slate roof, which is very slippery. Workers should install numerous roof brackets and boards (roof jacks) to keep themselves and roofing material on the roof. They also need to maintain an eave debris safety net and a roped-off zone away from the eaves. Falling slate can cause severe injury or death to persons under the eaves. Because of the weight, do not stack large quantities of slate on any one roof jack. Spread smaller quantities of slate out over numerous roof jacks.

Synthetic slate is made of fiber mat (usually fiberglass) that has been impregnated and coated with cement. It looks like real slate but is lighter in weight. Even though it is lighter than slate, it still requires strong roof framing for support. Synthetic slate is fireproof and can last 40 years or more. Synthetic slate is still a relatively expensive material and is typically used on structures that require historically appropriate materials. The synthetic slate material is difficult to cut and must be carefully fastened to avoid cracking the shingles.

3.3.0 Metal Roofing

Metal roofing is available in a great variety of materials and styles. Companies can purchase these materials with a baked enamel, ceramic, or plastic coating or without a coating so that they can apply a separate roof coating after installation. Common metal roofing materials include the following:

- Aluminum (plain or coated)
- Galvanized steel (plain or coated)
- Terne metal (heat-treated, copper-bearing steel hot-dipped in metal consisting of 80% lead and 20% tin)
- Aluminum- or zinc-coated steel
- Stainless steel

Although they are typically more expensive to install, metal roofs provide many benefits, including the following:

- *Longevity*—Have a life span of 50 years or more
- *Durability*—Can withstand severe weather events
- *Fire resistance*—Resist fire and smoke damage
- *Low maintenance*—Clean easily and require little maintenance
- *Energy efficiency*—Can save energy by reflecting the sun's heat
- *Eco-friendliness*—Made from a high percentage of recycled materials and are 100% recyclable
- *Moss and fungus resistance*—Resist the growth of moss, mold, and fungus
- *Maximum snow and water shedding*—Resist the accumulation of snow and ice

Besides the common residential panel roofing styles (*Figure 13*), many new engineered/preformed, architectural metal fascia/roofing systems are available for commercial or residential use in a wide selection of colors and in styles (*Figure 14*) that include shingles, panels, and tiles.

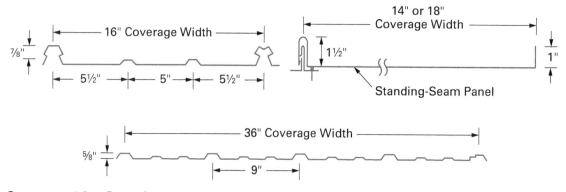

Figure 13 Common metal roofing styles.

(A) Residential Metal Roofs

(B) Metal Roofs Color Samples

Figure 14 Metal roofing options.
Sources: claylib/Getty Images (14A); Pornprasit/Shutterstock (14B)

3.4.0 Built-Up Roofing

Conventional built-up roofing (BUR) membrane (*Figure 15*) has been used for over 100 years on very low-slope roofs of residential and commercial structures. While it is still a viable form of roofing, it is gradually being replaced by pre-manufactured membrane roofing systems.

Built-up roofing, which is field fabricated, consists of three to five layers of heavy, asphalt-coated polyester or fiberglass felt embedded in alternate layers of hot-applied or cold-applied bitumen (coal tar based or asphalt based) that is the waterproofing material. Most commercial applications use hot asphalt as the waterproofing material. The top surface layer of bitumen is sometimes left smooth, but is most often covered with a mineral-coated cap sheet or embedded with a loose mineral surface consisting of small aggregates such as washed gravel, pea rock, or crushed stone.

Hot asphalt applications require special heating equipment along with some method of transporting the hot asphalt to the roof unless the heating equipment is lifted and positioned onto the roof. Installing this type of roof is labor-intensive and requires experienced roofers. Because the membrane is created in the field, its quality is subject to many variables, including the weather, application techniques, and the experience of the roofers. Most BUR is placed over one or more insulation boards bonded or fastened to the roof substrate.

The life span of a correctly applied, built-up roof is about 10 to 20 years, depending on the number of layers. Generally, specific damage to this type of

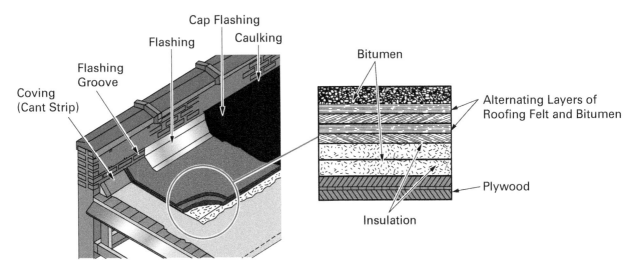

Figure 15 Conventional built-up roofing membrane.

roof can be easily repaired. However, the normal, gradual deterioration of the roof, which may result in deep splits over much of the surface or delamination of the layers, will generally require that the roof be completely removed and a new roof applied.

3.4.1 Modified Bitumen Membrane Roofing Systems

Modified bitumen roofing systems are classified as either styrene-butadiene-styrene (SBS) or atactic polypropylene (APP) modified bitumen products. The SBS products are usually a composite of polyester or glass fiber and modified asphalt coated with an elastomeric (rubberlike) blend of asphalt and SBS rubber. The APP products have a coating made with an elastomeric blend of asphalt and atactic polypropylene. Both products are the weatherproofing medium and are used in a hybrid built-up roof as a cap sheet over one or more base/felt plies that have been secured with hot asphalt or cold adhesive to a deck board covering the insulation or directly to the insulation.

Depending on the manufacturer, SBS products are either supplied with a pre-applied adhesive and the product is rolled for adhesion, or the product is secured with hot asphalt or a cold adhesive applied separately. APP products, called *torch-down roofing*, are usually secured to the layers below by heat welding. Using flame heating equipment, the back of the product roll is heated as the product is unrolled and pressed down. The back coating of the product is heated to the point where the bitumen coating acts as an adhesive, bonding the product to the layer below and the overlapped edges of the adjacent sheet. Flame heating equipment (*Figure 16*) can be used for BUR and torch-down roofing systems. Both types of systems require a mineral covering or protective coating to prevent ultraviolet destruction of the modified bitumen materials. In some cases, the product is available with various colors of ceramic roofing granules pre-applied to the exposed surface.

Figure 16 Flame heating equipment for BUR and torch-down roofing.
Source: Alessandro Mascheroni/Alamy Images

3.5.0 Single-Ply Roofing

There are two general categories of premanufactured membrane roofing systems: modified bitumen systems or single-ply systems. The single-ply membrane systems are wholly synthetic roofing materials that exhibit elastomeric properties to various degrees. Within these two general systems are a number of types. Today, there are many manufacturers of both systems, and it is important to note that each manufacturer requires specific, compatible materials and accessories for their versions of each type within each system. These materials and accessories include flashing, fasteners, drain and vent boots, inside/ outside corners, cant strips (coving), cleaners, solvents, adhesives, caulking, sealers, and tapes. To obtain the maximum performance from the system and for warranty purposes, use compatible materials and accessories and follow the manufacturer-specified application procedures. Normally, these systems are installed over insulation boards by roofers experienced with a specific manufacturer's product.

Single-ply roofing systems can be classified as either thermoplastic (plastic polymer) or thermoset (rubber polymer) systems. Within these classifications, a number of types exist. One of the most common thermoset polymer membranes is an ethylene-propylene-diene monomer (EPDM) product. This polymer, usually reinforced with polyester **scrim**, retains its flexibility over a wide range of temperatures and is very resistant to ozone and ultraviolet-ray damage. The membrane is usually supplied in large sheets or rolls and is spliced together with compatible adhesives or tapes.

Scrim: A loosely knit fabric.

Thermoplastic single-ply membranes have become very popular for commercial and industrial roofing. Two of the most common types of thermoplastic membranes available are polyvinyl chloride (PVC) and thermoplastic polyolefin (TPO). PVC membrane is usually reinforced with a polyester scrim and can be joined using solvent or hot-air welded seams that are extremely durable (*Figure 17*). PVC membrane is lightweight and aesthetically pleasing. It is very resistant to ozone and ultraviolet-ray damage. It is also puncture and tear resistant. TPO membrane combines the advantages of the solvent or hot-air seam-welding capability of PVC with the greater weatherability and flexibility benefits of the more traditional EPDM membranes. TPO membranes may also have a polyester scrim reinforcement. Both PVC and TPO membranes come in sheets or wide rolls.

Figure 17 Single-ply membrane installation.
Source: Doralin Samuel Tunas/Shutterstock

Single-ply roofing systems are clean and economical to install because they do not use hot-asphalt installation techniques common to BUR and modified bitumen membrane roofing systems. Most single-ply membranes have an insulation board or protective mat underlay along with a fireproof slip sheet, if necessary.

Single-ply membranes are usually anchored to a roof structure in one of the following four ways:

- *Loose laid/ballasted*—The perimeter is anchored with adhesives or mechanical fasteners, and the entire surface is weighted down with a round stone ballast or walking pavers (thin concrete blocks). This method is used only on very low-slope roofs capable of supporting the ballast load. It is fast and economical. See *Figure 18*.

- *Partially adhered*—The entire area of the membrane is spot-adhered to the roof with mechanical fasteners and/or adhesives. This method produces a membrane with a dimpled, wind-resistant surface.

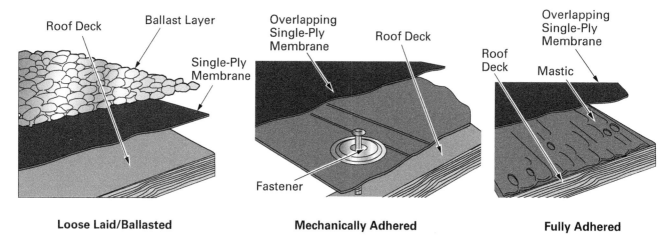

Figure 18 Typical methods of anchoring single-ply membrane to a roof.

- *Mechanically adhered*—The entire area of the membrane is spot-adhered to the roof with mechanical fasteners. This method produces a dimpled, wind-resistant installation and allows easy removal of the membrane during future roof replacement.
- *Fully adhered*—The entire area of the membrane is completely cemented down with an adhesive. This method produces a smooth, windproof surface.

3.6.0 Underlayment and Waterproof Membrane

Underlayment is available as 60 lb rolls of nonperforated 15 lb, 30 lb, and 60 lb asphalt-saturated felt (*Table 2*). The felt used under roofing materials must allow the passage of water vapor. This prevents the accumulation of moisture or frost between the underlayment and the roof deck. The correct weight of felt to be used is usually specified by the manufacturer of the finish roofing material. Roofs with a slope of more than 4 in 12 normally use 15 lb felt.

In areas of the country where water backup under the finish roof is a problem due to wind-driven rain, ice, and snow buildup, a waterproof membrane is available from a number of roofing material manufacturers under such names as Storm-Guard™, Water-Guard™, and Dri-Deck™. The membrane is usually made of a modified asphalt-impregnated fiberglass mat, coated on the bottom side with an elastic-polymer sealer that is also an adhesive. The membrane must be applied directly to the roof deck, not the underlayment. One of the top side edges is also covered with an adhesive so that the next course of overlapping membrane will adhere to the previous course. Any nails or staples driven through the membrane are sealed by the membrane, thus preventing water leakage.

Because the membrane is impermeable, any moisture from inside the structure condensing under the membrane will eventually damage any wood directly under the membrane. As a result, some manufacturers suggest that the membrane only be applied along the bottom edges of the eaves and up the roof to at least 24" above the outside wall, along the rake edges, up any valleys, around skylights, and on any saddles or other problem areas. The rest of the roof is covered with conventional

TABLE 2 Sizes, Weights, and Coverage of Asphalt-Saturated Felt

Approximate Weight per Roll	Approximate Weight per Square	Squares per Roll	Roll Length	Roll Width	Side of End Laps	Top Lap	Exposure
60 lb	15 lb	4	144'	36"	4" to 6"	2"	34"
60 lb	30 lb	2	72'	36"	4" to 6"	2"	34"
60 lb	60 lb	1	36'	36"	4" to 6"	2"	34"

underlayment. However, other manufacturers indicate that the membrane can be applied over the entire roof deck, if adequate ventilation exists under the deck and a vapor barrier is installed on any inside ceiling under the deck.

Installing waterproof membrane is a two-person job. After unrolling and cutting off a manageable strip, one person must hold the material in place while another peels away a protective sheet covering the bottom adhesive. The membrane adhesive remains tacky during installation and the membrane can be lifted off the roof deck and repositioned, if necessary. However, after a short time it will set up and cannot be removed without damaging the membrane.

3.7.0 | Drip Edge, Flashing, and Roof Ventilation

Figure 19 shows various types of drip edges that are available. Some are for reroofing applications only. Drip edges and any other flashing must be made of materials that are compatible with the roofing and any items that are being flashed. They must last as long as the finish roof material. Today, aluminum, galvanized steel, copper, vinyl, and stainless steel are the most common materials used for drip edges and flashing. Galvanized steel, copper, or stainless steel must be used with any cement-based roofing material or against masonry materials, due to the corrosive nature of cement. Copper flashing works best with slate roofing because of its long life and resistance to corrosion.

Wraparound end cap flashing covers the edges of old roofing layers.

This type of drip edge is designed to contain pea gravel on a built-up roof and is commonly called a gravel stop.

Often called style D or dripcap, this flashing adds a lip to the roof edge that overlaps the gutter or rake edge of the roof.

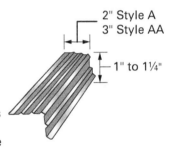

A canted strip edge of this variety carries the water away from the fascia. It is used on old roofs to hide old shingles or on new roofs. It provides a clean edge for new shingles. Style AA is used so that nails penetrate into wood when previous roof used a style A edging.

For roofing trimmed flush with the fascia, this type of end cap covers the edges of layers and keeps water and ice from backing up under the old shingles.

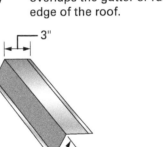

Angled gutter apron used at eaves to divert water into gutters to protect fascia.

Figure 19 Various types of drip edges.

For best results, coat any galvanized drip edge or flashing with an appropriate primer before installation. Also, fasten all drip edges and flashing with nails or staples made of a compatible material to prevent electrolytic corrosion between the fasteners and the flashing. Always follow the roofing material manufacturer's recommendations for flashing.

Figure 20 shows a W-metal (so named because its end profile looks like a letter W) or standing-seam **valley flashing**. This type of valley flashing is available in 8' to 10' lengths in widths of 20" to 24". If desired, it can be field-fabricated using flat roll flashing and a metal brake. In any case, it must be wide enough so that the finish roofing overlaps the metal by more than 6".

Valley flashing: Watertight protection at a roof intersection. Various metals and asphalt products are used; however, materials vary based on local building codes.

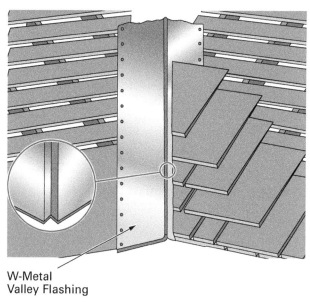

W-Metal
Valley Flashing

Figure 20 W-metal valley flashing.

The W-metal valley flashing is preferred with open valleys. Open valleys are defined as valleys where the valley flashing material will be visible after the finish roofing is applied. Open valleys are more difficult to install but accommodate higher rainfall rates than closed valleys. They can also be used with any type of finish roofing. Closed valleys can be used only on composition shingle roofs. The ridge in the middle of the W-metal valley (about $\frac{3}{4}$" to 1" high) prevents water rushing down the slope of one roof from washing under the shingles of the intersecting roof. On short valleys or relatively low-slope roofs, ordinary flat-roll flashing can be used in the valleys.

Proper attic ventilation is necessary to allow heat and moisture to escape so that damage to the roofing and roof deck does not occur. In the winter, ventilation keeps the roof deck cold and reduces the buildup of ice on the eaves. This helps prevent water penetration through the roof and subsequent water damage to the structure. It also carries away moisture so that it does not condense on the roof deck, which can cause rotting. In the summer, excess heat can cause overheating of composition shingles, resulting in early failure of the roof.

Residential attics are generally ventilated by convection vents in the form of gable vents or roof-mounted **ridge** or box vents. Sometimes, electric-powered or wind-powered turbine vents or fans are used (*Figure 21*). The air used for ventilation is provided by soffit vents at the eaves of the roof. The amount of soffit ventilation (in square feet or inches) must be equal to or greater than the amount of roof ventilation.

In residential structures, the proper amount of convection-type ventilation for an unheated attic space is usually defined as 1 ft^2 of ventilation for every 300 ft^2 of attic area, with 50% in the roof for exhaust and 50% in the eaves for intake. The attic area is calculated using the exterior foundation dimensions of the structure. For example, the amount of ventilation for a residence with an exterior foundation measurement of 40' × 60' would be calculated as follows:

$$40' \times 60' = 2,400\,ft^2 \text{ (attic area)}$$
$$2,400\,ft^2 \div 300\,ft^2 = 8\,ft^2 \text{ (total ventilation required)}$$
$$8\,ft^2 \times 144\,in^2 \text{ per } ft^2 = 1,152\,in^2$$
$$1,152\,in^2 \div 2 = 576\,in^2 \text{ for the ridge and } 576\,in^2 \text{ for the soffits}$$

Based on the calculated ventilation requirement, select and install appropriately sized roof and soffit convection ventilation devices.

Ridge vents, box vents, and turbine vents are available in a variety of styles and sizes for residential use. *Figure 22* illustrates different types of ridge vents. Metal or plastic ridge vents, sold in 10' lengths in several patterns and colors, can be installed over most roofing materials.

Ridge: The horizontal line formed by the two rafters of a sloping roof that have been nailed together. The ridge is the highest point at the top of the roof where the roof slopes meet.

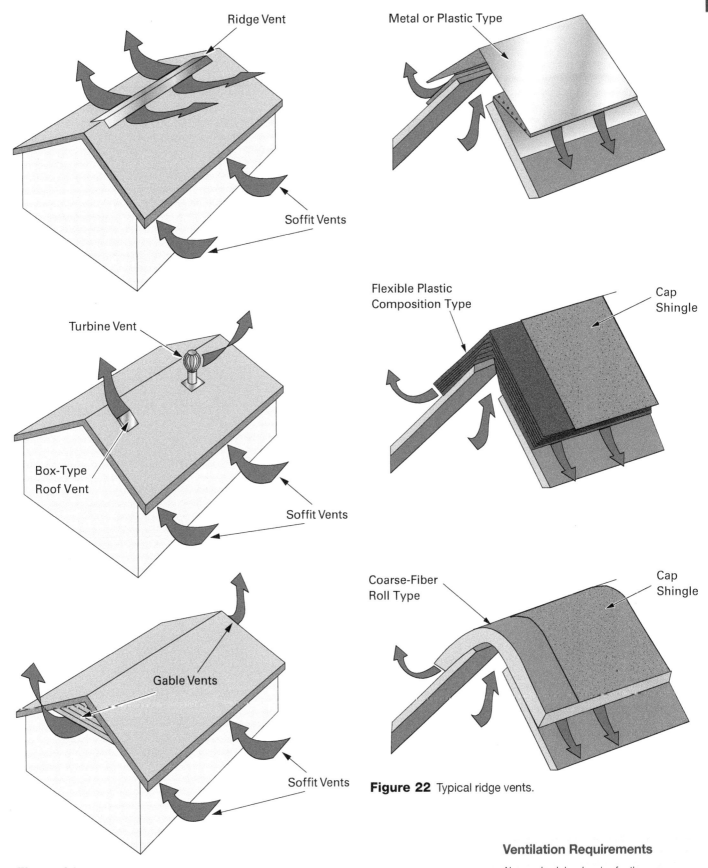

Ridge Vent

Metal or Plastic Type

Soffit Vents

Turbine Vent

Box-Type
Roof Vent

Soffit Vents

Flexible Plastic
Composition Type

Cap
Shingle

Gable Vents

Coarse-Fiber
Roll Type

Cap
Shingle

Soffit Vents

Figure 22 Typical ridge vents.

Figure 21 Residential roof vents.

The flexible plastic composition vent is available in 4' lengths. An inert, coarse-fiber vent is available in rolls. Flexible plastic and rolled coarse-fiber vents can only be used over composition shingles because both are designed to be covered with cap shingles that match the roof. Most residential customers prefer shingle-covered vents because they tend to blend into the overall roof.

Ventilation Requirements

Always check local codes for the proper ventilation requirements of a structure. In some areas, the amount of air exchange required may dictate fan-assisted ventilation if the capacity of free-air ventilation devices is not adequate.

Open-Flame Heat Welding

Today, most seam-sealing methods and equipment recommended by manufacturers of single-ply membrane roofing systems make use of hot-air welding methods that are quite safe. However, some contractors still use open-flame heating equipment for seam welding. Open-flame seam sealing can be hazardous and may cause damage to the membrane. For these reasons, open-flame seam sealing of polymer membrane roofing should be avoided.

The manufacturer's specifications for a ventilation device will provide details to determine the amount of free-air ventilation (in square feet or inches) that the device will supply. Ridge ventilators are probably the most efficient and are usually rated in square inches of free-air ventilation per linear foot of the product.

Waterproof Membranes

Waterproof membranes used on roof edges and valleys are self-healing. This means that if a screw or nail intentionally or accidentally penetrates the membrane, a modified asphalt coating will flow to seal the penetration when the sun heats the roof and membrane. Because they have an adhesive that secures them to the roof deck, waterproof membranes are virtually impossible to remove once the adhesive is heat-set by the sun. Reroofing a structure is very expensive if the membrane must be removed, because the roof deck must be replaced if it is wood or a wood product. Newer versions of membranes are available with a granular surface that allows the overlying shingles to be removed without damaging the membrane.

3.0.0 Section Review

1. Architectural shingles have a lifespan of _____.
   a. 10 to 25 years
   b. 25 to 40 years
   c. 40 to 70 years
   d. 70 to 100 years

2. For slate roofing installations, use nails and flashing made of _____.
   a. stainless steel
   b. copper
   c. galvanized steel
   d. aluminum

3. A metal roofing material with a coating that is 80% lead and 20% tin is called _____.
   a. composite-coated sheeting
   b. pot metal
   c. galvanized steel
   d. terne metal

4. Depending upon the number of layers, a correctly applied built-up roof should not need replacement for _____.
   a. 5 to 10 years
   b. 10 to 20 years
   c. 15 to 25 years
   d. 20 to 30 years

5. True or False: Thermoplastic single-ply membranes are a popular material used for roofing single-family residences.
   a. True
   b. False

6. Rolls of asphalt-saturated felt used for roofing underlayment weigh _____.
   a. 15 pounds
   b. 30 pounds
   c. 60 pounds
   d. 75 pounds

7. Metal drip edge and flashing must be installed with fasteners made of a compatible material to avoid _____.
   a. metal fatigue
   b. electrolytic corrosion
   c. distortion
   d. rusting

4.0.0 Roof Installation

Objective

Describe the installation techniques for common roofing systems.

a. Describe how to properly prepare a roof deck.

b. Explain how to install composition shingles.

c. Explain how to install metal roofing.

d. Discuss roof projections, flashing, and ventilation.

Performance Tasks

1. Demonstrate how to install composition shingles on a specified roof and valley.

2. Demonstrate the method to properly cut and install the ridge cap using composition singles.

3. Lay out, cut, and install a cricket or saddle.

4. Demonstrate the techniques for installing other selected types of roofing materials.

Roofing projects require planning prior to installation. As part of the planning process, check the manufacturer's recommendations for proper installation and fasteners. Prior to installing the finished roofing material, make sure to properly prepare the roof deck.

4.1.0 Preparing the Roof Deck

A typical roof installation is shown in *Figure 23*. Before applying the finish roofing, the roof deck must be flashed with a drip edge along the eaves and any valleys must be flashed. Then, install an underlayment and/or a

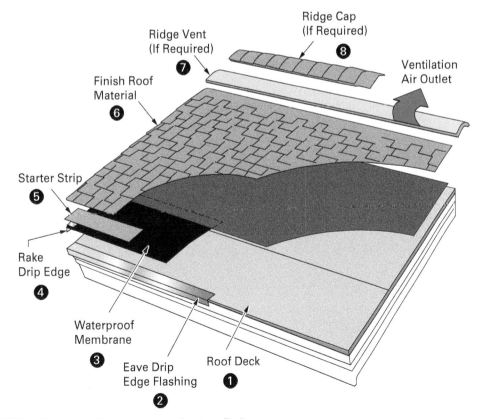

NOTE: Circled numbers represent the installation sequence.

Figure 23 Typical roof installation.

Protruding Nails and Debris

Before applying any roofing material to a roof deck, walk the nail pattern on the bare deck sheathing and check that all nails are driven flush with the surface. Also, make sure to remove all debris including small pebbles. After installing all flashing, underlayment, and drip edges, walk the nail patterns again to make sure all nails are driven flush. Be sure to remove all debris. Any protruding nails or debris under a composition shingle may eventually penetrate the shingle and cause leaks.

waterproofing membrane and cap the rake edges with metal drip edge flashing. On bare wood roof decks, the underlayment/membrane must be applied on dry wood as soon as possible. If the wood is moist due to rain or morning dew, allow it to dry before applying the underlayment/membrane. If the roof deck is damp, the membrane may not adhere to the roof deck, or the underlayment will buckle and cause the final roof to appear wavy. The underlayment/membrane prevents the finish roof materials from having direct contact with any damaging resinous or corrosive areas of the roof deck and helps resist or eliminate any water penetration into the roof deck. *Figure 24* and *Figure 25* show the recommended underlayment/waterproof membrane placement and drip edge installation.

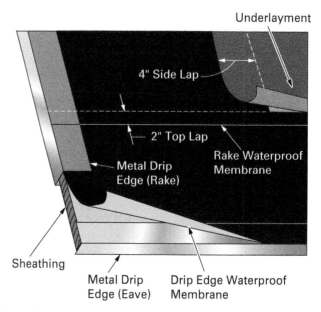

Figure 24 Drip edge and waterproof membrane placement.

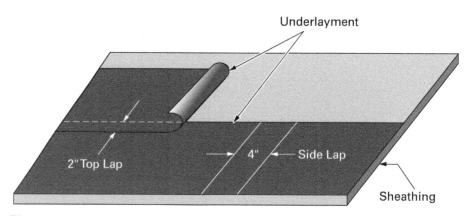

Figure 25 Underlayment or waterproof membrane placement over roof deck.

Overhang: The part of a structure that extends beyond the building line. The amount of overhang is always given as a projection from the building line on a horizontal plane.

Normally, install the drip edge along the length of the eaves first, followed by any valley flashing. Hold the drip edge against the fascia and nail it to the roof deck every 8" to 10". When installing valley flashing, the flashing should create an **overhang** at the upper and lower ends of the valley and be nailed every 6" to 8" on both sides, $\frac{1}{2}$" from the edges.

Residential Secondary Roof Systems

An alternative for roofs that cannot be insulated properly from inside the residence is an insulated, secondary roof system installed over the original roof deck. These systems have space between a rigid insulation layer and a second roof deck to allow free airflow over the insulation and under the second roof deck. This greatly reduces the melting of snow and the resulting ice dam on the second roof deck. These secondary roof systems are available from several manufacturers. In hot climates, they can reduce the heat load in a residence with similar insulation problems. The airflow of these systems can also reduce the heat load on the roofing materials used for the second deck.

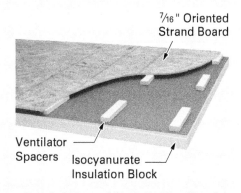

Bonded Second-Deck Roofing Panel

With the flashing secure, carefully trim both ends flush with the roof deck. After the eave drip edge is in place, cover the exposed nail heads with asphalt. Starting at the bottom of the roof, the underlayment and/or a waterproof membrane is rolled out and flattened with a roof roller before being tacked to the roof. In valleys, the waterproof membrane should extend over the flashing nails. The membrane will adhere and seal to the valley flashing; however, the underlayment should be trimmed to cover the flashing nails and should be cemented to the valley. In some cases, on lower-sloped roofs, the underlayment is half lapped and cemented with asphalt to create a better water barrier. After the underlayment/waterproof membrane is in place, the rake edges of the roof are capped with a drip edge that is nailed every 8" to 10" to the roof deck. The bottom end of the rake drip edge overlaps the eave drip edge, and the fascia flange is cut to interlock behind the fascia flange of the eave drip edge. The nail heads should be covered with asphalt cement.

After the roof preparation is complete, the finish roof materials can be lifted and distributed evenly over the roof deck.

4.1.1 Protection against Ice Dams

In areas subject to heavy snow, the snow will accumulate on the roof. Heat rising through the roof from inside the structure will melt the snow and cause ice to build up on the edge of the roof and in the rain gutters, creating an ice dam (*Figure 26*). As snow continues to melt, the ice dam will trap the water and force it under the shingles. Eventually, it will find its way into the building.

A combination of attic insulation, roof venting, and waterproof shingle underlayment can eliminate this ice dam problem. This underlayment comes in 36"-wide rolls. The material has a sticky side designed to stick to the roof deck, forming a tight seal against water penetration. It will seal around any nails that are driven through it.

Many sloped-roof residences in the North have ice-damming problems on the roof, usually at the eaves. This is especially true for those residences with finished attics where insulation and venting under the roof deck is limited to the rafter space. In most cases, when a new roof is installed, the application of a waterproof membrane from the eaves up the slope of the roof to a point that is at least 24" beyond the inside wall will prevent water backed up behind any ice dams from penetrating into the structure. However, for a problem residence with an existing roof that lacks a waterproof membrane underlayment, it may be desirable to install an ice edge at the eaves. On most roofing materials, ice dams cannot be easily removed. In addition, ice dams can damage some roofing materials, including slate.

An ice edge is an exposed metal sheeting mounted from the eaves up the slope of the roof from 18" to 36". It provides a shield against water penetration and allows any ice dams to be quickly shed from the roof during any brief thaws, thus reducing the chances of a large ice-dam buildup. The width of the ice

Figure 26 Ice dam.
Source: rviard/Getty Images

edging used is determined by how steep the roof is and how thick the ice usually becomes, which determines how much water backs up on the roof. Normally, continuous sheets of plain or tinted/painted aluminum flashing or special standing-seam aluminum panels are used as the ice edge (*Figure 27*); however, copper flashing can be used where corrosion is a problem. The standing-seam

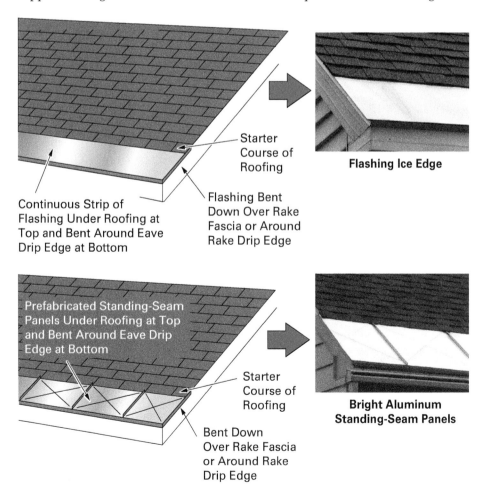

Figure 27 Types of ice edging.

panels are more expensive and are not subject to buckling due to temperature extremes. Their surface appearance is more uniform and does not appear wavy.

There are disadvantages to the use of ice edges. One is that they are generally not considered attractive unless colored to match the roof. The possible exception is plain copper used with slate roofing. Another disadvantage is that for ice edges to be effective in shedding ice, eaves troughs cannot be used on the building. This can lead to several severe hazards such as injury or death to people or damage to property or foliage, including damage to the siding of the residence caused by falling ice. In addition, water draining from the roof can collect and penetrate under the foundation and/or into a basement.

Ice Edge Hazards

Before ice edges are considered for use on the eaves of a building, the hazard of falling ice causing injury, death, or property damage under each eave must be carefully evaluated. With ice edging, a long, heavy ice dam (1,000 lb or more) along an entire roof eave can be released from the eave without warning. In addition, attempts to remove icicles may cause an ice dam along an entire eave to be released.

Source: marinatr/123RF

4.2.0 Installing Composition Shingles

Composition shingles come in various colors, textures, shapes, and sizes (*Figure 28*). The number of shingles in a square and the number of squares in a bundle can vary greatly. Before starting a roofing project, check the manufacturer's product information and specifications, which will provide installation instructions along with information on weight, dimensions, and recommended **exposure**.

Exposure: The distance (in inches) between the exposed edges of overlapping shingles.

Figure 28 Composition shingles.
Source: Natalija Datsko/Shutterstock

NOTE

The manufacturer provides a set of instructions with each bundle of shingles. Failure to follow these instructions may void the manufacturer's warranty.

The following general instructions pertain to a standard three-tab fiberglass shingle.

The instructions for the installation of all types of shingles, including wood, composition, and slate, use standard terminology to describe the placing of the shingles. This terminology is explained in *Figure 29*. Roof shingles can be placed from the left side of the roof to the right or the right side of the roof to the left, depending on the preference of the roofer. On wide roofs, the courses are sometimes started in the middle and laid toward both ends. This module uses the left-to-right convention.

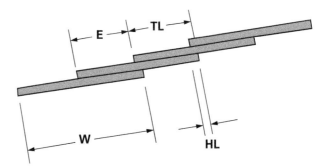

W = **Width:** The total width of strip shingles or the length of an individual shingle.

E = **Exposure:** The distance between the exposed edges of overlapping shingles.

TL = **Top Lap:** The distance that a shingle overlaps the shingle in the course below.

HL = **Head Lap:** The distance from the lower edge of an overlapping shingle to the upper edge of the shingle in the second course below.

Figure 29 Roofing terminology used in instructions.

Starter Strips

Precut starter strips are available from many manufacturers. In many areas of the country, they are available in two sizes: 5" wide for roofing over existing shingles and 7" wide for new and tear-off installations.

CAUTION

Failure to scatter unopened shingle bundles can result in a broken rafter or collapsed roof, due to the weight of the bundles.

All strip shingles start with a double first row, which may be made up of a starter row and a row of shingles or a double course of shingles in which two joints have been offset. A common practice is to place a starter row of shingles with the tabs cut off. The type of starter course used will depend upon the type of shingle being used, the availability of materials, and local building codes.

Unopened bundles of shingles are usually placed at various points on the roof for the roofer.

Under normal circumstances, the shingle can be fastened with four aluminum or galvanized roofing nails positioned at a nailing line from the bottom of the shingle, one at each end and the other two above and adjacent to each cutout (*Figure 30*). Depending upon the area of the country in which you live, this cutout may be referred to as a *notch* or *gusset*.

In areas with very high winds, the number of nails above the cutout can be increased to two, positioned about 3" apart and forming a triangle with the top of the cutout as a low point. In high-wind areas, staples are not normally used.

When laying a shingle, butt the shingle to the previous shingle in the course and align the shingle. Then fasten the butted end to the roof. Keep the shingle aligned and fasten across the shingle to the other end. The guide built into a roofing hammer or chalking lines can help with alignment.

4.2.1 Gable Roofs

This section describes the installation of long and short runs of standard shingles on gable roofs. It does not include instructions for metric or architectural shingles.

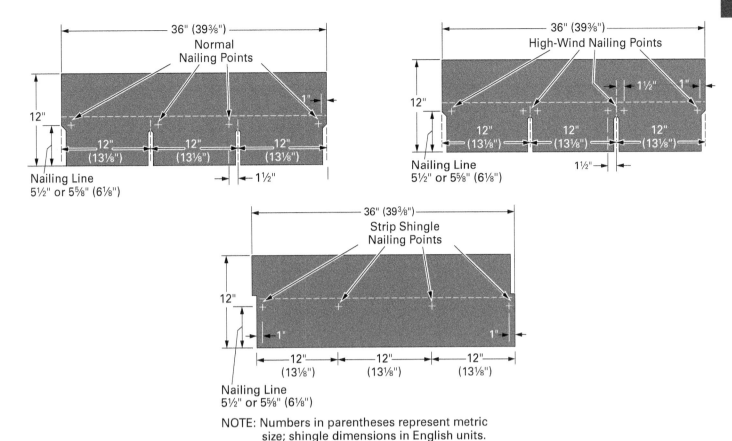

NOTE: Numbers in parentheses represent metric size; shingle dimensions in English units.

Figure 30 Nailing points.

4.2.2 Gable Roofs—Long Runs

On large roofs, start applying the shingles at the center of the long run. By beginning in the center, there is less chance of misalignment as you proceed in both directions. Shingle manufacturers suggest that you run shingles horizontally first rather than stacking them one row above another. The color of the shingles will blend better that way. The following procedure explains how to lay out and mark the roof.

To install a long run on a gable roof, proceed as follows:

Step 1   Measure along the length of the roof and find the halfway point. Do this along the ridge and along the eaves. Mark both of these places. Snap a chalk line vertically up the roof at these two marks.

Step 2   Measure 6" to the right and left of this centerline at the ridge and at the eaves. Snap a line vertically for these marks as well. This will provide three lines to start the rows on.

Step 3   Starting at the eaves at both ends of the roof, measure 6" up from the eave drip edge and place a mark. Then go up to 11" and make a mark there. Proceed up the tape and mark every 5" interval. With both sides marked, snap lines horizontally across the roof to align the shingles.

Step 4   For the starter shingles, cut the tabs off a tabbed shingle. Place the remaining part of the shingle so that the tar strip is nearest the edge of the roof. With standard shingles, you should have a starter row that is 7" wide by 36" long. If the starter strip is not done this way, then the first full row of shingles will not be sealed down. Place the starter strip on one of the vertical lines in the center of the roof. Make sure that the starter shingle stays on the first line snapped horizontally across the roof. If there are drip edges, position the starter strip with a 1/2" overhang on the eave and rake drip edges; otherwise, position the strip with a 3/4" overhang. Place and fasten the starter strip both ways from center.

Step 5   Take a full shingle and place it directly over the starter strip and 6" to the right or left of the vertical line used for the starter strip. This will ensure that the cutouts

NOTE

If you are going to stand on a scaffold, it should be erected in such a manner that you will be working approximately waist high with the eave line. This places you in a safe, comfortable position to install the critical double course of shingles along the eaves.

NOTE

Starting 6" over helps to minimize waste. Most of the time, the scrap you have left over on the right end will work on the left and vice versa.

will be spaced 6" away from the joints of the starter strip. This is considered the first full row of shingles.

Step 6   The second row is in the center, 6" to the right or left of the starting place where the first row was started. Proceed right and left from that point.

4.2.3 Gable Roofs—Short Runs

To install a short run on a gable roof, proceed as follows:

Step 1   Cut the tabs off the number of strip shingles needed to go across the roof. Mark a spot 6" up on both eaves. Snap a line across the roof at this location. This line will keep the starter strip running straight across the roof. With the first left starter strip shortened by 6", place the starter strip shingles so that the tar strip is nearest the eave. If drip edges have been installed, position the starter strip with a $\frac{1}{2}$" overhang on the eave and rake drip edge; otherwise, position the strip with a $\frac{3}{4}$" overhang. Nail the starter close to the top in four locations. After the starter strip is laid as far to the right as possible, return to the rake on your left and start to double up this first course by placing a full shingle directly over the top of the first upside-down shingle. See *Figure 31*. Continue this process with full shingles as you move to the right, and end before reaching the last starter strip shingle. You will observe that all cutouts and joints are covered with the full 12" tab. Make sure to nail the shingles correctly, as shown in *Figure 32*.

Step 2   Start the second course with a full strip minus 6". This layout, a 6" half-tab pattern, or a 6-up/6-off layout, means that half a tab is deliberately cut off and produces vertically aligned cutouts (refer to *Figure 31*). Overhang the cut edge at the rake by

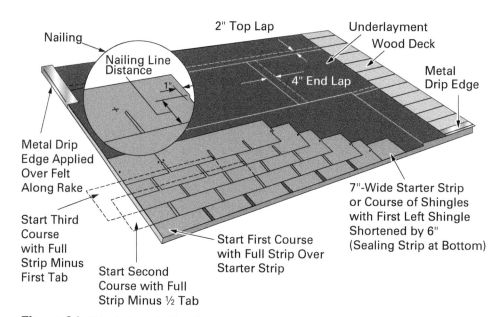

Figure 31 Shingle layout: 6" pattern.

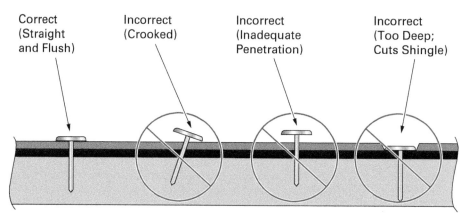

Figure 32 Correct and incorrect nailing.

the same ¼" margin as the first course. Nail this shingle in place and proceed to the right with full shingles. The gusset openings can be used as a checking procedure to obtain the proper 5" exposure. End the second course before reaching the end of the first course. This is called stair stepping and allows the maximum number of shingles to be placed before moving ladders or scaffolding.

Step 3   Start the third course with a full strip minus a full tab (12"). Follow the same procedure with the same rake overhang. Again, nail the shingles in place moving to the right and gauging your exposure by the 5" gusset.

Step 4   Returning now to the fourth course, start this row with half a strip (for example, with 18" removed). This will show a 6" tab. Nail this in place with the same overhang margin on the rake. Proceed with full shingles, nailing to the right and ending short of the previous course while constantly checking the alignment.

Step 5   Start the fifth course. This row starts with a full tab only (12"). Use the same overhang margin on the rake and full shingles as you move to the right. End before reaching the last shingle on the previous course.

Step 6   The sixth course starts with a 6" tab, the same overhang margin at the rake, and then full shingles. It continues to the right as the nailing proceeds. Again, end before reaching the last shingle on the previous course.

Step 7   Repeating the process, the seventh course starts with a full shingle. Each successive course shortens the shingles by an additional 6". This continues as previously described until the twelfth course.

Depending upon the individual or working team, two or more courses may be carried or nailed at a time as the shingling proceeds across the roof. When two people are working together, they usually work out their own system for speedy, accurate installation. The procedure described above uses the left rake, when facing the roof, as a starting point. Keep in mind that the entire application could also start from the right rake (this applies to gable roof construction). On small roofs, strip shingles may start at either end with a successful result because the roof measurement is usually symmetrical.

To obtain different variations of roof patterns using tabbed shingles, only a change of starting measurement is required. *Figure 33* shows one possibility using a course with a full shingle (36") followed by a second course using a reduced-size shingle (32"). The third course would be reduced again to a shorter measurement (28"). Repeat these three measurements starting with the fourth

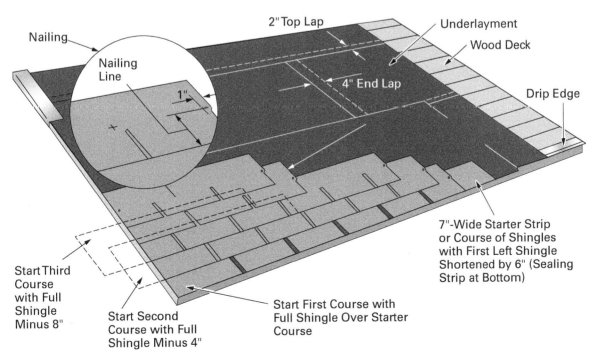

Figure 33 Shingle layout – 4" pattern.

course and continuing up the rake. This 4" pattern produces a diagonal cutout pattern.

Ribbon courses (*Figure 34*) are a way to add interest to a standard 6" pattern. After applying six courses, cut a 4"-wide strip lengthwise off the upper section of a full course of shingles. Fasten the strips as the seventh course is correctly aligned with the cutouts of the sixth course. Then reverse the 8"-wide leftover pieces of shingle and align them directly over the 4" strip. Fasten the 8" pieces to the roof deck at the top of the tabs. Cover both with a full-width seventh course of shingles. This creates a three-ply edge known as the ribbon. Repeat the pattern every seventh course.

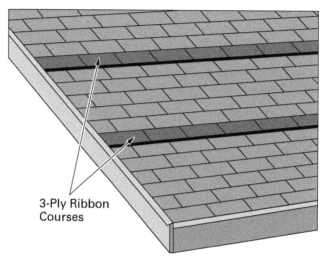

3-Ply Ribbon Courses

Figure 34 Ribbon courses.

Due to its simplicity, the first pattern mentioned (half tab = 6") is the most commonly used in the field. The full-strip asphalt shingle eliminates all pattern problems and alignment concerns on the vertical plane because it contains no gussets or cutouts.

Correct Nailing

Improper setting of nails and crooked nails can prevent the shingles from tabbing correctly and allow the wind to lift or tear the tabs of the shingles. Practice your fastening procedures so you drive the nail straight. The head should be flush with the surface of the shingle. Because your goal is to make the roof watertight, no pinholes or breaks are acceptable. If an accident should happen, a dab of asphalt cement spread with a putty knife will remedy the problem.

4.2.4 Hip Roofs

When you encounter a hip roof, the basic nailing procedures remain the same, but the shingle layout starting point must be at the center of the roof, as described for long gable roofs.

To begin, apply the starter strip as previously described for long gable roofs. Return to the vertical line and use a full shingle for the doubling of this starter course. Offset this shingle 6" on either side of this vertical line. This will automatically cover the seam and gussets underneath and seal the roof against water penetration. You have the option of continuing this dual shingle starting course in either direction until it terminates at the hip rafter. See *Figure 35*.

At this point, cut the shingles to match the angle of the hip rafter and cover with a hip cap (the same as a ridge cap), completing the installation of the shingles. The hip cap is centered on the hip rafter and usually consists of a 12" tab showing a 5" exposure.

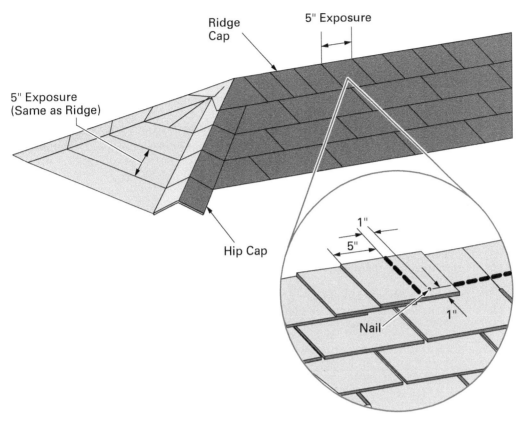

Figure 35 Hip and ridge layout.

Exposed nails in the last caps should be covered with roof sealant. The ridge cap should cover the hip cap to prevent leaks.

4.2.5 Valleys

If the building you are constructing is not a perfect rectangle, you may encounter an L or T shape, which calls for another variation of shingling procedures. Where two sloping roofs meet, this intersecting valley needs to be able to carry a high concentration of water drainage. Shingling becomes very critical, and the application must be done with extreme care.

4.2.6 Open Valley

With the valley flashing installed as previously described, snap two chalk lines the full length of the valley. They should be 6" apart at the ridge or uppermost point. This means they should measure 3" apart when measured from the center of the valley. The marks diverge at the rate of $\frac{1}{8}$" per foot as they approach the eaves. For example, a valley 8' in length will be 7" wide at the eaves; one 16' long will be 8" wide at the eaves. The enlarged spacing provides adequate flow as the amount of water increases, passing down the valley. See *Figure 36*.

The chalk line you have snapped serves as a guide in trimming and cutting the last shingle to fit in the valley. This ensures a clean, sharp edge and a uniform appearance in the valley. Clip the upper corner of each end shingle slightly on a 45-degree angle. This keeps water from getting in between the courses. Spread a 6" to 8" bed of plastic asphalt cement to secure the roofing material to the valley lining and to itself where an overlap occurs. Do not overdo it, and clean up all excess cement so no tar shows.

4.2.7 Closed-Woven Valley

Some roofers applying asphalt shingles prefer to use a closed-woven valley design, sometimes called a full-weave or laced valley. It is faster to install, and some feel it gives a tighter bond. Others believe closed valleys are inferior to

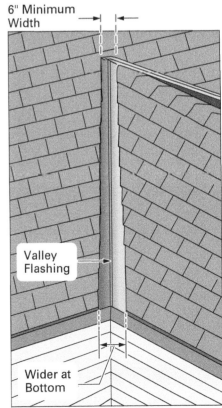

Figure 36 Open valley flashing (steep pitch).

open valleys because they do not shed high volumes of water very well. Only composition shingles work for this pattern. See *Figure 37.*

It is essential that a shingle be of sufficient width to cross the lowest point of the valley and continue upward on each roof surface a minimum of 12". Because of the skill required for this process, it is suggested that the two converging roofs be completed to a point 4' to 5' from the center of the valley. Then carefully complete the weaving process.

To create the 12" extension, it may be necessary to cut some of the preceding shingles in the course back to two tabs. To ensure a watertight valley, either a strip of 36"-wide, waterproof membrane, or 50 lb or heavier roll roofing over the standard 15 lb felt is placed in the valley, as shown in *Figure 37.*

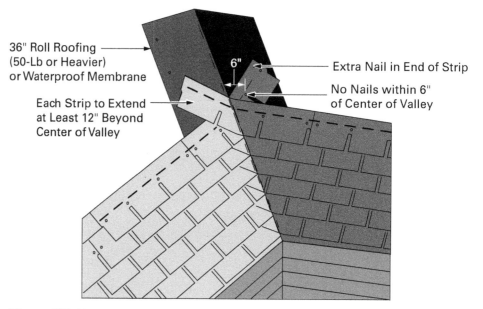

Figure 37 Closed-woven valley.

For the weaving process, place and fasten the first course in the normal manner. Note that no fasteners are located closer than 6" to the valley centerline. An extra fastener is placed at the high point at the end of the strip where it extends the extra 12". The first course on the opposite side is then laid across the valley over the previously applied shingles. Succeeding courses alternate, first along one roof area and then the other, as shown in *Figure 36.* Take extreme care to maintain the proper exposure and alignment. While weaving the shingles over each other, press them tightly into the valley to provide a smooth surface where the roof surfaces join.

Unequally Pitched Roof Intersections

If the roof valley is formed by an intersection of two unequally pitched roofs, the woven valley will creep up one side, making it nearly impossible to maintain the correct overlap of shingles. The open valley or the closed-cut valley should be used with unequally pitched roof intersections; however, the closed-cut valley will give a much neater appearance.

4.2.8 Closed-Cut Valley

In a closed-cut valley, sometimes called a half-weave or half-laced valley, the underlayment and valley flashing materials are the same as for the woven application.

To create a closed-cut valley, proceed as follows:

Step 1   Lay the first double-course of shingles along the eaves of one roof area up to and over the valley. See *Figure 38.* Extend it up along the adjoining roof section. The distance of this extension should be at least 12" or one full tab. Follow the same procedure when applying the next course of shingles. Make sure to press the shingles tightly into the valley.

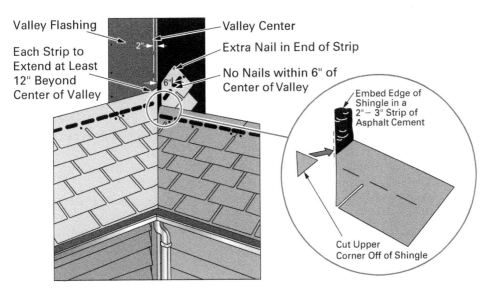

Figure 38 Closed-cut valley.

Step 2 Follow this procedure up the entire length of one side of the valley. If there is a high and a low slope, the first application should always be done on the low-slope side.

Step 3 When this roof surface is complete, you are ready to proceed with the intersecting roof surface, which will overlap the preceding application. Measure over 2" from the centerline of the valley in the direction of the intersecting roof. Carefully snap a chalk line from top to bottom. This will be your guideline for the trim cut on the shingles.

Step 4 Now apply the first course of shingles on the intersecting roof. Use extreme care to match your chalk line angle exactly. Also, trim off the upper corner of the shingle to prevent water from running back along the top edge. Embed the end of the shingle in a 2" to 3" strip of plastic asphalt cement, being careful to allow no tar to show on the original shingle opposite. Succeeding courses are applied and completed as shown in *Figure 37*, making the valley watertight.

Other Shingle Alignment Methods

Several other alignment methods can be used for exposure. One popular method begins by snapping a chalk line along the top edge of the shingle of the first course, or 12" above and parallel with the eave line. Snap several other chalk lines parallel with this first line. If you make the lines 10" apart, you can use these to check every other course by aligning the top of the shingle.

High-Wind Areas

In high-wind areas, do not use the fifth and sixth courses of the 6" pattern because of the possibility of the small starter tabs being torn off the roof. Use the fifth course tab (12" long) as a ridge cap and discard the sixth course tab (6" long). Double fastening at each cutout secures the shingles.

4.3.0 Installing Metal Roofing

With proper installation, metal roofs provide durability and energy efficiency to a building. A metal roof can last up to 50 years. This section covers the installation of different types of metal roofing systems. Carpenters must always follow the manufacturer's instruction and specifications during the installation process.

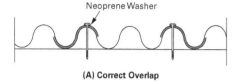

(A) Correct Overlap

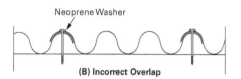

(B) Incorrect Overlap

Figure 39 Corrugated roofing.

4.3.1 Corrugated Metal Roofing

Corrugated metal roofing, or galvanized metal roofing, is roll formed with a wavy pattern. Only galvanized sheets heavily coated with zinc are recommended for permanent construction.

Galvanized sheets may be laid on slopes as low as a shallow 3" rise to the foot ($\frac{1}{8}$ pitch). If the roof requires more than one sheet to reach the top, the ends should overlap by at least 8". When the roof has a pitch of $\frac{1}{4}$ or more, 4" end laps are usually satisfactory. To make a tight roof, sheets should overlap by $1\frac{1}{2}$ corrugations at either side (*Figure 39A*).

When using roofing that is $27\frac{1}{2}$" wide with $2\frac{1}{2}$" corrugations and a corrugation lap of $1\frac{1}{2}$", each sheet covers a net width of 24" on the roof. For 26-gauge galvanized sheets, supports may be 24" apart. For 28-gauge galvanized sheets, supports should not be more than 12" apart. The heavier gauge has no particular advantage except its added strength, because the zinc coating is what gives this type of roofing its durability.

When $27\frac{1}{2}$" roofing is not available, sheets of 26" width may be used. When laying the narrower sheets, every other one should be turned upside down so that each alternate sheet overlaps the two intermediate sheets, as shown in *Figure 39A*.

For best results, fasten galvanized sheets with neoprene-headed nails, galvanized nails and neoprene washers, or screws with neoprene washers only in the tops of the corrugations to prevent leakage. To avoid unnecessary corrosion, use the fasteners specified by the roofing manufacturer.

Garages, storage buildings, and farm buildings often have corrugated metal roofing panels (*Figure 40*). They are available in widths up to 4'-0" and lengths up to 24'-0". Normally, these panels are used on roofs with slopes of 4 in 12 or steeper. They can be used on 2 in 12 roofs if a single panel reaches from the ridge to the eave.

The panels fasten to purlins. A purlin is a structural member, usually made of 2 × 4 wood stock, running perpendicular to the rafters. See *Figure 40*. The spacing should follow the directions specified by the manufacturer. The panels

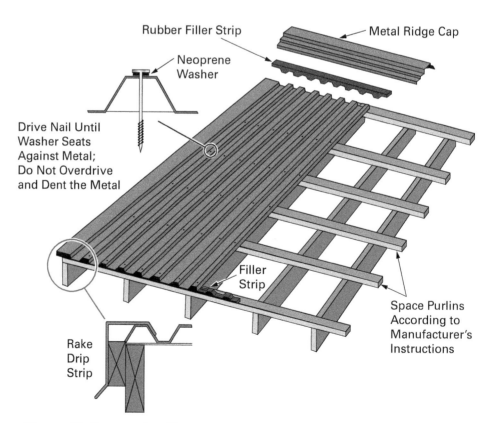

Figure 40 Corrugated roof layout.

come with filler strips that are set at the eave and the ridge. Normally, the panel overhangs the eave by 2" to 3". The installation of metal roofs should follow the manufacturer's specifications.

4.3.2 Simulated Standing-Seam Metal Roofing

The character of each standing-seam roof system dictates the amount and type of planning. Each system has different components and slightly different requirements for tools and equipment. In some cases that require seaming machines, supervisors may have to decide not only when to lease the machines, but how many machines to have on the project. A backup machine is always a good idea, especially if the system requires seaming shortly after panel installation.

Power-tool requirements also differ from system to system. Most systems use screw guns to install self-drilling and self-tapping screws, but some systems require impact wrenches or bulb rivet guns for fastening. It is important to have a sufficient number of power tools on the job to allow the work to move smoothly and efficiently.

It is also important that the assemblers carry oversized fasteners. Standing-seam roof systems minimize the number of through-the-panel fasteners by up to 90%. Therefore, it is crucial to install the fasteners correctly. If a fastener is stripped during installation, remove it immediately and replace it with a fastener of the next larger size.

The direction in which the sheeting takes place has to be considered. Some systems have strict requirements; others are more flexible. The approved construction documents will show the sheeting direction.

Competent assemblers should install the standing-seam roof system. Improper installation techniques cause most standing-seam roof failures, which underscores the need for special training.

Before roofing can begin, the structure must be plumb and level. The purlins must also be straight. Z-purlins, in particular, have a tendency to roll. If there are no purlin braces, use wood blocking. Most manufacturers suggest driving wood blocks tightly between purlins to ensure proper spacing, and recommend either 2 × 6 or 2 × 8 lumber, depending upon the purlin depth. Place at least one row of blocks in the center of the bay. The erection drawings contain the proper purlin spacing.

Purlins may also be straightened by adjusting the sag rods, if the structure has them. Many sag rods are cut to set a specific width automatically.

Until at least one run of panels has been installed, there is no safe place to work unless a work platform is constructed. A work platform should be made by stacking two panels on top of each other and placing walk boards in the center. Attach the panels to the structure with locking C-clamp pliers or some other means to prevent the panels from moving. This platform can store the insulation required for the first run, as well as the necessary tools and components.

The specific sequence of erection of standing-seam roof systems is determined by the manufacturer of the given system. What is recommended for some systems may not be recommended for others. This fact emphasizes the importance of knowing and understanding the particular system before installing it.

Usually, a blanket of insulation is stretched across the structural members prior to sheeting. The blanket width depends on the panel. Keep the stapling edge ahead of the panel, but no more than a foot ahead of it.

As previously mentioned, installing metal roofing requires special tools and experience. The following is an overview of the installation procedure for one type of metal roofing laid over a closed, fully sheathed roof deck. The precut panels are 12" to $16\frac{1}{2}$" wide and run from the eaves to the ridge of the structure. The joints between panels have a C-clip with a neoprene seal to weatherproof them.

Cover the roof with a 30 lb felt or waterproof membrane. The membrane or felt must overlap the edges of the eaves and rakes.

Screw the eave trim (*Figure 41*) in place before applying the panels. Then place the panels one at a time and secure them to the roof deck with a T-clip on one side of the panel.

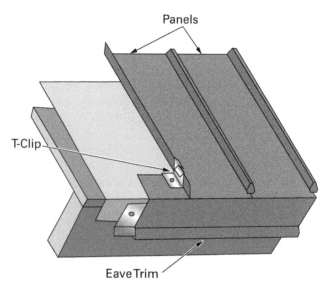

Figure 41 Eave trim.

The next panel is then inserted under the T-clip that is holding the first panel and secured on the opposite side with another T-clip (*Figure 42* and *Figure 43*).

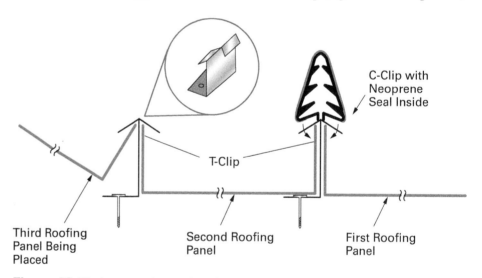

Figure 42 Placing, securing, and sealing panels.

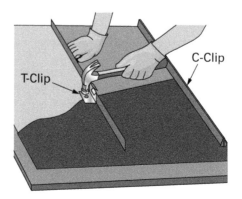

Figure 43 Nailing a T-clip to a roof deck.

At the joint of any two panels, the joint is weather-sealed with a C-clip running the full length of the joints. The T-clips are used every 12" in high-wind areas and every 18" elsewhere. After panels are placed under each side of the T-clips, the T-clip wings are bent down, and the C-clip is forced down on the seam (*Figure 42*). The neoprene flaps inside the C-clip seal the seam against water penetration, and the bent-down wings of the T-clip keep the C-clip from dislodging from the seam.

After the first panel is placed at one of the rakes, a rake edge is applied, as shown in *Figure 44*. If necessary at the opposite rake, the panel is trimmed off and Z-strips are sealed and secured to the panel before the rake edge is installed.

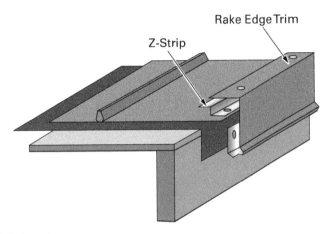

Figure 44 Rake edges.

At valleys, the panels are trimmed to the angle of the valley. Channel strips running parallel in the valley are sealed and screwed to the valley and hold the edges of the panels (*Figure 45*).

At the ridge, fasten and seal Z-strips to the panels between the seams as well as to the seams. Then seal and secure the cover flashing to the Z-strips (*Figure 46*).

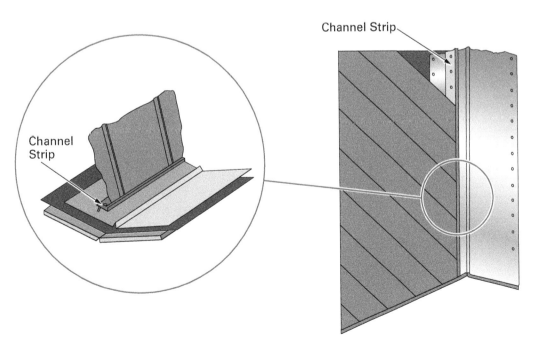

Figure 45 Channel strips and valley flashing.

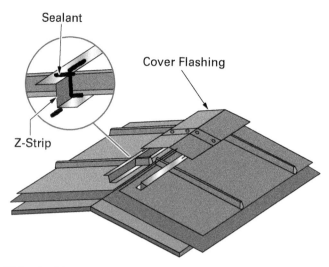

Figure 46 Ridge flashing.

4.3.3 Snug-Rib System

The snug-rib system uses a concealed fastener that is leak resistant and eliminates through fasteners. This combination of a V-beam industrial sheet and the snug seam joint makes for greater beam strength and deeper corrugation than other roofing profiles. Because of the greater strength of this type of roofing panel, the purlin spacing can be increased.

The snug-rib joint is a simple and highly efficient watertight joining system. The joint is created by engaging the hooked edges of two panels into a Y-shaped extruded spline, previously measured and anchored to the purlins with a self-templating clip. A neoprene gasket is then rolled into the extrusion between the panel edges, where it holds the panel edges securely in place and creates a watertight seal.

There is a $19\frac{1}{2}$" covering width. The material has a V-shaped corrugation that has a $4\frac{7}{8}$" pitch and a $1\frac{3}{4}$" depth. Lengths vary from 77" to 163", depending upon the gauge of the material. The end laps must be a minimum of 12", located over roof purlins, and staggered with the end laps in adjacent panels. Install this system according to the manufacturer's specifications.

NOTE

No primary fasteners penetrate the weatherproofing membrane.

Translucent Fiberglass Light Panels

Translucent fiberglass light panels allow daylight to show through the roof. They can cover the entire roof or just specific sections. Each manufacturer provides instructions for use and installation.

Source: Sanit Fuangnakhon/Shutterstock (left); 52Ps.Studio/Shutterstock (right)

Simulated Standing-Seam Roofing Systems

A number of different methods of joining metal roofing panels are employed for these types of systems. Older systems used galvanized steel or copper standing seams that were soldered together to form a weather seal. Other systems require crimping machines or use snap-type seals.

Source: LesPalenik/Shutterstock

4.4.0 Roof Projections, Flashing, and Ventilation

The following sections describe various types of roof projections and flashing.

4.4.1 Soil Stacks

Another roofing task is waterproofing around soil or vent stacks. Most building roofs have pipes or vents emerging from them. Most are circular and call for special flashing methods. Asphalt products combined with metals can serve this purpose. A soil pipe made of cast iron, copper, or other approved materials can be a vent for plumbing. Several types of **vent-stack flashing** are available to seal these openings. See *Figure 47.*

To apply vent-stack flashing, proceed as follows:

Step 1 First, apply the roofing up to where the stack projects. Use extreme care when cutting and fitting the shingles around the stack. See *Figure 48*.

Step 2 Slip the flange over the stack and place it down into a bed of carefully spread asphalt cement that is the same size as the flange. The flange, sometimes called a *boot* or *collar*, can be made of metal, plastic, or rubber. See *Figure 49*. Prefabricated boots that slip over the stack are available in different pitches.

Step 3 Mold the flange boot to the soil stack to ensure a snug fit. Use the manufacturer's recommended sealant to close up any opening. When the next course of

Vent-stack flashing: Flanges that are used to tightly seal pipe projections through the roof. They are usually prefabricated.

Figure 47 Vent-stack flashing.

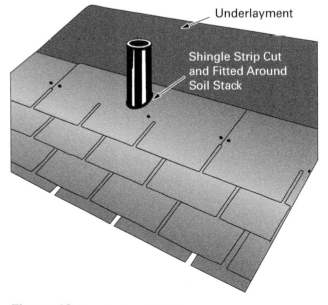

Figure 48 Layout around stack.

shingles is laid, it covers the upper portion of the flange. Prior to this course, a bed of cement can be spread on the top of the flange. The resulting seal waterproofs the vented stack opening. See *Figure 50*.

Step 4   After completing the installation, install the remaining shingles as previously described.

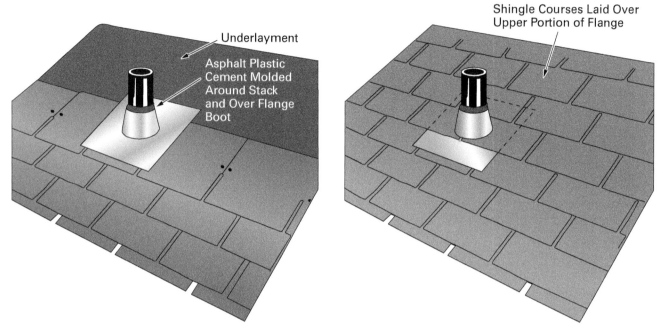

Figure 49 Placement of flashing.   **Figure 50** Covering flashing.

4.4.2 Vertical Wall Flashing

In the process of roof construction, there are times when the roof abuts a vertical wall horizontally or at an angle. This is a very critical spot to make watertight. Extreme care must be taken to follow correct procedures. An example of a sloped or angled abutment is shown in *Figure 51*. The initial step is to let the underlayment turn up on the vertical wall a minimum of 3" to 4". As an alternative, a strip of waterproof membrane may be applied to the roof deck and turned up on the wall.

This turn-up bend must be done very carefully to maintain the seal and overlap of material without creasing or tearing the felt or membrane. Regular shingling procedures are used as each course is brought close to the vertical wall. **Wall flashing** (step flashing) is used when the rake of the roof abuts the vertical wall.

Wall flashing: A form of metal shingle that can be shaped into a protective seal interlacing where the roof line joins an exterior wall. Also referred to as *step flashing*.

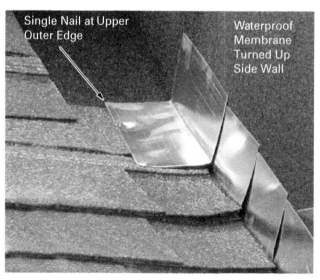

Installed Step Flashing

Figure 51 Wall (step) flashing.

Metal step-flashing shingles are applied over the end of each course of shingles and covered by the next succeeding course. The flashing shingles are usually rectangular. They are approximately 6" to 7" long and from 5" to 6" wide. When used with shingles laid 5" to the weather, they are bent so half of the flashing piece is over the roof deck with the remaining half turned up on the wall. The 7" length enables one to completely seal under the 5" exposure with asphalt cement and provide a 2" overlap up the entire length of the rake.

A careful study of *Figure 51* shows that each flashing shingle is placed just up the roof from the exposed edge of the shingle that overlaps it. It is secured to the deck sheathing with one nail in the top corner.

When the finished siding or clapboards are brought down over this flashing, they serve as a **cap flashing**, sometimes referred to as *counter flashing*. Usually, a 1" reveal margin is used and the ends of the boards are fully painted or stained to exclude dampness and prevent rot. With proper application of flashing and shingles, the joint between a sloping roof and a vertical wall should be watertight.

On a horizontal abutment (*Figure 52*), continuous flashing must be applied horizontally across the entire top of the abutting roof and against the vertical wall under the siding.

Continuous flashing can be formed with a metal brake or by hand, as shown in *Figure 53*. The flashing should be at least 9" wide and bent to match the angle of the joint to be flashed. Position the bend so that there will be at least 4" of flashing on the roof and 5" on the wall.

Cap flashing: The protective sealing material that overlaps the base and is embedded in the mortar joints of vulnerable areas of a roof, such as a chimney.

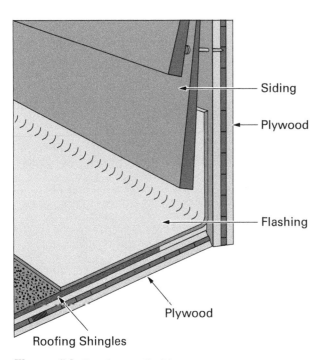

Figure 52 Continuous flashing.

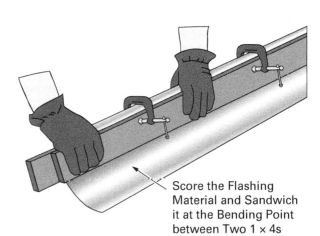

Score the Flashing Material and Sandwich it at the Bending Point between Two 1 × 4s

Figure 53 Bending continuous flashing.

Before applying the flashing, adjust the last two courses of shingles so that the last course, which will be trimmed to butt against the wall, is at least 8" wide. After this abutting course is installed, place roofing cement on top of the last course of shingles. Place the flashing against the wall (slipping it under any siding, if necessary) and press it into the cement. Do not nail the flashing to the wall or roof deck. If desired, apply several beads of roofing cement to the top of the flashing. Press the tabs cut from shingles into the cement to cover and hide the flashing (*Figure 54*). Position the tabs the same distance apart as the cutouts on the shingles and stagger them to match the pattern on the roof deck.

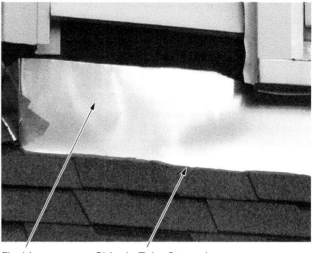

Flashing   Shingle Tabs Cut and
                Cemented to Flashing

Figure 54 Covering flashing.

4.4.3 Dormer Roof Valley

The installation of a dormer roof valley will require you to combine some of the procedures previously covered. *Figure 55* shows an open valley for a gable dormer roof. Note that the shingles have been laid on the main roof up to the lower end of the valley.

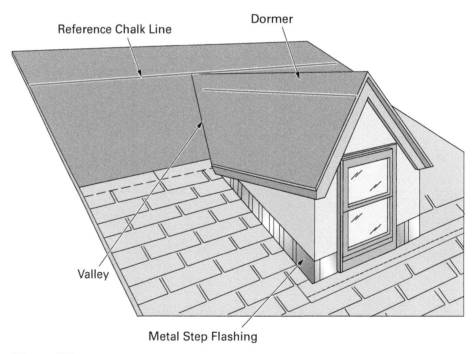

Reference Chalk Line   Dormer

Valley

Metal Step Flashing

Figure 55 Dormer flashing.

Gable Dormer Roof Valleys

Besides an open valley, a closed-cut valley can be used on a gable dormer that has a different roof pitch than the main roof. A woven valley may be used if the gable and main roof pitches are equal.

Extreme care must be used during the installation of the last course against the vertical wall to ensure a tight, dry fit. *Figure 56* displays the standard valley procedures for dormer flashing and shows how the valley material overlaps the course of shingles to the exposure line for a watertight seal.

Regular valley nailing procedures are used until work proceeds past the dormer ridge and resumes a full in-line course of shingles, as determined by a reference chalk line. See *Figure 57*.

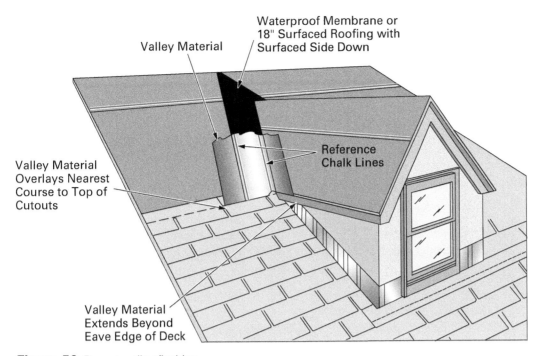

Figure 56 Dormer valley flashing.

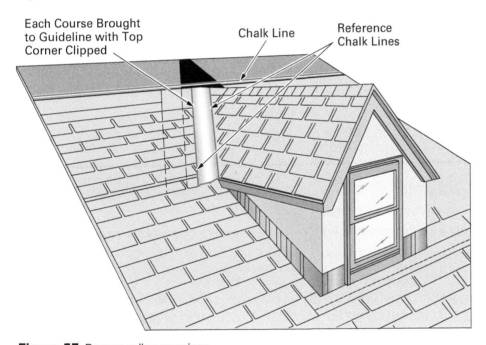

Figure 57 Dormer valley coverings.

4.4.4 Chimneys

Chimneys are subject to varying loads and certain opposing structural movements due to winds, temperature changes, and settling. Therefore, roofing materials and **base flashing** should not be attached or cemented to both the chimney and roof deck. The process of shingling around chimneys must be approached with extreme care. Due to the size of the opening in the roof and of the chimney itself, additional work must be done on the roof deck around chimneys prior to shingling. A cricket, also called a *saddle*, must be made. See *Figure 58*.

A cricket placed behind the chimney keeps rainwater or melting snow/ice from building up in back of the chimney. It steers flowing water around the chimney. The cricket is usually supported by a horizontal ridge piece and a vertical piece at the back of the chimney, as shown in *Figure 58*. The height of the cricket is typically half the width of the chimney, although these requirements will vary. Check local codes. The ridge, which is level, extends back to the roof

Base flashing: The protective sealing material placed next to areas vulnerable to leaks, such as chimneys.

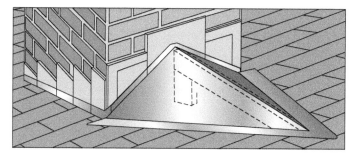

Figure 58 Simple chimney cricket.

slope. On wide chimneys, it may be necessary to frame the cricket, as shown in *Figure 59*. Either type of cricket may be covered with two triangular pieces of $^3/_4$" exterior plywood cut to fit from the ridge to the edge of the chimney and the roof slope. Heavy-gauge metal can also be used to form the cricket. The covering is then nailed to the support and roof deck.

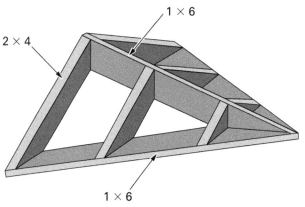

Figure 59 Cricket frame.

Flashing at the point where the chimney comes through the roof requires something that will allow movement without damaging the water seal. It is necessary to use metal base flashing. Then secure a metal counter or cap flashing to the masonry.

To apply the chimney flashing, proceed as follows:

Step 1   Apply shingles over the roofing felt up to the front face of the chimney (*Figure 60*).

Step 2   The base flashing for the front cut is applied first. The lower section is laid over the shingles in a bed of plastic asphalt cement. Bend the triangular ends of the upper section around the corners of the chimney.

Step 3   The sides of the chimney are base-flashed next. Either step flashing or continuous flashing can be used. Step flashing is applied as the shingles are applied up to the top side of the chimney (*Figure 61*). Note that the first piece of step

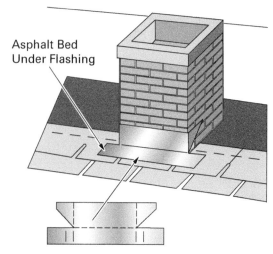

Figure 60 Front base flashing.

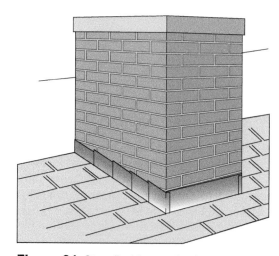

Figure 61 Step-flashing method.

flashing overlaps the front base flashing and is cut and bent around the front of the chimney. Like the front base flashing, the step flashing is fastened only to the roof deck with a nail or, if desired, roof cement. Bend and cement the flashing to the underlayment along the slope at the sides of the chimney, with the lower end overlapping the front base flashing. Bend the triangular end pieces around the chimney. Apply shingles up the roof to the top side of the chimney. Cement the shingles to the side base flashing to form a waterproof joint.

Step 4 It is now necessary to go to the top side of the chimney and complete the water-proofing operation by cutting and fitting base flashings over the cricket, known as *cricket flashing*, as shown in *Figure 62*.

Step 5 Using a pattern, cut the cricket base flashing.

Step 6 Bend the base flashing to cover the entire cricket, as shown at C in *Figure 62*. Extend the flashing laterally to cover part of the side base flashing previously installed. Set it tightly using plastic asphalt cement. Use care to spread the cement in the proper location and not extend out onto the finished shingles. Neatness is a must. Bend the ends around the chimney.

Step 7 Cut a rectangular piece of flashing, as shown at D in *Figure 62*, and make a V-cutout on one side to conform to the rear angle of the cricket. Center it over that part of the cricket flashing extending up to the deck. Set it tightly using plastic asphalt cement. This piece provides added protection where the ridge of the cricket meets the deck.

Step 8 Cut a second small rectangular piece of flashing. Cut a V on one side to conform to the pitch of the cricket, as shown at E in *Figure 62*. Place it over the cricket ridge and against the flashing that extends up the chimney. Embed it in plastic asphalt cement to the cricket flashing. Nail the edges of the flashing. In most cases, similarly colored pieces of roll roofing overlap the entire cricket and extend onto the roof deck. The roll roofing is cemented to the cricket flashing and sealed with cement at the chimney edge.

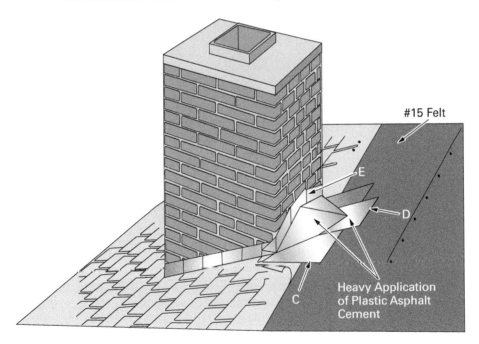

Figure 62 Cricket flashing.

Step 9 For completion, cap flashing (also called *counter flashing*) is installed. It is usually made of sheet copper 16 ounces or heavier. It can also be made of 24-gauge galvanized steel, which needs to be painted on both sides. *Figure 63* and *Figure 64* show metal cap flashing on the sides of the chimney and on the face. Cap flashing is secured to the brickwork, as shown in *Figure 65*.

Step 10 *Figure 65* shows a good method of securing the cap flashing. Cut a slot in the mortar joint to a depth of ¼" to ½". Insert a 90° bent edge of the flashing into the cleared slot between the bricks using an elastomeric sealant or mortar in the slot to secure the flashing to the masonry. When installed, the cap should lie snugly against the masonry. The front unit of the cap flashing should be one

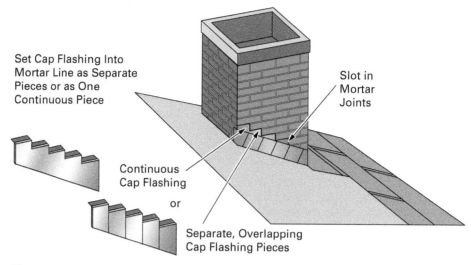

Figure 63 Cap flashing methods at sides.

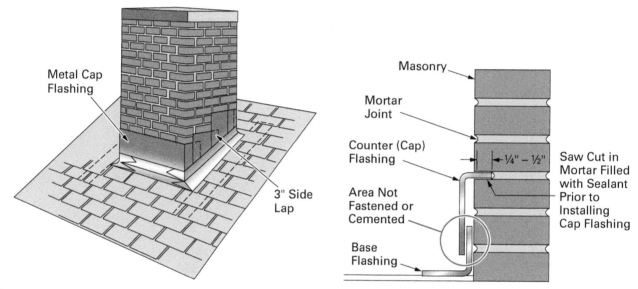

Figure 64 Flashing cap and lap.

Figure 65 Counter (cap) flashing installation.

Nailing Step Flashing

Step flashing should only be nailed to the roof deck, never to the wall sheathing. This will allow settling or shifting of the structure without tearing the flashing and roofing away from the roof deck.

Combustible-Material Spacing Requirements for a Cricket

Some building codes require that the wood framing and sheathing for a cricket must be spaced up to 1" from the chimney masonry. Always check your local codes for spacing requirements pertaining to combustible materials near chimneys.

continuous piece. On the sides and the rear, the sections are similar in size and conform to the locations of mortar joints and the pitch of the roof. If the sides are lapped, they must lap each other by at least 3". The slots are refilled with the brick mortar mix or elastomeric sealant and conform to the original brickwork. Patient installation of the flashing will provide the watertight seal necessary for a dry roof. Do not cement the counter (cap) flashing to the base flashing.

Step 11   Once the chimney flashing and shingling have been accomplished and the shingles next to the cricket are cemented under the edges to make a waterproof joint, the regular shingling process resumes on the next full course above the cricket. Another method of finishing the cricket is to extend the horizontal composition shingles of the roof deck up the pitch of the cricket. Then use step flashing and cement shingles parallel with the cricket ridge to form a half-weave valley at the edges of the cricket. This second method requires cementing cap shingles over the ridge of the cricket. The application of the shingles continues until the roof ridge is reached.

4.4.5 Hip or Ridge Row (Cap Row)

Special shingles are required to complete the hip or ridge rows. In some cases, ridge caps and hip caps are premanufactured.

Architectural shingles have a matching cap-row shingle (*Figure 66*) that must be used. Cap-row shingles cannot be cut from architectural shingles. Most of the time if the roof was shingled with standard three-tab shingles, cap rows are cut from the shingles and the 12" × 12" tab is used (*Figure 67*). The tab can be reduced to 9" × 12",

Figure 66 Cap shingles.
Source: SMJoness/Getty Images

Cut Tabs from Whole Shingles and
Taper Each Tab as Shown

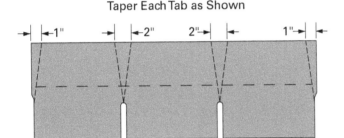

Figure 67 Cutting cap shingles.

but nothing less. Because the hip or ridge is a potential spot for water leaks, pre-cautions must be taken. If using ridge venting, do not apply the ridge caps.

To install a ridge or hip row, proceed as follows:

Step 1 Butt and nail shingles as they come up on either side of a hip or ridge. On a ridge, lay the last course and trim the shingles, as shown in *Figure 68*. On a hip, trim the shingles at an angle on the hip line.

Step 2 After cutting the cap shingles, bend them lengthwise in the centerline. In cold weather, warm the shingles before bending to prevent cracks. Begin at the bottom of any hips. Cut the first tab to conform to the dual angle at the eaves.

Step 3 Lap the units to provide a 5" exposure of the granular surface. See *Figure 69*. Secure with one nail on each side, 6" back from the exposed end and 1" from the edge. As each succeeding tab is nailed going up the hip, the nail penetrates and secures two tabs. This tight bond prevents the wind from getting underneath and lifting the tab.

Nailing the ridge row is similar to the procedure described for the overall hip. Nailing takes place from both ends of the ridge. A final cap piece joins the ridge together in the center, and the exposed nails are covered with roof cement. An exception to this may occur in a very windy area. In that case, the ridge cap would be started at the point on the roof opposite the wind direction. As each ridge shingle is placed, it automatically allows the wind to pass over it, and there is no possibility of shingles blowing off. The junction of the roof ridge and any hip ridges can be capped with a special molded cap or by a fabricated end cap. Bed the final ridge cap or hip/ridge caps in asphalt and secure with nails, as shown in *Figure 70*. Cover the nail heads with roof cement or sealant.

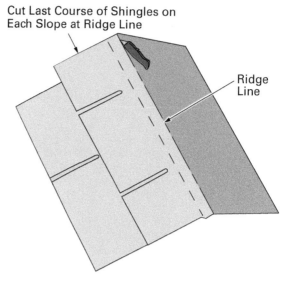

Figure 68 Applying the last course of ridge shingles.

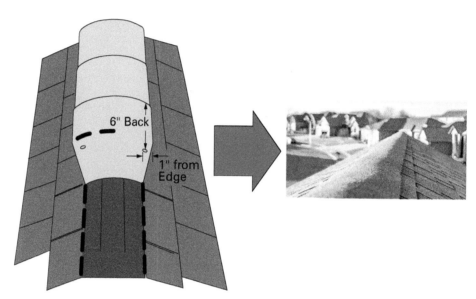

Figure 69 Installing a ridge cap.
Source: Lost_in_the_Midwest/Shutterstock (right)

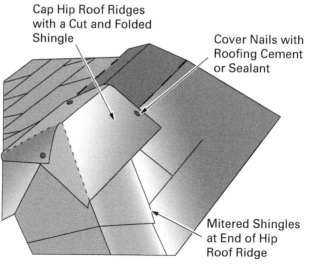

Figure 70 Hip and ridge end cap.

4.4.6 Installing Box Vents

Box vents (*Figure 71*) are easily installed on most roofing materials. On new roofs, the proper size hole is cut in the roof and the hole is surrounded with at least a 24"-wide waterproof membrane. When the roofing courses reach the area of the ventilator, the ventilator is installed with the lower edge of its flashing overlapping the course below it. The flashing is fastened to the roof at the top and sides. Then, roofing courses are applied over and cemented to the flashing at the top and sides of the ventilator.

On existing roofs, the hole is cut, and the upper flashing of the ventilator is slid under the courses above the hole and fastened to the roof. The side and bottom flashing is not fastened; however, the side flashing of the ventilator is completely cemented down to the roofing under the sides. The bottom flashing does not need cement.

Figure 71 Typical box vent.

Preformed Cap Flashing

Commercial preformed cap flashing may be installed on vertical brick, concrete, or block surfaces including chimneys. This flashing is made of aluminum-coated steel. A slot is cut in all sides of a chimney using a $1/4$"-thick diamond-impregnated steel wheel mounted in a small, high-speed electric grinder. The flashing is trimmed to shape, and the V-edge of the flashing is pressed into the groove. The flashing is formed to shape and sealed with an elastomeric sealant. When set, the sealant and flashing may be painted to blend with the roof.

| Preformed Cap Flashing | Cutting $1/4$" Groove | Completed Grooves |

| Seating V-Edge of Front Flashing in Groove | Fitting Side Flashing | Side Flashing Installed and Formed to Front Flashing |

| Sealant Applied to Groove and Flashing | Flashing Painted to Match Roofing |

4.4.7 Installing Ridge Vents

Use the following procedure to install ridge vents:

Step 1   Determine where the roof vent slots will be cut on the ridges and any hips (*Figure 72*). Slots cut along a ridge should start and stop approximately 12" from the gable (terminal) ends, any vertical walls, any higher intersecting roof, any roof projections at the ridge, any valleys, and any hip joints. Slots cut in hips should end 24" above the eaves and should be in 24" sections separated by 12" to maintain maximum roof strength. For appearance, the roof ridges and hips should be covered completely with the roof vent material to maintain a continuous roof line, as shown in *Figure 73*.

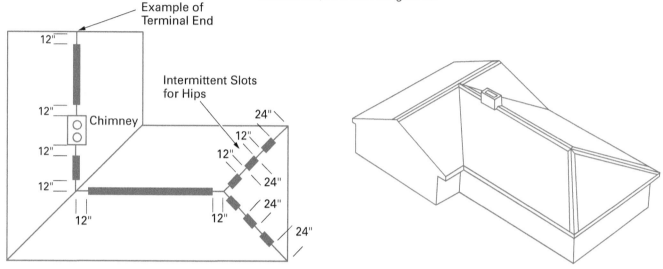

Figure 72 Example of slot cutout placement.

Figure 73 Example of a continuous roof line.

WARNING!

Always wear proper personal protective equipment (PPE), including fall protection equipment, when working on roofs.

Step 2   Next, refer to the manufacturer's specifications and note the width of the slot required for the vent being used. Determine the roof construction (ridge board or no ridge board). If a ridge board is used (*Figure 74*), add 1½" to the required slot width and divide the result by two to find the total slot width to be cut on each side of the peak. If no ridge board is used, only divide the required slot width by two to find the slot width to be cut on each side of the peak.

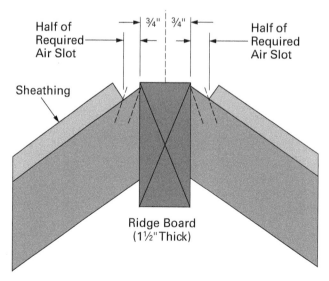

Figure 74 Determining total slot width.

Step 3   At the peak, measure down the roof on both sides for half of the required slot width as determined above and snap a chalk line along any ridges or hips, at both sides of the peak or hip (*Figure 75*). Mark the start and stop point for each slot, as previously specified.

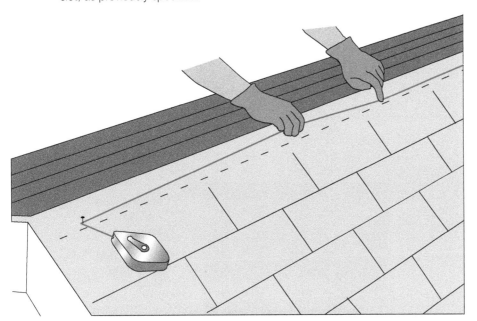

Figure 75 Snapping a chalk line for slot width.

Step 4   Using a knife, cut away the shingles along the chalk lines between each start and stop point.

Step 5   Using a power saw set to the thickness of the shingles plus the roof sheathing, carefully cut into the sheathing and along the chalk lines to remove the sheathing without damaging the rafters or, if present, the ridge board.

Step 6   Place a shingle ridge cap at each gable end, before and after each roof projection, at each vertical wall, at each higher intersecting roof joint, and at each eave end for any hips (*Figure 76*). This prevents any water intrusion at or under the exposed ends of the vent from penetrating into or through the sheathing.

Step 7   If packaged in rolls, unroll the vent material, and temporarily tack it in place over the ridges and hips, or secure the flexible plastic composition sections to the roof over the ridges and hips. Make sure that the vent covers the entire length of all ridges and hips (*Figure 77*). If rigid vent sections (*Figure 78*) are being used, make sure that vents are secured through the pre-punched holes.

Step 8   Cut and taper the tabs from a number of three-tab standard or architectural cap shingles for use as cap shingles.

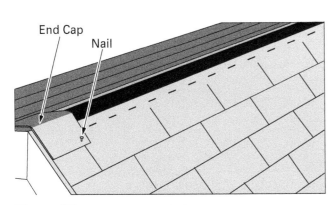

End Cap

Nail

Figure 76 Exposed-end shingle caps.

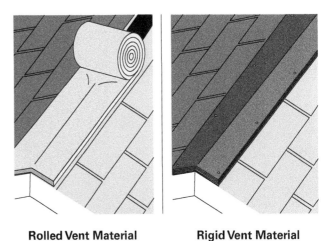

Rolled Vent Material   **Rigid Vent Material**

Figure 77 Positioning vent material.

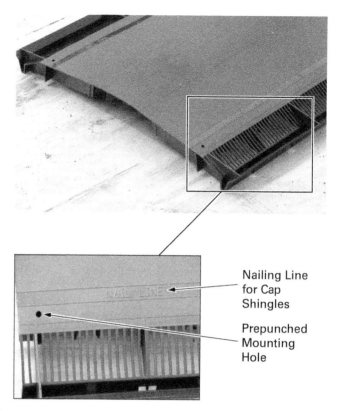

Figure 78 Typical rigid vent section.

Nailing Line for Cap Shingles

Prepunched Mounting Hole

Step 9 Starting from the exposed ends of a ridge or hip and using sufficiently long nails, fasten the cap shingles to the roof through the vent material nailing line (*Figure 79*). The cap shingles are applied and mated over the ridges and hips in the same manner as described in Section 4.4.5 *Hip or Ridge Row (Cap Row)*. However, make sure that the shingles are nailed snugly without compressing the vent material. It is advisable to place a bead of roofing cement under the overlapping sections of the shingles to help secure them.

Overlapping Cap Shingles

Figure 79 Applying cap shingles over a vent.

Gutters and Downspouts

Companies specializing in the field often measure, form, fabricate, and install gutters on residential and commercial buildings. Occasionally, however, carpenters install gutters and downspouts. Therefore, you should be familiar with available products and have a general knowledge of how they are installed. Refer to the manufacturer's product information and data sheets for installation guidelines. Some localities specify the size and capacity of gutters applied to commercial and/or residential structures, so always check local codes.

Vinyl and metal gutters and downspouts come in various colors, sizes, and shapes with smooth or patterned finishes. Galvanized metal gutters are usually unfinished and must be painted after they are in place.

Some gutters come with a debris guard that requires no cleaning and sheds roof debris, such as twigs and leaves, while directing water into the gutter. Another similar product is available for application to existing gutters.

The relationship between the gutter and the edge of the finish roofing material is important. Always center the gutter under the edge of the roof and below the slope of the roof to catch water that runs off, but not to catch snow and ice that slides off. The size of gutters and downspouts will depend on the intensity of rainfall and the amount of roof area needing drainage. Furthermore, the slope of the gutter and the number of outlets will affect the gutter size. The approved construction documents will designate the required sizes of the gutters and the location of the downspouts.

(A) Debris Guard

(B) Typical C-Style

(C) Typical K-Style

Sources: sheilaf2002/123RF (SA10A); brizmaker/Shutterstock (SA10B); Feverpitched/Getty Images (SA10C)

Plastic or Metal Ridge Vent

If using a rigid plastic or metal vent not intended to be capped with shingles, fasten the vent to the roof through the flanges using a noncorrosive fastener with a neoprene seal washer; then insert an end cap in the exposed ends of the vents, if required.

Precautions When Cutting a Ridge Vent

Use goggles and make sure your footing is solid along the ridges when cutting with a power saw. Make sure the power cord is laid out so that there is no chance of getting entangled with it during the cutting operation. Make sure that the saw depth is set so that the ridge board or rafters are not cut.

Alternate End Treatments of a Ridge Cap

Some contractors cut the ridge vent further in from the end of the roof so that the ridge cap is not flush with the end of the roof. This hides the end view of the cap from the ground for a more attractive appearance.

4.0.0 Section Review

1. Drip edge should be nailed at an interval of _____.
   a. 4" to 6"
   b. 6" to 9"
   c. 8" to 10"
   d. 8" to 12"

2. A row of shingles with the tabs cut off is often the first row laid at the eave edge of the roof. These shingles are known as a(n) _____.
   a. starter row
   b. initial row
   c. base row
   d. underlay row

3. To form a tight roof, adjacent corrugated metal roofing panels should be overlapped by _____.
   a. 1" corrugations
   b. 1½" corrugations
   c. 2" corrugations
   d. 2½" corrugations

4. To keep rainwater or melting snow/ice from building up in the back of the chimney, a _____ is placed behind the chimney.
   a. vertical wall flashing
   b. dormer flashing
   c. flash cap and lap
   d. cricket

5.0.0 Estimating Roofing Materials

Performance Tasks	Objective
There are no Performance Tasks in this section.	Describe the estimating procedure for roofing projects.

Regardless of the type of roofing to be installed, the amount of material required must first be estimated. After the amount of material has been determined and obtained, the roof deck must be prepared before the finish roofing material is raised to the roof deck and installed.

The approved construction documents for the structure will show the roof dimensions and material estimates. If these documents are not available, the carpenter can estimate the amount of material needed by measuring the length and width of each section of the roof. Then, the area of each section of the roof is calculated and a percentage is added for waste. The result is converted into the number of squares (100 ft^2) of material required.

Step 1   Measure the length and width of each triangular and rectangular roof section of the structure, including any overhangs (*Figure 80*).

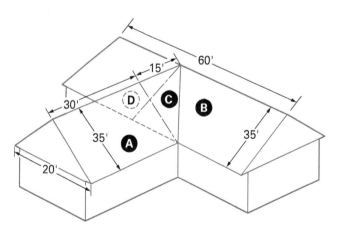

Figure 80 Roof example (including overhangs).

Step 2   Calculate the area for one-half of each roof section and add the areas together. Then, multiply by 2 to obtain the total roof area and subtract any triangular areas covered by roof intersections.

To calculate the total area of the roof shown in *Figure 80*, proceed as follows:

$30' \times 35' = 1,050.0 \text{ ft}^2$ (Area A)

$60' \times 35' = 2,100.0 \text{ ft}^2$ (Area B)

$15' \times 35' \times \frac{1}{2} = 262.5 \text{ ft}^2$ (Area C)

$1,050.0 \text{ ft}^2 + 2,100.0 \text{ ft}^2 + 262.8 \text{ ft}^2 = 3,412.5 \text{ ft}^2 (\frac{1}{2} \text{ roof area})$

$3,412.5 \times 2 = 6,825.0 \text{ ft}^2$

$35' \times 20' \times \frac{1}{2} = 350.0 \text{ ft}^2$ (Area D)

$6,825.0 \text{ ft}^2 - 350.0 \text{ ft}^2 = 6,475.0 \text{ ft}^2$ (roof area)

Step 3   Add an average of 10% for hips, valleys, and waste. For a complicated roof with several valleys or hips, more than 10% may be required. For a plain, straight gable roof, less is usually required. In addition, for wood or slate roofs, an additional 100 ft² is usually required for each 100 lineal feet of hips and valleys. For our example, a relatively simple roof with standard three-tab composition shingles, a 10% waste factor is included.

$$\text{Total material} = (10\% \times 6,475) + 6,475$$
$$= 648 + 6,475$$
$$= 7,123 \text{ ft}^2$$

Step 4   Convert the total square feet of material required to squares by dividing by 100 ft² (if using wood roofing, round up or down to the nearest bundle or square). For standard three-tab composition shingles, the rounding would be to the nearest $\frac{1}{3}$ or $\frac{2}{3}$ of a square (one or two bundles) or whole square.

$$\text{Squares} = 7,123 \div 100 = 71.27 = 71\frac{1}{3}$$

Step 5   The number of rolls of underlayment required is determined from the same total material requirement of 7,123 ft². Starter strips, eave flashing, valley flashing, and ridge shingles must be added to complete the estimate. All of these are determined with linear measurements.

5.0.0 Section Review

1. When estimating the material needed for a building with a hip roof, increase the calculated quantity by _____.
   a. 5%
   b. 7.5%
   c. 10%
   d. 15%

Module 27202 Review Questions

1. Any project where workers will be more than 6' off the ground or on an elevated platform requires a _____.
   a. job hazard analysis
   b. minimum of 3 workers
   c. fall protection plan
   d. nonconductive ladder

2. A scaffold that meets OSHA requirements and is safe to use is marked with a(n) _____.
   a. blue ribbon
   b. orange tape
   c. checkered flag
   d. green tag

3. For safe use, a ladder must extend above the edge of a roof by a distance of _____.
   a. 1'
   b. 2'
   c. 3'
   d. 4'

4. Workers should immediately distribute roofing bundles around the roof area because roof structures are designed to carry a specific _____.
   a. slope
   b. dead load
   c. balance
   d. sheathing

5. When a roof slope increases to 6 in 12, _____.
   a. bundles need to be spread out on the roof
   b. roofing brackets are no longer needed
   c. more strain is placed on the worker's feet and body
   d. ladders cannot be used to access the roof

6. The standard roofing hammer is also called a _____.
   a. shingle hatchet
   b. slater's hammer
   c. claw hammer
   d. cricket hammer

7. A slater's hammer can be used to _____.
   a. remove nails
   b. bend metal flashing
   c. punch nail holes
   d. cut tile

8. Power tools should be examined for damage _____.
   a. after each use
   b. only by a supervisor
   c. once per month
   d. before each use

9. Composition shingles require approximately _____ pound(s) of nails per square (100 ft^2).
   a. 1
   b. $1\frac{1}{2}$
   c. 2
   d. 3

10. Non-fibered forms of asphalt roofing cement are usually used in the _____.
    a. preparation of the roof deck
    b. installation of metal roofing
    c. installation of a hip roof
    d. lapped installation of underlayment

11. Organic-fiber composition shingles have been largely replaced by shingles made with _____.
    a. asbestos
    b. wood
    c. fiberglass
    d. synthetic resin

12. Fiberglass shingles have a typical life span of _____.
    a. 5 to 12 years
    b. 15 to 20 years
    c. 20 to 25 years
    d. 25 to 40 years

13. The most common type of composition shingle applied to residential structures is the _____.
    a. two-tab
    b. three-tab
    c. architectural
    d. strip

14. Per square foot, both tile roofing and slate roofing weigh about _____.
    a. 3 to 7 lb
    b. 5 to 9 lb
    c. 7 to 10 lb
    d. 9 to 12 lb

15. Some benefits of metal roofing include _____.
    a. longevity and cost
    b. low maintenance and noisiness
    c. energy efficiency and eco-friendliness
    d. energy efficiency and cost

16. The abbreviation BUR stands for _____.
    a. builder's union representative
    b. bundle-unitized roofing
    c. built-up roofing
    d. building unit resources

17. Premanufactured membrane roofing systems are grouped into _____.
    a. two categories
    b. three categories
    c. four categories
    d. five categories

18. One of the most common thermoset single-ply membranes is known as _____.
    a. APP
    b. SBS
    c. PVC
    d. EPDM

19. Two common thermoplastic single-ply membranes are _____.
    a. APP and BUR
    b. SBS and PVC
    c. PVC and TPO
    d. EPDM and TPO

20. The felt used under roofing materials must allow the passage of _____.
    a. air
    b. gas
    c. water vapor
    d. water

21. The attic ventilation area is normally divided between exhaust and inlet vents in the proportion of _____.
    a. 50% exhaust, 50% inlet
    b. 60% exhaust, 40% inlet
    c. 70% exhaust, 30% inlet
    d. 80% exhaust, 20% inlet

22. The width of ice edging is determined by the _____ and how thick the ice usually becomes.
    a. type of roofing material
    b. slope of the roof
    c. amount of insulation
    d. framing material

23. When applying shingles to a large gable roof, start at the _____ of the long run so there is less chance of misalignment.
    a. center
    b. bottom
    c. left side
    d. right side

24. The 6" pattern of laying three-tab shingles means that 6" is _____.
    a. added to each course
    b. subtracted from each course
    c. subtracted from every other course
    d. subtracted from the second course, 12" from the third, 18" from the fourth, and so on

25. An open valley is wider at the bottom than at the top _____.
    a. to provide a more visually pleasing joint
    b. for easier cleaning and maintenance
    c. to accommodate the higher volume of water that will be present near the bottom
    d. to allow for variations in the width of shingles

26. There are two types of closed valleys, cut and _____.
    a. stepped
    b. woven
    c. staggered
    d. interleaved

27. In corrugated metal roofing, only galvanized sheets coated with _____ are recommended for permanent construction.
    a. copper
    b. plastic
    c. aluminum
    d. zinc

28. If a fastener is stripped during the installation of a standing-seam metal roof, you should _____.
    a. remove the stripped fastener and replace it with a fastener of the next larger size
    b. remove the stripped fastener, seal the hole with roofing cement, and drill a hole 2" from the original for a new fastener
    c. leave the stripped fastener in place and put another fastener alongside it
    d. remove the stripped fastener and replace it with a fastener of the same size

29. The snug-rib system uses a neoprene gasket to hold the panel edges securely in place and to create a _____.
    a. purlin
    b. watertight seal
    c. closed-cut valley
    d. ridge vent

30. A(n) _____ is used to seal and waterproof pipe projections through a roof.
    a. channel strip
    b. ridge flashing
    c. vent-stack flashing
    d. rake edge

31. Step flashing is used on _____.
    a. valleys
    b. slopes against walls
    c. hips
    d. vent pipes

32. The additional structure required to properly apply roofing around a chimney is a _____.
    a. cricket
    b. dormer
    c. fascia
    d. valley

33. When installing a box vent on a new roof, the hole must be surrounded with at least a(n) _____.
    a. 18"-wide waterproof membrane
    b. 20"-wide waterproof membrane
    c. 24"-wide waterproof membrane
    d. 36"-wide waterproof membrane

34. A square of three-tab composition shingles consists of _____.
    a. one bundle
    b. two bundles
    c. three bundles
    d. four bundles

35. How many squares of material would be needed for a simple roof with standard three-tab composition shingles that has a total material of 6,316 ft^2?
    a. 63
    b. 63$\frac{1}{3}$
    c. 63$\frac{2}{3}$
    d. 64

Answers to Odd-Numbered Module Review Questions are found in *Appendix A*.

Answers to Section Review Questions

Answer	Section	Objective
Section One		
1. b	1.1.0	1a
2. a	1.2.1	1b
Section Two		
1. b	2.1.0	2a
2. d	2.2.1	2b
3. c	2.3.1	2c
Section Three		
1. b	3.1.0	3a
2. b	3.2.2	3b
3. d	3.3.0	3c
4. b	3.4.0	3d
5. b	3.5.0	3e
6. c	3.6.0	3f
7. b	3.7.0	3g
Section Four		
1. c	4.1.0	4a
2. a	4.2.0	4b
3. b	4.3.1	4c
4. d	4.4.4	4d
Section Five		
1. c	5.0.0	5

Steel Framing Systems

Objectives

Successful completion of this module prepares you to do the following:

1. Identify the tools and components of cold-formed steel framing systems and their safe use.
   a. Identify safety guidelines that should be followed when working with steel framing systems.
   b. Identify steel framing materials.
   c. List common steel framing tools and fasteners.
   d. Explain how to perform a material takeoff for a steel frame project.
2. Identify the steps required to lay out and install a steel stud wall.
   a. Describe basic steel construction methods.
   b. Explain how to frame nonstructural steel walls.
   c. Describe how to frame structural steel walls.
3. Identify other steel framing applications.
   a. Explain how to use steel framing members in floor and roof construction.
   b. Describe how to use steel framing members in ceiling construction.

Performance Tasks

Under supervision, you should be able to do the following:

1. Estimate the amount of materials required to complete an instructor-specified steel framing project.
2. Lay out a steel stud wall with openings to include bracing and blocking.
3. Demonstrate the ability to build headers (back-to-back, box, and L-header).

Overview

This module describes the uses and installation of cold-formed steel framing systems. In commercial and multifamily residential construction, it is common to use steel framing materials in place of wood studs to frame walls and partitions. It is also becoming more common in single-family residential construction as the price of lumber rises. Steel framing requires a carpenter to master tools and joining techniques different from those required in wood framing construction. Learn how to effectively complete various steel framing tasks, such as the installation of load-bearing steel assemblies to nonstructural walls, by familiarizing yourself with common steel framing materials and building methods.

NOTE

Codes vary among jurisdictions, so consult the most applicable code whenever regulations are in question. Referring to an incorrect or outdated set of codes can result in improper building processes that can negatively impact your project deadlines and budgets.

Digital Resources for Carpentry

Scan this code using the camera on your phone or mobile device to view the digital resources related to this craft.

1.0.0 Materials and Tools

Performance Task

1. Estimate the amount of materials required to complete an instructor-specified steel framing project.

Objective

Identify the tools and components of cold-formed steel framing systems and their safe use.

a. Identify safety guidelines that should be followed when working with steel framing systems.

b. Identify steel framing materials.

c. List common steel framing tools and fasteners.

d. Explain how to perform a material takeoff for a steel frame project.

Cold-formed steel: Sheet steel or strip steel that is manufactured by press braking of blanks sheared from sheets or cut lengths of coils or plates, or by continuous roll forming of cold- or hot-rolled sheet steel coils.

Designers, builders, and architects have long recognized steel for its strength, durability, and functionality, as well as its potentially environmentally friendly characteristics. **Cold-formed steel** framing helps with industry efforts to promote sustainable construction due to its long-lasting durability and reusability in steel production.

Going Green

Steel Recycling

Steel is manufactured with an average of 67 percent recycled materials, making it the world's most recycled material. In North America alone, millions of tons of steel are recycled or exported for recycling per year.

By virtue of its material characteristics and properties, steel offers the following advantages for building construction:

- Steel studs and joists are strong and lightweight, with the highest strength-to-weight ratio of any building material.
- Steel framing members are rigid and dimensionally stable, providing structural integrity across long spans of material.
- Steel is uniform and provides a flat surface for sheathing or other material attachment.
- Steel framing is naturally fire-resistant, resulting in safe and reliable foundational elements and reduced fireproofing costs to builders and owners.
- Steel can be engineered to meet the strongest wind and seismic ratings specified by building codes.

North American Codes and Standards

The use of cold-formed steel (CFS) members in building construction began around 1850. In North America, however, steel members were not common building materials until 1946, when the American Iron and Steel Institute (AISI) Specification was first published. This design standard was primarily based on research sponsored by AISI at Cornell University. Subsequent revisions to the document reflected technical developments that ultimately led to the publishing of the North American Specification for the Design of Cold-Formed Steel Structural Members.

AISI, along with the American National Standards Institute (ANSI), the American Society for Testing and Materials (ASTM) International, the International Code Council (ICC), and the Steel Stud Manufacturers Association (SSMA), govern the design, manufacturing, and use of cold-formed steel and framing. All framing members carry a product identification to disclose the minimum sheet steel thickness, coating designation, minimum yield strength, and manufacturer's name.

1.1.0 Cold-Formed Steel Framing Safety Guidelines

Working with cold-formed steel presents several safety issues that differ from working with wood framing members. Cold-formed steel framing safety guidelines are as follows:

- Wear cut-resistant gloves when handling steel framing members to protect your hands from sharp edges.
- Use caution when handling wet steel framing members because they may be slippery.
- Use proper personal protective equipment (PPE), including hearing protection, respiratory protection, and a full-face shield over wraparound safety glasses, when cutting steel framing members with an abrasive saw. An abrasive saw can produce high-decibel noise, flying metal fragments and debris, and fumes from the zinc coating applied to cold-formed steel framing members. Inhaling these fumes may result in a condition known as *zinc chills*, so task-specific safety plans may require a respirator mask as well.
- Avoid dropping or placing heavy steel framing members on electrical cords to reduce the chance of damaging them or creating an electrical hazard from exposed wires.

1.2.0 Steel Framing Materials

Framing components include **studs**, joists, and roof trusses. Accessories such as clips, web stiffeners, resilient channels (RC), fastening devices, and anchors are required for complete and proper installation of these members. Some manufacturers also offer specialized products that enable builders to shorten construction times for complicated curves, arches, and other unique architectural features. The vertical and horizontal framing members serve as structural load-bearing components for many low-rise and high-rise structures.

Studs: Cold-formed steel structural and nonstructural framing members, consisting of a web, two flanges, and two returns.

WARNING!

Cutting cold-formed steel framing members with an abrasive saw makes the metal very hot, creating zinc fumes. To prevent the inhalation of these fumes, always wear proper respiratory protection when cutting framing members with an abrasive saw.

Steel components are also noncombustible, lightweight, and resistant to heat and corrosion. The ASTM governs cold-formed steel manufacturing standards and tolerances to establish proper lengths and web depths to reduce the likelihood of members flaring, crowning, bowing, or twisting.

Cold-formed steel components are also coated to protect against corrosion according to ASTM standards. Hot-dipped zinc galvanizing is the most effective coating method. Depending on the thickness of zinc applied to the steel, and the environment in which the steel is placed, zinc coatings can protect the steel for years.

Steel framing is also compatible with all types of surfacing materials. A variety of load-bearing and nonbearing systems are available, but *Figure 1* shows the basic components of most steel framing systems. To meet custom material requirements, stud, **track**, and joist material can be cut within $\frac{1}{8}$" of specification. Length is restricted only by the mode of physical transportation—typically 40' for containers and flatbed trucks. Custom ordering allows for less field cutting, labor, and waste on the jobsite.

Track: A steel framing member consisting of two flanges and a web. Although similar in shape to studs, track has no returns. Track web depth is measured between the inside edges of the flanges.

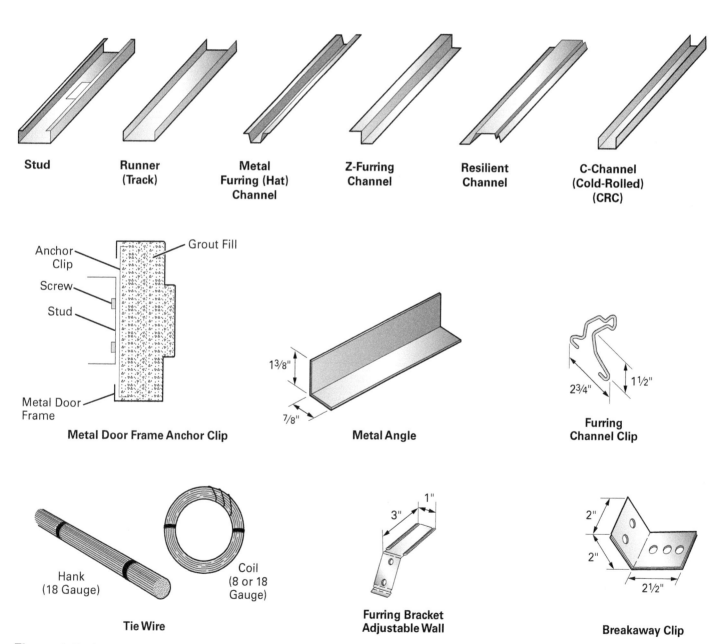

Figure 1 Basic components of a steel framing system.

Avoid using steel framing members that are not marked with the required identification. The identification may be etched, stamped, or labeled every 96", along the centerline of the **web** (*Figure 2*). Missing identification can result in project delays until materials are identified in accordance with appropriate standards.

Web: A portion of a framing member that connects the two flanges.

Figure 2 Steel product skid label or *inkjet*.
Source: Don Wheeler

Going Green

Modern Steel Production

Modern steel production relies on two technologies: a basic oxygen furnace (BOF) or an electric arc furnace (EAF). The BOF process uses 25 to 35 percent recycled steel to make new steel. It produces sheets for products requiring drawability, such as automotive fenders, steel framing members, refrigerator enclosures, and packaging. Conversely, the EAF process uses 95 to 100 percent recycled steel and is primarily used to manufacture more rigid products like structural beams, steel plates, and reinforcement bars. Regardless of the process, the resulting steel product has a minimum of 25 percent recycled content; however, the industry standard averages 67 percent recycled content.

During the production process, molten steel is poured into an ingot mold or a continuous caster, where it solidifies into large rectangular shapes known as slabs. These slabs are then passed through a machine with a series of rolls that reduce the steel to thin sheets of the desired thickness, strength, and other physical properties. Lastly, steel sheets undergo a hot-dipped galvanizing process before they are rolled into coils that weigh approximately 13 tons.

1.2.1 Identification of Framing Materials

Manufacturers of various steel framing products have published a wide range of codes for identifying steel components. To eliminate this confusion, The American Iron and Steel Institute (AISI) developed a universal designator system, known as the S-T-U-F system. This S-T-U-F system was once the industry standard for identifying common framing members, including studs (S), tracks (T), cold-rolled channels (U), and furring channel (F).

However, with the arrival of new products that didn't fall into the simplified acronym, AISI revised the system to the cold-formed steel (CFS) indicator system. This new system has the same indicator acronym for product type, but with an added L for L-headers. When ordering CFS materials for your project, an inkjet (as seen in *Figure 2*) will present an alphanumeric label that designates web depth, profile type, flange width, and minimum base thickness. To break this concept down further, we will use 250S125-43 as an example.

The first three numbers designate web depth, typically measured from the outside edges, except for tracks, which require inside-to-inside measurements for a tight fit. These measure in $1/100$", so the 250 and S represent a stud with a web depth of $2^1/_2$".

The next three numbers in the sequence designate flange widths, which are also measured in $1/100$". Therefore, the example stud has a flange width of $1^1/_4$". Lastly, the numbers after the dash represent base material thickness in mils ($1/1000$"). You can find more information on required material thicknesses in *Table 1*.

TABLE 1   Minimum Base Steel Thickness of Cold-Formed Steel Members

Designation Thickness (Mils)	Minimum Base Thickness (Inches)	Reference Gauge Number
18	0.0179	25
27	0.0269	22
30	0.0296	20—Drywall
33	0.0329	20—Structural
43	0.0428	18
54	0.0538	16
68	0.0677	14
97	0.0966	12
118	0.1180	10

1.2.2 Furring

Design requirements may specify that builders maintain separation between gypsum board (drywall) materials and studs by using steel furring channels to provide sound isolation. In this case, the gypsum board is usually furred out using a resilient channel (*Figure 3*).

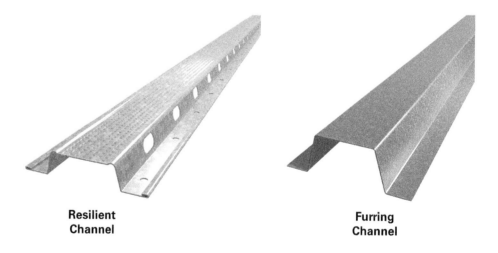

Resilient
Channel

Furring
Channel

Figure 3 Resilient channel and furring channel.
Source: Image provided by ClarkDietrich

Resilient furring channels attach over both metal and wood framing to provide a sound-absorbent spring mounting for gypsum board and should be attached as specified by the manufacturer. The channels not only improve sound insulation, but also protect fragile gypsum boards from cracking due to structural movement.

Resilient furring channels may also accommodate the application of gypsum boards over masonry and concrete walls. In wood frame construction, gypsum board can be screwed to resilient metal furring channels to provide a higher degree of sound control. Furring also offers the following benefits:

- Provides additional space for insulation
- Allows builders to match out-of-plane walls or walls of different thicknesses, creating a smooth and uniform vertical surface
- Provides additional space to conceal fixtures or structural elements within a wall

1.2.3 Slip Connectors

Slip connectors (*Figure 4*) are devices that allow for the vertical movement of a structure without imposing additional loads on cold-formed steel framing or other wall components. Use these connectors where a structural system other than steel framing is needed to carry loads of upper floors and the roof down to the foundation. Loads on the upper portions of a structure cause it to deflect, which in turn may induce loading on walls and wall components that are not designed to carry these loads. Slip connectors are designed to allow for this movement without creating unmanageable stress on components.

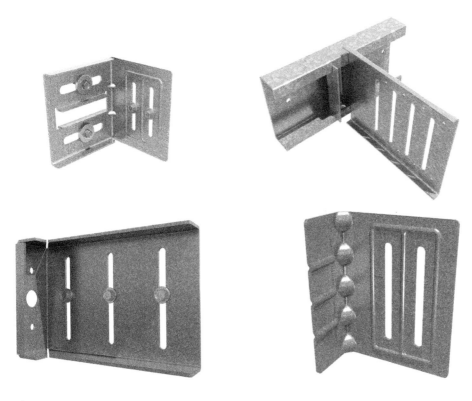

Figure 4 Slip connectors.
Source: Image provided by ClarkDietrich

Slip connectors are generally located at the top of a wall panel, where it meets the underside of a structural element, such as a floor slab or beam. Under gravity, seismic, or wind loads, this upper portion of the structure may deflect up or down. The connector is designed to allow this movement, restrain the wall system from out-of-plane movement, and prevent any additional axial loading on the stud.

Slip connectors are also useful at locations where a wall system is continuous, bypassing intermediate floors (*Figure 5*). Where this occurs, a slip connector extends from the side of the structure and supports the wall components laterally. Connectors are also installed at roof bypasses, where the wall system extends past a roof structure to form a parapet or high wall. At this location, the slip connector must permit movement of the structure either up, due to wind uplift, or down, due to gravity loads.

Figure 5 Slip connector application.
Source: Image provided by ClarkDietrich

1.2.4 Deep Leg Tracks

A deep leg track (*Figure 6*), or deflection track system, requires cold-rolled channel and angle clips passing through the uppermost stud punchout to align studs vertically along the wall plane and float within the track legs. This system allows for live load movement across the structural assembly without transferring loads to the wall studs. This system must be designed with lateral bracing to prevent rotation and should not be used in load-bearing spans above doors and windows.

Figure 6 Deep leg track.
Source: Image provided by ClarkDietrich

1.3.0 Steel Framing Tools and Fasteners

Steel framing typically requires more use of power tools than wood framing, so be sure to always practice power tool safety, including the grounding of all electrical tools. The following are some common tools used in steel framing:

- *Impact driver* — The impact driver (*Figure 7*) drives screws to connect steel members and attach sheathing material and gypsum wall board to steel. Impact drivers can offer significantly more torque than power screwdrivers, so it is important to avoid over-driving your fasteners.

- *Stud driver* — There are two types of stud drivers for applying fasteners into concrete slabs, structural steel, and foundations: powder-actuated tools (*Figure 8*) and rotary hammers (*Figure 9*). The holding strength of the fastener in concrete depends on the compressive strength of the concrete, shank diameter, and depth of penetration of the fastener, as well as spacing and edge conditions. Headed or threaded powder-actuated drive pins are available with knurled shanks to increase holding power in structural steel material. Fasteners loaded in tension may require washers to prevent the fastener from pulling through the steel track.

Figure 7 Impact driver.
Source: Courtesy of Stanley Black & Decker, Inc.

Knurled: Having a series of small ridges that provide a better gripping surface on metal and plastic.

Figure 8 Powder-actuated tool.
Source: Courtesy of Stanley Black & Decker, Inc.

Figure 9 Rotary hammer.
Source: Courtesy of Stanley Black & Decker, Inc.

NOTE

A power screwdriver is not just a drill with a screwdriver bit. Power screwdrivers typically have an adjustable depth control to prevent overdriving the screws. Many power screwdrivers have clutch mechanisms that disengage when the screw has been driven to a preset depth. Some power screwdrivers are designed to perform specific fastening jobs, such as fastening gypsum board to walls and ceilings.

CAUTION

If the screwgun is running too fast, the tip of the screw may burn out before it penetrates the steel. Once the screw is properly seated, the screwgun automatically stops spinning the screw, preventing the screw from stripping.

- *Powder-actuated tool* — Craft professionals must be properly trained in the use of powder-actuated tools (PATs) and must have an operator's certificate before operating them. The operator must wear a hard hat, as well as eye, ear, and face protection. Treat a powder-actuated tool as if it were a loaded gun. Before you handle it, determine whether it is loaded with a powder charge, a fastener, or both. Always point the muzzle away from yourself and others and ensure the tool is in a locked position when it is not in use.

Header: A horizontal structural framing member that supports and transfers weight loads over openings in floors, roofs, or walls. The most common location for headers is above windows and doors.

- *Rotary hammer* — Use this powerful tool to drill holes in concrete or masonry walls to secure steel frames in place. The rotary hammer's ability to combine the functions of a drill and a hammer allows for efficient and precise drilling. Both corded and battery-powered models are available with various drill bits to create accurate holes in tough surfaces.

- *Locking C-clamps and bar clamps* — These tools come in a variety of sizes and are designed to hold steel members together during fastening. C-clamps (*Figure 10*) prevent separation, also known as screw-jacking, when the first layer of steel climbs the threads of the screw. Use bar clamps to hold steel wall members together until permanent fastenings are applied or use them to hold a **header** in place until it is fitted into the top track. Steel framing should have regular end tips without pads to reach around the steel flanges.

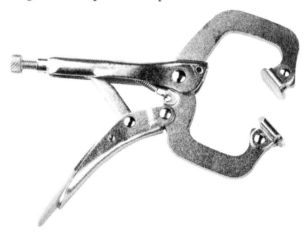

Figure 10 Locking C-clamps.
Source: Oleksandr Chub/Shutterstock

1.3.1 Cutting Tools

Prefabricated steel framing components often reduce the need for steel cutting in the field. However, if field cutting needs to be performed, the following are some common tools and methods:

- *Chop saw* — Chop saws use an abrasive blade for field cutting steel (*Figure 11*). While chop saws are very effective for square cuts and for cutting bundled studs,

Figure 11 Chop saw.
Source: Susan Law Cain/Shutterstock

they are very noisy and emit hot flying metal filings. Be aware that the edge produced by a chop saw is very rough, with sharp burrs left on the steel. There are several circular saws available with abrasive or carbide-tipped blades that are designed for cutting steel studs. Some of them have special guards to catch flying metal chips, which can be a hazard when using a high-speed saw to cut steel. Even with the appropriate guards in place, you should always wear task-specific PPE to protect your ears, eyes, face, and hands during cutting operations.

- *Swivel-head shears* — Shears are manufactured in both electric- and battery-operated models, although the battery-operated model shown in *Figure 12* is more common in the field. They are portable and can cut steel studs, track, sheet metal, and cold-formed steel up to 20-gauge thickness. This versatile tool also provides a 180-degree rotating cutting edge to accommodate hard-to-reach areas. Some drawbacks to shears are that they may be difficult to use for cutting a tight radius on CFS components, and blades are expensive to replace.

- *Circular saw/dry-cut metal-cutting saw* — Newer types of metal-cutting saws are available with blades made of aluminum oxide that produce a smooth edge.

- *Aviation snips* — Aviation snips (*Figure 13*) are hand tools that can cut cold-formed steel up to 18-gauge thick. They are useful when cutting and coping steel, for snipping flanges, and for making small cuts. Some brands of snips are color coded for left, right, and straight cutting angles. Always wear gloves and safety goggles when using snips.

Figure 12 Swivel-head shears.
Source: Courtesy of Stanley Black & Decker, Inc.

Figure 13 Aviation snips.
Source: Yanas/Shutterstock

- *Hand-held grinder* — Hand-held grinders (*Figure 14*) are commonly used for cutting and grinding metal pieces to size and shape. They can also be used to remove burrs, rust, and other imperfections from steel surfaces. For example, a craft professional might use a hand-held grinder to cut a piece of steel stud or track to fit between other framing members, or to smooth the edges of a metal bracket or angle. When using a hand-held grinder, it's important to take safety precautions, such as wearing eye and ear protection and using the correct type of abrasive disc for the task at hand. Always use both hands with one near switch and one on the side handle. Always use correct guard, closed

Figure 14 Hand-held grinder.
Source: Ridge Tool Company

Type 1 guard for cutting, Type 27 for grinding. Do not grind with the side of a cutting wheel. Be very careful not to twist or put a side load on a cutting wheel. Do not use a bonded abrasive wheel that is past the expiration (EXP) date marked near the center of the wheel.

1.3.2 Hole-Cutting Tools

Most openings in steel studs are prepunched at the factory. However, it may be necessary to punch additional openings for electrical or telecommunications cabling that may be installed later. Holes in webs of studs, joists, and tracks must conform to an approved engineered design or a recognized design standard.

A metal stud punch is available for punching additional openings in studs up to 20-gauge thick. Protect any electrical and telecommunication cabling that passes through stud openings. Grommets and other specialty devices are designed to protect cables from sharp edges and reduce conduit rattle (*Figure 15*).

Figure 15 Punch openings in steel studs.
Source: JPL Designs/Shutterstock

For small holes up to $1\frac{1}{2}$" in diameter, a hole punch or steel stud punch can be used for steel members up to 20-gauge thick. Insert grommets or rattle prevention to protect wiring from sharp edges or isolate copper or other dissimilar metals from the steel.

Hole saws are recommended for larger holes up to 6" in diameter and for material thicknesses greater than 20-gauge. Hole saws (*Figure 16*) are used on a drill motor to cut through the steel.

Figure 16 Hole saw.
Source: Courtesy of Stanley Black & Decker, Inc.

1.3.3 Screws

The right fastener will make a significant difference in the structural integrity and quality of your work. Consider the loads to be transmitted through the connection, as well as the thickness, strength, and configuration of the materials to be joined. Fasteners include screws, pins, clinches, and welds, as well as anchor bolts, rivets, powder-actuated fasteners, and expansion bolts. For applications requiring increased stability, or when working with steel thicker than 18 gauge, join the components by shielded metal arc welding to provide additional strength if this technique is accepted by the project Engineer of Record (EOR). Always follow manufacturer recommendations or engineered solutions for fastening systems during all phases of cold-formed steel construction projects.

Screws are the most common type of steel framing fasteners. Because pilot holes are not drilled in steel framing members, screws used to secure the members must be "self-tapping" and able to penetrate the target surface before they engage. Steel framing screws are installed with impact drills and screwguns and are available in a variety of head styles to fit a wide range of structural and cosmetic requirements.

Screws have three distinct thread thicknesses: coarse (threads spaced farthest apart), medium, and fine (threads spaced closest together). Steel framing screws usually have coarse threads for optimum penetration, but thicker steel, such as 12 gauge, requires the use of a fine-threaded screw. Screws are generally finished with zinc or cadmium plating to withstand environmental impacts and resist corrosion. The size and the type of screw needed will depend on the thickness of the sheathing material and steel.

The two main types of steel framing screws are self-drilling and self-piercing (*Figure 17*). Self-drilling screws are designed to drill through layers of steel before the screw threads engage. Self-piercing screws' sharp points typically penetrate 20- to 25-gauge material, which is why they are commonly used to attach plywood and gypsum wall board to thinner layers of steel.

Figure 17 Self-drilling and self-piercing screws.
Sources: Ruslan Maiborodin/Shutterstock (left); Nikolay132/Shutterstock (right)

Code and standard requirements state that screws must extend through the steel a minimum of three exposed threads, making $\frac{1}{2}$" or $\frac{3}{4}$" screws acceptable for most steel-to-steel connections. When applying plywood, gypsum board, or rigid foam insulation, proper screw length is determined by adding the measured thickness of all materials plus an extra $\frac{3}{8}$" allowance for the exposed threads.

Screw heads are available in many different forms. The head locks the screw into place and prevents it from sinking past the surface layer of the material. The head also contains the drive type (type of bit tip) to apply the screw. The head profile is selected to avoid interference with other building components.

Common self-drilling screws are shown in *Figure 18* and include the following:

- *Gypsum board (bugle-head) screw* — These screws are common fasteners for attaching gypsum wallboard to steel framing members. No. 6 sharp-point bugle-head drywall screws were designed for this application, but you must use a depth-setting nosepiece to avoid tearing the gypsum board or protective paper. For steel thicker than 20 gauge, use a self-drilling bugle-head drywall screw. Fine-threaded drywall screws may be acceptable fastener options for thinner-gauged studs.

- *Flat-head screw* — This screw is typical for wood flooring and facings because it is designed to countersink and seat flush with the target without causing splintering or splitting.

- *Wafer-head screw* — This screw is larger than the flat-head screw and is used to secure soft materials to steel framing members. The large head provides a greater bearing surface and seats flush to achieve a clean, finished appearance.

- *Hex washer-head screw* — The most popular head style for steel-to-steel connections with no sheathing or gypsum board is the hex washer. The hex washer face provides a surface for the driver socket, ensuring good stability in driving. If sheathing and gypsum board are applied over the fastener, the head style must have a very thin profile to prevent blowouts or bumps at the screw locations.

- *Trim-head screw* — Small trim-head screws are preferred when installing baseboard and other trim. The small head penetrates the trim, leaving a tiny hole that can be filled with putty afterward.

- *Oval-head screw* — This screw spans clearance holes and has a low-profile appearance for attaching accessory items to steel stud walls.

- *Pan-head screw* — This screw is used to fasten studs to runner tracks as well as connect steel bridging, strapping, or furring channels to studs or joists.

- *Winged screw* — Only use winged screws to attach plywood to metal framing members.

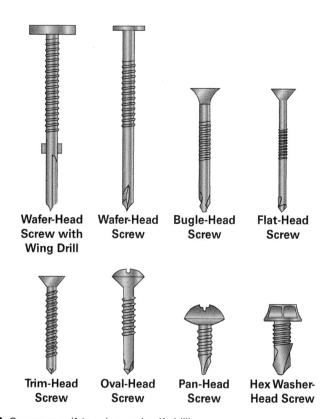

Wafer-Head Screw with Wing Drill   Wafer-Head Screw   Bugle-Head Screw   Flat-Head Screw

Trim-Head Screw   Oval-Head Screw   Pan-Head Screw   Hex Washer-Head Screw

Figure 18 Common self-tapping and self-drilling screws.

1.3.4 Pins

Pneumatic or cordless nailers are commonly used in steel framing to attach plywood, OSB, or siding to walls and roofs. Some contractors use only pins, while others use screws around the perimeter and pins in the field of the board.

Plywood or OSB sheathing material may be applied using pins (*Figure 19*). To be effective, the plywood must be held tightly against the steel member before the pin is driven because firing the pin does not tighten the plywood against the steel. Roof sheathing is more easily installed with pins because the carpenter usually stands on the plywood, keeping it tight to the framing member. An alternative method is to tack the plywood to the steel with screws along the perimeter first and then pin the field.

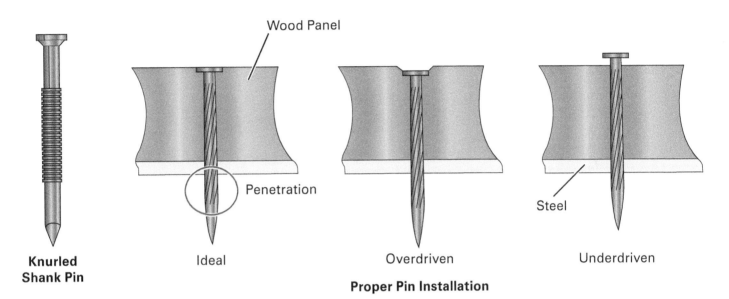

Knurled Shank Pin   Ideal   Overdriven   Underdriven

Wood Panel

Penetration

Steel

Proper Pin Installation

Figure 19 Proper pin installation.

Advancements in fastener technology include pins for steel-to-steel connections. These connections are accomplished by driving a pin with a knurled shank (*Figure 20*) into the layers of steel.

Figure 20 Knurled shank pins.
Source: Photo courtesy of Simpson Strong-Tie Company Inc.

1.3.5 Powder-Actuated Fasteners and Powder Loads

Powder-actuated fasteners are available in a variety of diameters and lengths and include drive pins and threaded studs (*Figure 21*). Drive pins are used for permanent installations. Electricians and HVAC professionals use threaded studs for applications where the material or equipment fastens to the concrete to allow for easier removal. Most drive pins and threaded studs have a plastic fluted washer that centers the pin or stud in the powder-actuated tool (PAT) during installation.

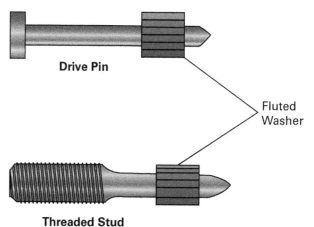

Figure 21 Drive pin and threaded stud.

WARNING!

Only use powder-actuated fasteners if you are trained and certified with a verified certification card on your person. Certified personnel must also select the proper powder load for the job in accordance with the manufacturer's instructions. If in doubt, check with your supervisor.

Powder load: A crimped and sealed metal casing that contains a powdered propellant.

A **powder load** provides the propellant for a powder-actuated tool. Powder loads resemble shell casings used in conventional firearms. PATs have a .22-, .25-, or .27-caliber bore diameter and require a load of the same caliber. These loads range in strength from #1 (lowest velocity) to #6 (high velocity), which are color-coded for easy identification (*Figure 22*).

Load Color	Powder Level	
Gray	1	Low Power
Brown	2	
Green	3	
Yellow	4	
Red	5	
Purple	6	High Power

Figure 22 Powder loads.

1.3.6 Bolts and Anchors

Bolts and anchors are also commonly used to fasten cold-formed steel framing to masonry, concrete, and other steel components. Except for some proprietary anchors, predrilling of holes is necessary. Bolts require the installation of a washer and must meet or exceed the requirements of *ASTM A307*. Expansion anchors are common for connections to concrete or masonry and require information from the manufacturer to determine the capacity and spacing requirements.

1.4.0 Material Takeoff

An estimator typically calculates material quantities for commercial construction projects and orders them to be delivered to the jobsite. For residential construction projects, a carpenter may take on the estimating role to order necessary building materials and supplies. Therefore, an understanding of basic material quantities and takeoffs is essential for craftworkers in any construction field.

Material takeoffs provide the planning framework to properly determine the quantities of materials required to complete a construction project. Special structural elements such as floor openings, cantilevers, and partition supports that affect material requirements should be considered during these early planning phases.

Begin the takeoff process by studying specifications to calculate different material types and their respective dimensions. Investigate project documents and blueprints and double-check the scales and dimensions of the various components needed. Once you organize a list of material types and dimensions, you can perform the material takeoff.

1.4.1 Bottom and Top Tracks

To determine the amount of material needed for the top and bottom tracks, use the following procedure:

Step 1 Determine the length of the walls in feet. Multiply the length by 2 to account for the top and bottom tracks. Skip this step when using a deep-leg track or slip track, as these materials are installed at the top, while bottom members will be standard tracks.

Step 2 Divide the result by 10 to determine the number of 10' pieces of stock material, then round up to the next full number to account for waste.

1.4.2 Studs

To determine the number of studs needed, use the following procedure:

Step 1 Determine the linear feet of all walls.

Step 2 Divide the linear feet total by the stud spacing (typically 16" or 24" OC), then add two studs for each wall end to account for uneven spacing.

For example, a 32' × 64' building has 192' or 2304" of exterior walls. If the stud spacing is 16" OC, your equation will be (2304 ÷ 16) + 8 (for 4 wall ends). Therefore, you will need 152 studs to complete this project. Round up all results to the next full number.

1.0.0 Section Review

1. What is the best form of PPE to protect against zinc chills?
   a. Respirator
   b. Cut-resistant gloves
   c. Waterproof jacket
   d. Safety glasses

2. The universal designator system for steel framing materials is called the _____.
   a. FMIS system
   b. CFS system
   c. SSID system
   d. STMF system

3. To attach OSB sheathing to steel framing, you would use a(n) _____.
   a. impact driver
   b. hammer drill
   c. screwdriver
   d. hammer and nails

4. The number of cold-formed steel studs, spaced 16" OC, required for a building measuring 50' × 75' is _____.
   a. 50
   b. 75
   c. 125
   d. 196

2.0.0 Layout and Installation of Steel-Framed Walls

Performance Tasks	Objective
2. Lay out a steel stud wall with openings to include bracing and blocking. 3. Demonstrate the ability to build headers (back-to-back, box, and L-header).	Identify the steps to lay out and install a steel stud wall. a. Describe basic steel construction methods. b. Explain how to frame nonstructural steel walls. c. Describe how to frame structural steel walls.

Historically, cold-formed steel framing has been widely accepted for use in nonstructural applications or partition walls. Today, it is commonly used for floor, structural wall, and roof assemblies as well. This section covers steel framing as it applies to nonstructural and structural walls in residential and commercial structures (*Figure 23*).

Figure 23 Example of a structural wall system.
Source: northlight/Shutterstock

2.1.0 | Basic Steel Construction Methods

Basic steel construction methods include in-line framing, web stiffeners, standard web holes and patches, and proper stud seating. Steel frame assemblies also require certain types of bracing for stability.

2.1.1 In-Line Framing and Web Stiffeners, Holes, and Patches

In-line framing, or direct alignment, is the most common framing method for providing a direct path to transfer a load through the framing system, from studs to joists to the ground. In this method, cold-formed steel framing members are aligned vertically so that the centerline of the joist web is within $3/4$" of the centerline of the structural stud member or web joist below.

Web stiffeners, or bearing stiffeners (*Figure 24*), are studs or tracks that prevent joists from crippling at the point where the load transfers from a stud into the floor joist under structural walls. Studs provide additional rigidity because they are formed with a web, two flanges, and two **returns**. Stiffeners are installed across the joist depth of the web and on either side of the web. Minimum stiffener thickness matches the width of the floor joist, and the length of the stiffener must equal joist depth minus $3/8$". Fasten the stiffener to the web with No. 8 screws, with at least three in a single row, or four in each corner. Refer to shop drawings and engineering specifications to ensure you are using the correct fasteners.

Returns: A pair of components of a C-shaped stud that extend perpendicularly from the flanges as stiffening elements. These may also be called *stiffening lips*.

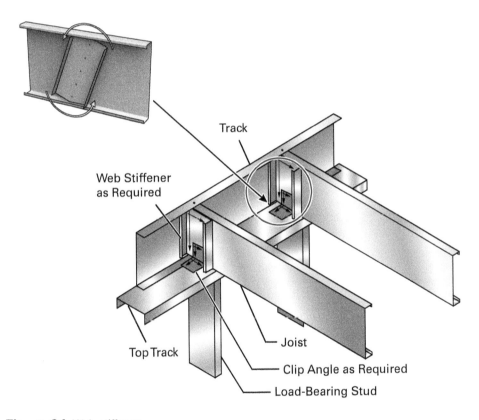

Track

Web Stiffener as Required

Top Track

Joist

Clip Angle as Required

Load-Bearing Stud

Figure 24 Web stiffener.
Source: Image provided by ClarkDietrich (top)

Framing members may be designed and manufactured with standard punchouts, or web holes (*Figure 25*). Holes in webs of studs, joists, and tracks must conform to an approved design unless you are using a pre-approved, engineered proprietary product. Install prepunched materials to align to allow conduit and piping to run through

Figure 25 Standard stud punchouts.
Source: Jit Lim/Alamy

in a straight line. Always refer to the manufacturer's recommendations regarding web hole and patch size to guarantee the structural integrity of the framing member.

2.1.2 Bridging, Bracing, and Blocking

Bracing improves the strength and function of individual framing members by restraining them from moving laterally or twisting. Bracing uses a combination of the following three common methods:

* Cold-rolled channels placed through the punchouts
* Steel strapping attached to the flanges with periodic solid **blocking** and/or X-bracing (*Figure 26*)
* Sheathing attached to the flanges

Blocking: A standard track, brake shape, or flat strap attached to structural members or sheathing panels to transfer shear forces.

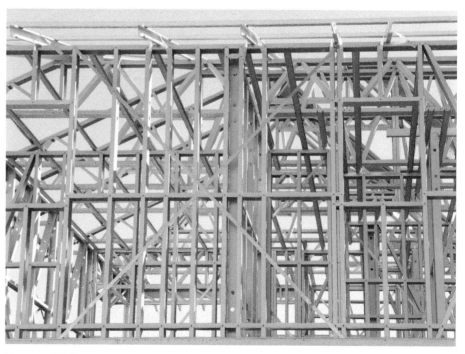

Figure 26 Example of X-bracing.
Source: Lev Kropotov/Alamy

2.1.3 Thermal Considerations

Some construction products may require additional insulation to meet energy codes. The *Thermal Design and Code Compliance for Cold-Formed Steel Walls,* published by the Steel Framing Alliance, provides designers and contractors with guidance on the thermal design of buildings using cold-formed steel framing members. The thermal performance of a steel-framed structure may also be improved with batt and other insulating materials within the wall cavity, while some thermal regions will require insulation foam board on the exterior of the frame. Designs should consider the effects of moisture when assessing the application of cavity and continuous insulation.

2.1.4 Protecting Piping and Wiring

Copper piping is commonly run through the punchouts of wall studs, creating the potential for direct contact with the galvanized steel, resulting in a galvanic reaction that will compromise the strength and performance of both the steel and the copper. Reduce contact between dissimilar metals by installing nonconductive grommets, plastic bushings, or other materials.

Plastic piping, on the other hand, does not require protection from contact with steel framing members, but consider installing nonmetallic brackets to isolate the pipe from the hole. This isolation will prevent noise and the potential for damaging the pipe through long-term movement.

Separate wiring sheathed with a nonmetallic coating from sharp edges typically found in punchouts of wall studs and joists (*Figure 27*). *The National Electrical Code® (NEC®)* states that nonmetallic sheathed cable must be protected by bushings or grommets securely fastened in the opening prior to the installation of the cable. Secure any cables that follow the length of framing members with tie-downs, such as nylon cable ties. Attach them to the studs at intervals conforming to local building codes.

> **NOTE**
>
> Use care when using pressure-preservative-treated or fire-retardant-treated wood products with cold-formed steel framing, as accelerated corrosion can occur. It is preferable not to use pressure-treated wood with steel framing, but if you do, specify a less corrosive pressure treatment, such as sodium borate. Always separate the steel framing material from the pressure-treated wood with a nonabsorbent closed-cell sill seal.

Figure 27 Proper protection of wiring.
Source: Image provided by ClarkDietrich

2.2.0 Framing Nonstructural (Nonbearing) Steel Walls

Framing members in interior systems may be nonbearing or designed as part of the structural system. This section primarily discusses nonstructural walls that function as space partitions within the exterior walls of a building.

Nonstructural walls are made up of studs, tracks, and accessories. The primary differences are the characteristics of the materials and the application of connectors and accessories.

Nonstructural framing members typically have a base steel thickness of 25, 22, or 20 gauge, compared with a minimum thickness of 20 gauge for structural studs. In addition, the minimum stud flange dimension for nonstructural framing members is $1\frac{1}{4}$", and the minimum return dimension is $\frac{1}{8}$", compared with $1\frac{5}{8}$" flange and $\frac{1}{2}$" lip for structural studs. Nonstructural members typically have a G40 galvanized coating weight, compared with G60 or higher for structural studs. The rules for nonstructural stud spacing are different from structural wall blueprints and may be found in the gypsum specifications, rather than the building codes.

2.2.1 Steel Curtain Walls

Since their introduction in the early 1900s, metal and glass **curtain wall** systems have become very popular in the architectural design of modern structures. Unlike interior partitions, wind-bearing curtain walls resist loads from exterior wind pressures which may exceed 60 pounds per square foot in some cases. Cold-formed steel curtain walls are made up of the following components:

- *Angle* — A **clip angle** and continuous angle are designed to connect framing members within the curtain wall system.
- *Clip angle* — A steel angle, generally 3" to 12" long, which connects a framing member to its supporting component.
- *Continuous angle* — A steel angle that connects a stud curtain wall to the primary frame. The angle is hot-rolled to an engineer-specified thickness between $\frac{3}{16}$" and $\frac{3}{8}$"; however, thinner-gauge materials can be adequate if the span and load requirements are relatively small.
- *Diagonal brace (or kicker)* — A sloping brace used to provide **lateral** support to a curtain wall assembly. When installed horizontally, this brace is known as a *strut*.
- *Embed* — A hot-rolled steel plate or angle, reinforced with shear studs or steel rebar, which is cast into a concrete floor or beam. Embeds allow for the welded attachment of steel supports.
- *Girt* — Horizontal structural member that supports wall panels and is primarily subject to bending under horizontal loads from wind and other forces.
- *Slide clip* — A connection device that permits deflection of the primary frame to which a stud attaches while it braces the stud against lateral forces.
- *Slip track* — A track section used for infill curtain wall applications (*Figure 28*). Slip tracks accommodate vertical movements of a primary frame (normally $\frac{1}{4}$" to $\frac{3}{4}$") while bracing the wall against lateral forces. Slip tracks may also be specified at the top of interior gypsum board partitions.

Curtain wall: A light, nonbearing exterior wall attached to the concrete or steel structure of the building.

Clip angle: An L-shaped piece of steel (normally with a 90-degree bend), typically used for connections.

Lateral: Running side to side; horizontal.

Figure 28 Slip track.
Source: Image provided by ClarkDietrich

The term *curtain wall* distinguishes this system from load-bearing framing. Load-bearing framing requires that the wall members carry the weight of the structure above. With curtain wall framing, the structure is usually already in place, and the wall framing is filled in between the floor slabs. The stud-to-track gap distance for a curtain wall is no more than $\frac{1}{4}$" unless it is otherwise specified in an approved design. This is different from load-bearing construction, which permits only a $\frac{1}{8}$" gap. The curtain wall framing system typically carries a gravity load from attached cladding or finish materials.

2.2.2 Construction Methods for Steel-Framed Curtain Walls

The four main methods for building assemblies in curtain wall construction are as follows:

- *Infill* — This method describes applications where studs are only one story tall, spanning from floor to floor. Infill framing requires less stud material and frame connections are often easier with the use of powder-actuated fasteners for various applications. In addition, the spans are often shorter than bypass conditions, so thinner steel or wider spacing could be acceptable. However, a common disadvantage of infill framing is that more track material is needed, and wall sections can be difficult to panelize.

- *Panelization* — You can use this building method if field measurements are made after the floor systems are in place, or if a slip connector or telescoping stud system is used. When using structural steel as the primary framing component, the spandrel beam can get in the way, forcing builders to make bottom-beam attachments. If framing takes place outside the spandrel beam, it can cause difficulty in supporting insulation.

- *Bypass framing (balloon framing)* — This common building method allows for a single-stud framing of two or more floors. Multiple spans can reduce moment stresses in members, allowing for wider stud bay spacing and thinner framing members, resulting in reduced material usage and lower project costs. In addition, the balloon framing method (*Figure 29*) makes it easier to prepanelize large sections, including multistory panels. However, some connections can be more difficult to make, such as bracing and support connections behind columns and spandrel beams.

Figure 29 Balloon framing.
Source: Jeremiah Zamora/Alamy

Depending on the condition, double connections may need to be made. This method requires clips or slip connectors between the structure and stud. In this case, use connectors that friction-fit inside the stud to reduce connection time. Finally, if slabs or other structural elements extend into the stud cavity, they may need to be chipped away, or stud framing may need to be altered to correctly install the bypass framing.

- *Stacked wall framing* — This method permits bypass framing while isolating slip connections to one- or two-story segments. As a result, multistory panels and prefinished panels can be fabricated and installed, including insulation. However, to prevent water infiltration, this approach requires special detailing at slip connections and exterior finishes. Because the entire panel weight goes to fixed connections, usually at every other floor, these fixed connections need to be strengthened to carry the added dead load of the taller panels. Also, some connections are more difficult to reach, such as bracing and support connections behind columns and spandrel beams.

In multistory construction, movement may occur at the floor below a curtain wall system, causing the entire wall to move down. In this case, the connector at the top of the wall must have sufficient capacity to allow the wall system to move down without creating tension on wall components.

Other than slip connections, the most typical connections used in metal-stud curtain wall design include the following:

- *Base connections* — Stud to track, and track to concrete or steel deck.
- *Head or sill to jamb connections* — Connections at the top and bottom of window or door openings.
- *Continuous angle connections* — Field-welded connections typically found at spandrel framing.
- *Clip angle connections* — Connections typically found at non-slipped bypass or spandrel framing connections for the intent of carrying lateral and self-weight forces.
- *Outrigger clips* — Short lengths of angles designed as an axially loaded strut to transfer lateral reactions back to the structure.
- *Wind girt connections* — These connections are not typically designed to carry gravity loads, but are practical alternatives for tall spandrel framing where diagonal bracing would be impractical.
- *Diagonal braces or kickers* (as opposed to X-bracing for a **shear wall**) — Stud diagonals that transfer bottom spandrel stud reactions back to the structure.
- *Stud-to-stud connections* — Either lapped or track-to-track. On occasion, these connections require a slip track or slip-pin detail to allow for movement.
- *Knee-wall base* — A moment connection at the bottom of a knee or stub wall, usually either a freestanding parapet or a long segment of wall under a continuous or ribbon window condition.

Shear wall: A wall designed to resist lateral forces such as those caused by earthquakes or wind.

2.2.3 Bracing for Curtain Walls

C-shaped CFS studs are the most typical cold-formed steel curtain wall member. Due to its shape and geometric properties, a stud tends to rotate under lateral load. This rotation causes unbraced studs to move out of plane in a condition known as *torsional-flexural buckling*. Mechanical bridging and/or sheathing materials can restrain the flanges and prevent this buckling. When using discreet bracing rather than sheathing, shorter bridging spaces will increase structural member rigidity and reduce buckling.

Member deflections are the primary serviceable issue with a typical steel-framed exterior curtain wall. You can find these deflection limits in the building code and engineering specifications.

2.2.4 Finishing for Curtain Walls

A wide range of finish systems may be applied to curtain wall frames, including the following:

- Brick veneer
- Split-faced block veneer
- Tile or thin-cast brick
- Exterior insulation finish systems (EIFS)
- Glass fiber-reinforced concrete (GFRC)
- Metal panel
- Modified Portland cement (stucco)
- Fiber-cement board or siding
- Dimensional stone, such as granite or limestone
- Wood siding
- Many other finish systems

2.2.5 Radius (Curved) Walls

A radius wall is an excellent example of a partition that might be easier and faster to build with steel framing members than wooden ones due to its curved, circular shape. Complex radius walls may require the use of plywood templates to complete the assembly. Curved walls can be framed using curved track for partitions or exterior load-bearing walls. The track can be bent at the jobsite by slitting the flanges. But ordering curved track from specialty companies can reduce the time needed to construct a radius wall in the field.

Some specialty companies use power equipment to bend the track without slitting the flanges, while other manufacturers produce a flexible track that can be ordered and formed on the jobsite based on the desired effect. This provides a clean, neatly bent track to an exact radius. Wall track is bent around the flanges (*Figure 30*).

Figure 30 Example of a radius track.
Source: Dennis Axer/Alamy

2.2.6 Other Nonstructural Wall Assemblies

Nonstructural steel walls are constructed to serve the following additional functions:

- *Fire-resistance-rated assemblies* — Building codes require that certain partitions provide fire-resistance-rated separation of interior spaces, with a specific fire-resistance rating that measures the amount of time an assembly maintains the ability to confine a fire and perform its structural function. Approved assemblies must be constructed precisely as described in the directories published by the rating agencies.

- *Area separation walls* — These nonloadbearing, parapeted wall types are also known as *maximum foreseeable loss walls* (MFLs). They separate the building into distinct areas, but they also act as a protective measure to reduce property loss in the case of a catastrophic fire event. Many fire-resistant MFLs are designed to remain standing after one of its sides has collapsed to keep the fire from spreading to other areas of the building.

- *Continuity head-of-wall conditions* — An assemblage of specific materials or products that are designed to resist the passage of fire through voids created at the intersection of fire barriers and the underside of nonfire-resistance-rated roof assemblies for a prescribed time.

- *Shafts* — An enclosed space extending through one or more stories of a building, a shaft connects vertical openings in successive floors and roofs.

- *Shaft wall systems* — Shaft walls, commonly known as *chase walls*, are special interior firewall systems constructed from one side only to enclose shafts for stairs; mechanical, electrical, and plumbing runs; and elevator hoistways. This building method allows either side of the structure to collapse to protect the shaft during extreme fire conditions. This characteristic is vital for protecting evacuation routes for occupants during an emergency. Higher sound and fire ratings may also be achieved by attaching additional layers of gypsum board to the outer or inner face. However, typical shaft wall construction requires the use of 1" shaftliner panels or coreboard, supported by C-T or C-H studs with one or two layers of $1/2$" fire-resistance-rated gypsum board. Horizontal shaft walls may also be used for a soffit or **plenum** where ratings are required.

Plenum: An enclosed space, such as the space between a suspended ceiling and an overhead deck, which is used as a return for heating, ventilating, and air conditioning (HVAC) systems.

2.3.0 Framing Structural Steel Walls

Structural walls support the weight of the building and protect occupants from wind loads and other forces of nature. A basic, cold-formed structural steel wall includes structural studs, tracks, fasteners, bracing, and bridging. Headers for window and door openings may be constructed using structural studs, joists, or proprietary cold-formed shapes. In addition, shear walls are typically framed within the stud wall assembly.

Structural steel wall framing members have a material base steel thickness of 20 to 10 gauge, with a minimum metallic zinc coating of G60. Structural steel studs are typically produced in sizes from $2^{1}/_{2}$" to 8", with flanges ranging from $1^{3}/_{8}$" to $2^{1}/_{2}$" or more. The size of the returns depends on the flange size. Structural track is sized to accommodate the web depth of wall studs, and the flanges are typically sized $1^{1}/_{4}$" to 3". Several proprietary and nonproprietary accessories are also available for structural wall assembly, including cold-formed channels for bridging, as well as clips, angles, and straps.

2.3.1 Layout

Lay out the wall studs accurately to align with other structural assemblies such as roof and joist framing. Steel framing is typically spaced at 12", 16", or 24" on center. Perform the following steps to complete panelized wall layout:

Step 1   Place the top and bottom track members on the straight edge of the panel table. Arrange them with the webs next to each other with temporary clamps and fit them tightly against the edge of the end stop (the straight edge at the end of the wall).

Step 2 Mark the layout of the wall studs on the flanges of the top and bottom tracks, starting with a wall stud at the end of the wall. Use highly visible ink, such as a black felt-tip marker. Place a line at the web location and an X on the side of the line to indicate the stud flanges.

Step 3 Mark the next stud location to match the first truss or **roof rafter** location from the end wall. Continue marking every 24" (or 16", depending on the specified layout) for the full length of the wall. Where the exterior corner walls intersect, the wall that runs to the edge of the foundation has an extra stud. This stud is 3" from the end and acts as a backer to screw to the shorter intersecting wall.

Step 4 Next, look through the architectural drawings and approved construction documents to find door and window locations that need rough openings. If the dimensions are not provided, initiate an RFI (request for information) for the engineer or architect to clarify. Mark the location for the center of the openings on the top and bottom tracks. Use a red felt-tip marker to distinguish these marks from the layout marks. Check the door and window sizes on the drawings and verify rough openings with actual window sizes.

Step 5 In structural framing layout, use a tape measure and center the dimensions over the red marks on the track and mark each end of the tape measure. This shows the location of the webs of the king studs. Draw an X on the side of the mark away from the window because the webs of the king studs will be on the rough opening side.

Roof rafter: A horizontal or sloped structural framing member that supports roof loads.

All load-bearing studs must be aligned with the trusses, joists, or rafters above or below the wall. Because these walls carry loads, it is important to fit each stud tightly in the track member to allow the stud to properly carry the axial (downward) load from above. The top and bottom wall tracks must be of equal or greater thickness than the studs. Panelized walls may be constructed on a concrete slab, floor deck, or panel table. For all systems of wall construction, make sure that the surface the wall is built on is level. Walls must not be out of plumb more than $\frac{1}{8}$" for every 10'.

Full-length walls may be framed up to 40' depending on available assistance at the jobsite. A large crew can spread out and help support the wall as the team raises it, whereas a smaller crew may struggle to keep the wall from twisting. Shorter sections need more plumbing and alignment, but they are easier to assemble and may be built and spliced together (*Figure 31*).

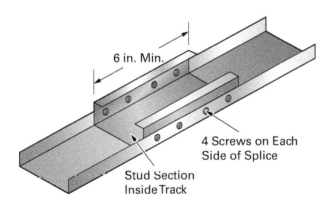

Figure 31 Track splicing.

2.3.2 Wall Assembly

The first step of structural wall assembly is ensuring that the foundation or bearing surface is free of all defects. The bearing surface should be uniform, with a maximum $\frac{1}{8}$" gap between the surface and track. After layout is complete, perform the following steps:

Step 1 Separate the top and bottom tracks and install a wall stud at each end of the wall between the top and bottom tracks.

Step 2 Clamp the stud flanges to the track flanges with locking C-clamps at each end. Tap the track on one end with a hammer to seat the studs as tightly as possible in the track. Fitting each stud tightly and perpendicular to the track keeps the wall straight.

Step 3   Screw one low-profile No. 8 screw through the flange of the track into the flange of the stud on either side of the stud. If an elevated panel table is used, the framer may be able to install the screw (from underneath) on the other flange as well. If not, once all the studs and headers are in place, and all screws are installed on one side, flip the wall over to install the screws on the other side.

When framing for a gypsum board covering, the framer often installs screws on one side of the wall. However, some local codes require screws in each flange, on both sides of a load-bearing wall to keep the studs from twisting, while also providing proper connections for in-line framing. Continue twisting the studs into the track. Install the studs with the open side of the stud facing toward the start of the layout. Align the punchouts in the studs to provide straight runs for bracing if needed. The studs should all align and face the same direction on parallel load-bearing walls.

WARNING!

Do not attempt field modifications to framing members without approval from a design professional and never install damaged materials that will compromise structural integrity.

Install king studs at the rough openings with the hard side of the stud facing the rough opening and punchouts aligned. Do not install studs at the markings between the king studs as these are reserved for cripple studs. Continue down the length of the wall until all the studs are twisted and screwed in place and refrain from removing the wall panel from the table until the headers and rough openings are completed.

During wall assembly, it is important to seat the studs into the track as tightly as possible. Studs must be seated with a gap of no more than $1/8"$ to ensure that building loads are transferred through the studs instead of the fastener (*Figure 32*).

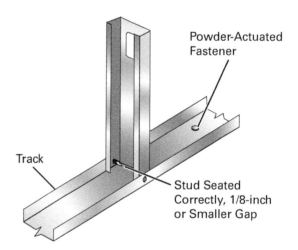

Figure 32 Stud seating.

Structural walls may be constructed using any of the following methods:

- *Stick building* — Walls are framed one stud at a time on a flat surface of the site, such as a concrete slab.
- *Panelization* — This assembly process reduces construction time and improves the overall performance of the steel framing assembly.
- *Pre-engineered method* — This assembly typically increases the size and spacing of structural steel members.

Although most cold-formed steel floor or roof assemblies call for screw attachments, welding may be used when specified. Other tasks may require powder-actuated fasteners to connect track to structural steel framing members or concrete. Expansion bolts are used at jamb locations and corners, while expansion anchors are typically used at shear wall locations.

Advantages of Steel Framing

Metal framing offers a great deal of flexibility in building design, especially for satisfying fire-resistance rating requirements for tall structures because steel is noncombustible and can be manufactured to long lengths.

Source: welcomia/Shutterstock

2.3.3 Wall Installation

If the foundation has anchor bolts, measure their locations and place holes in the bottom track of the wall panels to allow the walls to fit over the bolts. This step is not necessary if you use strap anchors. Place temporary bracing material near the foundation to prepare to raise the wall. You can use spare stud material at the jobsite, preferably 12' long, for temporary bracing. Perform the following steps to complete the task.

Step 1 Seal the concrete foundation with weatherproof caulking material and use foam closed-cell sill sealer beneath the track.

Step 2 Move the wall panel and set the bottom track on the foundation. Position it over the anchor bolt locations and tilt the wall up. Leaving the wall tilting slightly outward, clamp the temporary brace material to the wall studs in two or three locations (every 8' to 12' along the wall), depending on the length of the wall.

Step 3 Make sure the braces do not lap past the inside face of the wall. Before removing the clamps, secure a No. 10 hex-head screw through the brace into the stud.

Step 4 Secure the bottom of the brace with a stake driven into the ground or other solid surface, and then screw the stud to the stake to hold it in place. Repeat this process with all wall panels until all load-bearing walls are standing.

Some builders choose to frame walls in place, especially when working with smaller teams. In this case, the track should be cut for the full length of the wall. Mark the top and bottom track for layout, and then anchor the bottom track in place, securing the studs at each end of the track. Position the top track at the ends with intermediate studs and use a string line and level to position the remaining studs for the wall. Install headers, X-bracing, or wall sheathing with the wall standing.

Balloon framing is another acceptable framing method for some structures. This is categorized as a pre-engineered structure, where the carpenter and the engineer must work closely together to size the members and develop details for balloon framing.

WARNING!

Ensure that the braces do not hang past wall edges to reduce the risk of the wall assembly snagging on them during tilting operations. This could result in accidents and potential injuries.

Proper bracing is required to prevent a cold-formed steel framing system from twisting and buckling. In wall construction, there are several common methods for bracing, including the X-bracing already discussed, as well as lateral bridging (*Figure 33*) through punchouts and temporary bracing once the walls are in place.

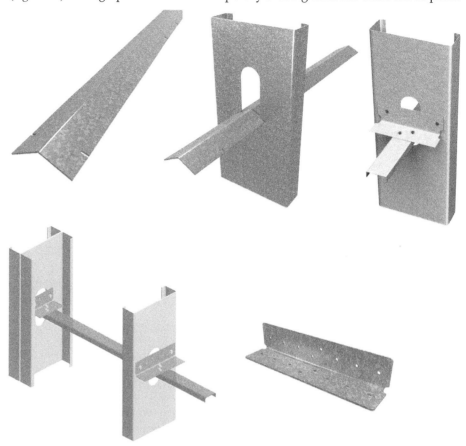

Figure 33 Lateral bridging with ClarkDietrich's Spazzer® bar, SwiftClip™, and EasyClip™.
Source: Image provided by ClarkDietrich

Racking: Being forced out of plumb by wind or seismic forces.

If a straight wall is constructed on a panel table, installing plywood or temporary bracing (*Figure 34*) prevents **racking** when the wall is removed from the table. Before taking the wall off the table, check for squareness by diagonally

Figure 34 Temporary bracing.
Source: eyemark/123RF

measuring the panel and make any necessary adjustments. Lay extra studs or truss material across the wall diagonally, and screw the bracing to the wall studs, especially at door openings where the bottom track is weak.

Panelized wall installation also requires diagonal temporary bracing to keep walls plumb and reduce the risk of the wall falling out of position until permanent bracing can be installed. This step improves structural function, while also protecting workers and equipment from walls falling over.

2.3.4 Backing

Wall-mounted fixtures, appliances, and chair rails require **backing** for proper support. Install solid wood backing or fasten light-gauge sheet metal between studs before the wall finish material is applied. Interior elevation drawings or specific backing drawings should contain information on proper backing placement. Some proprietary backing products such as ClarkDietrich's Danback™ flexible wood backing and backer bar have been shown to reduce time and labor required to accomplish this task (*Figure 35*).

Backing: A flat, horizontal wood or metal member that provides a supportive attachment surface between stud bays.

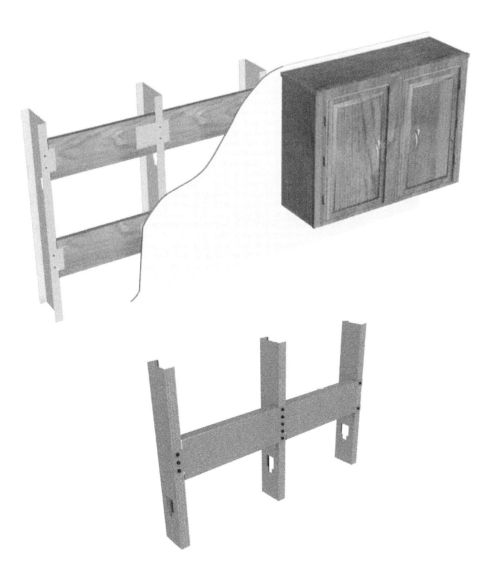

Figure 35 Wood and metal backing.
Source: Image provided by ClarkDietrich

2.3.5 Shear Walls

Systems of interconnected shear walls and floor and roof/ceiling diaphragms work together to transfer lateral (shear) loads from wind and seismic activity evenly through the structure. Create a **diaphragm** or shear wall by attaching wood structural sheathing, steel decking, X-bracing (*Figure 36*), or other materials to the floor, wall, ceiling, or roof framing. In addition to field-fabricated shear wall systems, there are now several proprietary, high-strength pre-engineered systems that may help shorten construction time (*Figure 37*).

Figure 36 X-bracing (flat strap).
Source: Image provided by ClarkDietrich

Figure 37 Pre-engineered shear wall system.
Source: Don Wheeler

Each multistory steel-framed structure should account for stacking and load path to ensure adequate shear transfer between roof or floor diaphragms and shear walls by using assemblies such as bar joists, long-span decks, and wood framing. Shear walls connect to the foundation with proprietary or engineered connectors and hold-downs.

The most common ways of applying shear-wall bracing are structural sheathing and X-bracing. Structural sheathing, such as wood structural panels (WSP) or oriented strand board (OSB), may be adequate to keep the wall from racking if the design does not call for excessive openings in the wall (*Figure 38*). Structural sheathing is most effective when installed in a vertical orientation with the long dimension perpendicular to stud framing. Secure WSP or OSB to the wall while panelizing, or after the wall is plumb and level.

X-bracing is an alternative way to obtain shear strength when not using structural sheathing. X-braces are diagonal steel straps attached to the walls with screws or welded connections (*Figure 39*). This bracing must be designed by an engineer, and the straps must be inspected for the correct number of fasteners. Do not tighten the straps until the walls are plumbed and aligned.

NOTE

All concealed cavities, such as headers in exterior walls, must be pre-insulated before they are installed.

Figure 38 Wood structural sheathing over steel framing.
Source: welcomia/Shutterstock

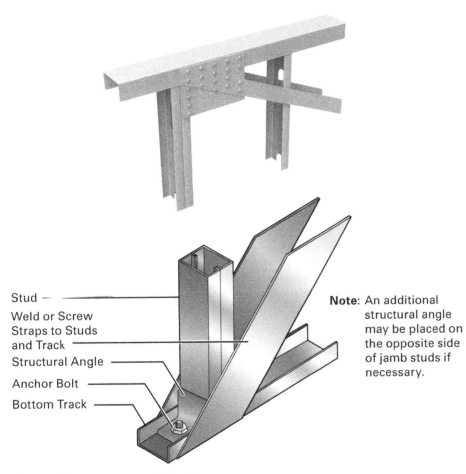

Stud

Weld or Screw
Straps to Studs
and Track

Structural Angle

Anchor Bolt

Bottom Track

Note: An additional
structural angle
may be placed on
the opposite side
of jamb studs if
necessary.

Figure 39 Attachment of X-bracing.
Source: Image provided by ClarkDietrich (top)

2.3.6 Header Assembly and Installation

Load-bearing headers are typically boxed, unpunched studs that are capped on the top and bottom with track sections (*Figure 40*). They must be engineered for bending and shear strength, along with web crippling (crushing) at the locations of the loads from above. The types of headers commonly built from standard studs include box, webbed, and back-to-back headers. Manufacturers also produce prefabricated headers with a wide selection of products for various loads and applications.

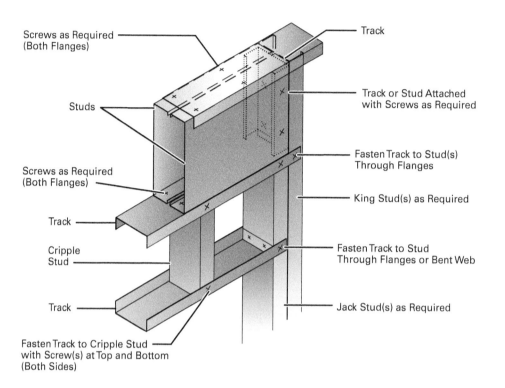

Figure 40 Job-built header.

Back-to-back headers are formed by placing two studs with the webs of the members touching each other. They are positioned in the top track of the wall and finished like a box header. The L-header consists of one or two large angle pieces that fit over the top track to span the top opening of windows and doors. Some craft professionals have moved away from job-built headers and use prefabricated systems (*Figure 41*), such as the ClarkDietrich REDHEADER PRO™ and HDS® headers. These manufactured options are cut to length and require fewer individual components, so they significantly reduce labor and material costs. Furthermore, many of these proprietary headers are hollow, allowing room for added insulation and improved energy performance.

If the rough opening (RO) requires a traditional header, you can accomplish this task by performing the following steps:

Step 1   Measure the RO width with a tape measure, adding 3" on each side for flanges. Make a mark on your selected track with a felt-tipped marker and use a speed square to extend straight lines around the track at these locations.

Step 2   Use a pair of tin snips to cut along your primary cutoff line and then snip the flange lines on each return to the right angle where the returns and web intersect. Then bend these flange sections toward the web to create an elongated U-shape.

Figure 41 ClarkDietrich REDHEADER PRO™ and HDS® headers.
Source: Image provided by ClarkDietrich

Step 3 Next you need to find the high spot in the doorway floor. Do this by setting up a laser level at an arbitrary elevation near your eyeline on a nearby post or stud that allows the laser line to fall on each side of the door jamb.

Step 4 Use your tape measure to find the distance between the floor near the jamb and the laser on each side of the RO. The smaller of the two measurements will be your high point and the location from which you will base all subsequent elevation measurements. Ensure that you keep your tape measure straight and plumb to guarantee accurate measurements.

Step 5 Measure up from the floor to the RO height on the high-point jamb and mark the bottom-of-header elevation. Double-check the approved architectural drawings to ensure that this elevation also accounts for the finish floor. Use a speed square to carry these lines around the jamb to guarantee a straight header installation.

Step 6 Reset your laser level to this final bottom-of-header elevation mark and mark the laser location to match the other jamb. You now have reference lines to set your traditional header with the flat web side facing down toward the RO.

Step 7 Once in place, you can use clamps to hold the header level and in place. Then, fasten the header flanges to both inside jambs with three screws, two on top and one on the bottom for each flange. The finished product should present a square and plumb opening to fit doors and windows without extensive troubleshooting (*Figure 42*).

Door openings do not require a bottom sill. However, the bottom track should run continuously at the bottom of doors to hold the wall together temporarily. The track can be cut out after the wall is plumb, level, and permanently braced.

NOTE

After header assembly, verify the position and height of the window in the wall to guarantee the window opening matches the drawings.

Figure 42 Completed exterior rough openings.
Source: B Brown/Shutterstock

2.3.7 Built-Up Shapes

C-shaped studs that are commonly used in cold-formed steel framing provide minimal resistance to twisting. Built-up shapes, such as a nested stud and track assembly, provide the increased stiffness of a closed section, as well as flat surfaces for attaching finish materials. Built-up shapes are commonly used for door and window jambs, headers (*Figure 43*), beams, and posts. The strength of built-up shapes is determined by the properties of the individual members and the chosen fastening method.

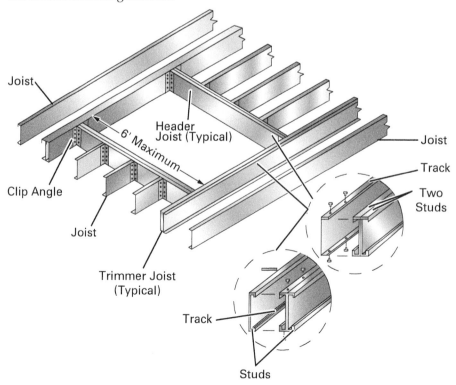

Figure 43 Floor opening constructed with built-up headers.

2.3.8 Jambs

A group of king and jack studs is known as a jamb. Load-bearing jambs require a minimum of two studs on each side of the framed opening (one trimmer plus one king), with more if required by the building code or by engineering analysis. These structural members must match the dimension and thickness of adjacent wall studs. The studs must be fastened together to act as one member, either by capping with track and screw fastenings or by stitch welding. Floor joists supporting jamb studs with multiple members require bearing stiffeners.

2.3.9 Window Sills and Head Tracks

Sills are horizontal track sections used to frame the bottom of window openings. They are clipped and screwed to the jamb studs. Sills are not to be confused with head tracks that span above the door and window openings. Construct sills with multiple track sections and cripple studs to support and evenly distribute excessive lateral load. Maximum spans of either structural member depend on wind loads and other assembly requirements.

2.3.10 Fire-Resistance-Rated Construction

A fire-resistance rating (FRR) represents the ability of a wall, floor, roof, or structural member to withstand fire for a specified period ranging from one hour to four hours. A rating of two hours, for example, means that a structural assembly such as a wall, floor, or roof would not allow flame or hot gases to pass through it for two hours. The fire rating also refers to the ability of a structure or material to withstand the force of water sprayed from a hose. Builders and architects select different fire-resistance-rated materials and systems to meet codes and protect the structural integrity of a building against fire hazards.

2.0.0 Section Review

1. In in-line framing, framing members are aligned vertically so the centerline of the joist web and centerline of the structural stud are within _____.
   a. $3/4$"
   b. 4"
   c. 8"
   d. 36"

2. When assembling a structural steel-framed wall, the maximum allowable gap between stud and track is _____.
   a. $1/8$"
   b. $1/4$"
   c. $1/2$"
   d. $3/4$"

3. In wall framing, the normal spacing for steel studs is _____.
   a. 22" on center
   b. 20" apart
   c. 18" apart
   d. 16" on center

3.0.0 Other Steel Framing Applications

Performance Tasks

There are no Performance Tasks in this section.

Objective

Identify other steel framing applications.

a. Explain how to use steel framing members in floor and roof construction.

b. Describe how to use steel framing members in ceiling construction.

While walls are the most common elements to be constructed with cold-formed steel framing members, other building elements may also be steel-framed. These elements include floors, roofs, and ceilings.

3.1.0 | Steel Floor and Roof Assemblies

Rim tracks: Horizontal structural members that connect to the ends of floor joists.

Cold-formed steel floor assemblies typically use standard CFS floor joists (*Figure 44*), proprietary floor joists, pre-engineered steel floor trusses, **rim tracks**, web stiffeners, clip angles, hold-down anchors, and fasteners. They can be installed on foundation walls and stem walls, as well as directly to interior structural walls. They are similar to conventional framing and use single- or multiple-span installation techniques.

Figure 44 Standard CFS floor joists.
Source: Peter Bennett, Citizen of the Planet/Alamy

The construction industry has witnessed a dramatic rise in the use of cold-formed steel framing members in roof assemblies in the last decade. They allow for easy and standardized assembly and are durable and noncombustible. In addition to the standard C-shape member, scores of proprietary shapes, fabrication methods, and installation requirements are available from truss manufacturers nationwide. *Figure 45* shows an example of custom-designed steel trusses.

Figure 45 Complex framing using steel trusses.
Source: Zoonar/Lev Kropotov/Alamy

3.2.0 Ceiling Systems

Ceiling systems typically consist of a suspended drywall grid with gypsum board. An alternative method uses furring and cold-rolled channel to provide a rigid framework for suspended gypsum board and other ceiling assemblies, with the steel members suspended (*Figure 46*). Furring channel is commonly clipped or wire-tied perpendicular to the underside of the U-channel at appropriate intervals for attaching gypsum board with screws.

Figure 46 Ceiling framework.
Source: Don Wheeler

Attach furring channels to other structural steel members with screws or directly to the bottoms of bar joists. In the latter case, the furring is installed perpendicular to the joists and wire-tied at appropriate intervals.

3.0.0 Section Review

1. Cold-formed steel floor assemblies can be installed on foundation walls, stem walls, and _____.
   a. compacted gravel
   b. interior structural walls
   c. nonloadbearing walls
   d. retaining walls

2. In a ceiling system using steel members, gypsum board is attached with _____.
   a. bolts
   b. wire ties
   c. drive pins
   d. screws

Module 27205 Review Questions

1. A steel framing component that has a web, two flanges, and two returns is called a _____.
   a. track
   b. stud
   c. Z-furring channel
   d. hat channel

2. When a stud has a designator of 800S162-54, the web depth of the stud is _____.
   a. 0.8"
   b. 8"
   c. 80"
   d. 800"

3. In the CFS designation system, the letter U in the acronym STUFL represents _____.
   a. channel
   b. track
   c. studs with returns
   d. furring channel

4. The organization that created the universal designator system for steel components is the _____.
   a. Specialty Steel Industry of North America (SSINA)
   b. Occupational Safety and Health Administration (OSHA)
   c. United Steel Alliance (USI)
   d. American Iron and Steel Institute (AISI)

5. Devices that allow for the vertical movement of a structure without imposing additional loads on cold-formed steel framing are called _____.
   a. grommets
   b. adjustors
   c. slip connectors
   d. angle clips

6. A tool that can cut steel components of up to 20 gauge in thickness and leave no abrasive edges is a _____.
   a. chop saw
   b. snip
   c. C-clamp
   d. swivel-head shear

7. When selecting a screw to attach gypsum board to steel framing parts, the screw length should be the thickness of all the material plus _____.
   a. $\frac{1}{8}$"
   b. $\frac{1}{4}$"
   c. $\frac{3}{8}$"
   d. $\frac{1}{2}$"

8. A common caliber for powder-actuated tools is _____.
   a. .18 caliber
   b. .27 caliber
   c. .38 caliber
   d. .45 caliber

9. When using bolts to fasten steel framing to concrete, you must install _____.
   a. washers
   b. spacer bar
   c. caulking
   d. cotter pins

10. The maximum centerline-to-centerline tolerance for in-line framing of cold-formed steel is _____.
    a. $\frac{1}{8}$"
    b. $\frac{1}{4}$"
    c. $\frac{1}{2}$"
    d. $\frac{3}{4}$"

11. Structural stud sizes typically range from _____.
    a. 1" to 6"
    b. $1\frac{5}{8}$" to 6"
    c. 2" to 5"
    d. $2\frac{1}{2}$" to 8"

12. For a structural wall, the maximum allowable gap between the foundation and the bottom track is _____.
    a. $\frac{1}{16}$"
    b. $\frac{1}{8}$"
    c. $\frac{1}{4}$"
    d. $\frac{3}{8}$"

13. Diagonal steel straps attached to the walls are used to _____.
    a. brace shear walls
    b. temporarily brace panelized walls
    c. provide structural sheathing
    d. support headers

14. Furring channels in a ceiling system can be attached to the bar joists and must be _____.
    a. welded in place
    b. parallel to the joists
    c. adjusted with slip connectors
    d. perpendicular to the joists

Answers to Odd-Numbered Module Review Questions are found in *Appendix A*.

Answers to Section Review Questions

Answer	Section	Objective
Section One		
1. a	1.1.0	1a
2. b	1.2.1	1b
3. a	1.3.0	1c
4. d	1.4.2	1d
Section Two		
1. a	2.1.1	2a
2. a	2.2.1	2b
3. d	2.3.1	2c
Section Three		
1. b	3.1.0	3a
2. d	3.2.0	3b

Thermal and Moisture Protection

Source: Pat Canova/Alamy Images

Objectives

Successful completion of this module prepares you to do the following:

1. Describe insulation and explain why it is used in buildings.
    a. Explain what insulation does.
    b. Describe insulation requirements.
2. Describe insulation materials and types.
    a. Define flexible insulation and explain how it is used.
    b. Define loose-fill insulation and explain how it is used.
    c. Define rigid or semi-rigid insulation and explain how it is used.
    d. Define reflective, spray, and foam insulation and lightweight aggregates.
3. Explain how to install different types of insulation.
    a. Describe how to install flexible insulation.
    b. Describe how to install loose-fill insulation.
    c. Describe how to install rigid or semi-rigid insulation.
    d. Describe how to install foam insulation.
4. Explain how moisture control is accomplished in the construction of a building.
    a. Explain how to design a structure to control moisture.
    b. Describe how ventilation and vapor retarders are used for moisture control and explain how they are installed.
5. Explain how water and air infiltration control are accomplished in the construction of a building.
    a. Describe how water-resistive barriers are used to resist liquid water.
    b. Describe how air barriers are used and installed.
    c. Explain how to protect openings, penetrations, and joints.

Performance Tasks

Under supervision, you should be able to do the following:

1. Install blanket insulation in a wall.
2. Install a vapor retarder on a wall.
3. Install water-resistive barriers and materials.

CODE NOTE

Codes vary among jurisdictions. Because of the variations in code, consult the applicable code whenever regulations are in question. Referring to an incorrect set of codes can cause as much trouble as failing to reference codes altogether. Obtain, review, and familiarize yourself with your local adopted code.

Digital Resources for Carpentry

Scan this code using the camera on your phone or mobile device to view the digital resources related to this craft.

Overview

A properly insulated building creates an environment comfortable to live or work in and economical to heat or cool. Without proper insulation, warm air will escape the building in cold weather, causing the heating system to operate constantly. This results in an increased use of energy. In hot weather, the air conditioning system will have to work harder, producing the same result. Proper insulation in walls, floors, and roof decks will minimize this problem. Vapor retarders help prevent moisture from penetrating the building. Moisture can cause serious problems, including wood decay and mold growth. A skilled craftworker knows how to select and install insulating materials and vapor retarders.

1.0.0 Insulation

Performance Tasks

There are no Performance Tasks in this section.

Objective

Describe insulation and why it is used in buildings.

a. Explain what insulation does.
b. Describe insulation requirements.

Four important considerations for the construction of any building include thermal insulation, moisture control and ventilation, waterproofing, and air infiltration control. This module covers these areas and presents materials and procedures that can ensure effective installations.

When insulating a structure, you need to pay close attention to these factors. Carpenters need to control moisture migration as well as the movement of heat through a wall. You will need to ensure that you comply with all relevant codes, such as the *International Energy Conservation Code*® (*IECC*®), *International Building Code*® (*IBC*®), and *International Residential Code*® (*IRC*®), as well as applicable local codes.

1.1.0 What Is Insulation?

Insulation reduces the transfer of heat through a building. Most materials used in construction have some insulating value. Air functions as an excellent insulator when confined to very small spaces and kept very still. Manufactured insulation material traps a large amount of air in many tiny spaces to resist the transfer of heat and sound. Double-pane and triple-pane windows also use this method to reduce heat loss.

The amount of insulation in a building directly affects heating and cooling costs. It also affects the value of the building. Some jurisdictions require a permanent certificate on the building's electrical box stating the insulative properties of material used in the structure. When required, the builder or designer completes this certificate.

Figure 1 Cavity Insulation.
Source: ScotStock/Alamy Images

There are two overall categories of insulation—cavity and continuous. **Cavity insulation** is placed between framing members (*Figure 1*). When you think of insulation, you may imagine fluffy pink fiberglass blankets tucked between wood studs. That is an example of cavity insulation.

Continuous insulation runs over a building's structural members continuously (*Figure 2*). It does not have thermal bridges except for fasteners and service openings. It is integral to the opaque surface of a building envelope and can be installed in the interior and exterior.

It is important to know how to use and install both types of insulation to meet code insulation requirements.

Cavity insulation: Insulating materials that are located between framing members.

Continuous insulation: Insulation that runs over a building's structural members seamlessly without breaks or gaps.

Figure 2 Continuous insulation.
Source: Sergii Petruk/Alamy Images

Source: stockcreations/Shutterstock

Types of Insulation

There are many different types of insulation. Each type has specific applications for which it is best suited. Examples include:

- Fiberglass insulation
- Loose-fill insulation
- Rigid or semi-rigid insulation
- Reflective (or foil-faced) insulation
- Sprayed-in-place insulation
- Lightweight aggregates
- Mineral wool insulation
- Kraft-faced insulation

1.1.1 Thermal Resistance

Carpenters must know how thermal resistance works so they can understand the importance of installing insulation properly. They must also follow the requirements of applicable codes, which dictate how much thermal resistance a building should have.

A law of physics states that heat will always flow (or conduct) through any material or gas from a higher temperature area to a lower temperature area. That means that in a building, heat will naturally flow from a warm area into a cool area. Have you ever wondered why the top floor of a house is the warmest on a cold day? That's because the interior heating moves up to the sky above the ceiling to the coldest air. Trying to keep a single room warm is a battle between the heat's natural tendency to escape into a cooler area and the protections builders have put in place to prevent its movement. Insulation keeps a building warm and prevents heat from escaping into the cooler air outside.

Thermal resistance is a measure of how a material will resist the flow of heat energy. Different types of materials have better thermal resistance than others, making them more conducive to insulation.

R-value measures a material or object's thermal resistance, or ability to resist **heat conduction**. As a rule, the higher the R-value, the greater the effectiveness of the insulation. The R-value depends on the material, type of insulation, density, and thickness, as well as temperature, moisture accumulation, and aging. Code requires different R-values for different parts of a structure (as outlined later in the module). Carpenters must know how to find R-values so they can meet these requirements.

R-value is expressed as:

$$R = 1/k \text{ or } 1/C$$

Where:

k = amount of heat in British thermal units (Btu) transferred in one hour through 1 ft^2 of a material that is 1" thick and has a temperature difference between its surfaces of 1°F; also called the *coefficient of thermal conductivity*

C = conductance of a material, regardless of its thickness; the amount of heat in Btus that will flow through a material in one hour per ft^2 of surface with 1°F of temperature difference

R = thermal resistance; the reciprocal (opposite) of conductivity or conductance

Thermal resistance: A measure of how a material will resist the flow of heat energy.

R-value: A measure of an object or material's amount of thermal resistance.

Heat conduction: The process by which heat is transferred through a material, which is caused by a difference in temperature between two areas.

The higher the R-value, the lower the conductive heat transfer. *Table 1* shows the R-values of a number of common building materials, including some common insulating materials.

Think About It

Calculating the Effective R-value

To calculate the Effective R-value of an entire assembly, add together the individual R-values of each of the unbroken layers of materials in the assembly.

TABLE 1 Approximate R-Values of Common Materials

Material	Thickness	R-Value °F/ft²/hr in Btus
Air film and spaces		
Air space bound by ordinary materials	½" to 4"	1.00
Masonry units		
Concrete masonry unit (CMU)	8"	1.11
Concrete masonry unit (CMU)	12"	1.28
Common brick	4"	0.80
Sheathing materials		
Fiberglass	1"	4.00
Fiberboard sheathing	½"	1.32
Plywood	½"	0.63
Extruded polystyrene (EPS)	1"	5.00
Foil-faced polyisocyanurate (ISO)	1"	7.20
Siding materials		
Bevel-lapped wood siding	½" × 8"	0.80
Hardboard	½"	0.34
Plywood	¾"	0.94
Brick	4"	0.44
Aluminum, steel, vinyl (hollow backed)		0.61
Stucco	1"	0.20
Common insulating materials		
Fiberglass batts	3½"	11.00
Fiberglass blown (attic)	1"	2.12–4.30
Rigid fiberglass (less than 4lb/ft³)	1"	4.00
Mineral wool batt	1"	3.14–4.00
Vermiculite	1"	2.13
Polyurethane foam	1"	5.88–6.25
Expanded polystyrene foam	1"	4.00
Interior finish materials		
Gypsum board	½"	0.39

Source: Arch Toolbox (https://www.archtoolbox.com/r-values/)

The total heat transmission through a wall, roof, or floor of a structure in Btu per ft² per hour with a 1°F temperature difference is called the total heat transmission or **U-factor**. Whereas the R-value takes into account the individual insulating materials, the U-factor considers the entire assembly. This includes the thermal bridging of framing and studs.

The U-factor is the inverse of the R-value. It is expressed as follows:

U-factor: A measure of the total heat transmission through a wall, roof, or floor of a structure.

$$U = \frac{1}{R_1 + R_2 + \cdots + R_N}$$

Where:

$R_1 + R_2 + \ldots R_n$ represents the sum of the individual R-values for the materials that make up the thickness of the wall, roof, or floor. As shown in *Table 2*, the whole wall R-value is significantly influenced by the percentage of windows and doors in the wall and their heat transmission.

Air Versus Inert Gas

The air trapped between the panes of double-pane and triple-pane windows makes a good insulator, but there are better alternatives. Inert gases that block more heat than air create a stronger barrier. That's why some window manufacturers fill the space between the panes with argon—an inert, nontoxic, nonflammable gas.

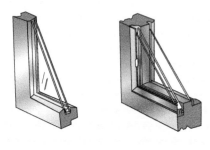

Radiation: Energy emitted from a source in electromagnetic waves or subatomic particles.

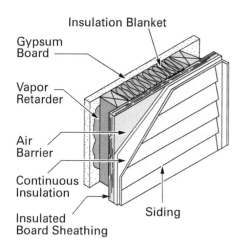

Gypsum Board
Insulation Blanket
Vapor Retarder
Air Barrier
Continuous Insulation
Insulated Board Sheathing
Siding

Figure 3 Typical wall construction.

Wood Framing Versus Steel Framing

Many modern constructions use steel framing. Steel can be a stronger, more durable material than wood that is more resistant to water, mold, air, and fire. However, steel framing conducts more heat than wood. This leads to increased energy costs and consumption while detracting from the interior comfort of the building. To counteract this effect, steel framed buildings require exterior continuous insulation.

TABLE 2 Example Wall Systems Based On the U-Factor Approach

2 × 6 Studs	Cavity Insulation		Continuous Insulation		Windows/ Doors		Whole Wall
Spacing	R	% Wall	R	% Wall	R	% Wall	R-Total
16	21	63	0	0	3	15	8.9
24	21	71	0	0	3	15	9.6
16	21	63	6	85	3	15	11.2
24	21	71	9	85	3	15	12.2
16	21	63	6	85	5	15	14.1
24	21	71	9	85	5	15	18.6

The lower the U-factor, the lower the heat transmission.

While the R-values provide a convenient measure to compare heat loss or gain, the total U-factor for a structure provides the calculations for sizing the structure's heating and cooling equipment (*Figure 3* and *Table 3*).

Doubling the R-value of a wall or roof can theoretically reduce the conductive heat loss or gain by half. However, as insulation thickness increases, the heat transmission (U-factor) decreases, but not in a direct relationship. Increases of insulation will continue to decrease heat loss, but at lower and lower percentages. At some point, it becomes economically useless to add more insulation. The same is true for double-, triple-, and quadruple-pane windows. Conductive heat loss or gain does not include heat gains or losses due to air leaks or **radiation** through windows or other openings.

Did You Know?

Over-Insulating

Installing excess insulation wastes money and may cause other problems. Moisture can collect inside an over-insulated building that lacks sufficient ventilation and water barrier protection. This promotes the growth of mold and fungus. It is even possible that cancer-causing radon gas trapped in the building can accumulate over time to dangerous levels.

TABLE 3 R-Values of Typical Wall Construction

2 × 4 Stud Wall with Rigid Board	
Type	**R-Value**
Air films*	1.00
¾" wood exterior siding	0.93
1" polystyrene rigid board	5.0
3½" batt or blanket insulation	11.0
Vapor retarder	0.0
½" gypsum board	0.39
Total Effective R-Value	18.32
2 × 6 Insulated Stud Wall	
Type	**R-Value**
Air films*	1.00
¾" wood exterior siding	0.93
3" expanded polystyrene foam	12.0
3" rigid fiberglass	12.0
Vapor retarder	0.0
½" gypsum board	0.39
Total Effective R-value	26.32

*Stagnant air film that forms on any surface

1.2.0 Insulation Requirements

Increasing energy costs and mandated government energy conservation have resulted in much higher R-value requirements for new construction. While building code and design standards for insulation have traditionally been based on average low-temperature zones and charts based on the range of low temperatures expected, the requirements change constantly. The *IECC*® recommends insulation values based on a location's **climate zone**, which is determined by local temperature and humidity levels (*Figure 4* and *Table 4*).

Climate zone: A geographical region based on climate criteria, as determined by the *IECC*®.

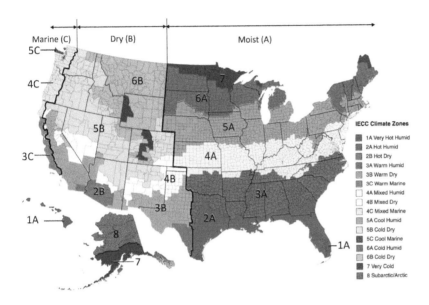

Figure 4 Climate zones in the United States.
Source: 2021 IECC - International Energy Conservation Code (https://basc.pnnl.gov/images/climate-zone-map-iecc-2021+AJ566)

Continuous insulation may be needed to ensure the highest level of insulation and meet code requirements. The code often gives the option of using cavity or continuous insulation (or a combination of both) with different required R-values. Cavity insulation generally requires higher R-values to account for the additional protection needed to counter the effects of thermal bridging. Cooler climates often require continuous insulation in addition to cavity insulation, as shown in *Table 4*.

Note that *Table 4* shows only a sample of the information available in Table C402.1.3 in Chapter 4 of the *IECC*® Commercial Provisions. The code also contains information on different types of roofs, walls, floors, and opaque doors.

In general, any exterior surface of a structure exposed to a thermal difference relative to its internal surface requires insulation. Such areas include the following:

- Roofs
- Above ceilings
- In exterior walls
- Beneath floors over crawl spaces
- Between two rooms that require different temperatures; that is, a conditioned vs. an unconditioned space
- Around the perimeter of concrete floors and around foundations

As you study the information in *Table 4*, you will notice that the ceiling insulation has the greatest R-value.

TABLE 4 Minimum R-Values of Commercial Insulation Requirements

Climate Zone	Floors: Mass		Walls: Wood Framed		Roofs: Attic and Other	
	All Other	Group R	All Other	Group R	All Other	Group R
1	NR	NR	R-13 + R-3.8ci or R-20	R-13 + R-3.8ci or R-20	R-38	R-38
2	R-6.3ci	R-8.3ci	R-13 + R-3.8ci or R-20	R-13 + R-3.8ci or R-20	R-38	R-38
3	R-10ci	R-10ci	R-13 + R-3.8ci or R-20	R-13 + R-3.8ci or R-20	R-38	R-38
4 except Marine	R-14.6ci	R-16.7ci	R-13 + R-3.8ci or R-20	R-13 + R-3.8ci or R-20	R-49	R-49
5 and Marine 4	R-14.6ci	R-16.7ci	R-13 + R-7.5ci or R-20 + R3.8ci	R-13 + R-7.5ci or R-20 + R-3.8ci	R-49	R-49
6	R-16.7ci	R-16.7ci	R-13 + R-7.5ci or R-20 + R-3.8ci	R-13 + R-7.5ci or R-20 + R-3.8ci	R-49	R-49
7	R-20.9ci	R-20.9ci	R-13 + R-7.5ci or R-20 + R-3.8ci	R-13 + R-7.5ci or R-20 + R-3.8ci	R-60	R-60
8	R-23ci	R-23ci	R-13 + R-18.8ci	R-13 + R-18.8ci	R-60	R-60

ci = Continuous insulation, NR = No Requirement
NOTE: Group R = Residential – A building or structure intended for sleeping purposes

1.2.1 Defining Climate Zones

Carpenters must consider the climate zone, or geographical region in which a construction project takes place, based on the climate criteria outlined in the *IECC®*. Regions have different zones with varying requirements due to climate differences in each region. The climate zone will affect everything from the scope of the work to the appropriate material and equipment to use. Consider building a house in Texas and building a house in Maine. These states have extremely different weather conditions that will affect the building needs and materials.

The *IECC®* breaks everything down by county. At the start of any project, check the code to make sure you design for the right zone.

Zones are broken into three categories—climate zones, moisture regimes, and warm-humid designations. The United States has 8 climate zones. For example, as shown previously in *Figure 4*, the northern part of the state of Illinois is in Zone 5. Cook county is in Zone 5A. The A, B, and C refer to the moisture regimes (A: Moist, B: Dry, and C: Marine). In the *IECC®*, a county marked with an asterisk is a warm-humid location. Cook county does not have an asterisk. Therefore, if you were installing insulation in a house in Chicago (in Cook County), you would need to consider that it is in Climate Zone 5A, which is not a warm-humid location.

1.2.2 Thermal Bridging

Ideally, every part of a building can perfectly align with the insulation. The insulation meets seamlessly between the wall and ceiling and around windows. But in reality, this is often not the case. Fixtures and other building elements stick out, or difficult junctions in the structural design make it difficult to install the insulation seamlessly. This can cause thermal bridging.

Thermal bridging occurs when a small area of floor, wall, or roof loses substantially more heat than the surrounding area. Thermal bridging can cause heat loss, condensation, moisture, and mold problems, as well as expansion and contraction. Any of these effects can lead to occupant discomfort or even illness based on the severity of the effect. They can also impact the longevity of the building and building materials. That's why it is critical for contractors to be aware of potential causes and impacts of thermal bridging.

Thermal bridging: When a small area of floor, wall, or roof loses substantially more heat than the surrounding area.

Thermal bridging can happen in any kind of building and can be an unintended consequence of poor installation or design. For instance, a brick wall with a structural angle supporting the brick creates a huge thermal bridge. Redesigning this structure would prevent or minimize thermal bridging. When installing insulation, thermal bridging is most commonly caused by gaps in the insulation, which lead to air leakage and heat escape. Carpenters need to be especially careful when installing insulation to ensure there are no gaps and that the insulation will remain structurally sound and in position over time to prevent gaps from appearing in the future.

ASHRAE® Standard 90.1-2022 includes requirements to address the impact of thermal bridging in building envelopes. Contractors will need to calculate the value of any thermal bridges in a structure to ensure it complies with all applicable code requirements, including state or local codes.

In recent years, there has been a push to increase energy conservation in buildings. Codes consider the operational energy of a building to ensure it meets energy conservation goals and has reduced energy consumption. Thermal bridging can impede energy conservation, causing a building to be less efficient and consume more energy.

Wood or metal studs create considerable breaks in the insulation, causing thermal bridges. As conductors, wood and metal channel heat out of the building. This poses a big problem for traditional insulation practices that have installed insulation between the wood or metal studs of a building.

Continuous installation (*Figure 5*) should prevent thermal bridging (including wood and metal studs) if done correctly. The requirement to install continuous installation or an equivalent amount of insulation in some zones is one of the most substantial changes in recent years to avoid thermal bridging and reduce energy consumption. This is a major shift away from the traditional method of installing cavity insulation only.

Figure 5 Installing continuous insulation.
Source: sima/Shutterstock

1.0.0 Section Review

1. The R-value is a measure of the ability of a material to _____.
   a. resist the passage of moisture
   b. resist heat transfer
   c. allow cold air to enter a building
   d. convert water vapor into a liquid

2. You are installing insulation for a building in Broward County, Florida. In the *IECC*®, it is marked as 1A Broward*. What does that indicate?
   a. It is in Climate Zone 2 and is marine.
   b. It is in Climate Zone 2 and is dry.
   c. It is in Climate Zone 1, is marine, and is a warm-humid location.
   d. It is in Climate Zone 1, is moist, and is a warm-humid location.

2.0.0 Insulation Materials and Types

Performance Tasks

There are no Performance Tasks in this section.

Objective

Describe insulation materials and types.

a. Define flexible insulation and explain how it is used.
b. Define loose-fill insulation and explain how it is used.
c. Define rigid or semi-rigid insulation and explain how it is used.
d. Define reflective, spray, and foam insulation and lightweight aggregates.

Insulation materials can be divided into four general classifications (*Table 5*). These materials are used in the manufacture of five basic categories of insulation: flexible, loose-fill, rigid or semi-rigid, reflective, and miscellaneous.

TABLE 5 Insulation Materials

Classification	Material	Comments
Mineral	Rock	Rock and slag are used to produce mineral wool by grinding and melting the materials and blowing them into a fine mass. It is available as loose-fill or blanket insulation.
	Slag	
	Glass fiber	
	Vermiculite	
	Perlite	
Vegetable (natural)	Wood	Many vegetable products are processed and formed into various shapes, including blankets and rigid boards.
	Sugar cane	
	Corn stalks	
	Cotton	
	Cork	
	Redwood bark	
	Sheep's wool	
	Straw	
	Hemp	Hemp insulation is not widely used in the US.
Plastic	Polystyrene	Polyurethane can be made into polyurethane foam insulation, which is a very common type of insulation.
	Polyurethane	
	Polyisocyanurate	
	Phenolic	
Metal	Foil	Metallic insulating materials are generally applied to rigid boards or papers and used primarily for their reflective value.
	Tin plate	
	Copper	
	Aluminum	

These materials can generally be used to insulate both wood and steel framed buildings. Installers need to pay close attention to the R-values of each material to make sure the R-value of the entire wall assembly meets applicable code requirements.

The R-value of the insulation is marked on the insulation itself or its packaging (*Figure 6*).

R-Value

Figure 6 Typical R-value identification.

2.1.0 Flexible Insulation

Flexible insulation is usually manufactured from fiberglass in blanket form (*Figure 7*) and fiberglass or mineral wool in batt form (*Figure 8*). Fiberglass insulation is a very common form of cavity insulation used in most homes in the United States. Mineral wool insulation has become more popular because it provides noise insulation and fire protection, resists mold, and can also be used as exterior continuous insulation. In some cases, flexible insulation can also be manufactured from wood fiber or cotton and treated for resistance to fire, decay, insects, and rodents.

Blankets of flexible insulation are available in 16" or 24" widths and the batts in 15" or 23" widths. Both come in thicknesses ranging from 1" to 12". The batts, packaged in flat bundles in lengths of 24", 48", or 93", may be unfaced or faced with asphalt-laminated kraft paper or fire-resistant foil scrim kraft (FSK) with or without nailing flanges. The blankets come in rolls often encased in an asphalt-laminated kraft paper or plastic film. In most cases, they have a facing with nailing flanges. Blankets and batts with kraft or film casing and/or facings are combustible and must not be left exposed in attics, walls, or floors.

Fiberglass batt insulation at a $3\frac{1}{2}$" thickness (standard rating or high rating) may be used on exterior walls between the studs. This has been the normal insulation thickness in the past, because of the use of $3\frac{1}{2}$"-wide studs spaced 16" on center (OC). However, in the northern parts of the country, some builders use 2 × 6 studs spaced 16" or 24" OC, which has allowed for an increase in the wall insulation to $5\frac{1}{2}$". This thickness is ample insulation for all parts of the United States.

Figure 7 Flexible blanket insulation (fiberglass).
Source: DonNichols/Getty Images

Flexible insulation: A type of insulation that is made from a flexible material.

Fiberglass: A material made of sand and recycled glass.

Batt: A flat, pre-cut piece of insulation.

Fiberglass insulation: A type of batt, roll, or loose-fill insulation that is made of extremely small pieces (or fibers) of glass.

Mineral wool insulation: A type of batt or loose-fill insulation that is made of natural stone fibers or slag.

Figure 8 Flexible batt insulation (mineral wool).
Source: ronstik/Shutterstock

Fiberglass Ingredients

The primary ingredients of fiberglass are silica (sand) and recycled (previously melted) glass. A chemical binder holds the spun glass together. The binder's ingredients include formaldehyde, phenol, and ammonia. The ammonia in the binder sometimes gives fiberglass a strong odor.

While the insulation itself does not burn, the binding material will burn off the glass fibers when the temperature rises high enough (about 350°F). For this reason, fiberglass insulation should not be used in applications that would subject the chemical binder to temperatures approaching its flashpoint.

2.1.1 Fiberglass Insulation Safety

Flexible fiberglass is probably the first thing that comes to mind when the average person thinks of insulation. Most of us don't realize that this common material must be handled carefully. The tiny strands of glass in fiberglass insulation can irritate skin, injure eyes, and cause a variety of respiratory problems.

While insulation installers must wear protective equipment, their responsibilities don't end there. If they do not properly remove debris from the installation or if they disturb existing insulation, fiberglass particles could spread through the building. Fiberglass that enters an HVAC system will move to all parts of the building. Always use care when handling fiberglass insulation. This protects you as well as the building's current and future occupants.

Heat Losses

At relatively cold temperatures, a building loses heat in many ways and through different areas. These include heat lost directly through its walls, ceiling, and roof. Heat also escapes through windows, doors, gaps or cracks, and thermal bridges in the structure. That is why it is so important to make sure every part of the building has proper insulation.

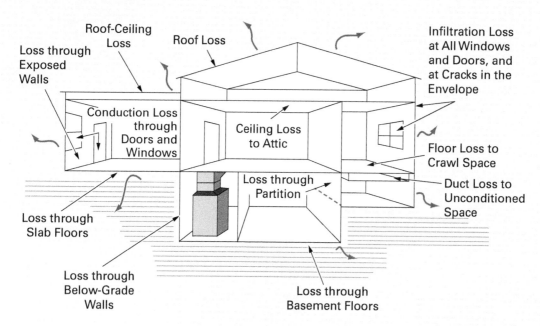

Unfaced insulation: Insulation that does not have a vapor retarder.

Vapor retarder: A material used to retard the flow of vapor and moisture into walls and prevent condensation within them. The vapor retarder must be located on the warm side of the wall.

2.1.2 Unfaced Versus Kraft-Faced Insulation

Unfaced insulation is fiberglass insulation that does not have a **vapor retarder**. It can be used for interior walls that do not face outside or in rooms that do not need moisture control, such as dining rooms and living rooms. It is generally noncombustible, which helps make the building more fire-resistant.

Kraft-faced insulation is fiberglass insulation that has a vapor retarder on one side. It can be installed on exterior walls and attic ceilings with the vapor retarder side facing out or in rooms that need moisture control. It generally comes in rolls or batts, which can be cut with a utility or insulation knife. The vapor retarder can be made of paper, vinyl, or foil. Make sure the type you use complies with applicable codes. Faced insulation must also be on the outside of the insulation stack to prevent moisture from building up inside the insulation.

Kraft-faced insulation: Insulation that has a paper, vinyl, or foil vapor retarder on one side.

2.2.0 Loose-Fill Insulation

Loose-fill insulation is a type of cavity insulation supplied in bulk form packaged in bags or bales (*Figure 9*). In new construction, it is usually blown or poured and spread over the ceiling joists in unheated attics. In existing construction that was not insulated when it was built, the material can be blown into the walls as well as the attic.

Loose-fill insulation: Insulation that comes in the form of loose material in bags or bales.

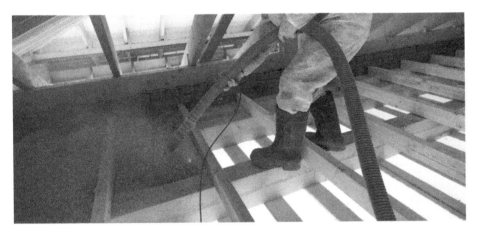

Figure 9 Loose-fill insulation.
Source: Kurteev Gennadii/Shutterstock

The materials used in loose-fill insulation include rock or glass wool, wood fiber, shredded redwood bark, cork, wood pulp products such as shredded newspaper (cellulose insulation), and vermiculite. All wood products, including paper, must be treated for resistance to fire, decay, insects, and rodents.

Shredded paper absorbs water easily and loses considerable R-value when damp. In addition to wall surface and/or ceiling vapor retarders, it is essential to install a waterproof membrane along the eaves to prevent water leakage.

The R-value of loose-fill insulation depends on proper application of the product. Follow the manufacturer's instructions to obtain the correct weight per square foot of material as well as the minimum thickness. Before installing loose insulation, calculate the area of the space to be insulated (minus adjustments for framing members). Then, determine the required number of bags or pounds of insulation from the bag label charts for the desired R-value.

Disadvantages of Loose-Fill Insulation

A disadvantage of loose-fill insulation is that it tends to settle over time. Insulation may have to be added to refill the cavities formed. Do not cover loose-fill insulation with materials that could pack or crush it.

Heat Gains

At relatively warm temperatures, exposed walls and roofs absorb heat from the sun. Heat also enters a building through windows, doors, and gaps or cracks in the structure.

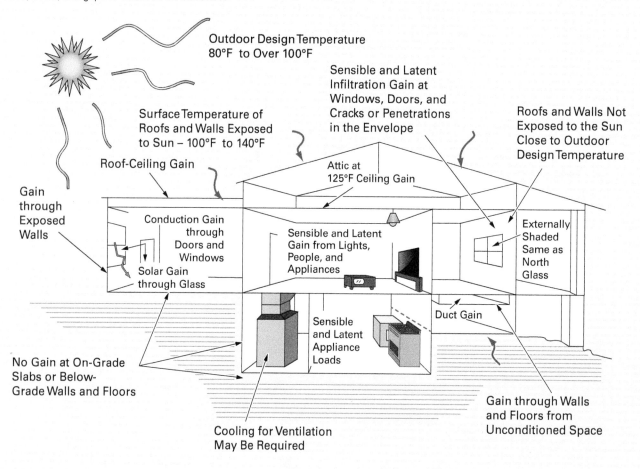

2.3.0   Rigid or Semi-Rigid Insulation

Rigid or semi-rigid insulation: A type of insulation that comes in formed boards made of mineral fibers.

Rigid or semi-rigid insulation is available in sheet or board form and is generally divided into two groups: structural and nonstructural. It is available in widths up to 4' and lengths up to 12'.

Structural rigid foam boards come in densities ranging from 15 to 31 pounds per square foot. They are used as sheathing, roof decking, and gypsum board. Their primary purpose is structural, while their secondary is insulation. The structural types are usually made of processed wood, cane, or other fibrous vegetable materials.

Nonstructural rigid foam board (*Figure 10*) or semi-rigid fiberglass insulation is usually a lightweight sheet or board made of fiberglass or foamed plastic such as polystyrene, polyurethane, polyisocyanurate, and expanded perlite. It can be used for cavity or continuous insulation. Most of these products are waterproof and can be used on the exteriors or interiors of foundations, under the perimeters of concrete slabs, over wall sheathing, and on top of roof decks. When used above grade with proper flashing, a protective and decorative coating is sometimes applied directly to the panel. This is known as an **exterior insulation finish system (EIFS)**.

Exterior insulation finish system (EIFS): Nonstructural, nonbearing, exterior wall cladding system that consists of an insulation board attached adhesively or mechanically, or both, to the substrate (*IBC*® 2021).

In other cases, it is sealed with an air infiltration film, and normal siding is applied. The foam boards range in thickness from 1" to 4" with R-values up to R-30. Because all foam insulation is flammable, it cannot be left exposed. It must be covered with at least $\frac{1}{2}$" of fireproof material.

Figure 10 Rigid foam board.
Source: Whiteaster/Shutterstock

Some manufacturers also provide rigid foam cores that can be inserted in concrete blocks or used with masonry products to provide additional insulation in concrete block or masonry walls.

The most common rigid and semi-rigid insulation types include the following:

- *Rigid expanded polystyrene* — This material has an R-value of R-4 per inch of thickness. Water significantly reduces this value because rigid expanded polystyrene is not water-resistant. It is not recommended for below-grade insulation. It is the lowest in cost. This material is also called *beadboard*.

- *Rigid extruded polystyrene* — The R-value of this material is about R-5 per inch of thickness. It is water-resistant and can be used below grade. When used in above-grade applications, it is subject to damage by ultraviolet light and must be coated or covered.

- *Rigid polyurethane and polyisocyanurate* — Initially, these boards have R-values up to R-8 per inch. However, over time the R-value drops to between R-6 and R-7 due to escaping gases. This is referred to as *aged R-value*. These products are subject to damage by ultraviolet light and must be coated or covered.

- *Semi-rigid fiberglass* — The R-value of this material is about R-4 per inch. Boards of this kind are used on below-grade slabs and walls to provide water drainage as well as insulation. They are also used under membrane roofs as insulation. On below-grade applications, the walls or floors must be water-proofed with a coating or membrane between the wall or floor and the insulation to block water penetration.

2.4.0 Other Types of Insulation

There are other types of insulation that do not fit the previous three categories.

2.4.1 Reflective insulation

Reflective insulation, also known as *foil-faced insulation* (*Figure 11*), usually consists of multiple outer layers of aluminum foil bonded to inner layers of various materials for strength. The number of reflecting surfaces (not the thickness of the material) determines its insulating value. Reflective insulation reflects light and can protect against heat in areas that get a lot of sunlight. When used with other materials, it can warm areas. To be effective, the metal

Reflective insulation: A type of insulation made of outer layers of aluminum foil bonded to inner layers of various materials.

Figure 11 Reflective insulation in an attic.
Source: Ozgur Coskun/Alamy Images

Using Sprayed- and Foamed-in-Place Insulation

Sprayed-in-place insulation materials work well for irregular surfaces. These include walls and ceilings that have curves or that have beams, pipes, or other equipment protruding from them. Foams and sprays can be built up in layers to the desired insulation thickness.

foil must face an open-air space that is $\frac{3}{4}$" or more in depth. In some cases, flexible insulation has reflective material bonded to it as the inside surface for both insulation and vapor seal purposes.

Sometimes foil is used independently of another material and is added to the other insulation. You can buy this foil insulation separately in rolls or batts. When installing it, do not push it into the insulation; it will work best with 1" to 2" of space between the foil and existing insulation. For obstacles like supports and braces, the foil can be cut, wrapped around the obstacle, and connected with foil tape. Seal any air or duct leaks before installing foil insulation to avoid condensation, especially in cold climates.

Did You Know?

Insulation Weight

Insulation materials have weight that must be considered when designing a building. One of the advantages of using lightweight insulation board, such as rigid polyurethane and polyisocyanurate, is that it permits greater freedom of design. The lighter the insulation, the less weight load-bearing members of the structure must support.

Reflective Insulation and Heat

Reflective insulation by itself can only block radiated heat. At relatively hot temperatures, it helps keep buildings cooler by deflecting heat from the sun. At relatively cold temperatures, it can do little to prevent heat from escaping the building.

2.4.2 Sprayed-in-Place and Foam Insulation

Sprayed-in-place insulation (*Figure 12*) can be applied to new or existing construction using special spray equipment. It can be injected between brick veneer and masonry walls; between open studs or joists; and inside concrete blocks, exterior wall cavities, party walls, and piping cavities. It is often left exposed for acoustical as well as insulating properties. Trained and certified contractors must apply sprayed-in-place insulation.

While foam is a very valuable form of insulation, carpenters must assess it from a fire standpoint. Spray foam—like other types of insulation materials—is flammable. Contractors must consider the fire rating of the foam when insulating a structure and ensure that the overall wall assembly fire rating complies with applicable codes. Foam plastics must have a thermal barrier that separates them from the building interior unless they have passed the code fire test. An acceptable barrier is $\frac{1}{2}$" gypsum board.

In the 1970s, urea formaldehyde foamed-in-place insulation was injected into many homes. However, due to improper installation, the foam shrank and gave off formaldehyde fumes. As a result, its use was banned in the United States and Canada. Later, it was allowed back on the market in certain areas of the United States. A urethane foam that expands on contact can also be used. It does not have a formaldehyde problem, but it does emit cyanide gas when burned. As a result, it requires fire protection and, like urea formaldehyde, it may also be banned in some areas of the country.

Another foamed-in-place product is a phenol-based synthetic polymer called Tripolymer® that is fire-resistant and does not drip or create smoke when exposed to high heat. This material does not expand once it leaves the delivery hose of the proprietary application equipment.

2.4.3 Lightweight Aggregates

Concrete, concrete blocks, or plaster often have added insulation material consisting of perlite, vermiculite, blast furnace slag, sintered clay products, or cinders to improve their insulation quality and reduce heat transmission.

Figure 12 Sprayed-in-place insulation.
Source: anatoliygle/123RF

2.4.4 Materials for Continuous Insulation

The following types of insulation materials are used for continuous insulation:

- *Nonstructural rigid foam board* — One of the most common materials used for continuous insulation, rigid foam board is a firm material that is easy to install on boards over the sheathing of a building. It is also possible to fasten rigid foam board to framing or sheathing as cladding without compressing the insulation, which meets the code requirement that continuous insulation be uncompressed. Types of nonstructural rigid foam board include:
  - Rigid expanded polystyrene
  - Rigid extruded polystyrene
  - Foil-faced polyisocyanurate (Polyiso)
- *Mineral wool (rockwool) board* — Mineral wool is a semi-rigid type of continuous insulation material. It is often made largely out of recycled and renewable material. Because mineral wool is made of bonded rock fibers, it is noncombustible and can resist temperatures of 2,000°F, making it very fire resistant. It is also vapor permeable and water resistant.
- *Spray foam* — Sprayed-in-place insulation is an effective type of continuous insulation. It can be sprayed continuously on the exterior of a building over the wood sheathing.

2.0.0 Section Review

1. Why is it important to wear protective equipment when handling fiberglass insulation?
   a. It can irritate skin, injure eyes, and cause respiratory problems.
   b. It can cause long-term heart problems.
   c. It can cause silicosis and chronic obstructive pulmonary disease (COPD).
   d. It is hot to touch and can cause burns.

2. Which type of insulation is typically blown or poured over ceiling joists in unheated attics?
   a. Flexible insulation
   b. Loose-fill insulation
   c. Semi-rigid insulation
   d. Sprayed-in-place insulation

3. Which of the following materials is likely to be used in structural rigid foam board insulation?
   a. Polystyrene
   b. Processed wood
   c. Fiberglass
   d. Expanded perlite

4. What is an acceptable thermal barrier that can be used to separate foam insulation from the building interior?
   a. Vapor retarder
   b. 2" fiberglass insulation
   c. $\frac{1}{4}$" of bonded aluminum foil
   d. $\frac{1}{2}$" gypsum board

3.0.0 Installation Guidelines

Performance Task

1. Install blanket insulation in a wall.

Objective

Explain how to install different types of insulation.

a. Describe how to install flexible insulation.
b. Describe how to install loose-fill insulation.
c. Describe how to install rigid or semi-rigid insulation.
d. Describe how to install foam insulation.

Before installing insulation, check building plans and applicable codes to determine the R-values and the types of insulation required or permitted for the structure. Then, calculate the required amount of insulation for the structure. Follow any specific instructions provided by the selected manufacturer when installing the insulation.

The estimator will determine the amount of insulation for the walls, ceilings, and floors of a structure. Installers do not typically need to be a part of the estimation process.

> **WARNING!**
>
> Wear proper eye protection, respiratory equipment, and gloves when handling and installing insulation.

3.1.0 Flexible Insulation Installation

Flexible insulation used in unfinished floors, walls, ceilings, attics, and crawl spaces can fit between joists, studs, and beams. It can insulate ducts and tank-style water heaters. Because it is not waterproof, it should not be used in foundations or other areas with high exposure to moisture.

3.1.1 Steps for Installing Flexible Insulation

Use the following procedure when installing typical flexible insulation:

Step 1 For walls, measure the inside cavity height and add 3". From the wall, lay the distance out on the floor and mark it. Unroll blanket insulation or lay batts on the floor. Use two layers or more. At the cut mark, compress the insulation with a board and cut it with a utility knife. On blanket or faced insulation, remove about 1" of insulation from the ends to provide a stapling flange at the top and bottom.

Step 2 If a separate interior vapor seal will be installed, install blanket or faced insulation so that the stapling flange is against the inside of the wall studs (*Figure 13*).

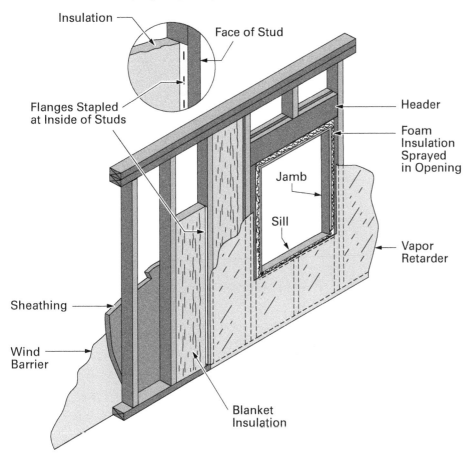

Figure 13 Blanket insulation without integral vapor seal.

- If the facing of the blanket or batt is the vapor seal, install the stapling flange against the inside of the stud. Do not go over the face of the stud (*Figure 14*). For faced or blanket insulation, use a power, hand, or hammer stapler to first staple the top flange to the plate.
- Align and staple down the sides.
- Staple the bottom flange to the sole plate. Pull the flanges tight and keep them flat when stapling. Follow the manufacturer's instructions for staple placement.
- For unfaced batt insulation, install the batt at the top and bottom first and push it tight against the plates.
- Evenly push the rest of the batt into the cavity (*Figure 15*).
- Spray foam insulation into the narrow spaces around windows and doors and cover it with a plastic or tape vapor seal. Be careful to use foam that doesn't expand so much that it distorts the window.

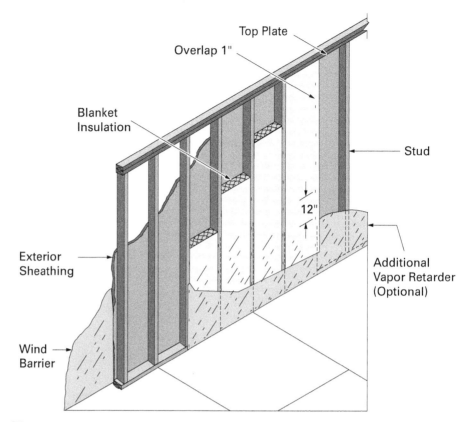

Figure 14 Blanket insulation with integral vapor seal.

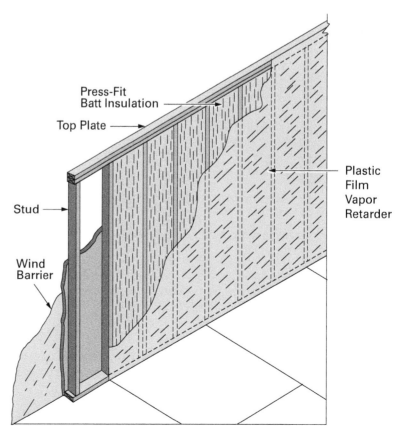

Figure 15 Batt insulation with separate vapor retarder.

> **WARNING!**
>
> Exercise caution when installing insulation around electrical outlet boxes and other wall openings or devices. Failure to do so may result in electrocution.

Step 3 Faced or blanket insulation for ceilings or floors is usually installed from the bottom in the same manner as the walls. Unfaced batts can be installed from either the top or the bottom.

- Make sure that ceiling insulation extends over the wall into the soffit area (*Figure 16*). Also make sure soffit baffles (*Figure 17*) are inserted over and cover the ceiling insulation. The baffles should be fastened to the roof deck to hold them in place so that they do not slide down into the soffit and block ventilation.
- For floors, ensure that the insulation is installed around the perimeter of the floor against the header (*Figure 18*). Floor insulation over a basement is installed with the vapor retarder facing down.
- Over a crawl space, the vapor retarder faces up. In either case, the insulation can be supported below by a wire mesh (chicken wire), if desired.
- Install a vapor retarder only if specified for the area. A vapor retarder on the inside face of the wall is not a good idea in certain geographical locations.

Think About It

Fiberglass Insulation

1. Are the flanges on faced fiberglass insulation always stapled to the inside of the stud?
2. Can you increase the effectiveness of fiberglass insulation by squeezing more into a smaller space?

Cathedral Ceilings

If a cathedral ceiling incorporates gypsum drywall attached to the bottom of the rafters, airflow must be maintained from the soffit to the ridge. Proper ventilation must also be maintained above and below skylights to prevent buildup of heat and moisture. Check your local code for the appropriate methods to use when working with cathedral ceilings.

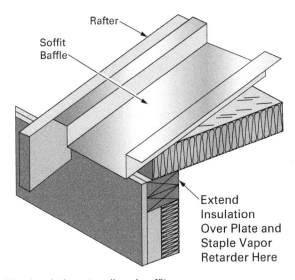

Figure 16 Ceiling insulation at wall and soffit.

Figure 17 Typical plastic soffit baffle (shown upside down).

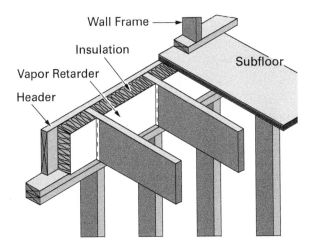

Figure 18 Perimeter floor insulation.

3.2.0   Loose-Fill Insulation Installation

For new construction, loose-fill insulation is used primarily for attic insulation. On older construction, it can also be blown into wall cavities through holes drilled at the center and tops of exterior walls. The following steps only cover attic or ceiling installation.

WARNING!

Wear proper eye protection, respiratory equipment, and gloves when handling and installing insulation.

3.2.1 Steps for Installing Loose-Fill Insulation

Use the following procedure when installing loose-fill insulation:

Step 1   Make sure that the finished ceiling below has been installed. Also, ensure that a separate vapor retarder has been installed to prevent moisture penetration of the insulation and to prevent the fine dust from the insulation from penetrating the ceiling in the event of future cracks (*Figure 19*). Make sure that soffit baffles and blocking have been installed to prevent the material from spilling into the soffits.

Step 2   Add markers throughout the area before pouring in the loose-fill insulation to indicate the initial installed thickness. This is a code requirement that allows an inspector to easily see the depth of the insulation. There should be one marker per 300 square feet. Markers can be attached to trusses or joists and should be marked with numbers that are at least 1 inch in height for visibility. The markers must face the access opening of the attic.

Step 3   Pour the insulation from bags or blow the insulation over the ceiling joists using special equipment. Using a straightedge, tamp the insulation and then level it to the required depth for the R-value desired (*Figure 20*).

Figure 19 Loose-fill insulation.

Polystyrene Forms

Structural forms made of polystyrene are sometimes used for residential and light commercial construction. Concrete is poured into the forms, which are left in place to provide insulation for the walls. The forms usually provide sufficient insulation by themselves, but check the local code for these requirements.

Source: Oleksandr Rado/Alamy Images

Figure 20 Leveling loose-fill insulation.
Source: Dorling Kindersley ltd/Alamy Images

3.3.0   Rigid or Semi-Rigid Insulation Installation

Rigid insulation panels can be fastened like sheathing over the studs or wood sheathing of a structure as exterior continuous insulation. Nails with large heads or washers or screws with washers prevent crushing or compressing the insulation (*Figure 21*).

This type of insulation is particularly important for steel-framed buildings to counteract the natural heat conduction of steel. Installing insulation outside of a steel frame prevents heat from escaping through the steel. Cavity insulation installed between steel framing members is often not sufficient to prevent heat from escaping through the steel studs.

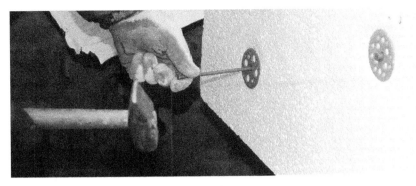

Figure 21 Installing rigid board insulation.
Source: Radovan1/Shutterstock

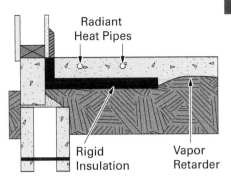

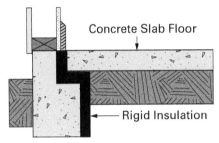

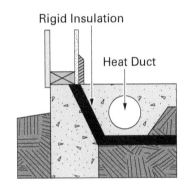

Figure 22 Rigid insulation installed under a concrete slab.

Rigid insulation panels may be installed on the exterior of a foundation. Typically, the exterior of the foundation is waterproofed first. Then, the panels are applied over special mastic and secured with concrete nails to hold them in place until the mastic sets. For existing construction, the panels may be installed on the interior of the foundation if the walls are adequately waterproofed.

Figure 22 shows typical methods of installing rigid insulation under surface slabs. Slabs lose energy when heat is conducted out and through the slab perimeter. Usually, insulation is only applied around the perimeter of the slab, which helps prevent this heat loss. It can be installed downward from the top of the slab on the inside or outside of the foundation wall. If a slab is separate from a foundation wall, insulation can be installed between the wall and the slab or on the exterior of the wall. A vapor retarder should be applied under the slab and over any insulation under the slab.

Climate Zones 4–8 require insulation below grade. It should extend vertically, under the slab, and perpendicularly by the amount indicated in Table N1102.1.2 in Section R402 of the *IECC®*. Insulation that extends away from the building must be protected by pavement or at least 10 inches of soil.

3.4.0 Foam Insulation Installation

Foam is sprayed between framing members rather than placed and attached like flexible and rigid insulation (see *Figure 23*). It is particularly useful for insulating small, tough-to-access spaces, such as the cavities around doors and windows, but it can also be used to insulate large spaces. Spray foam is messy and requires certain temperatures, so it is important to prepare fully before beginning the installation.

WARNING!

Remember that only trained contractors should install foamed-in-place and sprayed-in-place insulation.

3.4.1 Steps for Installing Foam Insulation

Installers should follow the manufacturer's instructions for using foam insulation. The following steps apply to most foam insulation installation processes:

Step 1 Make sure the foam cannisters are between 70°F and 85°F, though this may vary by manufacturer. Check the manufacturer's specifications for the best results. Some manufacturers may also recommend a certain temperature for the surface on which you will be spraying the foam.

Step 2 Prepare the area. Use plastic to cover everything that shouldn't be touched by foam. Make sure to cover outlets and loose cords as well.

Figure 23 Foam insulation.
Source: anatoliy_gleb/Shutterstock

Step 3 Wear the appropriate PPE. It is extremely important to avoid inhaling any foam and skin contact. Appropriate PPE includes the following:

- Disposable suit that covers the head
- Air respirator mask
- Eye protection
- Disposable gloves
- Shoe coverings

Step 4 Once you've prepared, you can begin spraying the foam. The R-value you need to meet will indicate how many inches of foam to apply. Remember that the foam will expand, so it's important to pay attention to the type of foam and how much it will expand over time. If you are combining foam insulation with another insulation material, you may only need a single layer of foam.

Step 5 Use a knife to remove any extra foam over the framing members.

Foam insulation can create continuous insulation when applied to the exterior sheathing of a structure. When applied on the exterior, it also acts as an air and water-resistive barrier. It may be more effective in this use because when applied in cavities, it is still being interrupted by thermal bridging (the studs). On the exterior of a structure, a continuous application creates an uninterrupted layer of insulation.

3.0.0 Section Review

1. When installing faced insulation in a wall that will have a separate vapor retarder installed, the staples should be fastened to the wall frame _____.
   a. on the faces of the wall studs, top plate, and sole plate
   b. on the inside surfaces of the wall studs, top plate, and sole plate
   c. only to the top plate and sole plate
   d. only to the faces of the wall studs

2. How many markers should be placed in loose-fill insulation per 300 square feet?
   a. 1   c. 3
   b. 2   d. 4

3. What is a type of fastener you can use when installing rigid insulation panels to prevent crushing the insulation?
   a. Screws with washers   c. Sheathing tape
   b. Nails with small heads   d. Wood glue

4. What should you do before installing foam insulation?
   a. Waterproof the exterior of the foundation.
   b. Install soffit baffles and blocking.
   c. Use plastic to cover everything that shouldn't be touched by foam.
   d. Install loose-fill insulation to the surrounding area.

4.0.0 Moisture Control	
Objective	**Performance Task**
Explain how moisture control is accomplished in the construction of a building. a. Explain how to design a structure to control moisture. b. Describe how ventilation and vapor retarders are used for moisture control and explain how they are installed.	2. Install a vapor retarder on a wall.

Water vapor contained in air can readily pass through most building materials used for wall construction. This vapor caused no problem when walls were porous because it could pass from the warm wall to the outside of the building before it could condense into liquid water (*Figure 24*).

Water vapor: Water in a vapor (gas) form, especially when below the boiling point and diffused in the atmosphere.

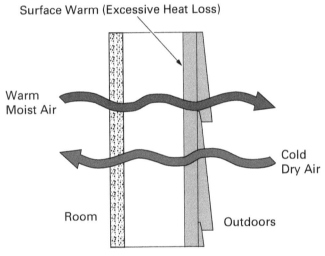

No Insulation or Vapor Retarder – Moist Air Passes through the Wall

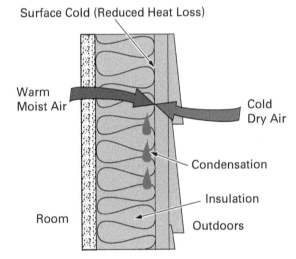

Insulation Only – Moist Air Passes into Wall and Condenses, Causing Damage to Insulation and Building Framework

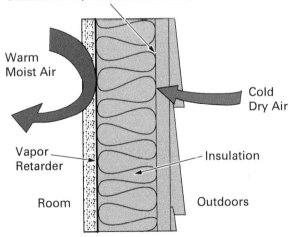

Insulation with Vapor Retarder – Moisture Cannot Enter Wall Space

Figure 24 Effects of insulation and vapor retarder.

When buildings were first constructed with insulation in the walls to cut down on heat loss, moisture in the air passed through the insulation until it reached a point cold enough to cause it to condense. The condensed moisture froze in very cold weather and reduced the efficiency of the insulation. The ice contained within the wall thawed as the weather warmed, and the resulting water in the wall caused studs and sills to decay over time.

For these reasons, it is important to keep cellars, basements, crawl spaces, exterior walls, and attics dry. Moisture in crawl spaces, basements, and attics also encourages wood-chewing insects such as termites, as well as the growth of mold. In the case of crawl spaces, moisture often rises from the ground into the crawl space during periods of heavy rain.

4.1.0 Designing a Structure to Control Moisture

To prevent the concentration of this damaging moisture, some precautions must be taken in the original design of the structure:

- The earth must slope down and away about 20' from the structure, carrying surface water away.
- The crawl space should be protected from moisture by a vapor retarder on the ground.
- The foundation walls should be penetrated with vents so that moisture will not be trapped in the crawl space.
- A vapor retarder should be installed between the insulation and the subfloor.

4.1.1 Moisture Control in Basements

Condensation: The process by which a vapor is converted to a liquid, such as the conversion of the moisture in air to water.

Basements usually have the most trouble with **condensation** in summer during humid weather. The earth under the concrete basement floor stays comparatively cool, causing the floor of the basement to be a cold surface. The hot air, saturated with moisture, condenses when it comes in contact with the cooler surfaces of the floor and walls. This problem is difficult to control. On a rough, porous concrete surface, the moisture will sink in and not cause a wetness problem. On a dense, smoothly finished floor, however, the tightly knit grains of concrete form a vapor retarder of sorts, and the water collects on the slab. Using dehumidification devices during the summer months can usually solve this problem.

Moisture seeping through the concrete floor is a different problem. In new construction, installing perimeter drainage and a vapor retarder under the concrete slab can help control this. When installing polyethylene film as an underslab vapor retarder, be careful not to tear, puncture, or damage the film in any way. Any passageways for moisture will defeat the purpose of the vapor retarder. Prior to pouring the concrete slab, make sure the polyethylene film is placed properly and free of punctures. Keep all construction debris away from the vapor retarder.

To keep moisture from rising up into the basement, place 6" of coarse gravel over the compacted earth to provide drainage to the perimeter drain before pouring the slab. A polyethylene film gets placed on top of the gravel to keep the concrete from penetrating into the gravel and possibly weakening the slab. Very wet areas or areas with a high water table may also require floor drainage in addition to a gravel bed.

4.1.2 Continuous Insulation

Continuous insulation helps control moisture in several ways. It seals off any gaps in the wall assembly, which prevents vapor from entering. This creates an air barrier that makes the structure airtight and helps reduce air leakage, making it easier to control air flow and, by extension, vapor flow. Continuous insulation should also provide an effective thermal barrier that regulates the temperature of the wall. When the air is hot on the outside of a structure, this should prevent condensation from building up inside the structure.

4.2.0 Interior Ventilation

One of the best ways to reduce or eliminate the chances of moisture damage in attics or in the space between the rafters and the finished roof is through proper ventilation. Ventilation provides a stream of outside air to remove trapped moisture before it can do any damage. In insulated attics, baffles (blocking strips) keep the insulation material from getting into the vented areas. With the increased use of blown-in insulation in attics, the code in many areas requires baffles.

The amount of ventilation required varies by climate and building codes. Louvers and vents provide ventilation in attics and gable and hip roofs, while eave vents and roof stacks provide ventilation in flat roofs (*Figure 25*).

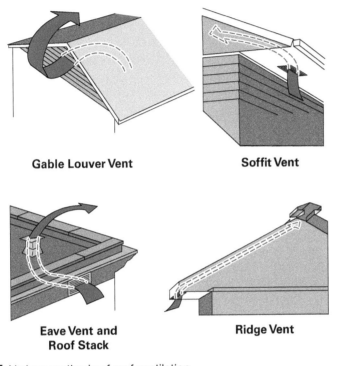

Gable Louver Vent

Soffit Vent

Eave Vent and Roof Stack

Ridge Vent

Figure 25 Various methods of roof ventilation.

Installing plenty of insulation and providing ample ventilation in the attic can help avoid ice dams.

Properly designed subroof ventilation is the best weapon for preventing water vapor infiltration into a steeply sloped roof, but it is less effective on roofs with low slopes because natural **convection** decreases with diminishing roof height. Moisture dissipation occurs through **diffusion** and wind-induced ventilation.

Convection: The movement of heat that either occurs naturally due to temperature differences or is forced by a fan or pump.

Diffusion: The movement, often contrary to gravity, of molecules of gas in all directions, causing them to intermingle.

Did You Know?

Mold

Moisture accumulating inside a building can damage the structure and promote the growth of mold. While it is not always harmful, this mold may cause allergic reactions or other respiratory problems in some people. Airborne mold spores can also cause infections, primarily in people with compromised immune systems.

Normally, the ventilation requirement for a gable roof is 1 ft^2 of free air ventilation for every 300 ft^2 of ceiling area if a vapor retarder exists under the ceiling. Without a vapor retarder, the requirement is 1 ft^2 for every 150 ft^2 of ceiling area. The total requirement must be split evenly between the inlet vents and the outlet vents.

Free air ventilation is the rating of the ventilation devices, taking into account any restrictions caused by screening, louvers, and other devices.

Source: LegART/Getty Images

Did You Know?

Ice Dams

In colder climates, ice dams become a problem. Ice dams form along the edge of a sloping roof when a building's attic does not have proper insulation and ventilation. Heat escaping through the roof melts accumulated snow, forming icicles along the edge of the roof. Over time, water collects under the outer layer of snow and gets trapped by the ice. This water backs up under the shingles and penetrates the roof, causing water damage and other problems.

In areas with a history of water damage to structures from ice dams at roof eaves, an ice barrier is required by code for added protection. The ice barrier consists of self-adhering polymer-modified bitumen material or two layers of cemented underlayment, which must extend from the eave to at least 24" inside the exterior wall line of the building.

Dew point: The temperature at which air becomes oversaturated with moisture and the moisture condenses.

Permeance: The ratio of water vapor flow to the vapor pressure difference between two surfaces.

Vapor retarder class: A measure of a material's ability to limit the amount of moisture that passes through it.

Vapor permeable: Permitting the passage of moisture vapor. A vapor permeable material, as defined by the *IBC®*, has a moisture vapor permeance of 5 perms or greater.

4.2.1 Vapor Retarders

Vapor retarders are an important part of moisture control. A vapor retarder, also known as a *vapor diffusion retarder (VDR)*, includes any material or substance that will not permit the passage of water vapor or will do so only at an extremely slow rate. When vapor reaches its **dew point**, which can occur inside the insulation or at the cool outer surface, it condenses. This results in a reduction or total loss of the thermal efficiency of the insulation, as well as dripping and damage. To prevent this process, select a vapor retarder that is easy to apply and resistant to jobsite abuse as well as compliant with any code requirements. A properly installed vapor retarder will protect ceilings, walls, and floors from moisture originating within a heated space.

As shown in *Table 6*, certain climate zones require different types of vapor retarders. It is important to check all applicable codes to ensure you are installing the right type of vapor retarder. The *IRC®* provides a rating system that separates vapor retarders into different classes based on their **permeance**. This helps clarify which types of vapor retarders work best in which climate zones. There are three classes, known as the material's **vapor retarder class** (see *Table 7*).

Vapor permeance refers to how a material allows water vapor to pass through it. A **vapor permeable** material permits the passage of moisture vapor and has a moisture vapor permeance of 5 perms or greater, as defined by the *IBC®*.

Some vapor retarder materials, such as kraft paper, are attached to blanket or batt insulation and installed with the insulation. Others, such as aluminum foil,

TABLE 6   *IRC®* Vapor Retarder Requirements by Zone

Climate Zone	Marine 4	5	6	7	8
Class I or II[1] vapor retarders required on the interior side of frame walls[2]	Yes	Yes	Yes	Yes	Yes
Class III[1] vapor retarder permitted for:					
Vented cladding over wood structural panels	Yes	Yes			
Vented cladding over fiberboard	Yes	Yes	Yes		
Vented cladding over gypsum	Yes	Yes	Yes		
Continuous insulation with R-value greater or equal to 2.5 over 2 × 4 wall	Yes				
Continuous insulation with R-value greater or equal to 5 over 2 × 4 wall		Yes			
Continuous insulation with R-value greater or equal to 7.5 over 2 × 4 wall			Yes		
Continuous insulation with R-value greater or equal to 10 over 2 × 4 wall				Yes	Yes
Continuous insulation with R-value greater or equal to 3.75 over 2 × 6 wall	Yes				
Continuous insulation with R-value greater or equal to 7.5 over 2 × 6 wall		Yes			
Continuous insulation with R-value greater or equal to 11.25 over 2 × 6 wall			Yes		
Continuous insulation with R-value greater or equal to 15 over 2 × 6 wall				Yes	Yes

[1] Class I: Sheet polyethylene, unperforated aluminum foil; Class II: Kraft-faced fiberglass batts; Class III: Latex or enamel paint
[2] Exception for basement walls, the below-grade portion of any wall, and construction where moisture or freezing will not damage any materials.

TABLE 7 Vapor Retarder Classes

Class	Definition	Examples
I	0.1 perms or less	Sheet polyethylene Nonperforated aluminum foil
II	Greater than 0.1 to less than or equal to 1.0 perms	Kraft-faced fiberglass batts Vapor retarder paint
III	Greater than 1.0 perms to less than or equal to 10 perms	Latex paint Enamel paint

2021 IRC Table R702.7(1) provides a list of materials that can be used for each class of vapor retarder and are deemed to comply with the test standard. No testing is required for these materials. All other materials are required to be tested.

may be applied to the back of gypsum board during its installation. Polyethylene film used as a vapor retarder is applied over studs and ceiling joists after installing the insulation. Vapor retarders are also installed under slabs, between the gravel cushion and the poured concrete.

The **permeability** of a substance measures its capacity to allow the passage of liquids or gases. Water vapor permeability is the property of a substance to permit the passage of water vapor and is equal to the permeance of a substance that is 1" thick.

The measure of water vapor permeability is the **perm**. This equals the number of grains of water vapor passing through a 1 ft² piece of material per hour, per inch of mercury difference in vapor pressure. All you need to remember is that a vapor retarder material has a perm rating of 1.0 or less and will not allow the passage of any appreciable or harmful amounts of water vapor. Any material with a rating higher than 1.0 is a breathable material that will permit the passage of water vapor to whatever degree its perm rating indicates. The higher the perm number, the greater the amount of water vapor that will pass through the material in a given time; 0.0 is totally impermeable.

A properly installed vapor retarder will protect ceilings, walls, and floors from moisture originating within a heated space (*Figure 26*). Note that the air flow in this image does not reflect all climates. This will vary depending on the climate zone.

An insulated wall will divide two temperature gradients. The area on the inside of the structure will normally be warmer than the air on the outside. Installing the vapor retarder on the warm side prevents moisture from moving through the insulation to the cool side and condensing.

Materials

Common vapor retarder materials include asphalted kraft paper, aluminum foil, and polyethylene film.

Asphalted kraft paper is usually incorporated with blanket or batt insulation. It serves as a means for attaching the insulation to the building framework and as a reasonably good vapor retarder when installed on the warm side of the wall or ceiling (*Figure 27*).

Aluminum foil may be incorporated with blanket or batt insulation in the same manner as kraft paper. It is also applied to the back of gypsum lath and gypsum board where it works as a relatively effective vapor retarder.

Polyethylene film is applied over the studs and ceiling joists after installing the insulation. When using gypsum board with polyethylene film or foil backing, the insulation will normally be plain batts or blankets that do not have an integral vapor retarder. As a vapor retarder, polyethylene film is stapled over the studs and covers the window frames. This helps to keep the window frames and sashes clean during application and finishing of the gypsum board. The film should be overlapped 2" to 4" and sealed with special mastic or tape.

Permeability: The measure of a material's capacity to allow the passage of liquids or gases.

Perm: The measure of water vapor permeability. It equals the number of grains of water vapor passing through a 1 ft² piece of material per hour, per inch of mercury difference in vapor pressure.

Smart Vapor Retarders

Naturally changing weather conditions present problems when installing vapor retarders. It is important to have a vapor retarder that is doing the right thing at the right time, depending on the conditions. In the past, it was advised to install a vapor retarder on the "warm side" of the wall, which meant the inside of the wall in colder climates or the outside of the wall in warmer climates. But this method can get confusing in places where sometimes it's warmer outside while other times of the year, it's warmer inside.

One solution is to use a smart vapor retarder. These vapor retarders have a permeance that varies based on humidity. They can achieve high permeance in summer to adapt to the higher humidity and low permeance in winter for low humidity, preventing condensation and moisture.

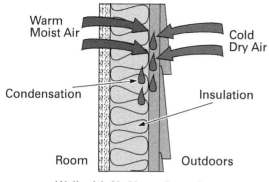

Wall with No Vapor Retarder

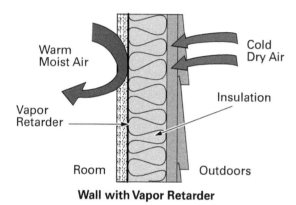

Wall with Vapor Retarder

Figure 26 Vapor retarder installation.

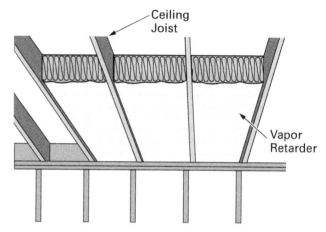

Figure 27 Installing insulation batts between ceiling joists with vapor retarder down.

4.2.2 Installing Vapor Retarders

There are different ways to install vapor retarders depending on the part of the structure.

Installation in Crawl Spaces

The exposed ground under an unventilated crawl space should be covered with a continuous Class I vapor retarder ground cover to protect the underside of the house from condensation (see *Figure 28*). Each joint of the vapor retarder should be taped or sealed and overlap by 6". The edges of the vapor retarder should be attached to the stem wall and extend at least 6" up the wall.

Vapor Retarder Backing

Many of the insulation materials produced today have a vapor retarder applied to the inside surface. Many interior wall surface materials are also backed with vapor retarders. When installed properly, these materials usually provide satisfactory resistance to moisture penetration. If the insulating materials do not include a vapor retarder, then one should be installed as a separate element.

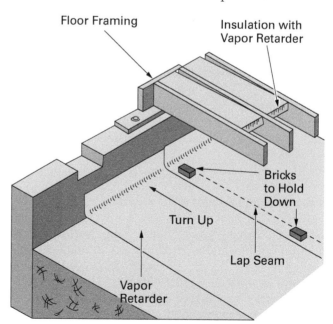

Figure 28 Vapor retarder installation for crawl spaces.

Besides installing vapor retarders, crawl spaces should be properly vented to permit the escape of moisture. Usually, this is accomplished by the use of a proper number of screened foundation vents installed in the above-grade foundation surrounding the crawl space. The normal requirement with a vapor retarder ground cover is 1 ft^2 of free air ventilation for every 150 ft^2 of crawl space area.

Installation in Slabs

When allowed to proceed unchecked, moisture will migrate from the ground upward through concrete and into the building, where it can cause moisture problems, damage, and higher energy costs. Even though the water table may be several feet below the slab, moisture vapor will migrate up to and through concrete slabs. Up to 80% of the moisture entering a structure does so by migrating from the ground beneath the structure. Moisture vapor passes through concrete more readily than liquid moisture.

Moisture in a building can cause deterioration of interior finishes, especially floors and equipment. Moisture can also add to energy costs by raising humidity and taxing cooling systems that require dehumidification.

Place a 6-mil (0.006 inch) approved vapor retarder between the base course and prepared subgrade. (Some codes may require a 10-mil vapor retarder.) The base course is sand, gravel, or crushed stone, concrete, or blast-furnace slag in a sieve on subgrade where the slab is below grade. Vapor retarders should be continuous under the slab. Take great care not to tear or puncture the retarder. Keep all construction debris away from the retarder location. Qualified contractors must install vapor retarders. When used in thickened-edge slab construction, as shown in *Figure 29*, a vapor retarder is placed between the gravel cushion and the poured concrete. The same arrangement is used for other types of slab-on-grade construction.

Figure 30 shows a method of constructing a finished floor over a concrete slab, which affords double protection against moisture. The sealer or waterproofer is placed on the slab itself, with the vapor retarder is suspended above the slab.

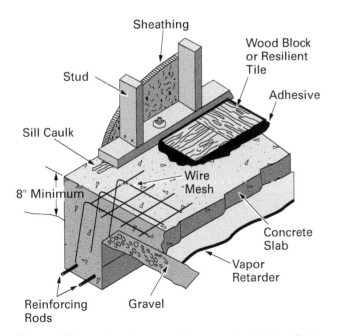

Figure 29 Thickened-edge slab vapor retarder installation.

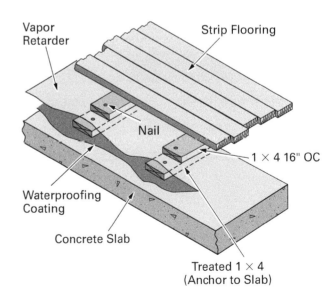

Figure 30 Surface-mounted vapor retarder on a slab.

There are some exceptions. *IECC*® does not require vapor retarders for the following:

- Any unheated accessory structures, such as garages and utility buildings
- Carports and unheated storage rooms of less than 70 ft^2

- Flatwork that is unenclosed and unheated, such as driveways, walks, and patios
- Areas approved by the building official as not needing a vapor retarder

Installation in Walls

A polyethylene sheet vapor retarder is easy to apply to frame walls where no integral barrier is provided or where a supplementary barrier is preferred. A flap would normally overlap both floor and ceiling barriers to seal the interior off completely. Adjacent sheets of the film are overlapped 2" to 4" and sealed with a special mastic or tape.

When applying vapor retarders to walls, pay particular attention to fitting the material around electrical outlet boxes, exhaust fans, light fixtures, registers, and plumbing. Considerable water vapor can escape through the cracks around the equipment, travel from the warm side of the wall to the cold side, and condense on the sheathing or siding. This is especially true if the insulation fits poorly at the top and bottom.

4.0.0 Section Review

1. The area of a house that typically has the most trouble with condensation during the summer is the _____.
   a. kitchen
   b. attic
   c. basement
   d. bathroom

2. The best way to prevent moisture damage in an attic is through _____.
   a. drainage channels on the roof deck
   b. moisture-absorbing materials in the insulation
   c. automated electric air dryers
   d. ventilation of the attic space

5.0.0 Water and Air Infiltration Control

Performance Task

3. Install water-resistive barriers and materials.

Objective

Explain how water and air infiltration control are accomplished in the construction of a building.
   a. Describe how water-resistive barriers are used to resist liquid water.
   b. Describe how air barriers are used and installed.
   c. Explain how to protect openings, penetrations, and joints.

Moisture can enter a building in numerous ways. It can penetrate the interior of a building as liquid water. Rain or snow will introduce water to the exterior of a building throughout the year in varying amounts depending on the local climate. Some places may have different waterproofing requirements depending on higher amounts of rain or snow. No matter where a building is, it is important to protect the interior from liquid water.

Outside air can also carry moisture into a building. When air leaks into a building, it can cause changes in temperature that create condensation on building surfaces.

Installers must take additional steps to control moisture by waterproofing a building, integrating water-resistive barriers, and preventing air infiltration.

5.1.0   Water-Resistive Barriers

The *IBC*® requires that exterior walls provide the building with a weather-resistant exterior wall envelope. **Flashing** and water-resistive barriers are essential components of this envelope. Concrete or masonry walls do not require a water-resistant exterior wall envelope.

A **water-resistive barrier** (WRB) prevents the passage of water into a building. It works together with vapor retarders to protect the building from moisture. WRB materials installed behind exterior wall coverings to the studs or sheathing should continuously cover expansion and control joints. When water does penetrate the exterior, the WRB should be ready to resist it from intruding deeper into the wall.

WRBs can be made of sheet materials, fluid-applied materials, and some types of insulating sheathing. Follow the manufacturer's instructions for installation. After applying WRBs in sheets, you should check your work for any holes or tears.

Various types of WRBs include the following:

- *Asphalt felt* — This material (*Figure 31*) is generally made from recycled paper products, such as corrugated paper and sawdust, which are coated with asphalt. It has a high permeance when wet and is able to soak up water—rather than repelling it—and let it dry.

- *Asphalt saturated grade D paper* — Similar to asphalt felt in composition, this WRB is made of asphalt impregnated into kraft paper. It generally is less expensive than asphalt felt but tends to rot if it stays wet. It comes in large rolls that should be applied over gypsum sheathing using 2-ply application. The time it can spend exposed is very limited and it needs to be covered as soon as possible. It should not be used below grade, horizontally, or for roofs.

- *Rigid foam plastic insulating sheathing* — Expanded or extruded polyisocyanurate or polystyrene panels can be applied over exterior framing and must have carefully taped and sealed joints. The material must have been tested as a water barrier.

- *Polyolefin water barriers* — This WRB is made of spun-bonded polyolefin and should be applied over sheathing. It can be used on vertical surfaces only and should not be used below grade.

- *Self-adhered water-resistive membrane* — While most paper WRBs are applied mechanically, self-adhered WRBs are applied with adhesive. This material is made up of rubberized asphalt compound laminated to an engineered film. It comes in large rolls and should be applied over sheathing. It can only be applied when the air and surface temperature is above 40°F and should be covered as soon as possible.

- *Fluid-applied* — This is an acrylic-based material that, when sprayed or rolled across a structure, forms a seamless membrane. It can also double as an air barrier. Fluid-applied has specific temperature requirements and should be applied only when the surface and air temperature is between 40°F and 100°F. It must be protected with cladding within 6 months. It should not be used below grade or for surfaces with standing water and should not be used to span joints in sheathing that are over $\frac{1}{8}$" wide.

In general, WRBs do not also act as air barriers; however, some materials, including liquid-applied WRBs and integrated sheathing systems, can act as both types of barriers.

Integrated sheathing systems use proprietary methods to combine sheathing panels with water-resistive barriers and air barriers. Some also have options to provide thermal barriers and fire-resistance ratings. Integrated sheathing panels range in size from 4' × 8' to 4' × 12', depending on the manufacturer.

Because each system is proprietary, the design of the system varies. While installation is similar to standard sheathing, each system includes its own accessories, such as tape, liquid flash, and corner seal, to create a continuous building envelope.

Flashing: Thin, water-resistant material that prevents water seepage into a building and directs the flow of moisture in walls.

Water-resistive barrier: One or more materials installed behind exterior wall coverings to prevent water from entering a building.

Figure 31 Asphalt roofing felt.
Source: Oleksandr Rado/Alamy Images

Popular integrated sheathing systems include the following:

- ZIP System® by Huber Engineered Woods, LLC, *https://www.huberwood.com/zip-system*
- ForceField® Weather Barrier System by Georgia-Pacific, *https://buildgp.com/forcefield/*
- Securock® Exoair® 430 System by USG® & Tremco®, *https://www.usg.com/content/usgcom/en/securock-exoair.html*
- LP® WeatherLogic® Air & Water Barrier by Louisiana-Pacific Corporation, *https://lpcorp.com/products/panels-sheathing/air-water-barrier*
- Tetrashield™ by Centria® (a Nucor® Company), *https://centria.com/products/insulated-metal-panels/insulated-sheathing-panels/tetrashield*

These integrated sheathing systems meet code requirements for water-resistive barriers, air barriers, and where applicable, fire-resistant-rated barriers. Always check the manufacturer's product information, data sheets, and installation instructions before using the product.

ZIP System®

Huber Engineered Woods, LLC, developed ZIP System® sheathing and tape as a streamlined weatherization product with an integrated air barrier and water-resistive barrier. ZIP System® sheathing panels consist of oriented strand board (OSB) with a built-in protective overlay to eliminate the need for building wrap. ZIP System® R-Sheathing panels consist of $^7/_{16}$" OSB laminated with an exterior water-resistive facer and a rigid foam insulation panel bonded to the interior face. During installation, the sheathing panels get sealed with ZIP System® flashing tape, stretch tape, or liquid flash to protect the wall from water intrusion. Panels come in 4' × 8', 4' × 9', 4' × 10', and 4' × 12' sizes and are compatible with both 16" and 24" OC framing and for vertical or horizontal installation.

Source: Huber Engineered Woods, LLC (https://www.huberwood.com/zip-system)

ZIP System® offers the following sheathing panels to meet various construction project needs:

- $^7/_{16}$", $^1/_2$", and $^5/_8$" *Sheathing Panels* — Provide an integrated air, water, and vapor management system
- $^7/_{16}$" *Long Length* — Designed for use in single-story and multistory projects to help reduce blocking and help eliminate horizontal seams
- $^7/_{16}$" *Wind Zone* — Designed with an extra $1^1/_8$" length to span from the sill plate to the top plate and protect against wind uplift
- $^1/_2$" *Long Length* — Designed for use in single-story and multistory projects that require extra strength
- *R-Sheathing* — Provide exterior insulation with integrated moisture, air, and thermal protection

The following table shows specification data for all ZIP System® sheathing panels.

ZIP System® Sheathing Panel Data

Panel	Vapor Transmission of WRB Layer	Air Barrier	R-Value
$^7/_{16}$", $^1/_2$", and $^5/_8$" Sheathing Panels	12 Perm ASTM E2178 Procedure B	ASTM E 2357 <0.037 L/(s*m²) @ 75 Pa for Infiltration ASTM E 2357 <0.012 L/(s*m²) @ 75 Pa for Exfiltration	N/A
$^7/_{16}$" Long Length	12–16 Perm ASTM E2178 Procedure B	ASTM E2178 <0.02 L/(s*m²) @ 75 pa ASTM E2357 <0.02 L/(s*m²) @ 75 pa	N/A
$^7/_{16}$" Wind Zone	12–16 Perm ASTM E2178 Procedure B	ASTM E2178 <0.02 L/(s*m²) @ 75 pa ASTM E2357 <0.02 L/(s*m²) @ 75 pa	N/A

Panel	Vapor Transmission of WRB Layer	Air Barrier	R-Value
$\frac{1}{2}$" Long Length	12–16 Perm ASTM E2178 Procedure B	ASTM E2178 <0.02 L/(s*m^2) @ 75 pa ASTM E2357 <0.02 L/(s*m^2) @ 75 pa	N/A
1" R-sheathing	12–16 Perm ASTM E2178 Procedure B	ASTM E2178 –0.0016 L/(s*m^2) (Air Barrier Material) ASTM E2357 0.037 L/(s*m^2) (Air Barrier Assembly)	3.6
1$\frac{1}{2}$" R-sheathing	12–16 Perm ASTM E2178 Procedure B	ASTM E2178 –0.0016 L/(s*m^2) (Air Barrier Material) ASTM E2357 0.037 L/(s*m^2) (Air Barrier Assembly)	6.6
2" R-sheathing	12–16 Perm ASTM E2178 Procedure B	ASTM E2178 –0.0016 L/(s*m^2) (Air Barrier Material) ASTM E2357 0.037 L/(s*m^2) (Air Barrier Assembly)	9.6
2$\frac{1}{2}$" R-sheathing	12–16 Perm ASTM E2178 Procedure B	ASTM E2178 –0.0016 L/(s*m^2) (Air Barrier Material) ASTM E2357 0.037 L/(s*m^2) (Air Barrier Assembly)	12.6

Source: Huber Engineered Woods, LLC, *https://www.huberwood.com/zip-system*

Additional ZIP System® product information, data sheets, and installation instructions are available from Huber Engineered Woods, LLC, *https://www.huberwood.com/zip-system.*

5.2.0   Air Infiltration Control

In addition to insulation, covering the exterior sheathing of a structure prevents wind pressure from causing infiltration of outside air into the structure. To achieve maximum energy efficiency in a structure, air infiltration must be strictly controlled. Installing air barriers in the interior of a structure can also help prevent air leakage.

5.2.1 Air Barriers

An **air barrier** in the wall assembly protects the structure from air leaks.

If a wall assembly has gaps, air can leak out of or into a structure. This can cause heat loss and temperature changes, which can lead to condensation and moisture damage. It can allow moisture to enter a structure in the air from outside. It can also allow rodents and insects to enter the structure. To keep the structure airtight, multiple air barriers can work together to seal the building. When a structure's air barrier works effectively, it reduces heat loss and energy consumption while preventing any potential moisture damage that could result from air leaks.

Air barrier: One or more materials joined together continuously to prevent or restrict the passage of air through a building's thermal envelope and assemblies.

An air barrier consists of one or more materials joined together continuously to prevent or restrict the passage of air through a building's thermal envelope and assemblies. It is often a good idea to have multiple air barriers to provide as much protection as possible. The *IRC*® requires installing a continuous air barrier in the building envelope and exterior thermal envelope. Carpenters should ensure the insulation aligns with the air barrier and seal any joints or breaks in the air barrier.

Air barriers should be:

- Continuous (no gaps)
- Impermeable to air leakage
- Durable (should last as long as the building)
- Repairable
- Rigid (supported from wind pressure)

Structures with one of the following systems do not require air barriers:

- Concrete masonry walls coated with block filler and two applications of a sealer or coating
- Portland cement/sand parge, plaster, or stucco that is at least $\frac{1}{2}$" thick

5.2.2 Interior Air Barriers

Gypsum board can serve as an effective interior air barrier when sealed continuously throughout a building. It must be sealed in the following locations:

- Around the rough openings of doors and windows
- To the first stud in the wall
- Along the top and bottom plates on exterior walls
- At the top plate of partitions when adjacent to an unconditioned space

Spray foam insulation, as well as other forms of air impermeable insulation, can also function as an air barrier.

Air barriers generally need to meet certain air permeance requirements that may vary by region. Make sure to check all applicable codes when installing air barriers.

5.2.3 Exterior Air Barriers

Different types of materials are used to protect different types of residential or commercial buildings. Some materials can act as both air barriers and WRBs.

Wraps

House wraps or building wraps are most commonly used to prevent water and air leakage in residential buildings (*Figure 32*). When properly applied and sealed, these wraps provide a nearly airtight structure regardless of the sheathing material used. Most versions of these wraps provide an excellent secondary barrier under all siding, including stucco and EIFS.

Building wrap can also be applied to commercial buildings (*Figure 33*).

Nails with large heads, nails or screws with plastic washers (*Figure 34A*), or 1" wide staples may be used to secure the wrap to wood, plastic, insulating board, or exterior gypsum board. Steel construction requires screws and washers. Special contractor's tape (*Figure 34B*) or sealants compatible with the wrap seal the edges and joints of the wrap.

WARNING!

Some building wraps are slippery and should not be used in any application where they can be walked on. Because the surface will be slippery, use pump jacks or scaffolding for exterior work above the lower floor. Take extra precautions when using ladders to prevent sliding on the wrap.

Figure 32 Residential building wrap.
Source: B Christopher/Alamy Images

Figure 33 Commercial building wrap.
Source: Peter Titmuss/Alamy Images

(A) Nails with Plastic Washers

(B) Contractor's Tape

Figure 34 Building wrap accessories.
Sources: Jonny White/Alamy Images (A); Alysson M/Shutterstock (B)

Always refer to the manufacturer's instructions for specific installation information. House or building wrap is generally installed as follows:

Step 1   Using two people and beginning at a corner on one side of the structure, leave 6" to 12" of the wrap extended beyond the corner to be used as an overlap on the adjacent side of the structure. Align the roll vertically and unroll it for a short distance. Check that the stud marks on the wrap align with the studs of the structure. Also check that the bottom edge of the wrap extends over and runs along the line of the foundation. Secure the wrap to the corner at 12" to 18" intervals.

Step 2   Continue around the structure, covering all openings. If a new roll is started, overlap the end of the previous roll 6" to 12" to align the stud marks of the new roll with the studs of the structure. Repeat these steps for the upper parts of the structure, making sure the top and bottom layers of wrap overlap by 6" to 12".

Step 3   At the top plate, make sure the wrap covers both the lower and upper (double) top plate, but leave the flap loose for the time being.

Step 4   At each opening, cut back the wrap using one of these methods:

- **Method 1**: Fold the three flaps around the sides and bottom of the opening and secure every 6". Trim off the excess. At the outside, install 6" flashing along the bottom of the opening, then up the sides over the top of the wrap. Install head flashing at the top of the opening under the wrap and over the side flashing. Tape the flap ends to the head flashing.
- **Method 2** (for windows and doors with flanges): Create a top flap of the wrap. Insert a head flashing under the flap and over the flange. Extend the flashing to the sides about 4" and tape the flap to the head flashing. On the remaining sides, trim the wrap to overlap the flange area and tape the edge to the flanges.

Step 5   Secure all the bottom edges of the wrap to the foundation with the recommended joint sealer, then fasten the lower edge to the sill. At the top plate, seal the edge to the upper plate with the sealer and fasten the edge to the plate.

Step 6   Seal all vertical and horizontal joints in the wrap with the recommended tape.

Liquid Applied Roofing

Liquid (also known as *fluid*) applied roofing comes in a liquid, acrylic-based form that cures quickly and becomes waterproof soon after installation, creating a seamless membrane (*Figure 35*). Using a sprayer, paintbrush, or roller, it is easily applied to small cracks and crevices on the surface of a building, which can be harder to fill with caulk or tape. It can also be applied to most types of building materials. Liquid applied tends to be more expensive than other products, so it can be used in combination with other waterproofing materials, particularly for those harder-to-fill places.

Rigid Foam Sheathing

Rigid foam sheathing is a form of exterior continuous insulation that also acts as an air barrier. The material is rigid plastic foam usually sold in boards of 4' × 8' or 4' × 10' that are often between 1" and 2" thick. The primary types of

Figure 35 Applying liquid applied roofing.
Source: Avalon/Construction Photography/Alamy Images

rigid insulation are expanded polystyrene (EPS), extruded polystyrene (XPS), and polyisocyanurate (polyiso). Foam sheathing is attached to the exterior side of the framing. To act as an effective air barrier, the seams between foam boards must be carefully taped with durable flashing and sheathing tapes.

5.3.0 Installing Air Barriers and WRBs over Openings, Penetrations, and Joints

Installing any form of air or water-restrictive barrier over flat walls is a fairly simple process. When you need to pass over a window or door, joint, or other penetration in the structure, it can get a little more complicated. Installers must pay close attention to these areas to ensure seamless barriers around them. Installers will also have to use different techniques to install barriers in these places.

When installing the air barrier and WRB around the structure initially, you should work around penetrations and joints or allow space for them so you can seal them later. For instance, when you put wrap over a wall, you'll need to cut out spaces for pipes or wires coming through the wall so that the wrap can go over a flat surface. Penetrations, joints, and openings should be sealed separately.

5.3.1 Window and Door Treatment

Pay close attention to windows and doors when installing any type of exterior air barrier around a structure. There are specific requirements for sealing windows and doors. If not done correctly, these penetrations can easily cause air leaks.

In general, the air barrier should be sealed to door and window frames using flexible flashing material. Flashing is a piece of thin, water-resistant material that prevents water seepage into a building and directs the flow of moisture in walls. It helps repel rainwater and keeps the exterior of the building dry.

The *IRC*® requires the application of corrosion-resistant flashing around the exterior of a building. Installers can use either approved self-adhered membranes or fluid-applied membranes as flashing. It should be applied in each of the following places:

- At roof and chimney intersections
- Around doors and windows
- Continuously above projecting wood trim
- At the intersections between walls and floors and exterior porches, stairs, and decks
- At built-in gutters
- At the top of foundation walls
- Behind the lowest course of brick veneer

Installers should follow any specific instructions for flashing provided by the door and window manufacturers. If instructions are not provided, you should install pan flashing at the sill of door and window openings. To allow drainage, seal or slope pan flashing to direct water to the water-resistive barrier or surface of the exterior wall finish. The head and sides of any window and door openings also need flashing or protection. A registered flashing design professional should oversee this process.

5.3.2 Joint Treatment

Joints have two purposes in a building. A non-movement joint connects one material to another. An example is two pieces of steel welded together to make a steel frame.

Movement joint: A type of joint that allows a building to move or relieve movement when weather or temperature changes cause any type of structural movement.

A **movement joint** allows a building to move or relieve movement when weather or temperature changes cause any type of structural movement. This includes the expansion of masonry and concrete over time. These joints should accommodate vertical and horizontal movement. They must maintain integrity during movement yet remain permanently waterproof and airtight. Therefore, it is important to seal any joint in an air or water-restrictive barrier with the proper joint treatment system to avoid problems with moisture penetration at joints.

An example of a movement joint is an expansion joint. These joints allow the building materials to expand and contract without distress. Expansion joints installed at intervals along a concrete structure (*Figure 36*) allow the concrete to expand over time. The joint will contract slowly, giving the concrete more space to move, preventing cracking, and increasing its longevity.

Figure 36 Worker pointing to an expansion joint in concrete.
Source: Sumith Nunkham/Getty Images

Section R703 of the *IRC*® indicates what type of joint treatment to apply for different types of siding material. In general, WRBs should be applied horizontally and overlap at least 6" where joints occur. Horizontal joints in panel siding must be lapped by at least 1" or shiplapped or flashed with Z-flashing over wood panel sheathing. Joints can be sealed with the following materials:

- *Caulk* — Caulk is generally better suited for filling smaller gaps and is typically applied from a tube.
- *Tape* — Approved types of joint bridging tapes can be very effective for filling joints. Strips of tape should be cut to the appropriate width.
- *Foam seals* — Foam seals should generally be used for joint widths with a maximum of 8". Open-cell foams are better suited for vertical applications because they allow flow-through of moisture. Closed-cell foams are watertight and are better suited for horizontal applications.

The sealant must be able to insulate the joint while allowing it to move over time up to its full range of movement. That means it should be flexible. Some sealants may work better for certain climates. Evaluate the project needs and code requirements before selecting and applying a sealant. Follow specific manufacturer requirements to fill and seal joints.

5.0.0 Section Review

1. Which type of water-resistive barrier should be applied only when the surface and air temperature is between 40°F and 100°F?
   a. Asphalt felt
   b. Foam plastic insulating sheathing
   c. Polyolefin water barriers
   d. Fluid-applied

2. Which of the following can serve as an effective interior air barrier?
   a. Gypsum board
   b. Building wraps
   c. Liquid applied
   d. Rigid foam sheathing

3. The *IRC*® required the application of corrosion-resistant flashing around which component of the exterior of a building?
   a. At the bottom of foundation walls
   b. Continuously around brick veneers
   c. Continuously on all sides of projecting wood trim
   d. At roof and chimney intersections

Module 27203 Review Questions

1. What is the main purpose of insulating a structure?
   a. To control the movement of heat through a wall
   b. To reinforce the structural members with additional stability
   c. To make a structure as soundproof as possible
   d. To finish the interior walls of the structure

2. The two overall categories of insulation are _____.
   a. continuous and rigid
   b. rigid and foam
   c. flexible and loose-fill
   d. cavity and continuous

3. Heat will always flow (or conduct) through any material or gas from _____.
   a. the interior to the exterior
   b. the exterior to the interior
   c. a higher temperature area to a lower temperature area
   d. a lower temperature area to a higher temperature area

4. Which type of insulation is specifically required by the *IECC*® to ensure sufficient insulation?
   a. Fiberglass insulation
   b. Reflective insulation
   c. Continuous insulation
   d. Lightweight aggregates

5. What is one place where insulation should be installed in a building?
   a. Garages
   b. Roofs
   c. In floors above the 1st floor
   d. In interior walls

6. You are installing insulation for a building in Juneau County, Alaska. In the *IECC*®, it is marked as 7 Juneau. What does that indicate?
   a. It is in Climate Zone 7 and moisture regime is irrelevant.
   b. It is in Climate Zone 7 and is dry.
   c. It is in Climate Zone 7 and is moist.
   d. It is in Climate Zone 7 and is marine.

7. Wood or metal studs are considered to be thermal bridges because they _____.
   a. create considerable breaks in the insulation
   b. channel cold air out of the building
   c. prevent expansion and contraction of building materials
   d. do not support continuous insulation

8. Which materials are used to make vegetable (natural) insulation?
   a. Corn stalks and cotton
   b. Glass and cotton
   c. Straw and perlite
   d. Foil and wood

9. Which material can be made into foam insulation?
   a. Slag
   b. Cork
   c. Polyurethane
   d. Copper

10. Which of the following is the best insulator?
    a. 4" thick concrete block wall
    b. Plastic film
    c. 3.5" thick fiberglass batt
    d. Building paper

11. What can the vapor retarder on kraft-faced insulation be made of?
    a. Slag, phenolic, or copper
    b. Fiberglass or plastic
    c. Cork, tin plate, or paper
    d. Paper, vinyl, or foil

12. How should kraft-faced insulation be placed to prevent moisture from building up in the insulation?
    a. The vapor retarder face should be on the inside of the insulation stack.
    b. The vapor retarder face should be on the outside of the insulation stack.
    c. It should be on the center of a stack with a waterproof membrane on either side.
    d. It should be between the sheathing and exterior air barrier with the vapor retarder facing out.

13. Because reflective insulation reflects light, it can be used to protect against _____.
    a. heat in areas that get a lot of sunlight
    b. cold in areas with severe temperature drops
    c. moisture in areas with a lot of rain
    d. cold in areas with little sunlight

14. Sprayed-in-place insulation must be applied _____.
    a. only when the temperature is above 70°F
    b. by trained and certified contractors
    c. on top of loose-fill insulation
    d. by the contractor who installs the framing studs

15. Which type of insulation can be made of sintered clay products or cinders?
    a. Foamed-in-place insulation
    b. Lightweight aggregates
    c. Rigid expanded polystyrene
    d. Mineral wool insulation

16. Mineral wool board is made of _____.
    a. plastic
    b. polystyrene
    c. phenol-based synthetic polymer
    d. bonded rock fibers

17. What should be checked before installation to determine required R-values and the types of insulation required or permitted?
    a. Building plans and applicable codes
    b. The architect's plans
    c. Building blueprints
    d. The job summary analysis

18. Flexible insulation is used in _____.
    a. unfinished floors
    b. windows
    c. foundations
    d. porches

19. What type of insulation can you use to fill the narrow spaces around windows and doors?
    a. Rigid insulation
    b. Loose-fill insulation
    c. Foam insulation
    d. Reflective insulation

20. What is an acceptable height for the numbers labelling markers in loose-fill insulation?
    a. $\frac{1}{4}$"
    b. $\frac{1}{2}$"
    c. $\frac{3}{4}$"
    d. 1"

21. For loose-fill insulation, use a _____ to tamp the insulation and level it to the required depth for the R-value desired.
    a. straight edge
    b. marker
    c. shovel
    d. baffle

22. When installing loose-fill insulation, ensure that a separate _____ has been installed to prevent moisture and dust penetration.
    a. soffit baffle
    b. sheathing
    c. vent
    d. vapor retarder

23. Insulation below grade is required in _____.
    a. Climate Zones 1–3
    b. Climate Zones 4–8
    c. Climate Zones 2–6
    d. Climate Zones 3–5

24. Foam insulation can be applied to the exterior sheathing of a structure to create _____.
    a. continuous insulation
    b. thermal resistance
    c. water vapor
    d. diffusion

25. What does condensed moisture do to insulation?
    a. It enhances the protection of insulation.
    b. It reduces the efficiency of insulation.
    c. It prevents insulation from freezing.
    d. It helps insulation reduce the building's heat loss.

26. Continuous insulation helps with moisture control by _____.
    a. blocking cold air from the building
    b. allowing hot air into the building
    c. sealing the gaps in the wall assembly
    d. allowing for movement and expansion of building materials

27. When a vapor retarder is used under the ceiling, proper free air ventilation for a gable roof is defined as 1 ft^2 for every _____ ft^2 of attic area.
    a. 150
    b. 160
    c. 300
    d. 320

28. When vapor reaches its _____, it condenses.
    a. permeability
    b. dew point
    c. convection
    d. U-factor

29. Class III vapor retarder is permitted for vented cladding over wood structural panels in which of the following climate zones?
    a. Zones 1 and 2
    b. Zones 3 and 4
    c. Marine 4 and Zone 5
    d. Zones 5 and 7

30. When installing a vapor retarder in a crawl space, each joint should be taped or sealed, and overlap by _____.
    a. 4"
    b. 6"
    c. 7"
    d. 8"

31. Fluid-applied water-resistive barriers should be applied only when the surface and air temperatures are between _____.
    a. 60°F and 80°F
    b. 30°F and 70°F
    c. 70°F and 100°F
    d. 40°F and 100°F

32. The vertical seams of building wrap are usually overlapped _____ at each corner of the building.
    a. 2" to 6"
    b. 3" to 4"
    c. 6" to 12"
    d. 12" to 18"

33. What does an expansion joint do?
    a. It connects exterior building material to interior building material.
    b. It allows building materials to expand and contract without distress.
    c. It prevents the expansion or movement of building materials.
    d. It controls cracking in masonry caused by material shrinkage.

34. In general, water-restrictive barriers should overlap by at least how many inches where joints occur?
    a. 2"
    b. 4"
    c. 6"
    d. 8"

35. Which of the following statements is true about the material used to seal a joint?
    a. It should be flexible.
    b. It should be rigid.
    c. It should be air permeable.
    d. It should be reflective.

Answers to Odd-Numbered Module Review Questions are found in *Appendix A*.

Answers to Section Review Questions

Answer	Section	Objective
Section One		
1. b	1.1.1	1a
2. d	1.2.1	1b
Section Two		
1. a	2.1.1	2a
2. b	2.2.0	2b
3. b	2.3.0	2c
4. d	2.4.2	2d
Section Three		
1. b	3.1.1	3a
2. a	3.2.1	3b
3. a	3.3.0	3c
4. c	3.4.1	3d
Section Four		
1. c	4.1.1	4a
2. d	4.2.0	4b
Section Five		
1. d	5.1.0	5a
2. a	5.2.2	5b
3. d	5.3.1	5c

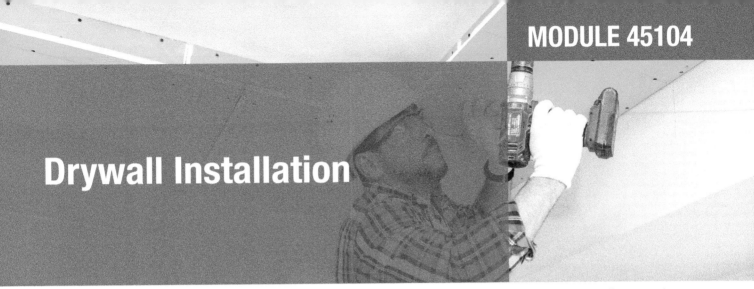

Drywall Installation

Objectives

Successful completion of this module prepares you to do the following:

1. Identify and describe the materials and tools required to install drywall.
   a. List characteristics of gypsum board and explain why it is used.
   b. Describe types of drywall fasteners and explain how they are used.
   c. Identify and describe types of adhesives and explain how they are used for drywall work.
   d. Identify and describe tools used to install drywall.
2. Explain the procedures for installing gypsum board on different types of ceilings and walls.
   a. Describe the procedure for single- and multi-ply installations of gypsum board on ceilings and walls.
   b. Identify and describe the types of drywall trims.
   c. Explain how moisture is controlled to prevent its effects on walls and ceilings.
   d. Explain guidelines for fire-resistance-rated and sound-rated walls.

Performance Tasks

Under supervision, you should be able to do the following:

1. Install gypsum drywall panels on wood joist ceilings and wood stud walls using the following:
   - Nails
   - Screws
2. Install gypsum drywall panels on steel grid ceilings and steel stud walls using screws.

Overview

Gypsum board is the most common drywall finish used inside residential and commercial construction. A variety of gypsum materials are used for these applications, and a number of construction methods are used to build standard ceilings and walls, as well as assemblies that meet building codes for fire resistance and sound control. Gypsum board drywall installation requires the knowledge and use of specialized tools, fasteners, and construction methods dictated by building codes, and therefore must be carefully considered.

Digital Resources for Carpentry

SCAN ME

Scan this code using the camera on your phone or mobile device to view the digital resources related to this craft.

1.0.0 Materials and Tools

Performance Tasks

There are no Performance Tasks in this section.

Objective

Identify and describe the materials and tools required to install drywall.

a. List characteristics of gypsum board and explain why it is used.
b. Describe types of drywall fasteners and explain how they are used.
c. Identify and describe types of adhesives and explain how they are used for drywall work.
d. Identify and describe tools used to install drywall.

Watching experienced workers install gypsum board walls inside a building, you might think the job is easy. They work in pairs, lifting a long panel and holding it firmly against the studs while one of them places the nose of the screw gun against the board and pulls the trigger. They move quickly and easily, covering the studs with perfectly straight, flat rows of panels.

In reality, expert installation of gypsum board takes a good deal of knowledge and practice. Gypsum boards are heavy, and it is not as easy as it looks to install them straight and tight against the studs and ceiling joists and to cut out around electrical outlets. Additionally, each construction project requires you to take the time to learn and understand the specifications for that project and how to meet them, as well as the building codes for the area in which the construction is taking place.

You are prepared to install gypsum board when you understand the following:

- The characteristics of gypsum board
- The fasteners, tools, and materials used to install gypsum board
- How to work safely with gypsum board
- The methods for applying gypsum board to different surfaces
- How to build fire-resistance-rated and sound-rated assemblies with gypsum board

1.1.0 Gypsum Board

Gypsum board: A generic term for paper-covered panels with a gypsum core; also known as *gypsum drywall*.

Gypsum board, also known as *gypsum drywall* when used inside buildings, is one of the most popular and economical methods of finishing the interior walls and ceilings of wood-framed and steel-framed buildings. Gypsum drywall gives a wall or ceiling made from many panels the appearance of being made from one continuous sheet.

Gypsum board is a generic name for products made with the mineral gypsum that feature a noncombustible core and a paper covering on the face, back, and long edges of each board. A typical residential (horizontal) board application is shown in *Figure 1* and a typical commercial (vertical) board application is shown in *Figure 2*.

Figure 1 Typical residential (horizontal) board application.
Source: Rachid Jalayanadeja/Shutterstock

Figure 2 Typical commercial (vertical) board application.
Source: ungvar/Shutterstock

The main difference between gypsum board and other sheet products such as plywood, hardboard, and fiberboard is its natural ability to resist fire because of its gypsum content. Gypsum is a mineral, calcium sulphate dihydrate, found in sedimentary rock formations. Human-made synthetic gypsum is also used in gypsum board production, mixed with gypsum rock mined from the earth. Gypsum is naturally noncombustible because it contains water in its crystal molecular structure.

Gypsum drywall is generally available in widths of 4' and lengths of 8', 10', 12', and 14'. Other lengths are available by special order. The 4' side of a gypsum board, the side that is not covered in paper, is called the **end** of the board. The longer side, which is covered with paper, is called the **edge** of the board. The ends of gypsum panels are square, but panels are manufactured for different purposes with a variety of different edges, including rounded, tapered, beveled, and square (*Figure 3*).

End: The side of a gypsum board perpendicular to the paper-bound edge. The gypsum core is always exposed.

Edge: The paper-bound edge of a gypsum board as manufactured.

Calcination: The process of heating gypsum rock enough to evaporate most of the water in its molecular structure, causing a chemical change in the material.

The Way it Was

Until the 1930s, walls were typically finished by installing thin, narrow strips of wood or metal known as *lath* between studs and then coating the lath with wet gypsum plaster. Skilled plasterers could produce a smooth wall finish, but the process was time-consuming and messy. Paper-bound gypsum board was introduced in the early 1900s. After World War II, it came into widespread use as a replacement for the tedious lath-and-plaster process.

1.1.1 Advantages of Gypsum Board

Gypsum board walls and ceilings have the following outstanding advantages:

- *Fire resistance* — Gypsum board is an excellent fire-resistive material because the gypsum in its noncombustible core contains molecules of water. Under high heat, this water is slowly released as steam, which retards the flow of heat through the assembly. Even after complete **calcination**, when all the water has been released from the gypsum in the board, gypsum drywall continues to provide a barrier to the spread of flames. Tests conducted in accordance with the *ASTM Test Standard E84* show that gypsum drywall resists flame spread and creates little smoke during a fire.

- *Sound attenuation* — A key consideration in the design of a building is how to control unwanted sound transmission between and within rooms. Standard gypsum board applied to walls and ceilings, because of its inherent mass and elasticity, effectively helps to control sound transmission between and within rooms to a certain extent. But wall and ceiling assemblies that use standard

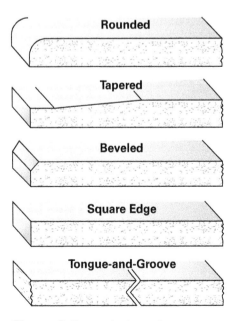

Figure 3 Types of edges of gypsum board.

gypsum board are non-rated for sound control. Enhanced gypsum board and specialized insulation are available for use in sound-rated assemblies.

- *Durability* — Walls and ceilings made from gypsum board are strong and have excellent dimensional stability. The wide range of gypsum products includes abuse-resistant and impact-resistant panels for high-traffic or high-contact settings. There are also gypsum board options that withstand damp conditions and resist moisture and mold.

- *Economy* — Gypsum board products are inexpensive to manufacture and relatively easy to apply, and so they may be installed at relatively low cost. They are the least expensive choice among wall surfacing materials that offer a fire-resistant interior finish.

- *Versatility* — Gypsum board products satisfy a wide range of architectural requirements. They are readily available and easy to apply, repair, decorate, and refinish. This combination of traits is unmatched by any other wall surfacing product.

- *Sustainability* — Gypsum board is considered a sustainable material because, in addition to naturally mined gypsum, it uses flue gas desulfurization (FGD) gypsum, a synthetic byproduct of coal-burning power plants. Additionally, gypsum board manufacturers have for decades used recycled **face paper** on their gypsum board products.

Face paper: The paper bonded to the surface of gypsum board during the manufacturing process.

Going Green

Recycling Gypsum Board Waste

The gypsum board waste that results from construction projects can be used for many purposes, including as a filler in plastics and cement, and as a soil conditioner that has been shown to reduce the levels of toxic soluble reactive phosphorus in waterways plagued by fertilizer runoff. With the proper separation of waste materials on the worksite—a practice that is growing in popularity—gypsum board without any fasteners or tape can be reclaimed by manufacturers to make new gypsum board. Leading gypsum board brands are now offering panels containing some recycled material. Companies such as USA Gypsum partner with building contractors to recycle gypsum board waste. Since 1998, USA Gypsum has recycled millions of pounds of gypsum board.

Gypsum Drywall—A Versatile Finish Material

It is common to think of gypsum drywall as a finish for flat walls and ceilings. These photographs show how it can be used in much more complex designs. Manufacturers have developed highly flexible gypsum boards, called Hi-Flex, to facilitate applications requiring complex designs.

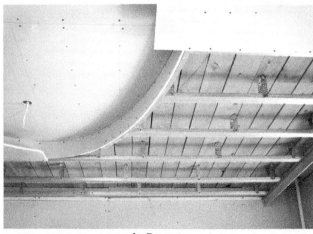

In Process

Finished

Sources: Sever180/Shutterstock (left); Jirawatfoto/Shutterstock (right)

1.2.0 Drywall Fasteners

Nails and screws are commonly used to attach gypsum board to framing members in both **single-ply** (one-layer) **construction** and **multi-ply** (multiple-layer) **construction**. Special drywall adhesives can be used to secure single-ply gypsum board to framing, furring, masonry, and concrete, or to laminate a face ply to a base layer of gypsum board or other base material. Adhesives must be supplemented with mechanical fasteners, such as nails or screws.

Installing gypsum board requires that you understand how to **countersink** drywall nails and screws. Countersinking means driving fasteners into the board until the head of the fastener rests just below the surface of the face paper on the board, without breaking through it. The face paper on a gypsum board, bonded as it is to the gypsum core, is a vital part of the structure of the board, so take particular care to never break it when driving a nail or screw. A fastener that has broken through the face paper is not holding the board properly to the framing member. The gypsum core around the nail or screw is now likely to crumble, causing the board to loosen or even detach from the framing member. If you do break the face paper on a board while driving a nail or screw, drive another fastener properly near the same place.

When you countersink a nail or screw correctly, with the head seated just below the surface of the panel, you create a uniform depression (see *Figure 4*). This depression leaves room to apply **joint compound**, also called *mud*, during the **finishing** process, to mask the presence of the nail or screw and create a smooth surface.

All gypsum board fasteners must be used with the correct safety shield, tool guard, or attachment recommended by the manufacturer.

1.2.1 Nails

Both **annular nails** and **cupped-head nails** are acceptable for applying gypsum board to wood framing or wood furring strips (see *Figure 5*). Annular nails have rings around the shank, the long part of the nail, which is why they are also called *ring shank nails*. Annular nails grip wood studs and create good **withdrawal resistance**, compared to nails with smooth shanks. Cupped-head nails have concave heads designed to countersink properly into gypsum board. Any nails used to install gypsum board should have thin-rimmed flat or concave heads. The heads should be between $\frac{1}{4}$" and $\frac{5}{16}$" in diameter to provide adequate holding power without cutting the face paper when the countersunk nail creates the proper depression.

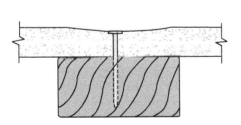

Figure 4 Uniform depression.

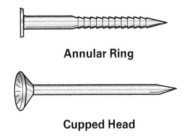

Annular Ring

Cupped Head

Figure 5 Nails.

Nails must be long enough to go through the gypsum board and far enough into the supporting wood construction to provide adequate holding power. The nail penetration into the framing member should be at least $\frac{7}{8}$" for smooth shank nails and $\frac{3}{4}$" for annular nails, which provide more withdrawal resistance and therefore require less penetration. For fire-resistance-rated assemblies, greater penetration is required (generally $1\frac{1}{8}$" to $1\frac{1}{4}$" for one-hour assemblies). Follow the specifications in the fire test report when choosing nails for a fire-resistance-rated assembly.

Drywall nails should be driven into gypsum board with a hammer specifically designed for installing gypsum board, such as the hammer end of a

Single-ply construction: A wall or ceiling installation built with one layer of gypsum board.

Multi-ply construction: A wall or ceiling installation built with more than one layer of gypsum board.

Countersink: To drive a nail or screw through a gypsum board and into the framing member until the head of the fastener rests just below the surface, without breaking the face paper.

Joint compound: A mixture of gypsum, clay, and resin applied wet during the finishing process to the taped joints between gypsum boards to cover the fasteners, and to the corner bead and accessories to create the illusion of a smooth unbroken surface. Sometimes called *mud* or *taping compound*.

Finishing: The application of joint tape, joint compound, corner bead, and primer or sealer onto gypsum board in preparation for the final decoration of the surface.

Annular nails: Nails with rings around the shank, which provide a stronger grip and higher withdrawal resistance than nails with smooth shanks.

Cupped-head nails: Nails with a concave head (shaped like a cup) and a thin rim.

Withdrawal resistance: The amount of resistance of a nail or screw to being pulled out of a material into which it has been driven.

CAUTION

Casing and common nails have heads that are too small in relation to the shank; they easily cut into the face paper and threaten the integrity of the board and its attachment to the stud. Nail heads that are too large are also likely to cut the paper surface if the nail is driven incorrectly at a slight angle.

Tooth: A textured surface created mechanically to help a covering material adhere more effectively.

Thread: The protruding rib of a screw that winds in a helix down its shank.

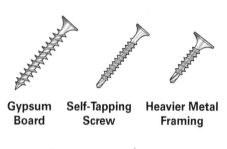

Gypsum Board   Self-Tapping Screw   Heavier Metal Framing

Light-Gauge Metal   Wood

Figure 6 Drywall screws.

Furring strip: A flat, narrow piece of wood attached to wood framing to create a level surface in preparation for installing gypsum board.

Pitch: The number of threads per inch on a screw shank, with more threads producing a finer pitch and fewer threads producing a coarser pitch.

Self-tapping screw: A screw that bores an internal screw thread in the material into which it is driven, creating a strong hold between the material and the screw.

Furring channel: A long, narrow piece of metal bent into the shape of a hat (which is why it is also called a *hat channel*), with two flanges (the brim of the hat) on either side of a channel (the crown of the hat), used to create a level surface on uneven masonry or metal framing in preparation for installing gypsum board.

Self-piercing screw: A screw with a point sharp enough to penetrate steel.

drywall hatchet, or with another hammer with a special wide curved head that forms the required depression around the nail head without breaking the face paper. The head of a drywall hammer has a waffle texture, which transfers to the depression around the nail head. This textured surface, or **tooth**, helps joint compound adhere to the depression during the finishing phase. Particular care should be taken when driving nails into gypsum board not to crush the core by striking it too hard with the hammer. Nails are not an appropriate fastener for applying gypsum board to steel framing.

Special Fasteners

Application of gypsum board requires special fasteners. Common nails and ordinary wood or sheet metal screws are not designed to penetrate the board without damaging it, or to hold the board tightly against the framing, or to permit correct countersinking for proper concealment of the fasteners during finishing.

1.2.2 Screws

Drywall screws (*Figure 6*) are used to attach gypsum board to wood or steel framing or to other gypsum board. They have a pronounced **thread** and a Phillips head design that is intended to be used with a drywall screw gun. The bugle-shaped head of a drywall screw is specially designed to countersink into the gypsum board and pull it tightly against the framing member. When a drywall screw is properly driven into a gypsum board, it creates a uniform depression that is free of ragged edges and fuzz.

When choosing the right screws for attaching gypsum panels to framing members, your main consideration will be what type of material you are attaching them to, as shown in *Table 1*.

TABLE 1 Drywall Screw Types

Threads	Application
Coarse	Drywall to wood
Fine, self-tapping	Drywall to steel: 18 mil (25 gauge) or 27 mil (22 gauge)
Fine, self-drilling	Drywall to steel: 30 mil (20 gauge)
Coarse	Drywall to drywall

The screw threads determine which of the following applications they are best suited for:

- A coarse screw is designed to fasten gypsum board to wood framing or a wood **furring strip**. This type of screw has a coarse **pitch**, which means the threads are relatively far apart. The point is diamond-shaped to pierce both gypsum board and wood, and the coarse threads hold the gypsum panels securely to the wood framing members.
- A fine, **self-tapping screw** is designed to fasten gypsum board to a standard 18 mil (25 gauge) or 27 mil (22 gauge) steel stud or steel **furring channel**. It has fine, close threads, so there are more of them on the shank of the screw. This type of drywall screw is a **self-piercing screw**, which means it is made with a sharp tip designed to easily pierce steel framing members. Easy penetration is important, because steel studs are often flexible and tend to bend away from the screws, and the screws tend to strip easily. These screws are also self-tapping: as they bore into the steel framing member, they tap the steel stud, which means they carve internal screw threads into the steel, creating a tight hold between the screw and the steel framing member similar to that between a nut and a bolt.

- A fine, **self-drilling screw** is recommended when attaching gypsum panels to steel framing that is 30 mil (20 gauge). The tip of this type of screw is shaped like a drill bit for cutting through thicker steel studs with ease. Self-drilling screws are also self-tapping.

- A coarse screw with a deeper, special thread design to create a tight grip between boards is used for fastening gypsum board panels to gypsum backing boards in multi-ply construction. The most common length of screws used for multi-ply construction is $1\frac{1}{2}$", but other lengths are available. However, in two-ply construction, in which the face layer is screw-attached to the base layer, additional holding power is developed in the base layer, which permits a reduced penetration in the face layer of a minimum of $\frac{1}{2}$".

Self-drilling screw: A screw with a point shaped like a drill bit to penetrate steel.

Did You Know?

Steel Framing Member Size

The thickness of the steel framing member determines its use. For drywall applications, the size of the steel framing members also determines the type of fasteners needed.

Thickness, in Millimeters	Reference Gauge No.
18	25
27	22
30	20 Drywall
33	20 Structural
43	18
54	16
68	14
97	12
118	10

Source: Adapted from the Steel Framing Industry Association (SFIA) specifications: https://sfia.memberclicks.net/assets/TechFiles/SFIA%20Tech%20Spec%202015%20updated%207.12.17%20v.2.pdf

The thickness of the gypsum board you use will depend on the building codes and the project specifications. The most common board thicknesses for single-ply interior wall assemblies are $\frac{1}{2}$" and $\frac{5}{8}$". For ceilings, many installers use $\frac{5}{8}$" board exclusively, because other thicknesses tend to sag or ripple over time. The correct screw length is based on the thickness of the gypsum panels and the framing material. For example, the recommended minimum penetration into wood studs for single-ply construction is $\frac{5}{8}$". That means if you are attaching $\frac{5}{8}$" board, you need screws that are at least $1\frac{1}{4}$" long.

Keep in mind that the required fastener penetration changes when constructing fire-resistance-rated assemblies with steel framing, in which case nails or screws must be at least 1" long and may need to be specially treated. Always read the architectural specifications and local building codes for an individual project and follow them carefully.

Selecting Drywall Screws

It is important to use the correct screws for each job. The numbers on the box will indicate the diameter (6, 8, 10, etc.), the length ($1\frac{1}{4}$", $1\frac{5}{8}$", etc.), and the thread (coarse or fine). A #6 screw, which has a relatively small diameter, is appropriate for applying gypsum board to wood and steel framing in single-ply construction and is the most common size screw used by gypsum board installers in the field.

Specialty panels may require specialty fasteners. Follow manufacturers' guidelines for installing specialty products.

1.3.0   Adhesives

Adhesives are used to bond single layers of gypsum board directly to the framing, furring, masonry, or concrete. They can be used to laminate gypsum board to base layers of backer boards, sound deadening boards, rigid foam, and other rigid insulating boards. The adhesive must be used in combination with screws, which provide supplemental support.

When choosing an adhesive, either to apply gypsum board directly to framing members or to apply a face layer to a base layer in a multi-ply assembly,

NOTE

Check the temperature range for the adhesive you plan to use to make sure it is compatible with the expected operating temperatures of the building.

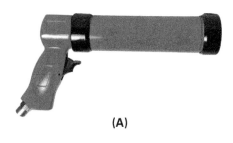

(A)

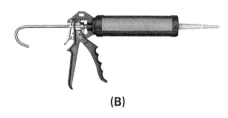

(B)

Figure 7 Adhesive applicators.
Sources: (A) darksoul72/Shutterstock;
(B) donatas/1205/Shutterstock

Bead: An application of adhesive or other construction material in a sphere or line not less than $3/8$" in diameter.

make sure it is recommended by the manufacturer for the specific conditions of your project. There are many all-purpose drywall adhesives available, but some of them may not be effective to hold gypsum board tightly to all types of framing members.

Whichever drywall adhesive you choose, follow all manufacturer directions for use. If the adhesive has a solvent base, follow appropriate precautions to keep the jobsite safe. Do not use solvent-based adhesives near an open flame or in poorly ventilated areas.

The most common type of drywall adhesive is applied with an electric, pneumatic, or hand-operated caulking gun (*Figure 7*) in a continuous or semicontinuous **bead**.

Drywall contact adhesive is a type of adhesive that can be used to apply a face layer of gypsum board to a base layer in multi-ply installations. It can also be used to apply gypsum board to steel studs. Contact adhesive is applied by roller, spray gun, or brush in a thin, uniform coating to both surfaces to be bonded. For most contact adhesives, some drying time is usually required before surfaces can be joined and the bond can be developed.

WARNING!

Observe all safety data sheet (SDS) precautions for adhesives. Extreme caution must be taken when using contact cement because it is highly flammable. It also produces toxic fumes and should be used in a well-ventilated area, as the fumes can quickly overcome a worker. Always wear appropriate PPE and follow manufacturer's instruction when using adhesives.

1.4.0 Tools Used to Install Gypsum Board

The following tools are commonly used to install gypsum board:

Figure 8 T-square.
Source: Photo Win1/Shutterstock

Figure 9 Utility knife.
Source: Andrei Kuzmik/Shutterstock

- *4' T-square* — This tool is indispensable for making accurate cuts across the narrow dimension of gypsum board. See *Figure 8.*
- *Utility knife* — This is the standard knife used for cutting gypsum board. It has replaceable blades stored in the handle. Knife blades need to stay sharp to cut the gypsum boards without needing to apply too much pressure. See *Figure 9.*
- *Rasp* — The rasp (*Figure 10*) is used to quickly and efficiently smooth rough-cut edges of gypsum board. The tool has both a file for finishing and a rasp for rough shaping.
- *Circle cutter* — The circle cutter (*Figure 11*) has a calibrated steel shaft that allows accurate cuts up to 16" in diameter. The cutter wheel and center pin are heat-treated.
- *Drywall saw* — This saw, also known as a *keyhole saw,* is used for cutting small openings and making odd-shaped cuts (*Figure 12*). A power cutout

Figure 10 Drywall rasp.
Source: malaha/Shutterstock

Figure 11 Circle cutter.
Source: Dmitry Naumov/Shutterstock

Figure 12 Drywall saw.
Source: Stocksnapper/Shutterstock

tool, also called a *drywall router,* (*Figure 13*) can be used for the same purpose. The power cutout tool comes with a depth gauge you can set before cutting, thereby ensuring that you cut only through the gypsum board. The depth gauge can help beginners get used to how the power cutout tool works, but more advanced drywall installers often prefer to work without it.

Figure 13 Power cutout tool.
Source: Keith Homan/Shutterstock

Measuring and Marking

When measuring gypsum board, use a soft lead pencil to mark the panels. Marks made by a ballpoint pen or marker may bleed through the joint compound and paint.

- *Gypsum board foot-operated lifter* — A foot-operated gypsum board lifter slides under a gypsum panel being installed just above floor level. When you press down on the lever with your foot, it lifts the panel up and pushes it forward into place so that you can easily attach it to the framing members. The lifter can be used for either parallel or perpendicular board applications.
- *Drywall hatchet* — This tool includes a drywall hammer with a wide symmetrical convex head designed to compress the gypsum panel face and leave the desired depression without breaking the face paper. The blade end can be used to cut boards roughly down to size, but then the ends must be smoothed before the boards are applied to the wall assembly. See *Figure 14.*
- *Screw gun* — Electric drywall screw guns, such as those shown in *Figure 15,* are designed to drive steel screws through gypsum board and into the framing member to a precise depth. The best screw guns for installing drywall on wood or steel studs have powerful brushless (meaning almost frictionless) motors that run at a minimum of 4,000 revolutions per minute (rpm). Screw guns feature an adjustable depth setting, located on the nose cone of the tool, for use with gypsum board of different thicknesses. When the screw being driven into the board reaches the preset depth, a clutch mechanism in the screw gun disengages and stops the screw from being driven any deeper. Manually loading a single-shot screw gun is as easy as placing the head of a drywall screw onto the magnetic bit tip, which will hold the screw in place until you press the screw against the board in the preferred location and pull the trigger. Many screw gun models offer a **collated magazine** that attaches to the nose cone and automatically feeds collated screws—screws arranged side by side on a long plastic strip—through the gun, allowing the installation process to move more quickly and easily.
- *Drywall lift* — This special device is designed to raise and support drywall panels during ceiling or high wall installations.

Figure 14 Drywall hatchet.
Source: Dorling Kindersley ltd/Alamy Stock Photo

Collated magazine: An attachment for a screw or nail gun that automatically feeds collated fasteners (fasteners arranged side by side on a strip of plastic) into the chamber of the gun for quick application.

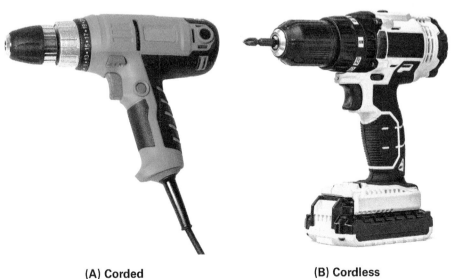
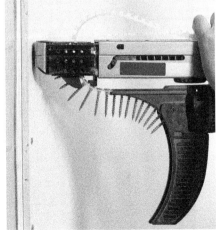

(A) Corded **(B) Cordless** **(C) Collated**

Figure 15 Screw guns.
Sources: (A) krolya25/Shutterstock; (B) Nakornthai/Shutterstock; (C) Ilja Enger-Tsizikov/Alamy Stock Photo

NOTE

Attaching gypsum boards properly to framing members takes attention to detail. For best results, hold the screw gun perpendicular to the work surface. Adequate pressure must be exerted to engage the clutch and prevent the screw from slipping (also known as walking). The gun should be triggered continuously until a fastener is seated or sunk through the gypsum board and sufficiently into the framing member. A one-piece socket makes driving easier and more efficient than a separate socket and extension pieces, because it provides a more rigid base and firmer control.

1.0.0 Section Review

1. Because of its contents, _____ has a natural ability to resist fire.
   a. plywood
   b. hardboard
   c. gypsum board
   d. fiberboard

2. Fine, self-tapping screws are used to apply _____.
   a. drywall to standard 18 mil (25 gauge) or 27 mil (22 gauge) steel
   b. drywall to wood
   c. drywall to 30 mil (20 gauge) steel framing
   d. drywall to drywall

3. The most common type of drywall adhesive is applied with a _____ in a continuous or semicontinuous bead.
   a. caulking gun
   b. sprayer
   c. brush
   d. roller

4. A _____ comes with a depth gauge you can set before cutting to ensure you cut only through the gypsum board.
   a. circle cutter
   b. drywall hatchet
   c. power cutout tool
   d. utility saw

2.0.0 Installation

Performance Tasks

1. Install gypsum drywall panels on wood joist ceilings and wood stud walls using the following:
   - Nails
   - Screws
2. Install gypsum drywall panels on steel grid ceilings and steel stud walls using screws.

Objective

Explain the procedures for installing gypsum board on different types of ceilings and walls.

a. Describe the procedure for single- and multi-ply installations of gypsum board on ceilings and walls.
b. Identify and describe the types of drywall trims.
c. Explain how moisture is controlled to prevent its effects on walls and ceilings.
d. Explain guidelines for fire-resistance-rated and sound-rated walls.

Any successful gypsum board installation project begins with a set of plans that dictate the materials and methods you will use to complete the project. Before you begin the gypsum board installation phase of the project, make sure you understand the project specifications and standards that apply to your portion of the project.

The responsibility for installing and finishing gypsum drywall varies from job to job and from one locale to another. In some situations, carpenters install, tape, and mud the drywall, and painters apply primer and surface **decoration**, such as paint or texturing. In other situations, professional drywall workers complete the installation, finishing, and decoration. The smaller the project, the more likely it is that the carpenter will install, finish, and decorate the drywall. Especially when you are first learning to install gypsum board, work with at least one other person to measure, cut, lift, and secure the panels tightly to the framing members.

Decoration: The application of the final surface covering on gypsum board walls.

2.1.0 Installing Gypsum Board

Gypsum board panels can be applied over any firm, flat base such as wood or steel framing or furring. Gypsum boards can also be applied to masonry and concrete surfaces, either directly or to surfaces that have been furred out with wood furring strips or steel furring channels.

The most common type of interior wall and ceiling assembly is the standard gypsum board system, which consists of a series of panels affixed with drywall nails or screws to wood or steel framing members. Installation of the boards leaves a series of vertical and horizontal **joints** between the gypsum panels. A butt joint is created where the short ends of two sheets of wallboard meet. A flat joint is the intersection of two bevel-edged wallboards. Once the panels have been attached to the framing members, the **outside corners** are reinforced with trim. Then the wall is finished, meaning the nail or screw depressions, the **inside corners**, and the joints between the panels are covered with **joint tape** and joint compound. Joint compound is applied to certain types of drywall trim as well.

Joints: Places where two pieces of wallboard meet.

Outside corners: Locations where two walls meet and face away from each other.

Inside corners: Locations where two walls meet and face each other.

2.1.1 Single-Ply and Multi-Ply Construction

In structures that are not required to be fire-resistance-rated or sound-rated, single-ply gypsum board systems are commonly used (*Figure 16*). Multi-ply systems, as shown in *Figure 17*, have two or more layers of gypsum board to increase sound control and fire-resistive performance.

Joint tape: Wide tape applied to the joints between gypsum boards and then covered with joint compound during the finishing process.

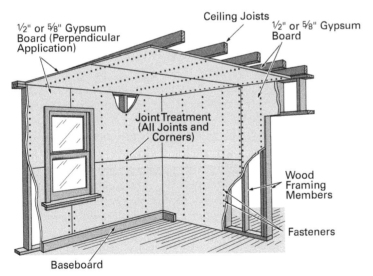

Figure 16 Single-ply construction.

The surface quality of multi-ply assemblies is often smoother than that of single-ply assemblies because face layers are often adhered to base layers, which means that fewer nails or screws are required to hold the face layer firmly in place. Additionally, the base layer of gypsum board reinforces the joints of the face layer, and the face layer masks any imperfectly aligned joints in the base layer. Because the face layer in a multi-ply assembly often requires fewer mechanical fasteners, problems with fastener pop occur less frequently.

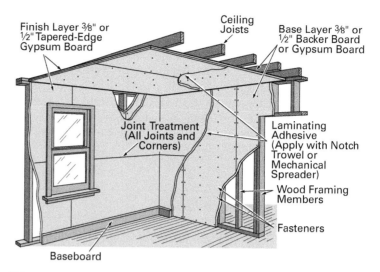

Figure 17 Multi-ply construction.

Satisfactory results can be assured with either single-ply or multi-ply assemblies by requiring the following:

- Proper framing details, consisting of straight, correctly spaced, and properly cured lumber
- Proper job conditions, including controlled temperatures and adequate ventilation during application
- Proper measuring, cutting, aligning, and fastening of the wallboard
- Proper joint and fastener treatment
- Special requirements for proper sound isolation, fire resistance, thermal properties, or moisture resistance

2.1.2 Jobsite Preparation

Materials for any gypsum board installation project should arrive at the jobsite in the manufacturer's originally sealed packaging. The delivery of the board itself should coincide as closely with the installation start date as possible. Gypsum board must be protected from direct contact with the elements, including sunlight and precipitation. If the gypsum board is stored outside, it must be stored under cover. Do not store gypsum board directly on the ground. All materials should remain stored in their original wrappers or containers until ready to use on the jobsite.

Because gypsum board can bend if stored improperly, keep unopened packages stacked flat wherever you are storing them. There should be no more than 40 to 100 boards in one stack, depending on the thickness of the boards. If the project requires more than 100 boards, it is acceptable to pile the stacks in tiers separated by risers or other supports that are at least 4" wide. Make sure the supports between the tiers are carefully aligned from bottom to top so that each tier rests on a solid bearing rather than on the boards themselves, as shown in *Figure 18*. Excessive weight on any stack of gypsum board can cause individual boards to break. Avoid stacking gypsum boards vertically, because the heavy boards are not stable in this position and can cause serious injury if they fall.

Avoid stacking long lengths of gypsum board on top of short lengths to prevent the longer boards from bending or breaking. When moving boards, carry them or transport them with a dolly. Never drag gypsum boards along the ground or floor, which can damage the edges and make taping and mudding the joints more challenging. Avoid leaning boards against framing members for a prolonged period of time or during periods of high humidity, as the boards can warp.

Storing and Handling Gypsum Drywall

Figure 18 shows gypsum drywall as it would be stored in a warehouse or building supply store. Just before installation, the drywall panels would be distributed along interior walls and stood on their long side, or edge.

Drywall is heavy. Two $\frac{1}{2}$" × 4' × 8' panels weigh about 110 pounds, while a pair of $\frac{5}{8}$" × 4' × 8' panels weigh close to 150 pounds. The panels need to be handled carefully so they don't break under their own weight. They should be lifted and carried by the edges rather than the ends. Individual panels are heavy and awkward to lift, so it is important that you use proper lifting procedures when handling them in order to prevent serious back injury. That means keeping the board close to your body and lifting with your legs rather than your arms or back.

Figure 18 Gypsum board storage.
Source: Minute of love/Shutterstock

Construction materials respond to the environment around them. Job conditions such as temperature, humidity, and airflow can affect the performance of gypsum panels and joint treatment materials, the appearance of the joint, and the behavior of adhesive and finishing materials after project completion. While there are types of specialty gypsum board that are moisture and mold resistant, standard gypsum board is not appropriate for use in damp or wet conditions.

Ideally, the temperature at the worksite where you are installing gypsum boards will be around the same temperature at which the site will normally be kept once the work is completed. If it is not possible to keep the temperature at that level, then the site must be kept at a minimum of 40°F while you are installing the board mechanically, meaning with nails or screws. The boards themselves should be kept in the area where they will be installed for at least 48 hours before installation begins, so that they can acclimate to the environment. Gypsum boards not stored under these conditions may shrink or swell after installation causing cracks, ridges, and indentations.

If you are using adhesive to attach the gypsum board to the framing members or to a base layer of gypsum board, the temperature must be at least 50°F at the worksite for 48 hours before beginning the work and until the adhesive has dried completely. It must also be at least 50°F for 48 hours before finishing and decorating drywall and after completion until the joint compound, primer, or paint has dried completely. Whether you are applying gypsum board mechanically or with adhesive, the temperature at the worksite should never exceed 95°F.

Before beginning installation, review the project specifications and gather all the tools and materials you will need. Maintain a safe worksite by keeping pathways clear of materials and tools, and by cleaning up debris at the end of each workday. Wear proper personal protective equipment (PPE): Gypsum board is heavy, so you will need work boots to protect your feet from dropped boards. You will also need work gloves for lifting gypsum boards and to protect your hands against cuts when you are scoring boards with a utility knife. Installing gypsum board ceilings requires that you wear a hard hat to protect your head from falling boards and safety glasses to protect your eyes from dust and debris. If you are applying gypsum board with adhesive, maintain proper ventilation at the worksite and review SDS sheets for additional precautions.

2.1.3 Planning Your Installation

Any gypsum board installation should be carefully planned. The project specifications will be your guide to ensure a successful drywall installation.

Moving Drywall

A drywall dolly like the one shown here is specially designed to transport drywall panels on the jobsite. When loaded, always use two people to move a drywall dolly. Do not overload the dolly.

Source: Dmitry Markov152/Shutterstock

When installing gypsum board on both ceilings and walls, you will always install board on the ceilings first.

Before you begin installing the drywall, verify the following information with the project specifications:

- Check that all insulation and other internal features of the wall and ceiling assemblies have been installed, keeping in mind that some insulation is applied inside ceilings after gypsum board installation.
- Make sure that the spacing of the framing members is appropriate for the thickness of the gypsum board you are applying. Standard spacing for framing members in wall and ceiling assemblies is 16" **on center (OC)** or 24" on center. Generally, the thicker the gypsum board, the farther apart the framing members can be and still provide sufficient support. Effective support lessens the possibility that the panels will sag, crack, or ridge in the finished assembly.
- Determine the size of the gypsum board you will be using.
- Determine whether your application will be a **perpendicular installation** (horizontal) or a **parallel installation** (vertical).

ASTM C840, Section 9 designates the industry standard for single-ply drywall installations (*Table 2*) that you will see in the project specifications.

On center (OC): The distance between the center of one framing member or fastener to the center of an adjacent framing member or fastener.

Perpendicular installation: Applying gypsum board so that the edges are at right angles to the framing members, meaning the board is oriented horizontally.

Parallel installation: Applying gypsum board so that the edges are parallel to the framing members, meaning the board is oriented vertically.

TABLE 2 Single-Ply Maximum Framing Spacing

Gypsum Board Thickness	Application	Framing Members On-Center Spacing
Ceilings		
3/8"	Perpendicular	16"
1/2"	Parallel	16"
5/8"	Parallel	16"
1/2"	Perpendicular	24"
5/8"	Perpendicular	24"
Walls		
3/8"	Perpendicular or Parallel	16"
1/2"	Perpendicular	24"
5/8"	Parallel	24"

For multi-ply construction on ceilings and walls, there are many possible combinations of board thickness, application direction, and fastener spacing when considering the base ply and the face ply. These details differ based on framing member spacing and whether mechanical fasteners, adhesives, or both are being used between the face ply and the base ply. Check the project specifications to make sure you have the correct materials. If the project plans do not address specific details, consult the tables in *ASTM C840, Section 9* for guidance. These ASTM guidelines apply to multi-ply construction for both wood and steel stud framing.

When you apply gypsum board to ceilings, there is always the risk that the boards will sag after installation. To protect against this risk, make sure that the gypsum board is completely dry and has been allowed to reach the temperature of the worksite. Control the humidity at the worksite carefully, especially if you are using portable heaters, which tend to make the air more humid. One way to control humidity is to provide good ventilation on the worksite.

Once you have determined which boards you will use on the ceilings and walls, measure the framing space where you will be applying the gypsum board. Accurate measuring will usually reveal any irregularities in framing or furring. Poorly aligned framing should be corrected before applying the panels.

Choose the orientation of the boards—perpendicular or parallel—that meets project specifications, professional standards, and building codes and that

NOTE

Gypsum board measuring 1/4" thick is not appropriate for single-ply application on wood or steel framing for ceilings or walls, but it can be used as a base ply or a face ply in multi-ply construction. It can also be applied directly to an existing surface of wood paneling, plaster, concrete, or masonry.

results in the fewest joints possible in the completed assembly. Joints between board ends and edges must be finished with tape and joint compound, so the fewer joints in an assembly, the less work it will take to finish the walls.

When ceilings are 8' high, perpendicular installation is generally preferred when using gypsum board with 4' ends, because two boards applied horizontally fit the height of the wall with little or no cutting to fit. In new construction, 9' ceilings have become the standard, and 54" wide gypsum panels are now more widely available, so that two boards applied horizontally fit the height of the wall with little or no cutting.

For rooms with ceilings taller than 9', parallel application is more practical and results in fewer joints. For certain projects, parallel application may be required for 8' ceilings in order to meet fire ratings.

For ceiling application, select the method that results in the fewest joints between boards. When ceilings are to receive water-based spray texture finishes, pay special attention to the spacing of framing members, the thickness of the board used, ventilation, vapor barriers, insulation, and other factors that can affect the performance of the system and cause problems, particularly sag of the gypsum board between framing members.

Parallel Versus Perpendicular Installation

Perpendicular installation often has the following advantages over parallel installation:

- There are fewer joints.
- Less measuring and cutting are required.
- Joints are at a convenient height for finishing.
- A single panel ties together more framing members to increase the strength of the assembly and hide framing irregularities.
- Panels applied perpendicularly are less likely to sag after finishing.

Plan to keep joints between boards at least 12" away from doors, windows, electrical outlets, and other openings in the assembly. Also plan to leave a gap of at least $\frac{1}{4}$" at the bottom of the wall assembly, because standard gypsum board is not water or mold resistant and can absorb moisture held by concrete or other masonry. Placing a shim in the gap between the wallboard and the floor can also prevent the board from wicking moisture from the floor. Leave a gap anywhere else the gypsum board wall assembly abuts masonry materials.

Control Joints

In certain settings, it is necessary to install **control joints**, also called *expansion joints*, in wall and ceiling assemblies. Control joints are small gaps left between gypsum panels in an assembly to help prevent damage to the boards under everyday conditions, including the expansion and contraction of the boards that come with temperature fluctuations and the shifting of the boards that comes with normal building movement. Movement of the structure can impose severe stresses and cause cracks, either at the joint between boards or in the **field**, or inner area, of a board. U- or V-shaped inserts—often called *control joints, control strips,* or *expansion joints*—fill in the gaps between gypsum panels and allow for the desired smooth finished surface. See *Figure 19*.

When applying gypsum board to long stretches of framing, professional standards require that you create a control joint every 30' in wall partitions and 30' in ceiling assemblies, not to exceed 900 ft^2 between control joints without **perimeter relief**. In ceilings with perimeter relief, control joints are necessary every 50', not to exceed 2,500 ft^2 between control joints. In an assembly in which ceiling framing members change direction, either control joints or intermediate blocking is required. Without control joints, gypsum panels applied in an uninterrupted straight plane for long stretches are vulnerable to **centerline cracking**.

Control joints are appropriate in other locations as well. They should be installed over window and door openings, which are weak points in a structure

Control joints: Deliberate gaps left between long stretches of gypsum panels to allow them to expand, contract, or shift without cracking.

Field: The inner area of a gypsum panel.

Perimeter relief: A gap left between a ceiling assembly and a wall assembly to keep the two assemblies separate and to allow the gypsum board in the ceiling assembly to move freely.

Centerline cracking: A crack in a finished drywall joint that can occur as the result of environmental conditions or poor workmanship.

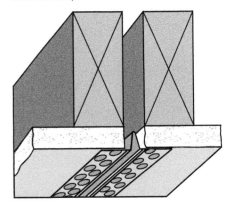

Figure 19 Typical control joint.

Ridging: A defect in finished gypsum board drywall caused by environmental or workmanship issues.

and areas of greater stress. Use control joints, metal trim, or other means to isolate gypsum panels from structural elements such as columns, beams, and load-bearing interior walls, and from dissimilar wall or ceiling finishes that might move differently. During planning, always confirm control joint locations with the architect or designer.

The modern trend toward less rigid structures in high-rise and commercial buildings has created a challenge for gypsum board installers. Because the structural members in these buildings naturally flex, expand, and contract more under everyday conditions, more of the load can be passed along to nonbearing walls and lead to drywall cracking and **ridging**. If you work on such a project, detailed designs are available for perimeter relief of nonbearing partitions to mitigate these risks. One solution is to use relief runners to attach nonbearing walls to ceiling and column members (*Figure 20*).

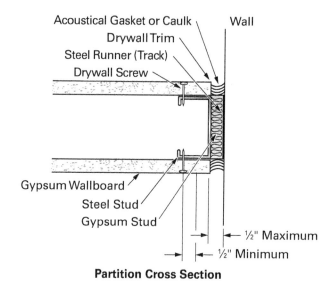

Partition Cross Section

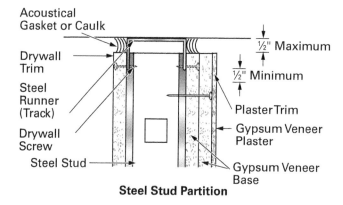

Steel Stud Partition

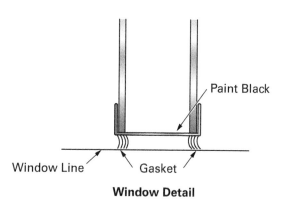

Window Detail

Figure 20 Designs for perimeter relief.

Building plans supplied by the project architect or designer should specify if and where control joints are to be incorporated in a wall or ceiling assembly. Installing control joints during the initial gypsum board installation will keep you—or someone else—from having to install them later, when cracks develop. If cracks do develop in a completed gypsum board installation in new construction, it is wise to wait until at least one heating season has passed before repairing or refinishing. During planning, always confirm control joint locations with the architect or designer.

Control Joints

Control (expansion) joints are used in large expanses of wall or ceiling drywall to compensate for the natural expansion and contraction of a building. Control joints help prevent cracking and joint separation. They are common in commercial construction, especially where exterior concrete walls contain expansion joints.

If control joints are called for in a structure where fire-resistance rating and/or sound control are important, a seal must be installed first where the control joint will occur in the wall or ceiling. The control joint has a $\frac{1}{4}$" slot that is covered by plastic tape. The tape is removed after the joint is finished, leaving a small recess.

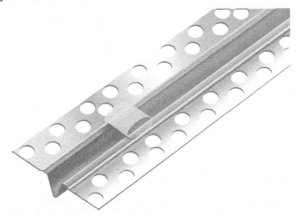

2.1.4 Cutting and Fitting Procedures

It is crucial to a gypsum board installation project that you measure, cut, and fit the boards accurately. Beginning with the ceiling panel installation, mark cutting lines on the panels using a 4' T-square and a pencil, such as a carpenter's pencil. Do not use a pen or a marker, which can bleed through primer applied during the finishing phase and cause project delays.

Step 1 Cut each board by first scoring through the paper down to the core with a sharp utility knife, working from the face side.

Step 2 Snap the board back, away from the cut face.

Step 3 Cut through the back paper with your utility knife. Less commonly, workers use a drywall saw to cut gypsum board.

Step 4 Cut away any jagged edges or burrs from the board with your utility knife. Use a rasp to smooth all cut edges and ends. Smooth ends and edges are vital to form neat, snug joints when the boards are installed. If burrs on the cut ends are not removed, they will form a visible ridge in the finished surface.

The wall or ceiling assembly you are building will include fixtures, such as electrical boxes, pipes, vents, and light fixtures, that must be marked and cut out of the gypsum board you are applying. As a beginning installer, take special care to mark and score the locations of these fixtures accurately on both the face and the back of the board before cutting out with a drywall saw or power router. Failure to do so can result in cutting mistakes that force you to scrap boards or to remove and replace boards you have already attached to the framing members.

The following are different ways to approach cutting holes in the panels for these fixtures:

- Some installers, especially more experienced workers, prefer to measure the location of a fixture with a measuring tape, mark the outline of the fixture on the board, cut it out with a utility saw or router, and then apply the board to the framing members.

- Other installers prefer to mark the location of the fixture on the board, attach the board to the framing members, and then cut out the space for the fixtures. The preferred method is to first tack the board to the framing members just enough to keep it in place, using a few fasteners at least 24" away from the cutout location. Do not screw the board tightly to the members until after you have completed the cutouts, because the pressure caused by the utility saw or router on a tightly attached board will break the board.

Marking cutouts accurately on gypsum panels can be challenging, and some experts recommend methods other than simply measuring with a measuring tape and marking boards with a pencil. This approach, if not done with care, can lead to mistakes and wasted boards.

Whichever marking and cutting method you use to accommodate wall and ceiling fixtures, make your cuts neatly and use a rasp to smooth the edges. Leaving rough edges and burrs on your cutouts makes it tricky to finish for a smooth surface.

To ensure a sound installation of gypsum drywall, use the following general practices:

- Install the ceiling boards first, then the wall panels.

- Fit the panels easily into place without force, maintaining **moderate contact** between them.

- Always match edges with edges and ends with ends. For example, put tapered end adjacent to tapered end, and match square-cut end to square-cut end.

- Plan to span the entire length of a ceiling or wall with single boards, if possible, to reduce the number of end joints, which are more difficult to finish than edge joints.

- Stagger end joints and locate them as far from the center of the ceiling or wall as possible so they will be inconspicuous.

- In a single-ply application, the board ends or edges parallel to the framing members should fall on these members to reinforce the joint. Do not locate joints in the same place on two sides of a wall assembly. When a joint occurs on a stud on one side of the assembly, do not also locate a joint on the same stud on the other side of the assembly.

- Mechanical and electrical equipment, such as cover plates, registers, and grilles, should be installed to provide for the final wall thickness when applying the trim.

- Keep gypsum boards clean and dry throughout the installation to facilitate the finishing process.

Applying Gypsum Board to Wood Stud Framing with Nails or Screws

Before you install gypsum board on wood stud framing or ceiling joists, carefully inspect the framing itself. Acceptable wood framing consists of properly cured, straight, correctly spaced lumber. Framing faults prevent solid contact between the gypsum board and the framing members, weakening the assembly structure. Moist, misaligned, or twisted supporting framing members and improperly installed blocking or bracing are common causes of face paper fractures when installing gypsum board on wood stud assemblies, as shown in *Figure 21*.

The moisture content of wood framing members should not exceed 15 percent at the time of gypsum board application. Because lumber shrinks across the grain as it dries, applying panels to wood studs with a higher moisture content

Moderate contact: The contact between the edges and ends of abutting gypsum boards in a wall or ceiling assembly, which should not be tight or too widely spaced.

NOTE

The depth of electrical boxes in a wall assembly should not exceed the framing depth, and boxes should not be placed back-to-back on opposite sides of the same stud. Electrical boxes and other devices should not be allowed to penetrate completely through the walls. This is detrimental to both sound control and fire resistance. Before installing gypsum board, make sure all the wires in the electrical boxes are tucked as far back as possible into the boxes to avoid nicking them when using a drywall router to cut out the gypsum board around the box.

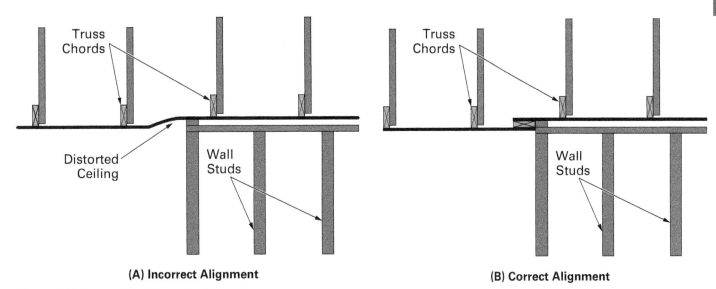

(A) Incorrect Alignment   **(B) Correct Alignment**

Figure 21 Incorrect and correct alignment.

will eventually expose the shanks of nails driven into the edges of the framing members. If shrinkage is substantial or the nails are too long, separation between the gypsum board and its framing member can result in fastener pops and loose boards. Also, a framing member that sticks out farther than the framing members on either side of it will result in a wall that is not even. No framing member should protrude more than $\frac{1}{8}$" farther than the adjacent members.

If you notice defective supports, you must correct them prior to the application of the gypsum board. Wood framing with a high moisture content should be allowed to dry further in appropriate environmental conditions. Protruding framing members should be trimmed or reinstalled. If fixing the framing or reframing is not possible, you can use wood shims to create a level plane for gypsum board application. If other options are unavailable, two-ply construction can also minimize framing defects.

When you are applying gypsum board to wood ceiling joists, some project plans will call for you to install wood cross furring perpendicular to the joists to provide better support for the ceiling panels.

For single-ply wall installation, when you are applying gypsum boards horizontally (perpendicular) to wood framing members, place all the ends of the boards on the members or other solid backing. When applying gypsum boards vertically, place all board edges on framing members.

You can apply gypsum board to wood studs and joists with drywall nails or drywall screws. The length of the fastener depends on the thickness of the board. See *Table 3* to determine appropriate nail length when using annular or smooth shank nails. The nail penetration into the framing member should be at least $\frac{7}{8}$" for smooth shank nails and $\frac{3}{4}$" for annular nails.

TABLE 3 Nail Length for Single-Ply Standard Gypsum Board Application on Wood Framing or Furring

Gypsum Board Thickness	Minimum Nail Length, Annular Nail	Minimum Nail Length, Smooth Shank Nail
$\frac{1}{4}$" (6.4 mm)	Only used in multi-ply assemblies	Only used in multi-ply assemblies
$\frac{3}{8}$" (9.5 mm)	$1\frac{1}{8}$" (28 mm)	$1\frac{1}{4}$" (32 mm)
$\frac{1}{2}$" (12.7 mm)	$1\frac{1}{4}$" (32 mm)	$1\frac{3}{8}$" (35 mm)
$\frac{5}{8}$" (15.9 mm)	$1\frac{3}{8}$" (35 mm)	$1\frac{1}{2}$" (38 mm)

Hold each gypsum board firmly against the framing members and drive each nail perpendicular to the board. This method will help you avoid **fastener pops**, which are protrusions of nail or screw heads above the surface of the gypsum board. The results of correct and incorrect nailing are shown in *Figure 22*.

Fastener pops: Protrusions of nails or screws above the surface of a gypsum board, usually caused by shrinkage of wood framing or by incorrect board installation.

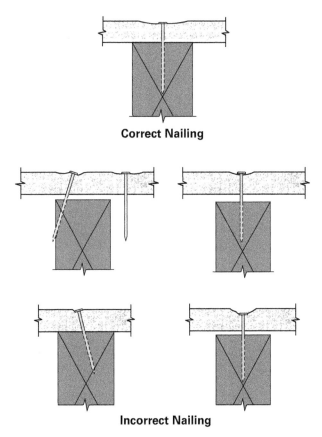

Correct Nailing

Incorrect Nailing

Figure 22 Correct and incorrect nailing.

Perimeter: The outer boundary of a gypsum panel.

Floating angle method: A drywall installation technique used with wood stud framing in which no fasteners are used where ceiling and wall panels intersect in order to allow for structural stresses.

Begin fastening in the inner area of the board, the field, starting in the middle and proceeding outward toward the board edges and ends, the **perimeter**. Around the perimeter of each board, place fasteners at least $\frac{3}{8}$" and not more than 1" from the edges and ends—except where the **floating angle method** is appropriate. Use the appropriate hammer to countersink each nail without breaking the face paper of the board.

In single-ply construction, space nails a maximum of 7" on center on ceilings and 8" on center on walls. See *Figure 23*. In multi-ply construction or when installing a single layer of board over rigid foam insulation, make sure to use nails that are long enough to penetrate the wood framing members at least $\frac{7}{8}$".

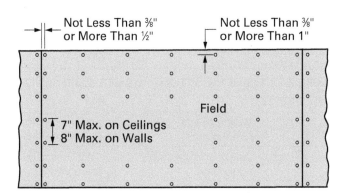

Not Less Than $\frac{3}{8}$" or More Than $\frac{1}{2}$"

Not Less Than $\frac{3}{8}$" or More Than 1"

Field

7" Max. on Ceilings
8" Max. on Walls

Note:
If screws are used in place of nails, spacing is as follows:

Framing Spacing	Walls	Ceilings
16" OC	16"	12"
24" OC	12"	12"

Figure 23 Single nail spacing.

Drywall screws are more commonly used than nails in gypsum board instal-lation. While drywall nails are less expensive, screws provide a stronger hold, so you use fewer of them. Modern drywall screw guns come with an adjust-able depth setting, which makes it relatively easy to consistently countersink your screws. These characteristics make screws and drywall screw guns more popular than hammers and nails among workers who install gypsum panels. Additionally, you can only use drywall nails to apply gypsum board to wood stud assemblies. In comparison, you can use drywall screws to apply gypsum board to both wood and steel stud assemblies.

When using screws to apply gypsum panels to wood stud framing, use #6 drywall screws long enough to penetrate the framing members at least $\frac{5}{8}$". See *Table 4*.

TABLE 4 Screw Length for Single-Ply Standard Gypsum Board Application on Wood Framing or Furring

Gypsum Board Thickness	Drywall Screw, Minimum Length
$\frac{1}{4}$" (6.4 mm)	Only used in multi-ply assemblies
$\frac{3}{8}$" (9.5 mm)	1" (25 mm)
$\frac{1}{2}$" (12.7 mm)	$1\frac{1}{8}$" (28 mm)
$\frac{5}{8}$" (15.9 mm)	$1\frac{1}{4}$" (35 mm)

These professional recommendations are important to consider, but it is also crucial to know that gypsum board installers operate somewhat differently in the field, where the standard screw used to apply gypsum board to wood or steel framing is a #6 drywall screw $1\frac{1}{4}$" long.

The spacing of screws in a single-ply gypsum board assembly on wood stud framing depends on the spacing of the framing members. When wall or ceil-ing framing members are 16" apart on center, space screws 12" apart on center for ceilings and 16" apart on center for walls. When the wall or ceiling fram-ing members are 24" apart on center, space screws not more than 12" apart on centers for both walls and ceilings. For such widely spaced framing members, double screws are recommended.

When using screws—as when using nails—hold the gypsum board tightly against the framing members while applying the fasteners to avoid fastener pop. When installing multi-layer gypsum board or single-layer board over rigid foam insulation, make sure to use screws long enough to penetrate the wood framing members at least $\frac{5}{8}$".

Floating Angle Method

When you are installing one layer of gypsum panels to wood stud framing, it is necessary to take preventive measures to minimize the possibility of fastener pop in areas adjacent to wall and ceiling intersections and to minimize cracking in gypsum boards due to structural stresses. The floating angle method, which is used only with wood stud framing, omits fasteners where walls and ceilings intersect. This method may be used for either single-ply or multi-ply gypsum board wall assemblies with wood framing, and it is applicable whether you are using nails or screws. *Figure 24* shows a typical single-layer application.

Where the ceiling framing members are perpendicular to the wall/ceiling inter-section, the ceiling fasteners should be located 7" from the intersection for single nailing and 11" to 12" for screw applications. Where ceiling joists are parallel to the wall/ceiling intersection, nailing should start at the intersection. See *Figure 25*.

Applying Gypsum Board to Wood Stud Framing with Adhesive

Drywall adhesives should be applied with a caulking gun in accordance with the manufacturer's recommendations. When you are using adhesive to make a joint between gypsum panels on a stud, apply two parallel straight beads of adhesive approximately $\frac{1}{4}$" in diameter, one near each edge of the stud. When you are apply-ing the field of the panel, rather than an end or edge, to a stud, apply one straight

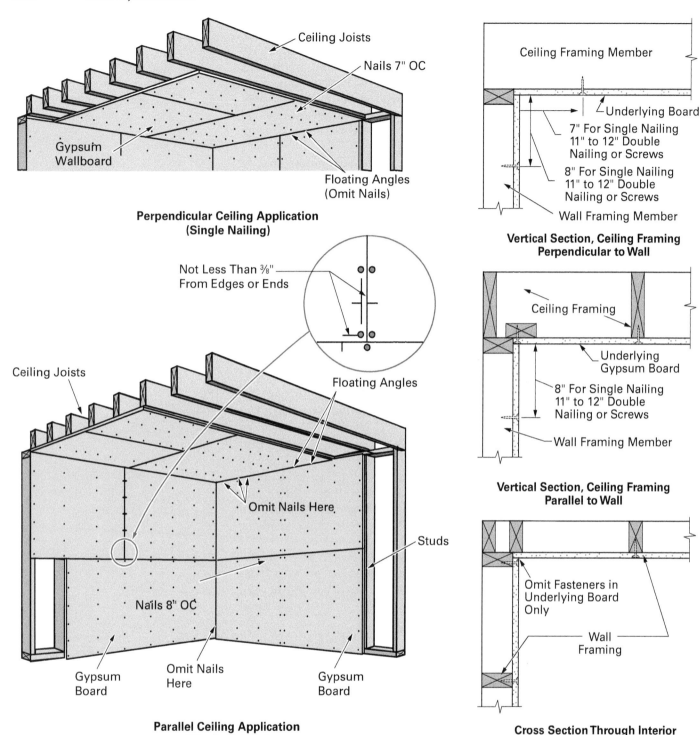

Perpendicular Ceiling Application (Single Nailing)

Ceiling Joists

Nails 7" OC

Gypsum Wallboard

Floating Angles (Omit Nails)

Not Less Than ³/₈" From Edges or Ends

Floating Angles

Ceiling Joists

Omit Nails Here

Studs

Nails 8" OC

Omit Nails Here

Gypsum Board

Gypsum Board

Parallel Ceiling Application (Single Nailing)

Ceiling Framing Member

Underlying Board

7" For Single Nailing
11" to 12" Double Nailing or Screws

8" For Single Nailing
11" to 12" Double Nailing or Screws

Wall Framing Member

Vertical Section, Ceiling Framing Perpendicular to Wall

Ceiling Framing

Underlying Gypsum Board

8" For Single Nailing
11" to 12" Double Nailing or Screws

Wall Framing Member

Vertical Section, Ceiling Framing Parallel to Wall

Omit Fasteners in Underlying Board Only

Wall Framing

Cross Section Through Interior Vertical Angle

Figure 24 Floating angle method.

Figure 25 Fastener patterns for the floating angle method.

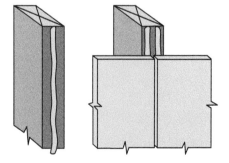

Figure 26 Adhesive applied to the edges of a stud.

bead of adhesive, approximately ¹/₄" in diameter, to the middle of the stud face. See *Figure 26*.

Do not expect adhesives alone to hold gypsum boards to wood framing members. Single-ply gypsum board systems attached with adhesives require supplemental perimeter fasteners. Place fasteners 16" on center along the edges or ends on boards that you have applied with adhesive, whether they are perpendicular or parallel to the supports. Ceiling installations require supplemental fasteners in the field as well as on the perimeter. They should be placed 24" on center. Follow the floating angle method for supplemental fasteners used with adhesive. See *Figure 27*.

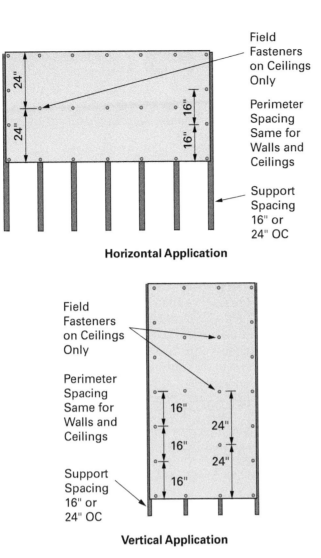

Horizontal Application

Vertical Application

Figure 27 Supplemental wall and ceiling fasteners.

Adhesive is not required at top or bottom plates, bridging, bracing, or fire stops. Where fasteners at vertical joints are undesirable, gypsum panels may be prebowed, as shown in *Figure 28*.

Using Adhesive on Drywall

When applying gypsum board with adhesive, choose a product that meets the requirements of *ASTM C557*. Allow the adhesive to dry for 48 hours before finishing the joints. Using adhesive does not eliminate the need for fasteners; it simply reduces the number of fasteners needed. Check the job specifications or local codes for details.

Adhesives cannot be used to attach drywall panels to studs if the building has an inside moisture barrier.

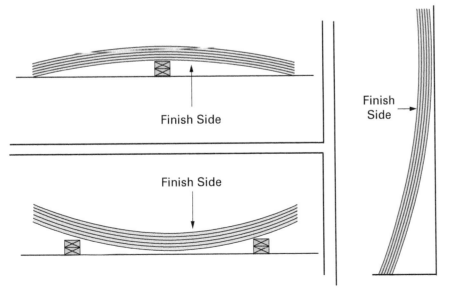

Figure 28 Prebowing of gypsum panels.

Prebowing puts an arc in the gypsum board, which keeps it in tight contact with the adhesive after the board is applied. Supplemental fasteners (placed 16" on center) are then used at the top and bottom plates.

Gypsum board may be prebowed by stacking it, face up, with the ends resting on 2' × 4' lumber or other blocks, and with the center of the boards resting on the floor. Allow it to remain overnight or until the boards have a permanent bow.

Applying Gypsum Board to Steel Stud Framing

For single-ply installation on steel stud framing, attach gypsum board to 18 mil (25 gauge) or 27 mil (22 gauge) steel framing and furring using #6, fine, self-tapping drywall screws. If the steel framing is 30 mil (20 gauge) or thicker, #6, fine, self-drilling drywall screws are required.

The recommended minimum screw penetration into steel studs for single-ply construction is $3/8$" or 3 threads. This length is less than the penetration required for wood studs because the hold between drywall screws and steel framing creates more withdrawal resistance than the hold between screws and wood framing. Keep in mind that these guidelines refer to the length of the threaded part of the screw. Self-drilling screws have a longer tip than other screws, and it is a good idea to allow an additional $1/4$" when choosing these screws to get the recommended penetration into the framing members.

Gypsum board installers applying gypsum board to steel studs most commonly use #6, fine, self-tapping or #6, fine, self-drilling screws that are $1\frac{1}{4}$" long for single-ply construction and $1\frac{5}{8}$" long for multi-ply construction. Choose the length of the screws you use for a particular project based on the guidelines in *Table 5*, by the length designated in the project specifications, or by the length commonly used in the field.

TABLE 5 Screw Length for Single-Ply Standard Gypsum Board Application on Steel Framing or Furring

Gypsum Board Thickness	Minimum Screw Length, Single-Ply Construction
$1/4$" (6.4 mm)	Only used in multi-ply assemblies
$3/8$" (9.5 mm)	$3/4$" (19 mm)
$1/2$" (12.7 mm)	$7/8$" (22 mm)
$5/8$" (15.9 mm)	1" (25 mm)

As with wood stud framing, the spacing of the screws when installing single-ply gypsum panels to steel stud framing depends on the spacing of the studs: When the framing is 16" on center, space screws on ceilings no more than 12" on center and on walls no more than 16" on center. When the framing of the steel studs is 24" on center, space screws on both ceilings and walls no more than 12" on center.

Most ceiling framing in commercial construction is built with a grid system of main beams supported by wires and perpendicular steel cross tees that click into place. These ceiling systems provide an excellent structure for applying gypsum panels, because the boards can be fastened all the way around the perimeter, compared to boards fastened to parallel ceiling joists, which, without installing furring, can only be fastened to the framing members on two sides.

When applying gypsum board to steel framing with adhesive, follow manufacturer's directions carefully. You must supplement adhesive with mechanical fasteners on intermediate beams or cross tees on ceilings as well as at the perimeters of gypsum panels on both walls and ceilings. For ceilings in particular, as shown in *Table 6*, the framing spacing varies both according to the load and the type of board being used.

TABLE 6 Maximum Spacing of Ceiling Framing

Gypsum Board (Thickness)		Application to Framing		Maximum OC Spacing of Framing
Base	Face	Base	Face	
$3/8$"*	$3/8$"	Perpendicular	Perpendicular or Parallel	16"
$1/2$"*	$3/8$" or $1/2$"	Perpendicular or Parallel	Perpendicular or Parallel	16"
$5/8$"*	$1/2$"	Perpendicular or Parallel	Perpendicular or Parallel	16"
$5/8$"*	$5/8$"	Perpendicular or Parallel	Perpendicular or Parallel	24"

*In two-ply construction, the threaded portion of gypsum drywall screws must penetrate at least $1/2$" into the base layer board. To ensure this minimum penetration, it is wise to allow approximately $1/4$" for the screw point by choosing a screw $3/4$" longer than the width of the gypsum board you are installing.

2.1.5 Resurfacing Existing Construction

Gypsum board may be used to provide a new finish on existing above-grade walls and ceilings of wood, plaster, wallboard, masonry, or concrete. If the existing surface is structurally sound and provides a sufficiently smooth and solid backing without shimming, $1/4$" gypsum board can be applied directly to the surface with nails, screws, or adhesives. If you are using drywall nails to apply gypsum panels to existing paneling or plaster, make sure the nails penetrate the framing beneath the old surface material by $7/8$". When using power-driven screws, make sure the threaded portion of the screw penetrates the framing by at least $5/8$".

Applying Gypsum Board Directly to Concrete and Masonry

Before beginning gypsum board installation, remove and set aside any surface covers for mechanical and electrical equipment, such as switch plates, outlet covers, and ventilating grilles. Electrical boxes should be reset prior to the installation of new gypsum board. Then address any irregularities on the surface of the wall. Rough or protruding edges and excess joint mortar should be removed, and any depressions filled with mortar to make the wall surface smooth and level. Base surfaces should also be cleaned of any curing compound, loose particles, dust, and grease to ensure an adequate bond. New concrete should be allowed to cure for at least 28 days before gypsum board is adhered directly to it.

Another acceptable way to create a flat plane for applying gypsum boards is to **fur out** the concrete or masonry surface with wood furring strips or steel furring channels. Furring also provides a separation between existing exterior walls and gypsum panels to help manage possible moisture issues. When applying gypsum panels to wood furring, use fasteners of a length that will penetrate the wood without coming into contact with the concrete or masonry surface beneath it.

When installing $1/4$" gypsum board directly to concrete or masonry with adhesive, choose an adhesive recommended by the manufacturer and follow all manufacturer's instructions and precautions carefully. In addition, use supplemental mechanical fasteners spaced 16" on center to hold the gypsum board in place while the adhesive is developing a bond.

Gypsum board should not be installed directly on below-grade exterior walls. If a project requires you to install gypsum board in such a setting, you should first fur out the walls and apply a vapor barrier and insulation before attaching the panels. Gypsum board can also be adhered directly to exterior cavity walls if the cavities are properly insulated to prevent condensation and the inside face of the cavity is properly waterproofed.

Fur out: To attach furring strips or furring channels to masonry walls, wood framing, or steel framing to create a level surface before applying gypsum board.

NOTE

An alternative to anchoring furring to a masonry wall is to build a $1\frac{5}{8}$" metal stud wall in front of the masonry wall for the application of gypsum board.

2.2.0 Drywall Trim

When you have completed the process of applying gypsum panels to a wall assembly, all edges of wallboard exposed to view will need drywall trim. Adding drywall trim—a long, narrow strip of metal, vinyl, or paper tape—to these joints creates the illusion of a smooth unbroken surface once the drywall has been finished. Trim comes in a variety of shapes, lengths, and widths, each one having a particular function. It can be made of metal (often galvanized metal) or vinyl (see *Figure 29*).

Figure 29 Drywall trim.
Source: Gintare Stackunaite/Shutterstock

Corner bead: A metal or plastic angle used to protect and finish outside corners where drywall panels meet.

Casing: A type of drywall trim that is used around windows and doors.

Flange: The rim of an accessory used to attach it to another object or surface.

Corner bead is a type of drywall trim made to mask the joints at outside corners of a room. Corner bead protects and reinforces outside corner joints, which are prone to wear and tear. It also provides a straight guide for finishing. **Casing** is a type of drywall trim used around windows and doors. Choose corner bead or casing with a wider **flange** if you need to hide flaws in problem areas in a drywall installation.

Corner bead or casing is relatively easy to apply with mechanical fasteners, drywall compound, or adhesive. Some metal corner bead comes with predrilled holes for mechanical fasteners or a perforated or mesh flange that can be effectively embedded in joint compound during application for a strong hold. The disadvantage of using metal corner bead is that if it is not handled carefully before installation, it can be easily damaged. Vinyl corner bead is less easily damaged, and, like metal bead, it can be applied with mechanical fasteners, drywall compound, or adhesive. Some vinyl corner bead, however, requires special compound or adhesive. Follow the manufacturer's directions carefully when applying any type of trim.

Metal and vinyl drywall trim also come in paper-faced options. This paper face, laminated over metal or vinyl bead or casing, blends in well with the paper face of the gypsum board, is easier to finish, and is more durable than trim covered with corner bead and joint compound alone. United States Gypsum (USG) makes a metal paper-faced corner bead, and No-Coat makes a vinyl paper-faced bead.

Some joints between gypsum boards are especially challenging to finish, such as arches and splayed angles. For these difficult areas, consider using paper tape fortified with plastic mesh or metal strips. This flexible tape can be applied using joint compound to inside and outside corners as well as in places that feature decorative elements where extra flexibility is needed.

Casings come in different shapes made for use in specific locations, usually where gypsum board meets other materials, such as a wooden window jamb or a fireplace. The cross-section of L-bead casing is shaped like the letter L, with one flange wider than the other. Use L-bead to cover the joints around window and door jambs. The cross-section of J-bead casing (sometimes called U-bead) is shaped like the letter J. Use J-bead to cap the end of a piece of gypsum board before applying it to the wall assembly where the wall meets another type of surface, such as a countertop or fireplace.

Reveal trim is a special insert that fits into a deliberate space left between gypsum boards to create a decorative effect in a wall or ceiling assembly. Reveals might be used as moldings around room perimeters, door frames, windows, archways, and other architectural features. They might also be applied at regular intervals in a wall or ceiling assembly to create a decorative pattern.

Some reveal trim must be installed at the time the gypsum panels are hung. If reveal trim is part of your drywall installation project, follow the project plans carefully. Drywall reveal is available in different shapes designed for different purposes. A standard drywall reveal, for example, is meant to be installed between two gypsum panels. It features a U-shaped main channel with two tapered flanges on either side. When the standard reveal is placed between the two boards, the flanges sit on top of the boards and are fastened to them with drywall fasteners through predrilled holes in the flanges.

In comparison, drywall F reveals have one tapered flange and are used to create a recessed molding around the perimeter of a room, where the wall meets the ceiling. The top of the U-shaped main channel of an F reveal lies flush against the gypsum board on the ceiling, and the flange is attached to the gypsum board on the wall below the reveal with a drywall fastener. F reveals are finished with joint tape and joint compound.

When you are ready to begin installing drywall trim, measure the length of the surface carefully. Then use aviation snips to cut the trim to fit.

Achieving a Rounded Appearance

A smooth, rounded finish appearance can be obtained by using a bullnose corner molding and cap such as the ones shown here.

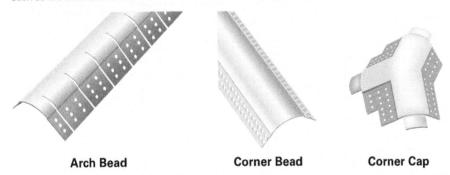

Arch Bead   **Corner Bead**   **Corner Cap**

2.3.0 Moisture Resistance

When a project calls for gypsum board installation in areas such as in laundry rooms, kitchens, bathrooms, and basements, give special consideration to the appropriate use of gypsum panels. In certain wet or humid settings, standard or moisture-resistant gypsum board might be appropriate, while in others its use is not recommended. At times, project managers and consumers must decide whether to follow manufacturer's product recommendations, professional standards, or local building codes. Sometimes local building codes are not as stringent as professional standards.

2.3.1 Moisture-Resistant Construction

Many brands of gypsum board offer moisture-resistant panels, and some local building codes allow these panels to be used as a base for applying tile with adhesive in wet areas, such as tubs, showers, and saunas. However, ASTM specifications (*ASTM C840*) recommend using both standard and moisture-resistant gypsum board only in dry areas where tile will be applied. In wet settings where a base is needed for tile application, cement backer board is the better option. Leading gypsum board manufacturer USG does not recommend using its gypsum board products in wet settings but instead offers its own water-durable cement backer board.

If the project you are working on does require you to use gypsum board in moist conditions, do not use foil-backed board or apply board directly over a vapor barrier, because the vapor barrier will trap moisture within the gypsum core of the board. In areas of high humidity or where water vapor is present, such as in basements, exterior below-grade walls or surfaces should be furred out and protected with a vapor barrier and insulation in order to provide a suitable base for attaching the moisture-resistant gypsum board.

If project plans call for you to install moisture-resistant gypsum board in a tub/shower enclosure as a backer for tile, make sure the shower pan or tub has been installed before you begin. Shower pans should have an upstanding lip or flange located at a minimum of 1" higher than the entry wall to the shower. It is recommended that the tub be supported. If necessary, fur out the framing members before applying gypsum board so the upstanding leg of the pan (*Figure 30*) will be on the same plane as the face of the board.

Suitable blocking should be provided approximately 1" above the top of the tub or pan. Between-stud blocking should be placed behind the horizontal joint of the board above the tub or shower pan. For ceramic tile applications, use studs that are at least $3\frac{1}{2}$" deep and placed 16" on center. Appropriate blocking, headers, or supports should be provided for tub plumbing fixtures and to receive soap dishes, grab bars, towel racks, and similar items.

Install tile backer boards with nails or screws spaced not more than 8" on center. When the ceramic tile to be applied is more than $\frac{3}{8}$" thick, space nails or screws no more than 4" on center. When it is necessary to treat joints and fastener heads with joint compound and tape, either use waterproof nonhardening caulking compound or seal joints and fastener heads with a compatible sealer prior to the tile application.

> **NOTE**
>
> The caulking compound or sealer used to seal joints and fastener heads must be compatible with the adhesive to be used for the application of the tile. Follow the adhesive manufacturer's instructions carefully.

Reinforce corners with rigid supports. Caulk the cut edges and openings around pipes and fixtures flush with a waterproof, nonhardening, silicone caulking compound or an adhesive complying with the American National Standard for Organic Adhesives for Installation of Ceramic Tile.

Tile Application

Ceramic wall tile application to gypsum board should meet the American National Standard Specifications for Installation of Ceramic Tile with Water-Resistant Organic Adhesive. The adhesives used should meet the American National Standard for Organic Adhesives for Installation of Ceramic Tile.

The surfacing material should be applied down to the top surface or edge of the finished shower floor, return, or tub, and installed to overlap the top lip of the receptor, subpan, or tub.

> **NOTE**
>
> Different types of waterproof boards have different applications and limitations, so it is always necessary to check the manufacturer's product data sheets, installation instructions, and local building codes for the type of board to be used.

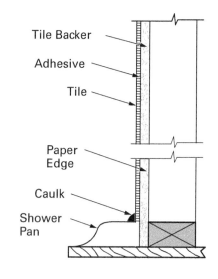

Figure 30 Pan is on the same plane as the face of the board.

2.4.0 Fire-Resistance-Rated and Sound-Rated Assemblies

The construction of ceilings, exterior walls, and interior partition walls in certain settings is driven by the fire resistance and acoustical performance requirements specified in local building codes. In buildings in which fire and sound rating are not required—such as in single-family residential homes—wood or steel stud frame wall and ceiling assemblies with one layer of $\frac{1}{2}$" gypsum drywall are satisfactory. In contrast, multi-ply ceiling and wall assemblies are necessary in construction that requires a one- or two-hour fire-resistance rating. For an even higher fire-resistance rating, such as in offices located next to the manufacturing spaces in a factory where there is a hazard of fire or explosion, it may be necessary to build multi-ply ceiling and wall assemblies that begin with a concrete block (CMU) wall, followed by rigid and/or fiberglass insulation, followed by fire-resistant gypsum board, as shown in *Figure 31*.

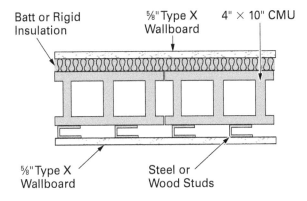

Batt or Rigid Insulation $\frac{5}{8}$" Type X Wallboard   4" × 10" CMU

$\frac{5}{8}$" Type X Wallboard   Steel or Wood Studs

Figure 31 High fire/noise resistance partition.

Wood and steel stud framing are both approved for use in ceiling and wall assemblies in single-family residential construction, but steel stud framing is the standard in multi-family residential and commercial construction, where fire-resistance and sound rating are required by building codes. The type and thickness of the insulation and gypsum board applied to the steel stud framing depend on the fire-resistance rating and acoustical performance requirements for the project. Multi-family residential construction is required to be one-hour fire-resistance rated. Sound control requirements are based on the amount of privacy required for the building's intended use. For example, high-rise condos, executive offices, hospitals, and homes for the elderly may require more privacy than general offices.

The requirements for sound attenuation and fire resistance can significantly affect the thickness of a wall. For example, a steel stud wall in an assembly with high sound and fire-resistance ratings might have a total thickness of nearly $6\frac{1}{2}$", while a wall in an assembly with a lower rating might have a total thickness of only $3\frac{1}{2}$".

2.4.1 Fire-Resistance-Rated Construction

Building codes for certain types of structures, such as multistory apartment and office buildings, require specific wall and ceiling assemblies. A **fire-resistance-rated assembly** can withstand fire for a certain period, measured in hours, as determined under laboratory conditions. A fire-resistance-rated assembly can effectively keep fire from spreading while maintaining the integrity of the assembly for a specified length of time. A single-family residential home may use a **non-rated assembly**, which may provide some resistance to fire, but not enough to qualify for any fire-resistance rating.

Fire-resistance-rated assembly: Construction built with certain materials in a certain configuration that has been shown through testing to restrict the spread of fire.

Non-rated assembly: A ceiling or wall assembly that does not exhibit enough fire-resistant properties to qualify for a fire-resistance rating.

There are many different construction methods for so-called party walls, or walls common to two adjoining buildings or rooms. Each is designed to meet different fire and sound control standards. The wall is generally at least 3" thick and contains several layers of gypsum board and insulation. A fire-resistance-rated wall may abut a non-rated partition or wall. Ensure that the rated wall is carried through to maintain the fire-resistance rating, as shown in *Figure 32*.

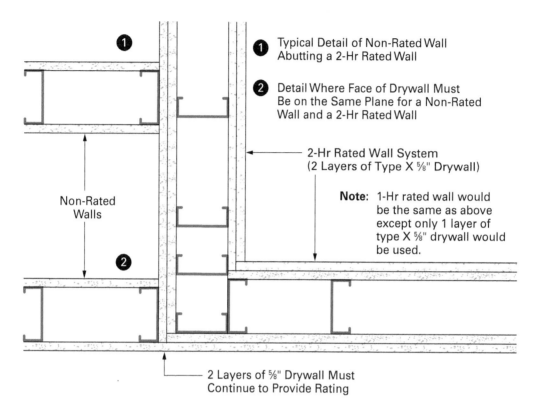

1 Typical Detail of Non-Rated Wall Abutting a 2-Hr Rated Wall

2 Detail Where Face of Drywall Must Be on the Same Plane for a Non-Rated Wall and a 2-Hr Rated Wall

2-Hr Rated Wall System (2 Layers of Type X ⅝" Drywall)

Note: 1-Hr rated wall would be the same as above except only 1 layer of type X ⅝" drywall would be used.

Non-Rated Walls

2 Layers of ⅝" Drywall Must Continue to Provide Rating

Figure 32 An example of a fire-resistance-rated wall abutting a non-rated wall.

Fire-Resistance-Rated Gypsum Board

Type X gypsum board is the most common type used in fire-resistance-rated assemblies. It contains noncombustible glass fibers in its gypsum core, which keep the board from shrinking during the calcination that happens during a fire.

Type C gypsum board is an enhanced, more expensive version of gypsum board that is also approved for fire-resistance-rated assemblies. In addition to glass fiber, it contains unexpanded vermiculite in its gypsum core, which expands when heated to keep the board from shrinking and causing a gap in the fire barrier.

Firestopping

Firestopping means cutting off the air supply so that fire and smoke cannot readily move from one location to another.

In frame construction, a firestop is a piece of wood or fire-resistant material inserted into an opening such as the space between studs. This firestop acts as a barrier to block airflow that would feed and carry a fire to the upper floors.

In commercial construction, firestopping material is used to close wall penetrations such as those created to run conduit, piping, and air conditioning ducts. If such openings are not sealed, fire will travel through the openings.

Avoid Back-to-Back Fixtures

Medicine cabinets; electrical, telephone, television, and intercom outlets; and plumbing, heating, and air conditioning ducts should not be installed back-to-back. Any opening for such fixtures, piping, and electrical outlets should be carefully cut to the proper size and caulked.

To meet the fire-resistance-rating standards established by the building and fire codes, all openings must be sealed. The firestopping methods used for this purpose are classified as mechanical and nonmechanical.

Mechanical firestops are fire-resistant devices used to seal the space around wiring or piping. There are many different options, such as lighting covers, sleeve kits, cableways, cable transits, pathways, grommets, plugs, and collars.

Nonmechanical firestops are fire-resistant materials, such as caulks and putties, that are used to fill the space around the conduit or piping. You may be required to install various nonmechanical firestopping materials when working with fire-resistance-rated walls and floors. Holes or gaps affect the rating of a floor or wall. Properly filling these penetrations with firestopping materials maintains the rating. Firestopping materials are typically applied around all types of piping, electrical conduit, ductwork, electrical and communication cables, and similar devices that run through openings in floor slabs, walls, and other fire-resistance-rated building partitions and assemblies.

Nonmechanical firestopping materials are classified as intumescent or endothermic. Both are formulated to help control the spread of fire before, during, and after exposure to open flames. When subjected to the extreme heat of a fire, intumescent materials expand (typically up to three times their original size) to form a strong insulating material that seals the opening for three to four hours. Should the insulation on the cables, pipes, etc., passing through the penetration become consumed by the fire, the expansion of the firestopping material also acts to fill the void in the floor or wall to help stop the spread of smoke and other toxic products of combustion.

Did You Know?

Firestopping Versus Fireproofing

Firestopping and fireproofing are not the same thing. Firestopping is intended to prevent the spread of fire and smoke from room to room through openings in walls and floors. Fireproofing is a thermal barrier that causes a fire to burn more slowly and retards the spread of fire.

Endothermic materials block heat by releasing chemically bound water, which causes them to absorb heat.

Firestopping materials are formulated in such a way that when activated, they are free of corrosive gases, reducing the risks to building occupants and sensitive equipment.

Firestopping materials are made in a variety of forms, including composite sheeting, caulks, silicone sealants, foams, moldable putty, wrap strips, and spray coatings. They come in both one-part and two-part formulations. The installation of these materials must always be done in accordance with the applicable building codes and the manufacturer's instructions for the product being used. Depending on the product, firestopping materials can be applied via spray equipment, conventional caulking guns, pneumatic pumping equipment, or a putty knife.

2.4.2 Sound-Rated Construction

The first step for sound isolation of any assembly is to close off air leaks and flanking paths. A flanking path is when sound leaks around or through building components. Noise can travel through the air over, under, or around walls; through the windows and doors; through air ducts; and through floors and crawl spaces. All these paths must be correctly treated.

Enhanced gypsum board and specialized insulation are available for use in a **sound-rated assembly**. One type, for example, has a core of viscose polymer between two layers of gypsum plaster, a combination that helps to block sound transmission between rooms and dampen sound in large rooms.

Buildings are generally required to meet a **sound transmission class (STC)** rating. The STC is a numeric rating representing the effectiveness of the construction in isolating airborne sound transmission. The higher the STC rating, the better the sound absorption. Hairline cracks and other openings can have an adverse effect on the ability of a building to achieve its STC rating, particularly in higher-rated construction. Where a very high STC performance is needed, air conditioning, heating, and ventilating ducts should not be included in the assembly.

Failure to observe special construction and design details can destroy the effectiveness of the best assembly. Improved sound isolation is obtained by the following:

- Separate framing for the two sides of a wall
- Resilient channel mounting for the gypsum board
- Using sound-absorbing materials in wall cavities
- Using multi-layer gypsum board of varying thicknesses in multi-layer construction
- Caulking the perimeter of gypsum board partitions, openings in walls and ceilings, partition/mullion intersections, and outlet box openings
- Locating recessed wall fixtures in different stud cavities

The entire perimeter of sound-isolating partitions should be caulked around the gypsum board edges to make it airtight, as detailed in *Figure 33A* and *Figure 33B*. The caulking should be a nonhardening, nonshrinking, nonbleeding, nonstaining, resilient sealant.

Sound-control sealing must be covered in the specifications, understood by all related tradespeople, supervised by the appropriate party, and inspected carefully as the construction progresses.

Separated Partitions

A staggered wood stud gypsum partition placed on separate plates will provide an STC between 40 and 42. The addition of a sound-absorbing material between the studs of one partition side can increase the STC by as much as 8 points. With $\frac{5}{8}$" Type X gypsum board on each side, an assembly has a fire-resistance classification of one hour. Separated walls without framing can also be constructed by using an all-gypsum, double-solid, or semi-solid partition.

Steel or wood tracks fastened to the floor and ceiling hold the partitions in place. For the attachment of kitchen cabinets, lavatories, ceramic tile, medicine cabinets, and other fixtures, a staggered stud wall rather than a resilient wall is recommended. The added weight and fastenings may short circuit the construction acoustically.

Resilient Mountings

Resilient attachments acting as shock absorbers reduce the passage of sound through the wall or ceiling and increase the STC rating. Further STC increases can result from more complex construction methods incorporating multiple layers of gypsum board and building insulation in the wall cavities.

A **resilient furring channel** is attached with the nailing flange down and at a right angle to the wood stud, as shown in *Figure 34*.

To install furring channels, drive $1\frac{1}{4}$" coarse screws or 6d coated nails through the prepunched holes in the channel flange. With extremely hard lumber, $\frac{7}{8}$" or 1" fine, self-tapping screws may be used. Locate the channels 24" from the floor, within 6" of the ceiling line, and no more than 24" on center. Extend the channels into all corners and fasten them to the corner framing. Attach $\frac{1}{2}$" × 3" gypsum board filler strips to the bottom plate directly over the studs by overlapping the ends and fastening both flanges to the stud. Apply the gypsum board horizontally with the long dimension parallel to the resilient channels using 1" fine, self-tapping screws spaced 12" on center along the channels. The abutting edges of boards should be centered over the channel flange and securely fastened.

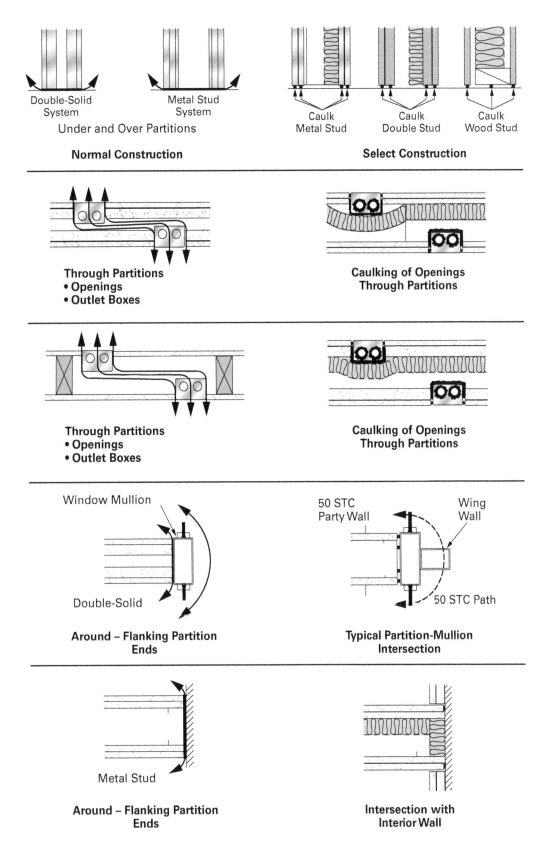

Figure 33A Caulking of sound-isolation construction (1 of 2).

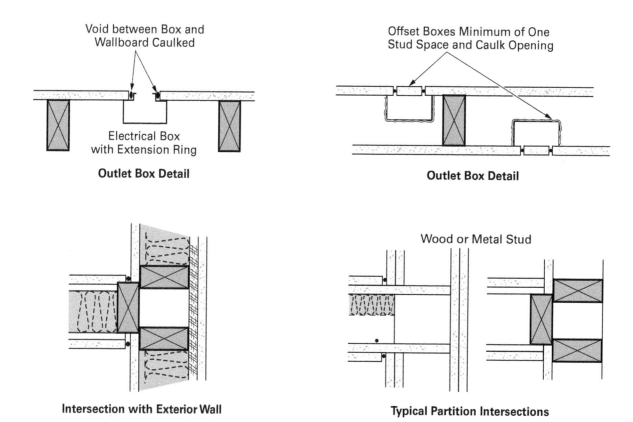

Pre-Design Construction
Simulating Laboratory Conditions

¼" Perimeter Relief and Caulking
to Seal Against Leaks

Wood Stud

Metal Stud

Gasket Impedes Structural
Flanking Through Floor

Typical Floor-Ceiling or Roof Detail

Void between Box and
Wallboard Caulked

Electrical Box
with Extension Ring

Outlet Box Detail

Offset Boxes Minimum of One
Stud Space and Caulk Opening

Outlet Box Detail

Wood or Metal Stud

Intersection with Exterior Wall

Typical Partition Intersections

Figure 33B Caulking of sound-isolation construction (2 of 2).

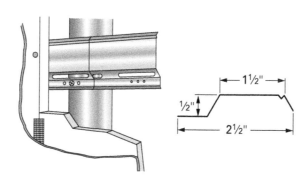

1½"

½"

2½"

Figure 34 Attached resilient furring channel.

Sound-Isolating Materials

Sound-isolating materials include the following:

- Mineral fiber (including glass) blankets and batts used in wood stud assemblies
- Semi-rigid mineral or glass fiber blankets for use with steel studs and laminated gypsum partitions
- Mineral (including glass) fiberboard
- Rigid plastic foam furring systems
- Lead or other special shielding materials

Mineral wool or glass fiber insulating batts and blankets may be used in assembly cavities to absorb airborne sound within the cavity. They should be placed in the cavity and carefully fitted behind electrical outlets and around any cutouts necessary for plumbing lines. Insulating batts and blankets may be faced with paper or another vapor barrier and may have flanges or be of the unfaced friction-fitted type.

Gypsum board may be applied over rigid plastic foam insulation. It is applied on the interior side of exterior masonry and concrete walls to provide a finished wall and to protect the insulation from early exposure to fire originating within the building. Additionally, these systems provide the high insulating values needed for energy conservation.

In new construction or for remodeling, these systems can be installed with as little as 1" dimension from the inside face of the framing or masonry ($\frac{1}{2}$" insulation and $\frac{1}{2}$" Type X gypsum board).

When applying gypsum board over rigid foam insulation, the entire insulated wall surface should be protected with the gypsum board, including the surface above ceilings and in closed, unoccupied spaces.

Single-ply or double-ply, $\frac{1}{2}$" or $\frac{5}{8}$" gypsum board should either be screw-attached to steel wall furring members attached to the masonry or nailed directly into wood framing, as shown in *Figure 35*. Follow the insulation manufacturer's instructions.

Rigid Foam Insulation

Gypsum board applied over rigid plastic foam insulation in the manner described may not necessarily provide the finish ratings required by local building codes. Many building codes require a minimum fire protection for rigid foam on interior surfaces equal to that provided by $\frac{1}{2}$" Type X gypsum board when tested over wood framing. The flammability characteristics of rigid foam insulation products vary widely, and the manufacturer's literature should be reviewed.

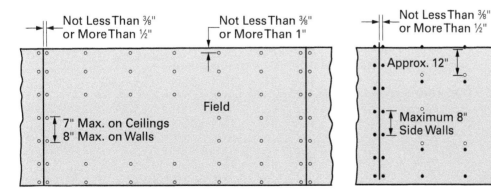

Figure 35 Nailing patterns for installation over rigid foam insulation.

Furring members should be designed to minimize thermal transfer through the member and to provide a $1\frac{1}{4}$" minimum width face or flange for screw application of the gypsum board.

Furring members should be installed vertically and spaced 24" on center. Provide blocking or other backing as required for attachment and support of fixtures and furnishings. Furring members should also be attached at floor/wall and wall/ceiling angles (or at the termination of the gypsum board above suspended ceilings) and around door, window, and other openings. Single-ply gypsum board should be applied vertically, with the long edges of the board located over furring members. The installation should be planned carefully to avoid end joints. The fastener spacing should be the same as required for single-ply application over framing or furring.

In multi-ply applications, the base ply should be applied vertically. In horizontal face ply applications, the face ply and end joints should be offset by at least one framing or furring member space from the base ply edge joints.

The fastener spacing should be as required for multi-ply application over framing or furring, as discussed previously.

In wallboard applications, mechanical fasteners should be of such a length that they do not penetrate completely to the masonry or concrete. In single-layer applications, all joints between gypsum boards should be reinforced with tape. In addition, gypsum board joints should be finished with joint compound. In two-layer applications, the base layer joints may be concealed or left exposed.

2.0.0 Section Review

1. To increase sound control and fire-resistive performance, _____ have two or more layers of gypsum board.
   a. steel framing members
   b. single-ply assemblies
   c. exterior walls
   d. multi-ply assemblies

2. Gypsum board should be kept in the area where it will be installed for at least _____ hours before installation.
   a. 24
   b. 72
   c. 12
   d. 48

3. A general practice to ensure a sound installation of gypsum board is to _____.
   a. rotate every other piece of gypsum board 180°
   b. start in the middle of a wall and work outward
   c. always match edges with edges and ends with ends
   d. always install wall panels before ceiling boards

4. Corner bead is a type of drywall trim made to mask joints at the _____.
   a. floor
   b. ceiling
   c. inside corner of a room
   d. outside corner of a room

5. In wet settings where a base is needed for tile application, _____ is the better option.
   a. Type X gypsum board
   b. cement backer board
   c. Type C gypsum board
   d. moisture-resistant gypsum board

6. To meet fire-resistance-rating standards for firestopping established by the building and fire codes, _____.
   a. openings must have airflow
   b. mechanical firestop devices must be used on all openings
   c. nonmechanical firestop materials must be used on all openings
   d. all openings must be sealed

Module 45104 Review Questions

1. Gypsum board retards the spread of fire because _____.
   a. its core is exposed on the end
   b. its field and edge are covered with paper
   c. its core is composed of all man-made materials
   d. it contains water in its crystal molecular structure

2. One of the advantages of gypsum board is that it is _____ because it is readily available and easy to apply, repair, decorate, and finish.
   a. economical
   b. sustainable
   c. versatile
   d. durable

3. Gypsum board is considered a(n) _____ material because in addition to naturally mined gypsum, it uses flue gas desulfurization (FGD) gypsum, a synthetic byproduct of coal-burning power plants.
   a. durable
   b. sustainable
   c. versatile
   d. economical

4. If you break the face paper on a gypsum board while driving a nail or screw, _____.
   a. remove the gypsum board
   b. remove the nail or screw
   c. drive the nail or screw farther into the gypsum board so it is invisible
   d. drive another nail or screw properly near the same place

5. When you are installing fire-resistance-rated wallboard, the nails should penetrate the studs by at least _____.
   a. $\frac{1}{2}$"
   b. $\frac{3}{4}$"
   c. $\frac{7}{8}$"
   d. $1\frac{1}{8}$"

6. The head of a drywall hammer has a _____, which transfers to the depression around the nail head and helps joint compound to adhere.
   a. flat surface
   b. waffle texture
   c. curved hook
   d. Phillips head attachment

7. _____ are recommended when attaching gypsum panels to steel framing that is 30 mil (20 gauge).
   a. Self-drilling screws
   b. Self-piercing screws
   c. Annular nails
   d. Coarse screws

8. _____ are designed to fasten gypsum board to wood framing or furring strips.
   a. Fine thread screws
   b. Self-tapping screws
   c. Self-drilling screws
   d. Coarse thread screws

9. When using any type of drywall adhesive, be sure to _____.
   a. apply multiple coats
   b. keep the work area hot to speed up drying time
   c. follow the manufacturer's directions for use
   d. avoid all solvent-based adhesives

10. You would use a _____ to smooth rough-cut edges of gypsum board.
    a. sanding block
    b. rasp
    c. drywall saw
    d. T-square

11. The point at which the edges of two panels of drywall meet is known as a _____.
    a. bedding seam
    b. joint
    c. gypsum lath
    d. depression

12. Gypsum drywall should only be installed with adhesive when the building temperature is greater than _____.
    a. 50°F
    b. 60°F
    c. 70°F
    d. 80°F

13. Fasteners should be applied to wallboard working from _____.
    a. top to bottom
    b. edge to edge
    c. center to edge
    d. corner to corner

14. In floating interior angle construction where the framing is perpendicular to the wall/ceiling intersection, the ceiling fasteners should be located _____ from the intersection.
    a. 1"
    b. 7"
    c. 10"
    d. 12"

15. Drywall trim needs to be applied to all _____.
    a. edges of wallboard exposed to view
    b. multi-ply construction
    c. single-ply construction
    d. wall board that has been secured with adhesive

16. If the project you are working on requires you to use gypsum board in moist conditions, _____.
    a. use multi-ply construction
    b. do not apply board directly over a vapor barrier
    c. use drywall contact adhesive to secure the board
    d. use parallel construction

17. _____ ceiling and wall assemblies are necessary in construction that requires a one- or two-hour fire-resistance rating.
    a. Wood framed
    b. Sound-rated
    c. Single-ply
    d. Multi-ply

18. When subjected to the extreme heat of fire, _____ materials expand to form a strong insulating material that seals an opening for three to four hours.
    a. endothermic
    b. intumescent
    c. moisture resistant
    d. multi-ply

19. Each of the following is a construction method used to control noise except _____.
    a. caulking around outlet box openings
    b. placing air conditioning ducts back-to-back
    c. using separate framing for the two sides of a wall
    d. mounting gypsum board in resilient channels

20. A low STC rating indicates _____.
    a. excellent fire resistance
    b. excellent sound isolation
    c. poor fire resistance
    d. poor acoustical performance

Answers to Odd-Numbered Module Review Questions are found in *Appendix A*.

Answers to Section Review Questions

Answer	Section	Objective
Section One		
1. c	1.1.0	1a
2. a	1.2.2	1b
3. a	1.3.0	1c
4. c	1.4.0	1d
Section Two		
1. d	2.1.1	2a
2. d	2.1.2	2a
3. c	2.1.4	2a
4. d	2.2.0	2b
5. b	2.3.1	2c
6. d	2.4.1	2d

Interior Finish, Trim, and Cabinets

Objectives

Successful completion of this module prepares you to do the following:

1. Describe the job preparation and safety planning required to complete a finish carpentry project.
   a. Identify tool and equipment safety guidelines for finish carpentry tools and materials.
   b. Explain how to perform a material takeoff for window, door, and ceiling trim.
2. Identify the different types of standard moldings and materials.
   a. Identify the different types of base, wall, and ceiling moldings.
   b. Identify the different types of window and door trim.
3. Explain how to install different types of molding and trim.
   a. Describe common techniques for cutting and fastening trim.
   b. Explain the process for installing door and window trim.
   c. Describe the steps for installing base and ceiling molding.
4. Identify common types of cabinets and countertops.
   a. Identify common types of wall, base, and island cabinets.
   b. Describe common types of countertops.
5. Identify common cabinet components and hardware.
   a. Describe various cabinet components, such as doors, drawers, and shelves.
   b. Identify common fasteners and hardware used in cabinet assembly.
6. Explain how to lay out and install a basic set of cabinets.
   a. Describe common surface preparation techniques required before cabinet installation.
   b. Explain how to install wall cabinets.
   c. Describe the process for installing base cabinets and countertops.

Performance Tasks

Under supervision, you should be able to do the following:

1. Make square and miter cuts to selected moldings using a hand miter box and a compound miter saw.
2. Make a coped joint using a jigsaw or coping saw.
3. Install interior door, window, base, and ceiling trim using a finish nailer.
4. Identify and install selected hardware on cabinet doors.
5. Identify and lay out various types of base and wall units following a specified layout scheme.
6. Install a selected type of cabinet and countertop.

CODE NOTE

Codes vary among jurisdictions. Because of the variations in code, consult the applicable code whenever regulations are in question. Referring to an incorrect set of codes can cause as much trouble as failing to reference codes altogether. Obtain, review, and familiarize yourself with your local adopted code.

Digital Resources for Carpentry

Scan this code using the camera on your phone or mobile device to view the digital resources related to this craft.

Overview

Skilled finish carpenters are always in demand for installing wall and ceiling moldings, as well as finish trim around doors and windows. These skilled craft professionals may also be called upon to install cabinets and other casework components. Quality trim and cabinetry work require an exceptional ability to measure accurately, calculate angles, and make precise, clean cuts. These skills are vital because even small errors will be visible in the finished product. Therefore, an accomplished finish carpenter must develop a reputation for reliable results and remain vigilant throughout the project to quickly address minor mistakes and make adjustments to this highly detailed work.

NCCER Industry-Recognized Credentials

If you are training through an NCCER-accredited sponsor, you may be eligible for credentials from NCCER. The ID number for this module is 27215. Note that this module may have been used in other NCCER curricula and may apply to other level completions. Contact NCCER at 1.888.622.3720 or go to **www.nccer.org** for more information.

You can also show off your industry-recognized credentials online with NCCER's digital credentials. Transform your knowledge, skills, and achievements into credentials that you can share across social media platforms, send to your network, and add to your resume. For more information, visit **www.nccer.org**.

1.0.0 Interior Finish and Trim Safety and Job Preparation

Performance Tasks

There are no Performance Tasks in this section.

Objective

Describe the job preparation and safety planning required to complete a finish carpentry project.

a. Identify tool and equipment safety guidelines for finish carpentry tools and materials.

b. Explain how to perform a material takeoff for window, door, and ceiling trim.

Trim: Millwork placed around wall openings and at wall intersections at floors and ceilings.

Moldings: A defined, decorative element that outlines the edges of a structure and covers gaps at intersections of walls, ceilings, floors, and openings.

Interior finish and trim installation are a few of the finishing touches that complete the building process. Because trim and cabinets are some of the most visually striking interior components, a highly trained, detail-oriented craft professional must install every trim piece with straight lines and tight-fitting joints to create a clean and professional final product.

This module covers the materials, equipment, and procedures used to install interior trim, cabinets, and countertops.

Trim is installed to create an attractive finished look and hide joints at the intersection of walls, floors, ceilings, and various wall coverings. Trim also protects vulnerable edges. Trim should blend with the overall design of the building or room. For modern residential and commercial buildings, the interior trim typically uses simple, modern moldings to create a clean aesthetic, while traditional architectural styles tend to use moldings with elegant and complex shapes and grooved profiles (*Figure 1*).

Figure 1 Interior trim.
Source: Vadym Andrushchenko/Alamy

Personal Protective Equipment

Proper personal protective equipment (PPE) is required when installing trim. Always follow general safety guidelines, including the use of proper respiratory, hearing, and eye protection.

Respiratory protection is required for sanding and finishing operations to reduce inhalation hazards. For sanding small amounts of trim in a fairly open area, a dust mask might be the only respiratory protection required. Other means of respiratory protection will likely be required to apply finish sealer to trim in an enclosed area. Check OSHA (Occupational Safety and Health Administration) guidelines and establish a site-safety plan to protect workers and occupants from different types of hazards. For more general safety guidelines, study NCCER Module 00101, *Basic Safety (Construction Site Safety Orientation).*

1.1.0 Tool and Equipment Safety

A craft professional must always be aware of general safety guidelines and site-specific hazards in every phase of the construction process. Accidents occur during finishing work because building interiors nearing completion can create a false sense of security. The following simple guidelines will help keep you and your team safe in this essential building phase:

- *Maintain your work area* — Cluttered work areas create tripping hazards (*Figure 2*), which increase the risk of life-threatening falls and injuries. Aside from mitigating workplace hazards, maintaining a clean and tidy work area reduces the work for other trades using the space after you, and it shows general contractors and project teams that you are a detail-oriented professional with pride in your work.

- *Protect and maintain power tools and equipment* — Familiarize yourself with each tool's operating manual and ensure that all necessary guards are in place. Do not use power tools in damp or poorly lit locations and avoid using them in the presence of flammable liquids or gases.

Figure 2 Clean versus disorganized work area.
Sources: ungvar/Shutterstock (left); Martin Deja/Alamy (right)

- *Use proper eye and respiratory protection* — Wear safety glasses, goggles, or a face shield while operating power tools, and use a dust mask or respirator to reduce the risk of breathing in dust and fumes. This rule extends to other craft professionals working near interior cutting stations that create dust, especially when working in rooms with limited ventilation.

- *Do not overreach* — Keep proper footing and balance at all times, especially when working from a ladder. Maintain at least three points of contact with a ladder and never allow your core to extend beyond the vertical rails.

- *Maintain tools with care* — Keep tools sharp and clean for safe performance and extended longevity. Follow instructions for lubricating and changing accessories and make necessary adjustments when the power tool is unplugged. Inspect tool cords, moving parts, and protective guards in accordance with the manufacturer's maintenance schedule. If damaged, take the tools out of service and have them repaired by an authorized service facility. Keep handles dry, clean, and free from oil and grease (*Figure 3*).

- *Stay alert* — Plan your work and remain focused on the task at hand. Do not operate a power tool when you are tired or under the influence of medication, alcohol, or drugs. Construction is a fast-paced, high-stress environment, but take time to plan your work and remain focused on the task at hand.

Figure 3 Common power tools.
Source: mihalec/Shutterstock

Many accidents occur due to rushed work processes or craft professionals working in a distracted state of mind. Remember that every craft professional is responsible for site safety, and it is your duty to call for work stoppage when you witness hazards or unsafe work practices. If you see something, say something! Your actions could save a life.

1.1.1 Finish Nailer Safety

Finish nailers, staplers, and pin nailers (*Figure 4*) are some of the most common tools for installing trim. Finish nailers speed up the trim installation time and often result in less damage to the trim pieces than hand-nailing.

Figure 4 Battery-operated finish nailer, stapler, and pin nailer.
Source: Ridge Tool Company

Finish nailers and staplers are designed to fire only when the trigger is engaged, and the tool is pressed firmly against the material being fastened. This safety feature reduces the risk of misfires, while also ensuring that nails are driven straight at the appropriate depth. Practice the following finish nailer safety guidelines to protect yourself and your co-workers:

- Review the manufacturer's instruction manual prior to using a finish nailer and follow all necessary safety, operation, and maintenance guidelines.
- Always wear appropriate personal protective equipment, including proper eye and hearing protection when using a finish nailer.
- Never aim a nail gun near your body or in the direction of other people. Although smaller than traditional framing nails, finish nails can easily puncture skin, causing severe pain and potential work-loss injuries (*Figure 5*).
- Check for pipes, electrical wiring, vents, and other materials behind the trim before nailing.
- Do not exceed the manufacturer's recommended pressure for finish nailers.

1.1.2 Router Safety

Routers (*Figure 6*) may be used when installing trim to create recesses for hardware and other accessories. A sharp router bit presents a cutting hazard, so follow these guidelines to ensure safety when using a router:

- Always disconnect a router from its power source when inserting and tightening bits, as well as when making depth adjustments.
- Always wear appropriate PPE, including proper eye, face, and hearing protection, and a face shield if you are working with a material that produces excessive dust and debris.
- Remove all electrical cords and other impediments from the cutting path.
- Keep both hands on the router when using it and don't let go until the motor comes to a complete stop (*Figure 7*).

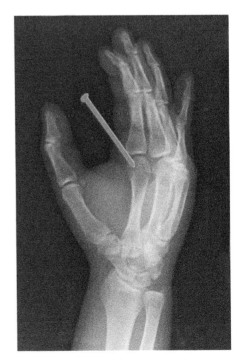

Figure 5 X-ray of a nail gun injury.
Source: Scott Camazine/Alamy

Figure 6 Plunge router and router bit.
Sources: Ridge Tool Company (left); KPixMining/Alamy (right)

Figure 7 Proper router safety.
Source: devilmaya/Alamy

1.1.3 Miter Saw Safety

Miter saws and compound miter saws (*Figure 8*) are common tools for cutting trim before installation. These tools speed up the process of cutting trim to angles and bevels in locations where a mitered joint is necessary to produce a seamless appearance. Mitered joints reduce gaps at outside corners of baseboards and crown moldings and top corners of trim around openings.

These power saws are typically attached to a sturdy table or stand to expedite work and prevent back injuries due to excessive bending (*Figure 9*). The main difference between the two is that miter saws chop straight down on materials at whatever angle they are locked into, while compound miter saws have a bevel adjustment that can cut these same angles with a vertically tilted blade. General safety guidelines for using a compound miter saw include the following:

- Always wear appropriate PPE, including eye and hearing protection.
- Keep your hands and other materials out of the blade path.
- Allow the blade to reach full speed before cutting materials and do not reach within the blade path before it has stopped rotating.
- Hold materials firmly against the rear fence for accurate, stable cuts.

Figure 8 Compound miter saw.
Source: Ridge Tool Company

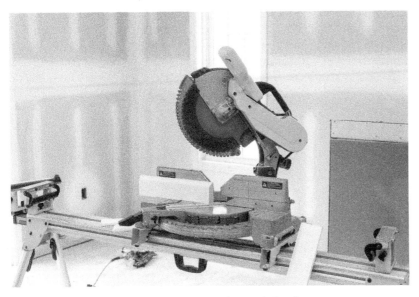

Figure 9 Compound miter saw attached to elevated stand.
Source: valentyn semenov/Alamy

1.1.4 Jigsaw, Table Saw, and Band Saw Safety

Jigsaws and stationary band saws (*Figure 10*) have thin reciprocating blades that allow a craft professional to cut tight corners, curves, and complex shapes. The main difference between the two is the jigsaw is typically hand-held for a versatile, portable option for finish carpenters needing detailed cuts on the jobsite, while heavier table saws and stationary band saws are used primarily in shops and designated cutting stations. Jigsaws are common tools for cutting countertops and sink cutouts, while table saws (*Figure 11*) and stationary band saws are excellent options for ripping long pieces of trim or carving intricate details.

Figure 10 Portable jigsaw and stationary band saw.
Sources: Ridge Tool Company (left); alfredhofer/Shutterstock (right)

Figure 11 Portable table saw.
Source: Ridge Tool Company

Safety guidelines for these tools are similar to those for a traditional table saw and circular saw, and include the following:

- Wear your PPE, such as cut-resistant gloves, and protect yourself against dust inhalation by equipping your tools with vacuum accessories and working in a well-ventilated area.
- Use a material roller for support when pushing long material through a table saw or stationary band saw alone. This will allow you to feed the material through your cut without the ends falling off the other side of the table.
- To keep your hands away from the fast-moving blades, use a manufactured or job-built material holder to push thin rips through table saws and stationary band saws.

- Hold your material firmly with clamps to reduce the vibrations caused by the reciprocating blades and keep both hands on the tool as you work. Otherwise, your material can move and increase the risk of cutting hazards. If you must hold the material with your nondominant hand, ensure that you hold the material edge and keep your fingers from the penetrating blade path (*Figure 12*).

Figure 12 Proper portable jigsaw safety.
Source: Ridge Tool Company

1.2.0 Performing a Material Takeoff for Interior Finish

Although the task of performing a material takeoff will typically fall to your project supervisor or estimator, it is an important skill that you should develop to ensure your team has the required amount of materials to finish your project scope. Various estimating software options, such as Procore® and Bluebeam®, can expedite and automate the process, but understanding the basic concepts of this project-planning task will allow you to develop a deeper understanding of material estimation. In short, the more you study blueprints and manufacturer instructions, the easier it is to see how each building component fits together to produce the final product (*Figure 13*).

To perform a material takeoff, begin by checking the architectural drawings, project specifications, and finish schedule. These construction documents will note specific types of materials used for each application, including doors, windows, baseboards, and ceilings. Then study the plans to determine how many linear feet of each type of trim will be required to complete the project.

Trim is divided into two categories (*Figure 14*):

- *Running trim* — This type of trim is installed in random lengths for such applications as baseboards, crown molding, and chair rails.

- *Standing trim* — This trim is specified in custom lengths for door and window casings or other special applications where splicing is not acceptable.

Estimating running trim is relatively simple, as seen in the following example of base trim:

Step 1   Find the sum of the perimeter of all rooms, minus the widths of the doorways.

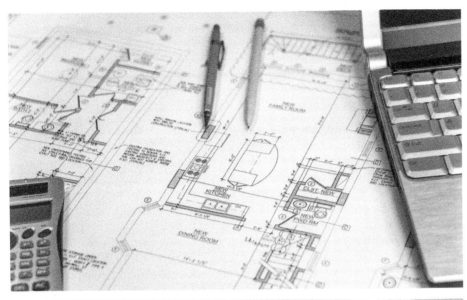

Figure 13 From blueprint to final product.
Sources: StudioDin/Shutterstock (top); Andy Dean Photography/Alamy (bottom)

Figure 14 Running and standing trim.
Source: Andreas Prott/Alamy

Step 2 Add a small percentage for waste. Although the standard percentage is 10 percent, this waste allowance will fluctuate depending on the complexity of the room layouts and your team's experience level. For example, if your team is composed of veteran finish carpenters and you are working on a housing development project with identical layouts, your waste percentage can be decreased significantly because you can estimate that fewer mistakes will be made.

Step 3 Order the longest standard mill lengths to cover the perimeter and minimize splicing.

Estimating standing trim requires a few more steps than running trim, as shown in the following door casing example:

A typical 6'-8" door that is 2'-10" wide requires four 7' pieces of side casing and two 42" head casings, all of which include an extra 4" for each end that requires mitering. Therefore, five 7' pieces of casing are required for each door. Because standard lengths are usually 6', 8', 10', 12', 14', and 16', you will want to order 14' lengths to minimize waste. Three 14' lengths (cut in half) will case one door with a 7' length left over for the next door. For every two doors then, only five 14' lengths are required. If all the doors are the same, divide the number of doors by two and multiply the result by five to obtain the total number of 14' lengths to order.

1.0.0 Section Review

1. Hearing protection, such as earplugs, should be worn when working with _____.
   a. handsaws
   b. coping saws
   c. compound miter saws
   d. block planes

2. You can find most information required to perform a material takeoff for trim in the _____.
   a. manufacturer instructions
   b. architectural drawings
   c. RCP plan
   d. structural drawings

2.0.0 Types of Molding and Trim Materials

Performance Tasks

There are no Performance Tasks in this section.

Objective

Identify the different types of standard moldings and materials.
   a. Identify the different types of base, wall, and ceiling moldings.
   b. Identify the different types of window and door trim.

2.1.0 Common Molding Materials

Wood moldings are sometimes referred to as *woodwork, trim, finish,* or *millwork.* To manufacture various molding profiles, preselected lumber thickness is sawn (or ripped) lengthwise into strips of various widths, which are then sorted and run through a band saw to produce pieces known as *blanks.*

Many wood moldings are made of pine because this soft wood type can easily be milled into intricate profiles. Premium pine is clear, knot-free, and milled to receive stain or other finishes. Paint-grade pine moldings, including **finger-jointed stock,** represent the next step down. Finger-jointed moldings (*Figure 15*) are assembled from shorter lengths, and two additional steps take place between the ripping and resawing operations.

Strips of clear wood are passed through a finger jointer, which cuts small fingers at the ends of each piece. These short, fingered pieces are then glued and joined into longer lengths. Finger-jointed moldings are only used for paint or opaque finishes because joints are highlighted by a stained or natural finish.

Finger-jointed stock: Paint-grade moldings made in a mill from shorter lengths of wood joined together.

Figure 15 Finger-jointed molding.
Source: Rawf8/Alamy

A common downside to working with pine moldings is that they can be damaged relatively easily. Premium hardwoods like poplar, oak, cherry, and maple are often the better option for high-wear areas. However, they are often reserved for custom projects because they are more expensive and harder to work with.

MDF Moldings

Prefinished moldings are often made of medium-density fiberboard (MDF), a relatively stable product made of glued and pressed wood fiber (*Figure 16*). MDF materials are denser than plywood and don't split, warp, and cup like natural wood. However, they are more susceptible to water damage and less resistant to wear and tear.

Figure 16 MDF molding.
Source: Nikola Spasenoski/Alamy

Standard moldings come in a variety of widths and are normally available in even lengths of 8', 10', 12', 14', and 16'. Some are available in odd lengths, such as moldings used for door casing, which are made in 7' lengths to reduce waste.

PVC Moldings

Polyvinylchloride (PVC) or cellular PVC trim (*Figure 17*) are some of the most affordable and versatile trim options available on the market today. The material is flexible and easy to install like MDF but doesn't absorb water, making it an excellent choice for trimming high-moisture areas of the home, such as bathrooms, kitchens, and laundry rooms. Installation processes vary between

Figure 17 PVC molding samples.
Source: Bobo_Custura/Shutterstock

different products, but many PVC and plastic moldings can be attached to a wall surface with a manufacturer-approved adhesive.

PVC moldings are also termite-proof, fire-resistant, and can hold up to normal wear and tear around high-traffic areas. You'll often find PVC trim used in commercial buildings, apartment complexes, and tract housing developments because they are easy to install and maintain, and they are considerably cheaper than wood or MDF products when they are bought in bulk.

2.1.1 Base Moldings

Base moldings (*Figure 18*) are used at the floor level to conceal the joint between a wall and the floor. They protect the lower wall surfaces from damage by vacuum cleaners, brooms, furniture, and foot traffic. Base moldings come in a variety of sizes and a multitude of contemporary and traditional profiles. Some of the most common profiles of base moldings are colonial, craftsman, and modern (flat).

Figure 18 Typical base moldings.
Source: Destina/Alamy

In general, base moldings have an unmolded lower edge for a cleaner connection to the floor. A base shoe molding can be added near the bottom of base molding to conceal any unevenness between the trim and floor or to hide edges of carpeting and other floorings (*Figure 19*). The top profiles of modern base moldings are often decorative, making the use of additional base cap molding on top unnecessary.

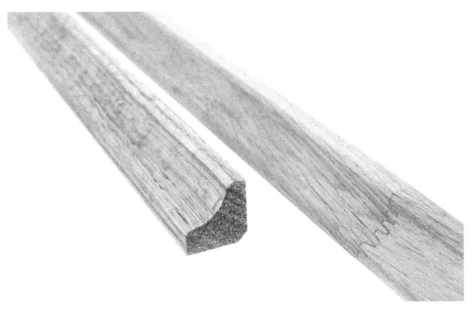

Figure 19 Shoe molding.
Source: JENYA/Alamy

Quarter-round moldings (*Figure 20*) are common alternatives to base shoe moldings and are excellent trim accessories for creating a seamless transition and reducing bottom gaps where dirt and dust could accumulate. Quarter-round moldings have a pie-shaped quarter-circle profile with most sizes between $\frac{3}{8}$" and $\frac{3}{4}$".

NOTE

Wider base moldings tend to make a room look smaller, so it is best to avoid their use unless the room is spacious with high ceilings.

Figure 20 Quarter-round molding.

2.1.2 Wall Moldings

Wall moldings provide a decorative and functional purpose by adding character to interior designs and protecting wall materials from damage. Extensive wall molding designs in classical, Victorian, colonial, craftsman, and art deco architectural styles all use a combination of outside-corner moldings, wainscoting, and chair rail to develop a cohesive visual style.

As the name suggests, outside-corner moldings protect outside corners, especially in high-traffic areas. This molding type is not as common today, but you may come across it in traditional, craftsman architectural styles (*Figure 21*).

Figure 21 Dining room with chair-rail, inside-corner, outside-corner, and wainscot-cap moldings.
Source: Artazum/Shutterstock

Wainscoting, Chair Rail, and Picture Rail Moldings

Wainscoting: A decorative wall panel covering that protects lower elevations of an interior wall.

Wainscoting is a wall finish consisting of panels applied partway up the wall, whereas a wainscot cap covers exposed end grain (*Figure 22*). A wainscot cap may include a lipped profile that covers the top of the wainscoting. The main difference between wainscot caps and back band is that caps finish the top of the wainscoting structure, while back band trim is a purely independent decorative element that is used to increase the thickness and aesthetic appeal of casing around doors and windows.

Chair-rail molding is installed at chair-back heights between 30" and 35" above the floor to protect the wall surface. It may also be used as a cap for wainscoting or as a horizontal dividing line between two surfaces such as wallcovering and paint.

You may also find horizontal moldings that resemble chair rails installed between 7' and 9' above finished floor (AFF) in houses built before World War II. These are called picture-rail moldings, and although their purpose in the early 20th century was providing a surface for hanging pictures and paintings, they are now used in modern design as an artistic wall delineation, adding a classic contrast to flat walls and an eye-catching reference line near the tops of window and door openings.

Figure 22 Wainscoting.
Source: Ursula Page/Shutterstock

2.1.3 Ceiling Moldings

Crown and cove moldings are installed around the perimeter of a room to hide the joints between walls and ceilings. Cove moldings' concave profiles (*Figure 23*) are the simplest and most common moldings for the wall-to-ceiling connection.

Crown moldings have decorative profiles that provide a more traditional appearance. Crown moldings come in several profile variations (*Figure 24*) with the top-third portion typically being concave, while the lower side of a crown molding has a wider surface to fit against the wall. You may also find these decorative elements on the upper edge of antique furniture or cabinetry.

Figure 23 Ornamental crown molding.
Source: Bilanol/Shutterstock

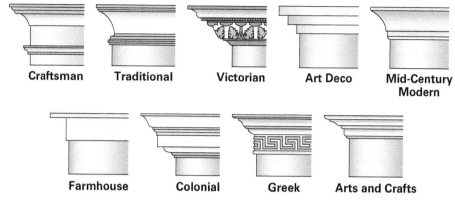

Figure 24 Popular crown molding profiles.
Source: NCCER, based on photo from FIXR

2.2.0 Window and Door Trim

Casing moldings (*Figure 25*) are used as trim around windows and doors to hide gaps between the frames (jambs) and the wall. Common profiles include colonial, craftsman, and modern, which are often selected to match base moldings.

Figure 25 Casing moldings.
Source: Krista Abel/Shutterstock

Both windows and doors have side and head casings that frame the face of the openings, with side-jamb extensions returning toward the openings on the sides, and head-jamb extensions on the top. Interior doors and doorless hallway openings do not have thresholds, to allow easy access from one room to the next. Windows have a lower trim piece to complete the window casing frame. Many molding styles use two built-out sections of molding known as a stool and an apron to add depth to the window trim (*Figure 26*).

2.2.1 Stools and Aprons

Stool: The bottom horizontal trim piece of a window that lays flat above the apron, commonly known as a *windowsill*.

Apron: A piece of window trim, sometimes known as *undersill trim*, that is installed under the stool of a finished window frame.

The **stool** (sometimes known as a *sill*) is a shelf-like, horizontal trim piece that terminates at the bottom of the window opening and often runs long on either side of the side casing to form a lip known as a *horn*. A stool covers the wall-to-window transition and can be accompanied by a decorative, supporting molding called an **apron**. These trim components are popular in craftsman residential interiors but are not as common in modern design, which typically uses flat casing around each side of a window opening to resemble a picture frame (*Figure 27*).

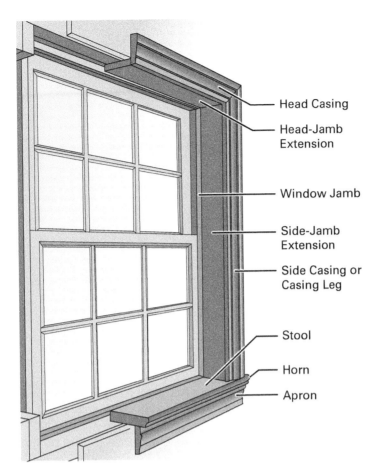

Figure 26 Window trim components.

Labels: Head Casing · Head-Jamb Extension · Window Jamb · Side-Jamb Extension · Side Casing or Casing Leg · Stool · Horn · Apron

Figure 27 Various styles of window trim at residential entryway.
Source: Artazum/Shutterstock

2.2.2 Door Stop Moldings

Some doors have casing stops or door stop moldings (*Figure 28*), which conceals gaps between the door frame and wall, provides additional support to the door frame, and acts as a stop for the door when it is closed in the jamb or a protective cushion that prevents the door from swinging open past a certain point.

Figure 28 Doorstop molding.

This protective characteristic prevents the door from swinging hard into the adjacent wall and creating dents and holes at the doorknob or handle. An additional doorstop can be installed at the location where the door strikes the wall surface or baseboards, or you can install it at a position on the floor that allows the door to open to the optimal swing.

Trim Relief

The back side of most baseboard, casing, and chair-rail trim molding is relieved to allow the trim to carry over any minor drywall/plaster irregularities. This allows the edges of the trim to lie flat against most surfaces. If major irregularities exist, they must be sanded down to allow the trim to lie as flush as possible. For paint-grade trim, any minor gaps can be filled with a flexible sealant.

Source: Andrii Biletskyi/Alamy

2.0.0 Section Review

1. A type of molding that trims the floor-to-wall connection is known as _____.
   a. crown molding
   b. wainscot molding
   c. casing molding
   d. base molding

2. The flat window trim that is installed below a stool is called a(n) _____.
   a. apron
   b. head casing
   c. base jamb
   d. stool cap

3.0.0 Installing Molding

Objective

Explain how to install different types of molding and trim.
   a. Describe common techniques for cutting and fastening trim.
   b. Explain the process for installing door and window trim.
   c. Describe the steps for installing base and ceiling molding.

Performance Tasks

1. Make square and miter cuts to selected moldings using a hand miter box and a compound miter saw.
2. Make a coped joint using a jigsaw or coping saw.
3. Install interior door, window, base, and ceiling trim using a finish nailer.

Because trim is one of the most visible features of the interior of a house or building, it must be treated carefully during preparation and installation. Accurate cuts must be made at the intersection or termination of trim when mitered to other trim members, flooring, or wall-to-ceiling connections as seamlessly as possible. Miter joints must be made precisely so the adjoining pieces perfectly match (*Figure 29*). This section provides some general installation guidelines for cutting and installing trim moldings.

Figure 29 Precise molding corner connections.
Source: rangizzz/Shutterstock

3.1.0 Cutting and Fastening Trim

Trim carpentry requires craft professionals to use special saws, such as a jigsaw or compound miter saw, to make different types of cuts to moldings. While most cuts are made using the power saw options, handsaws may also be an excellent option for cutting trim in areas where you must limit noise and dust. The most common handsaws used for cutting moldings are the backsaw, dovetail saw, and coping saw (*Figure 30*). These saws have fine-toothed blades that produce precise and clean cuts for several high-detail finish carpentry tasks.

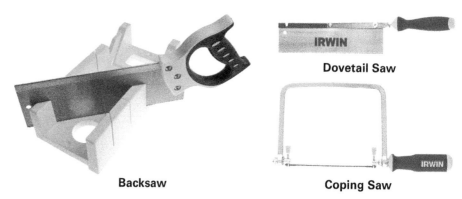

Dovetail Saw

Backsaw

Coping Saw

Figure 30 Special trim saws.
Sources: Aleksandr Ugorenkov/Alamy (left); Courtesy of Stanley Black & Decker, Inc. (right)

Coped joint: A joint made by cutting the end of a piece of molding to fit seamlessly to the contoured profile face of adjoining molding.

Coping saws and jigsaws are some of the most efficient tools for cutting curves and irregular lines along the profile of trim molding to make a **coped joint** (*Figure 31*).

Figure 31 Coped molding.
Source: NCCER, based on photo from Popular Woodworking

A coping saw has a steel frame with a narrow, fine-toothed blade that can rotate to cut small curves. A jigsaw will mimic this motion with a thin, high-speed reciprocating saw blade, and is available in portable battery-operated and stationary options.

3.1.1 Making a Coped Joint

When installing baseboard and ceiling moldings, coped joints create a more seamless appearance than miter joints on inside corners, because mitered joints tend to open at the inside corners.

You can make a coped joint with the tracing method by performing the following steps:

Step 1 Cut the first piece of molding to fit flush, butting into the intersecting wall. Install it with finish nails.

Step 2 Use a tape measure to find the length of the trim piece that will butt against the installed molding. Add a few inches to your measurement to account for the coping cutoff. It's better to have material that is too long than a piece that's too short.

Step 3 Flip the uninstalled piece face down and trace the profile of a scrap piece of molding on the back side to use as your cut line reference. Ensure that you stay to the outside of your mark when cutting. You can always sand or file imperfections to get a tight fit.

Step 4 Use a coping saw or jigsaw to trim off the excess material, tilting the blade to bevel cut more material off the back side than the front face profile. You can produce more intricate cutting patterns with a portable jigsaw by using a dome-shaped Collins® coping foot attachment. The smooth rounded base allows more flexibility and control as you make intricate angled cuts along your reference line (*Figure 32*). Test fit the two pieces and make any necessary adjustments before installing it flush against the installed piece.

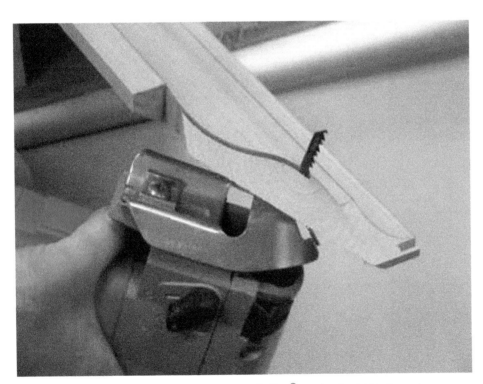

Figure 32 Portable jigsaw equipped with a Collins® coping foot attachment.
Source: Collins Tool Company

3.1.2 Making Square and Miter Cuts

Square cuts are made to straighten uneven square ends of moldings and to make butt joints at walls, door casings, and similar trim features. In basic trim work, 45-degree miter cuts are commonly made at the tops of doors and windows at the joints where the side and head casing meet, at outside corners of baseboards, and when scarf joints are required (*Figure 33*).

You can craft **scarf joints** by overlapping two pieces of molding, one with an open miter cut and the other with a 30-degree closed miter cut. Use a scarf joint when you need to join two sections of molding on a wall or ceiling to achieve the required length (*Figure 34*).

Whether you are using a handsaw and miter box or a powered compound miter saw, remember to cut materials with their face side or edges up, or toward you, to hide potential splinters on the hidden side.

Scarf joints: End joints made by overlapping two pieces of molding with angle cuts.

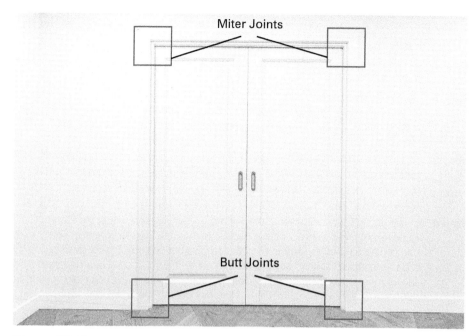

Figure 33 Typical butt and miter joints.
Source: Serjio74/Shutterstock

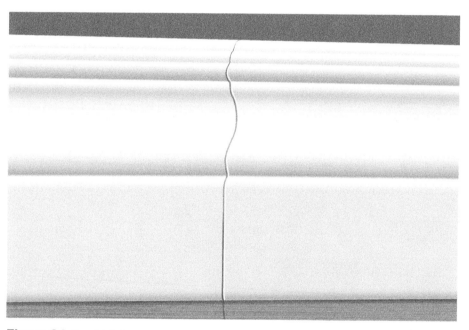

Figure 34 Scarf joint.

Flat miters, such as those made for door and window casings, are cut by holding the molding with its face side up and its thicker edge against the fence (back side) of the miter saw (*Figure 35*). Some moldings, such as the base, base cap and shoe, or chair rail, are held right side up. In this instance, place the bottom edge against the miter saw table with the back side against the fence (*Figure 36*). Once the molding is properly positioned, you can make quick and precise cuts.

When working with crown and cove moldings, some craft professionals position the molding upside down in the miter saw, at the same 45-degree angle at which it will be installed (*Figure 37*). The lower side of the molding fits against the wall and has a wider surface than the top side, which rests against the ceiling. When placing the molding in the miter saw, position it as if the fence were the wall, and the table were the ceiling. You can also use a jig to support the back side of the molding and keep it in the proper position while cutting.

Figure 35 Performing a flat miter cut.

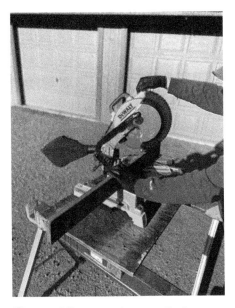

Figure 36 Performing a vertical miter cut.

Figure 37 Cutting crown molding at an angle.

When using a compound miter saw to cut crown molding, the molding's broad back surface is laid flat on the saw table, and the saw-bevel and miter-angle controls are set to achieve the desired cut. Most miter saw manufacturers provide a chart in the operator's manual that shows the bevel and miter settings to use with a compound miter saw for cutting different inside and outside corner cuts. An example cutting chart is shown in *Table 1*. Always make test cuts on scrap material to ensure the bevel and miter settings will produce the desired results before making compound miter cuts to crown molding.

TABLE 1 Compound Miter Saw Miter and Bevel Settings for Crown Molding (52° Top-Rear Angle and 38° Bottom-Wall Angle)

Type of Cut	Bevel Setting	Miter Setting	Remarks
Left side, inside corner	33.8°	Right, 31.6°	Position the top of the molding against the miter box fence.
			The left side of the molding is the finished piece.
Right side, inside corner	33.8°	Left, 31.6°	Position the bottom of the molding against the miter box fence.
			The left side of the molding is the finished piece.
Left side, outside corner	33.8°	Left, 31.6°	Position the bottom of the molding against the miter box fence.
			The right side of the molding is the finished piece.
Right side, outside corner	33.8°	Right, 31.6°	Position the top of the molding against the miter box fence.
			The right side of the molding is the finished piece.

3.1.3 Fastening Trim

Traditionally, interior trim was typically fastened in place using a lightweight, smooth-faced claw hammer and a nail set to drive finishing nails. Today, trim is commonly installed using pneumatic or battery-operated finish nailers (*Figure 38*).

Figure 38 Fastening trim.
Source: Courtesy of Stanley Black & Decker, Inc.

Finish nails are available in three diameters: 15-gauge, 16-gauge, and 18-gauge. The thinner the nail, the less likely it is to split the wood. Therefore, a 16-gauge or 18-gauge nailer is commonly used for installing door and window casing, window aprons, and other thin trim elements. Stools, hardwood trim, and similar heavier materials are installed using a 15-gauge nailer. Set your nail-driving depth with a flush stop or adjust the drive pressure. You can also fix nail depth by hand with a hammer and nail set (*Figure 39*).

Figure 39 Nail set.
Source: Courtesy of Stanley Black & Decker, Inc.

3.2.0 Installing Trim

Install interior door trim after the finished floor is laid, but before the base trim is installed. Use the same door casing material as you used to trim windows to create a uniform appearance throughout the room.

NOTE

The procedures in this section assume that the doors are installed and are plumb and square with the wall.

Matching a Base Molding and Door Casing

The door casing must be thicker than the base molding so that the base molding can butt into the casing without leaving the base-molding end grain visible. Note that the door casing is typically $1/16$" to $1/8$" thicker than the base molding. In those instances where the casing is not thicker than the base molding, a plinth block can be used, as shown on the doors here.

Plinth Blocks

Source: KUPRYNENKO ANDRII/Shutterstock

Step 1 Use a combination square or tape measure and mark the distance from the jamb edge to the desired **reveal** on the inside face of the side and top door jambs (typically $1/4$").

Step 2 Hold a section of case molding in place at the top of the door, then mark the length of the head casing. This is the distance between the reveal marks on the door-side jambs.

Step 3 Using the points marked on the head casing as a guide for where the heels of the miter cuts begin, cut left-hand and right-hand miters on the ends of the head casing.

Step 4 Set the head casing back from the edge of the door jamb the same distance as the reveal and nail it in place at the jamb edge and at the outer edge of the casing. Drill holes near miters to avoid splitting the casing.

Step 5 Cut two pieces of case molding slightly longer than needed for the door's two side-case moldings. Cut a left-hand miter on one side-case molding and a right-hand miter on the other side-case molding.

Step 6 Turn one of the side-case moldings upside down so that its mitered end rests on the floor, then mark the other end at the top edge (miter toe) of the head casing. Cut it off square, allowing the pencil mark to remain. If the finished floor is to be carpeted, locate the casing above the subfloor to allow for the carpet and pad height (*Figure 40*).

Reveal: The distance that the edge of a casing is set back from the edge of a jamb.

Figure 40 Base molding installed above carpet flooring.
Source: Artazum/Shutterstock

Step 7 Apply wood glue on the miters, set the side casing back the distance of the reveal from the edge of the jamb, and nail it in place.

Step 8 Repeat Steps 6 and 7 for the remaining side casing.

Step 9 Set all finish nails and, if necessary, sand the high edge at mitered corners without sanding across the grain.

Guidelines for Fastening Trim

- Remove any pencil marks left along the edge of a cutline before fastening the trim. Pencil marks in interior trim corners are difficult to remove after the pieces are fastened into position. Pencil marks can also show through a stained or clear finish, making the joint appear open. When marking interior trim, make light, fine pencil marks to reduce your cleanup time.
- Ensure that all joints fit tightly by measuring, marking, and cutting carefully. Take pride in your work and never leave an interior trim project with misaligned trim and ill-fitting corners. If you make a mistake, take the time to replace trim pieces until you are satisfied with the final product.
- Repair any dings or dents left by finish nailers or hammers.
- Set finish nails below the surface of the trim and fill nail holes with putty matching the paint or stain finish. The set depth for finish nails should equal the nail head width. Prefinished molding often requires color-matching nails that are driven flush.

NOTE

The procedures in this section assume that the windows are installed and are plumb and square with the wall.

3.2.1 Installing Window Trim

Start with the window and door casing when you begin trimming out a room. Two basic methods are used to trim windows: conventional and picture-frame (*Figure 41*). As shown, the conventional method uses casing at the top and both sides. The bottom of the side casing rests on the stool. A horizontal member called an apron is installed below the stool on the face of the wall. In the picture-frame method, the top, bottom, and sides of the window are all trimmed with casing.

The first step to trimming a window is ensuring that the edge of the jamb is flush with the inside wall surface. If the jamb projects beyond the surface, plane it or sand it flush. If the jamb is recessed, wood strips called *jamb extensions* must be nailed over the jamb edge to bring it flush with the inside wall

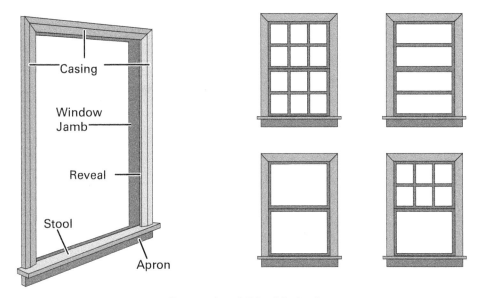

Conventional Trim Method

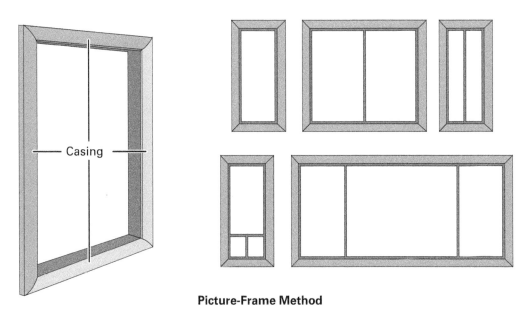

Picture-Frame Method

Figure 41 Window trimming methods.

surface (*Figure 42*). If the wall projects slightly beyond the jamb, the inside wall may require sanding to allow the casing to lie flush against the jamb and wall surface.

Jamb extensions may be supplied by the manufacturer with the window unit, they can be ordered from the mill, or they can be made on-site. Jamb extensions are not typically installed with the window, but later when the window is trimmed. If the extensions are supplied with the window, store them in a dry, safe place until you need them. Manufactured jamb extensions are usually precut to length and only require width adjustments. This is done by ripping the extension to the required width and cutting a slight back-bevel on the inside edge.

As with most carpentry procedures, installing interior trim can be accomplished in more than one way. Trimming a window using the picture-frame method is done in basically the same way as the conventional method, except

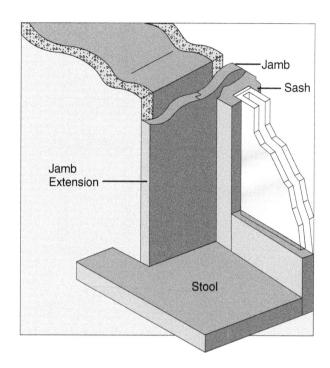

Figure 42 Installation of jamb extensions.

instead of using a stool and apron, a piece of casing is installed across the bottom of the window. Complete the following steps to cut and install a stool:

Step 1   Determine the length of the stool. The stool length is typically equal to the distance between the inside edges of the window jambs, plus the reveal on both sides (typically $\frac{1}{4}$" on each side), plus twice the casing width, plus twice the casing thickness (*Figure 43*). The reveal is the distance that the casing edge is set back from the edge of a jamb. For example, if the jamb-to-jamb measurement of the window is 36" and you are using $2\frac{1}{4}$" $\times$ $\frac{3}{4}$" casing, the length of the stool would be $42\frac{1}{2}$" (36" + $\frac{1}{2}$" + $4\frac{1}{2}$" + $1\frac{1}{2}$").

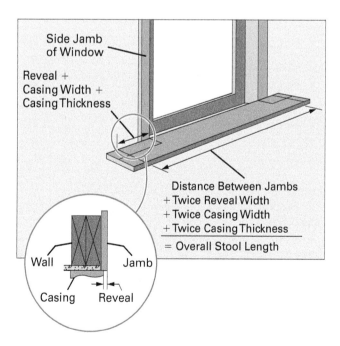

Figure 43 Method of determining the length of the stool.

Step 2   If you want to cover the wood grain at the stool ends with returns to provide a finished appearance, lay out and mark the length of the stool, but do not cut the stool. If return ends are not being used, the stool can be cut to length.

Step 3   Hold the stool trim level with the sill (centered on the window opening) and mark the inside edges of the side jambs on the top of the stool, using a square and pencil. Then, mark a line on the face where the stool fits against the wall surface.

Step 4   Carefully cut out the locations that will butt flush against the wall and slide the stool into position (*Figure 44*).

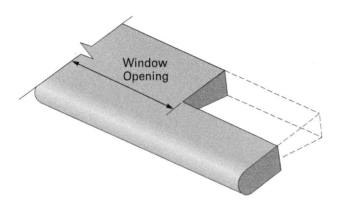

Figure 44 Window stool showing cutout for one side of the window opening.

Step 5   Lower the sash carefully on top of the stool and scribe the cutoff line so the sash clears the stool and cut off the excess material. When finished, the stool should fit flush against the wall with $^1/_{16}$" clearance between the back edge of the stool and the bottom rail of the window sash. This clearance is necessary to prevent the sash from binding when raising or lowering it.

Step 6   If you use returns on the ends of the stool to provide a finished appearance, mark and cut a return at each end of the stool. The returns are formed by making 45-degree miter cuts at each end of the stool, as shown in *Figure 45*. Glue and nail the return pieces to the ends of the stool.

Step 7   Sand all sawed edges before nailing the stool in place, ensuring that it is level and perpendicular. Some installation processes require that the underside of the stool be embedded in caulking compound before being nailed in place.

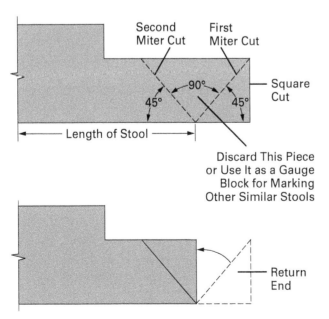

Figure 45 How to mark and cut a return.

Complete the following steps to cut and install the side and top (head) casing for your windows:

Step 1   On the inside face of the side and top jambs, measure and mark the distance from the jamb edge to the desired reveal setback. A $\frac{1}{4}$" reveal setback was used in this example.

Step 2   Hold a section of case molding at the top of the window and mark the length of the head casing. This mark illustrates the distance between the reveal marks on the window side jambs (*Figure 46*).

Figure 46 Checking the head casing for accurate miter cuts.
Source: joserpizarro/Shutterstock

Step 3   Using your marks on the head casing as a guide for where the heel of each miter cut begins, cut left-hand and right-hand miters on the ends of the head casing.

Step 4   Set the head casing back the distance of the reveal from the edge of the jamb and nail it into place. Drill holes near miters to prevent the casing from splitting.

Step 5   Cut two pieces of case molding slightly longer than needed for the two side-case moldings. Cut a left-hand miter on one side-case molding and a right-hand miter on the other.

Step 6   Turn one of the side-case moldings upside down to rest the mitered end on top of the stool, then mark the other end at the top edge (miter cut toe) of the head casing. Cut it off square, allowing the pencil mark to remain.

Step 7   Apply wood glue on the miters, set the side casing back the distance of the reveal from the edge of the jamb, and nail it into place.

Step 8   Repeat Steps 6 and 7 for the remaining side casing.

Simple aprons are installed below the stool flat and horizontally aligned with the outside edge of the side casing. However, custom window trim with molded casing might require mitered returns that fit seamlessly with the apron's face and hides the cut-off edge (*Figure 47*).

Figure 47 Custom window casings.
Source: Chris Haver/Alamy

Follow these steps to cut and install an apron with returns:

Step 1 Cut your apron piece to a length that matches the outside-to-outside window side casing. Use a scrap piece of molding as a template and draw its profile flush with the apron ends. Cut out the profile with a jigsaw or coping saw before sanding the ends. Glue and nail the returns in place at both ends (*Figure 48*).

Step 2 Nail the apron in place under the stool but refrain from forcing the stool upward to avoid pushing the piece out of level.

Step 3 Finally, set all finish nails in the apron and casing. If necessary, sandpaper the high edge at mitered corners with the grain.

Figure 48 Finished apron return.
Source: David Berlekamp/Shutterstock

Marking Gauge

A commercial marking gauge can be set to scribe a mark for the reveal setback on the edges of window or door jambs. If a commercial gauge is not available, a gauge can be made on the jobsite by rabbeting the four edges of a square block of wood for the desired amount(s) of reveal. This template block is placed inside the jamb with the correct reveal width overlapping the edge of the jamb. Use a pencil to mark the reveal while sliding the block around the side jambs and head jamb.

Source: Jeffrey B. Banke/Shutterstock

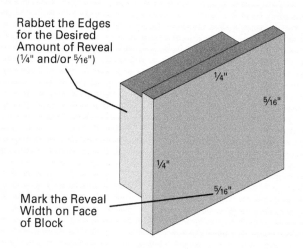

Rabbet the Edges for the Desired Amount of Reveal (¼" and/or ⁵⁄₁₆")

Mark the Reveal Width on Face of Block

Figure 49 Typical base molding corners.
Sources: hiv360/Shutterstock (top); shablovskyistock/Shutterstock (bottom)

3.3.0 Installing Molding

Base molding runs continuously around the bottom edges of a room and is one of the last items of interior trim to be installed because it must be installed around door casings. Inside baseboard joints are coped for intricate profiles and mitered for modern, plain baseboards. Most outside corners are mitered (*Figure 49*).

Before installing moldings, locate the studs within the room, and mark their locations on the wall above the baseboard height or on the subfloor if the walls are finished. Then, select and place lengths of base molding around the sides of the room. Select lengths that will allow you to make complete runs without joints; if you cannot do so, cut the joint over a stud so you can properly nail the material.

There is more than one way to complete any task, but typical procedures for installing base and related shoe molding are described here. Shoe molding is installed in the same basic manner as base trim, except that it is nailed into the floor instead of the baseboard. This prevents the joint under the shoe molding from opening if it shrinks. Also, because shoe molding is a smaller molding with solid backing, both inside and outside corners are typically mitered. Wherever it meets a door casing with nothing to butt against, back-miter and sand the end until smooth.

Typically, the installation of the base trim starts on a long wall of the room and/or on the side of the room opposite the door. Starting on a long wall makes it easier to get a good fit with a long piece of base trim (*Figure 50*).

Figure 50 Starting trim along a long wall, opposite the door.
Source: Artazum/Shutterstock

Figure 51 shows an example of a simplified baseboard installation, with the numbers 1 through 7 representing a typical sequence of molding installation. As shown, baseboard installation begins on the unbroken rear wall facing the door, because this wall can be fitted with a single length of baseboard. Installation proceeds according to the following steps:

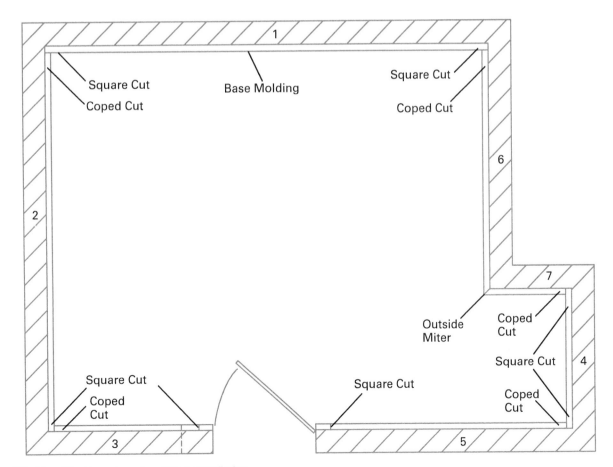

Figure 51 Simplified baseboard molding installation.

Step 1   Begin by measuring the wall at floor level.

Step 2   Next, select a straight length of baseboard molding and make a square cut on one end. Measure and mark the required length before making a square cut on the other end. Make the length slightly longer than required so that the cut will allow a spring fit.

Step 3   Nail the piece in place at the stud locations along the wall. If the finished floor is carpet, fasten the base trim above the subfloor to accommodate the pad and carpet, which will tuck underneath the base. Skip this step when installing molding after the carpet.

Step 4   Next, install the second piece of baseboard on the adjoining (left-hand) wall. This piece of baseboard is joined to the first with a coped cut at the inside corner. If necessary, file the contour to get a tight fit.

Step 5   Once the coped cut is made, measure from the bottom of the first base piece to the next corner and square-cut the second piece at the other end to butt tightly into the intersecting wall surface. Nail the second piece of trim in place. Keep working your way around the room in this manner until all baseboard molding is cut and nailed in place.

Guidelines for Finishing Trim

- If the trim is to be stained, make sure all traces of glue are removed from exposed surfaces, using a damp cloth. Dry glue will seal the surface and not allow the stain to penetrate, resulting in a blotchy finish.

- Sand all interior trim pieces smooth after they are cut and fitted, and before they are nailed in place. Always sand with the grain, never against the grain. Sanding of interior trim provides a smooth base for the application of stains, paints, and clear coatings.

- All sharp, exposed corners of trim should be rounded slightly.

As shown in *Figure 51*, all inside corners have coped joints and the outside corner has a mitered joint. Use square-cut butt joints in any location where the baseboard molding meets the door casing. Be careful when fitting the baseboard to the door casing. Do not make the joint so tight that you push the casing out of position when you spring the baseboard into place.

When making the outside miter joint for paint-grade moldings, make the coped cut for the inside corner on one piece first, then hold the baseboard in position and mark the back edge for the miter cut, as shown in *Figure 52A*. Make a 45-degree miter cut and fasten the piece to the wall, as shown in *Figure 52B*. Repeat the procedure for the second piece, as shown in *Figure 52C* and *Figure 52D*.

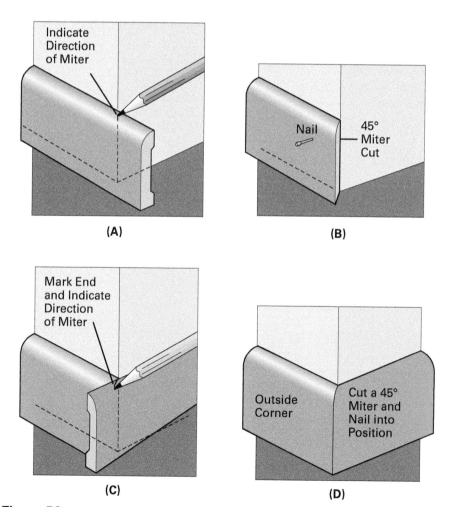

Figure 52 Procedure for fitting and cutting an outside miter joint.

If you need two spliced baseboard lengths to span a long wall, make the splice using a scarf joint by following these simple steps.

Step 1 Set the first piece of molding in place and mark the center point of the stud nearest to the end of the piece. Subtract half the thickness of the molding and cut the end to a closed 45-degree miter.

Step 2 Install the first piece, but do not nail into the last stud where the piece is mitered. Cut a 45-degree open miter at the end of the second piece. Then, measure from the face of the miter on the first piece to the wall's corner before cutting the second piece to length.

Step 3 Set it in place with the closed miter on the first piece overlapping the open miter on the second piece (*Figure 53*). Apply glue to the joint and nail both pieces into the stud, then continue nailing the second piece to the corner. Many craft professionals use a biscuit joint cut with a biscuit (plate) joiner to splice wide moldings (*Figure 54*). Biscuit joints are an excellent alternative to scarf joints but you will see them more commonly used in cabinetry than molding installations.

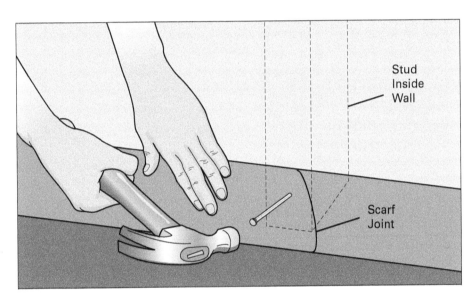

Figure 53 Scarf joint, cut and nailed directly over a wall stud.

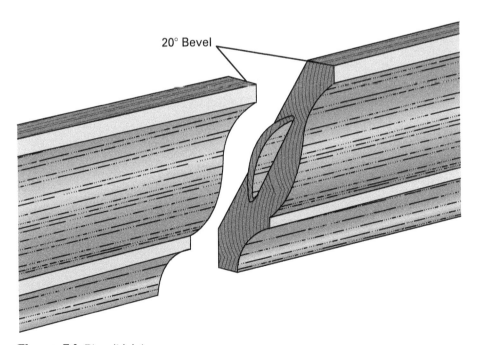

Figure 54 Biscuit joint.

Finding Outside Angles

Few outside corners are exactly 90 degrees, so it is best to use a sliding T-bevel or a digital angle-finder tool to find the exact angle made by the outside corner. Divide this angle by two and cut a closed miter at each piece of joining base to that angle.

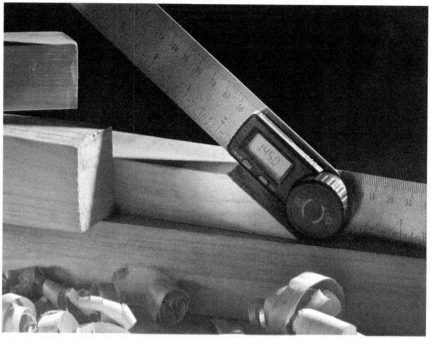

Source: Mariyana M/Shutterstock

Irregular or Wavy Baseboard Mounting Surfaces

When installing baseboard trim against a final finished floor without shoe molding, it is important to eliminate any noticeable gap between the bottom of the base trim and the floor. Complete this task by placing the final cut length of baseboard for each side of the room on the floor against the wall, checking that it is flush with the floor along its entire length. If the floor is warped or sagged, use a pair of dividers set to the widest gap width for any baseboard, and scribe a mark along the bottom edge of each wall baseboard. Trim the bottom of the base molding to the line with a jigsaw or band saw, or sand to the line with a belt sander. If the walls behind the baseboards are uneven, and the base molding is paint grade, high spots on the wall can be sanded down. Any gaps may be caulked after baseboard installation if they are not more than $\frac{1}{8}$". If the gaps are larger, or the baseboard is stain-grade, use a two-piece baseboard with a small cap and force-nail the cap into the gaps; otherwise, float the wall using drywall compound to level the gaps before nailing the cap and/or the base molding.

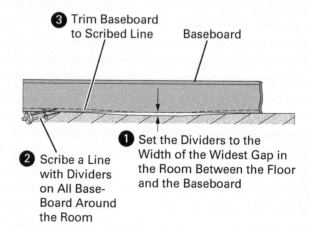

3.3.1 Installing Ceiling Molding

The general procedures for installing ceiling trim are basically the same as installing baseboard trim. However, when using crown/bed moldings, the moldings are not applied flat against the wall. Instead, they cover the wall/ceiling joint at a 45-degree angle (*Figure 55*).

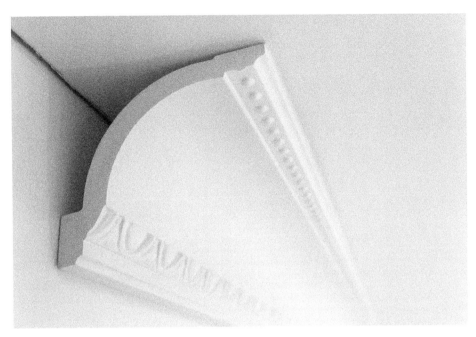

Figure 55 Ceiling molding installation angle.
Source: Andrii Biletskyi/Alamy

If the ceiling has only inside corners, install each piece of trim with a square cut at one end and a coped cut of matching profile at the other end, if possible (*Figure 56*).

As with base trim in rooms where a long wall is opposite the door, run square-cut crown molding from corner to corner and cope the molding of the

Figure 56 Ceiling molding with only inside corners.
Source: Ric Jacyno/Shutterstock

Backing Boards

When installing wide wood crown molding (4" or more), install wide backing boards (nailers) above the edges of ceilings that are parallel with the ceiling joists (A). These boards must be fastened to the top of the top plate to supply nailing support for the upper crown-molding edge. If wide backing boards haven't been installed, you can apply triangular wood-support blocks behind the crown molding, which secures to the wall studs (B). In this case, you must nail the crown molding at the bottom edge roughly two-thirds of the way up from the bottom. If a narrow nailer exists, you may fasten two pieces of baseboard trim at right angles and secure the assembly to the wall studs and narrow nailer (C). The crown molding is now ready to be installed and nailed to the base molding pieces. Other solutions involve formal cornice construction using both crown and bed molding (D).

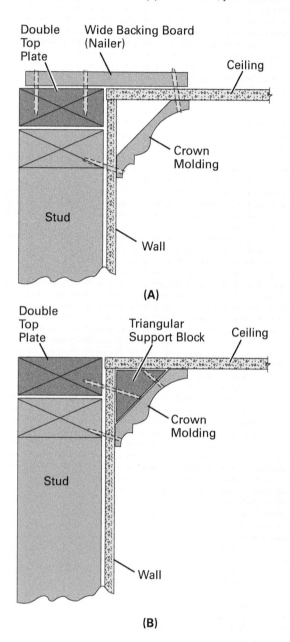

(A)

(B)

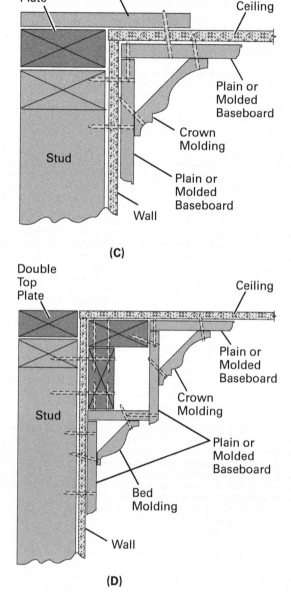

(C)

(D)

adjoining walls. This will help make the coped joints less noticeable if they separate due to wood shrinkage. Unfortunately, one side may require close fitting and coping of both ends of the molding. Like base trim, always cut crown molding slightly longer than required and spring fit it into place when installing and nailing. If outside corners are required, use compound miter cuts.

When installing crown or cove molding, you must find the distance from the top edge (ceiling) to the bottom edge (wall) where the molding is installed. You can accomplish this by holding a scrap piece of molding against the ceiling and wall to determine the distance from the ceiling to its bottom edge. After you've measured this distance, all walls can be marked at both ends, and lines snapped between the marks to represent installation points for the bottom edge of the molding around the room. You can also find this distance by measuring the molding with a framing square. Use the tongue and blade of the framing square to represent the wall and ceiling and align a piece of crown molding in the corner of the square before measuring the tongue distance (*Figure 57*).

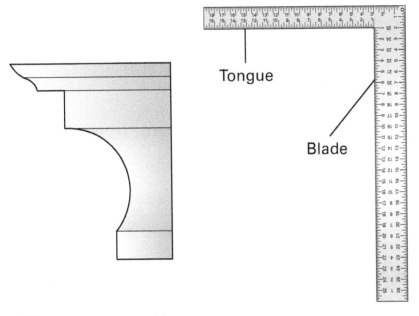

Tongue

Blade

Figure 57 Measuring crown molding.

Sagging or Uneven Ceilings

Note that most ceilings are not perfectly flat and level. When crown moldings or cornices are installed in a room, these uneven gaps may appear more noticeable on long runs. In the case of sagging ceilings, gaps will occur at both ends of the ceiling treatment. If the gaps are $\frac{1}{8}$" or less and the ceiling treatment is paint grade, the gaps may be caulked after installation. If the sag is less than $\frac{1}{4}$", and the ceiling trim is a molding, you may be able to bend the molding around the sag or nail the molding tight to the gap. Another solution is to scribe the gap or sag onto the top edge of the molding with a pair of dividers, trimming the molding to fit the ceiling. Finally, the best solution for fitting molding or cornices is to install the ceiling treatment straight and level and then float the gaps out using drywall compound or caulk the gaps if paint will be applied to the trim material.

3.0.0 Section Review

1. On the inside corners of baseboard and ceiling moldings with grooved profiles, it is best to cut your trim to form _____.
   a. coped joints
   b. scarf joints
   c. butt joints
   d. miter joints

2. The wood strips that must be nailed over the jamb edge to bring it flush with the inside wall surface are known as _____.
   a. stool projections
   b. jamb extensions
   c. casing planers
   d. jamb caps

3. When installing base molding, which areas of the room most commonly use miter joints?
   a. At the inside corners
   b. At the outside corners
   c. Where it meets the door casing
   d. When splicing two lengths of molding

4.0.0 Cabinets and Countertops

Performance Tasks

There are no Performance Tasks in this section.

Objective

Identify common types of cabinets and countertops.
   a. Identify common types of wall, base, and island cabinets.
   b. Describe common types of countertops.

This section introduces common types of cabinets and countertops. This background information will allow you to knowledgeably discuss cabinetry with customers, cabinet dealers, and manufacturers. The remainder of this module focuses on the basic procedures for identifying and installing stock cabinets and countertops at the jobsite.

4.1.0 Common Types of Cabinets

Cabinets are categorized into three major grades: stock, semicustom, and custom units.

Stock cabinets, also called modular cabinets, are mass-produced and sold off-the-shelf at home centers and builder supply centers. Stock cabinets are made in three forms: disassembled, assembled but not finished, and assembled and finished. Disassembled cabinets are known as ready-to-assemble (RTA) (*Figure 58*).

Semicustom cabinets are produced in a mill or cabinet shop. These cabinets are available in various materials, sizes, and finishes. They are usually sold by home centers, independent distributors, and kitchen and bath dealerships. In today's market, stock and semicustom cabinets will be the most common selections because they are easy to install and can be customized to several configurations that meet the client's needs and aesthetic preferences.

Figure 58 Ready-to-assemble cabinet unit.
Source: bane.m/Shutterstock

As the name suggests, custom cabinets are hand-built in a cabinet shop, mill-work plant, or on the jobsite to satisfy a specific application and design aesthetic. To create these units, a specialized cabinetmaker will measure the cabinet area, consult with the home or building owner, and build each cabinet unit to the desired specifications (*Figure 59*).

Figure 59 Custom cabinets.
Source: Artazum/Shutterstock

4.1.1 Wall Cabinets

Wall cabinets are typically the first units to be installed in a kitchen. A wall cabinet (*Figure 60*) is formed by fastening a series of individual wall cabinet units together in a desired configuration.

Figure 60 Wall cabinets.
Source: Papakah/Shutterstock

Wall cabinets are typically 12" deep and are available in single-door and double-door units with the same widths to match the base cabinets. They are made in several standard heights, with 30" being the most common. Shorter cabinets that are 24", 18", 15", and 12" in height are designed to be installed above sinks, refrigerators, and stoves (*Figure 61*). In kitchens without soffits, 36"- and 42"-high cabinets are frequently used to provide more storage space. Wall cabinets are typically available with adjustable or stationary shelves.

Figure 61 Short cabinet above sink.
Source: Hendrickson Photography/Shutterstock

The vertical distance between the countertop on the base cabinet and the bottom of the wall cabinets typically ranges from 15" to 18". However, building codes normally require that there be at least 24" to 30" of clearance between the cooking unit and any overhead wall cabinet, as well as a minimum of 22" between the sink and any overhead wall cabinet.

4.1.2 Base Cabinets

The lower units of cabinet configurations are known as base cabinets, and these are formed by fastening a series of individual base cabinet units together to achieve the desired configuration (*Figure 62*).

Figure 62 Base cabinets.
Source: Pixel-Shot/Shutterstock

Most base cabinets are manufactured $34\frac{1}{2}$" high and 24" deep. As shown, a countertop is installed on top of the base cabinet. By adding the usual countertop thickness of $1\frac{1}{2}$", the work surface of the base cabinet is at a standard height of 36" from the floor and the countertop is typically $25\frac{1}{2}$" deep.

Standard base cabinets come in several widths that vary in 3" increments, with single-door cabinets typically ranging between 9" and 24" wide. Most consist of one door, one drawer, and an adjustable shelf, although some cabinets have stationary shelves. Double-door base units range from 24" to 48" in width. A variety of floor-mounted tall cabinets are also made for oven, utility, and pantry units (*Figure 63*).

Figure 63 Kitchen casework with tall pantry units.
Source: suedanstock/Shutterstock

4.1.3 Island Cabinets

Kitchen islands are one of the most popular interior design elements because they provide additional counter space, storage, and seating in the center of a floorplan. These useful structures can be portable, like the stainless steel commercial islands used by professional chefs, or built in place, as you might find in luxury apartments and residential homes. For residential islands, the side facing the kitchen casework is often designed with cabinets and drawers for storing extra cookware and appliances, while the other sides provide an open space to accommodate bar seating (*Figure 64*). This allows the kitchen island to become a focal point of the home by providing functionality and an area to entertain guests while you cook.

Figure 64 Kitchen island with bar seating.
Source: Breadmaker/Shutterstock

The placement of these islands is critical to accommodate foot traffic, as well as to meet local code and potential ADA (Americans with Disabilities Act) requirements. You will find acceptable clearances in the approved construction set. You can also learn more information by researching the ICC A117.1-2017 standard, which outlines technical requirements that make buildings safer and more accessible. In most cases, there should be at least 36" between the outside edge of the island and any other cabinet or appliance, but 42" is preferable. Ensure that the island does not interfere with door swings or block access to large appliances.

Other applications call for the center island to be a prefabricated drop-in unit. Layout carpenters will mark the location on the slab or floor. Piping and electrical wiring may also be provided. Depending on the installation requirements and the manufacturer's instructions, cabinet installers may attach a 2 × 4 frame to the floor as an island base and attachment point (*Figure 65*).

(A)

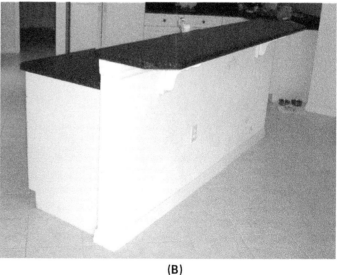

(B)

Figure 65 Island layout.

This base frame is not always necessary. For instance, if ceramic tile or hardwood flooring is installed against the island, the flooring material, combined with the weight of the island, will secure the island in place (*Figure 66*).

Figure 66 Kitchen island.
Source: Sheila Say/Shutterstock

Vanity Cabinets

Vanity cabinets used in bathrooms are constructed in the same way as kitchen base-cabinet units but are typically smaller to provide storage without dominating the bathroom layout. Most vanity cabinets are $31\frac{1}{2}$" high and 21" deep, with widths ranging from 18" to 36" in increments of 3", then 42", 48", and 60". They are available in a wide variety of door, drawer, and shelf configurations.

Source: Hendrickson Photography/Shutterstock

4.2.0 Common Types of Countertops

Numerous types of countertops can be used with cabinets. These include countertops covered with plastic laminate or those made from natural solid-surface materials. Some of the most popular materials used in modern construction are quartz, stainless steel, butcher block (wood), and even concrete (*Figure 67*).

Figure 67 Concrete countertop.
Source: Gcapture/Shutterstock

4.2.1 Plastic-Laminate Countertops

High-pressure laminate (HPL):
Composite material produced by pressing layers of plastic material under intense heat and pressure.

Plastic laminate, or **high-pressure laminate (HPL)**, is one of the most widely used materials to cover countertops. Made of layers of resin-impregnated kraft paper with a top layer of colored melamine, plastic laminates provide a durable, affordable, and easy-to-clean surface that can mimic the appearance and texture of wood, stone, and other natural materials (*Figure 68*). Plastic-laminate material is available in a wide range of colors and patterns in widths of 30", 36", 48", and 60", and lengths of 6', 8', 10', and 12'.

Figure 68 Laminate countertop with material samples.
Source: Chitaika/Shutterstock

A plastic-laminate countertop is usually $1\frac{1}{2}$" thick and overhangs the front and sides of the cabinets. Counter overhangs vary, but 1" is common for standard kitchen cabinets that are 24" wide. Tops for peninsulas or islands can be much wider. Bathroom vanity tops range from 19" to 22" wide.

Some plastic-laminate countertops are factory-built. These preformed countertops are made of plastic laminate bonded to a core material, generally medium-density fiberboard (MDF), plywood, or particleboard. The front edge of the countertop is rounded (bullnosed), and the rear edge will butt into the backsplash that protects the rear wall surface and moisture-absorbent MDF from water damage (*Figure 69*).

Backsplash: A protective and decorative wall covering typically installed behind kitchen and bathroom sinks, stoves, and countertops to prevent water damage to the wall surface.

Figure 69 Kitchen counter with subway tile backsplash.
Source: Hendrickson Photography/Shutterstock

These countertops are available in standard widths and are cut to length by the supplier or by the carpenter at the jobsite. Note that some L-shaped countertops are manufactured in two sections with a mitered corner. These sections are joined with hardware supplied with the countertop, along with preglued laminate endcaps.

A typical base for a job-built plastic-laminate-covered countertop is built up with two layers to achieve a total thickness of $1\frac{1}{2}$". The top layer (core) is a sheet of solid $\frac{3}{4}$" MDF, particleboard, or plywood (*Figure 70*). The second layer is made up of $\frac{3}{4}$" × 3" strips of plywood or particleboard attached to the front, back, and side edges. These strips reinforce and thicken the top. Cross strips

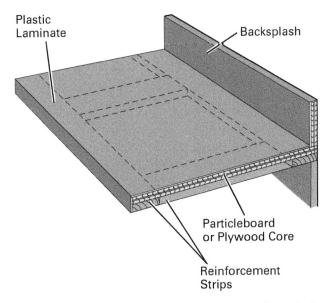

Figure 70 Typical construction of a plastic-laminate-covered countertop.

installed at 2' intervals at sink and appliance locations also provide additional strength. Any seams in the core material are reinforced underneath with large squares of plywood or another material.

After the countertop base is built, oversized strips of plastic laminate are cut and adhered to the exposed edges of the base using contact cement. The strips are then trimmed flush with a laminate trimmer or router, allowing an oversized top piece of laminate to be cemented in place. After trimming the top piece flush, the sink cutout(s) can be made (*Figure 71*). You may also need to scribe the countertop to the wall to ensure a tight and nearly seamless fit.

Figure 71 Worker installing sink into countertop cutout.
Source: Pawel_Brzozowski/Shutterstock

A backsplash attached to the back of the countertop is made as a separate piece. Plastic laminate is glued onto the backsplash core in the same way as for the countertop core and then is fastened to the back edge of the countertop with screws. When the countertop is installed, the backsplash is fastened to the countertop and adhered to the wall with construction adhesive. The joint between the countertop and backsplash is sealed with matching caulk or a waterproof sealer.

4.2.2 Solid-Surface Countertops

Countertops made of solid-surface materials are popular because they provide an attractive appearance and reliable durability. These countertops are custom-made in a variety of colors and finishes, including matte, satin, semigloss, and high gloss. Light colors, speckled patterns, and matte or satin finishes are excellent options for kitchen counters because they hide stains and normal wear and tear.

Marble, granite, quartz, and other stone countertop surfaces are typically installed by a specialty contractor; however, a finish carpenter may be tasked to build the supporting frame. Craft professionals are in high demand to produce concrete and butcherblock counters because these styles are popular for creating modern, industrial aesthetics and the production tasks align with a carpentry skill set (*Figure 72*).

Some manufacturer installation instructions require that dabs of silicone adhesive be applied to the top of the base cabinet frames before the countertop is placed on top of it. Others require that 1" wood strips be attached to the cabinets with the countertop glued to the strips. Some guidelines for a quality installation of a solid-surface countertop include the following:

- Level the counter with shims rather than hiding any gaps with silicone.
- Reduce the appearance of seams and ensure they are directly supported underneath.

- Inside corners should be cut on a radius and sanded to form a perfect curve, rather than a right angle, because a radius withstands stress better than sharp angles.
- Finishes should be consistent across the entire countertop surface. Examine the surface from several angles and under different lighting to identify any inconsistencies.

Figure 72 Butcher block countertop.
Source: Sheila Say/Shutterstock

4.0.0 Section Review

1. The three main categories of cabinets include base, wall, and _____.
   a. Eurostyle
   b. island
   c. custom
   d. shaker-style

2. Countertops made from marble, quartz, and stone are known as _____.
   a. MDF countertops
   b. ready-to-assemble countertops
   c. solid-surface countertops
   d. straight-surface countertops

5.0.0 Cabinet Components and Hardware

Objective

Identify common cabinet components and hardware.
   a. Describe various cabinet components, such as doors, drawers, and shelves.
   b. Identify common fasteners and hardware used in cabinet assembly.

Performance Task

4. Identify and install selected hardware on cabinet doors.

This section provides a basic overview of cabinets and countertops, including their surface preparation requirements and common installation techniques. In addition to the cabinets and countertops, a carpenter should be able to install cabinet doors, drawers, and hardware. This section presents the various cabinet components and hardware used to finish cabinets (*Figure 73*).

Figure 73 Typical configuration of base, wall, and island cabinets.
Source: Sheila Say/Shutterstock

Note that these are basic guidelines, and you should always follow manufacturer installation guidelines to ensure a stable and functional final product. Skipping steps or cutting corners could result in subpar performance and voided warranties, so remember that as a finish carpenter, one of your greatest strengths should be a laser-focused attention to detail.

5.1.0 Cabinet Construction and Components

All cabinets, whether designed for commercial or residential use, consist of a case fitted with shelves, doors, and/or drawers. Two types of cabinet construction are frameless and face-frame.

Frameless cabinets, also called European or Eurostyle cabinets, have no supporting framework or face frame (*Figure 74*). Support for the shelves and work surface is provided by the cabinet panels, which are heavy enough to carry the weight of the assembly and items stored in the cabinet. This style is common to achieve a contemporary look and provide more storage capacity because they are built without space-consuming framework. For example, a frameless four-drawer base cabinet contains about 1 cubic foot more usable space than a face-frame model of the same size.

Face-frame cabinets are similar in construction to frameless but are usually made of lighter stock materials because the face frame supports the doors. The edges of the cabinet front are covered with a solid lumber face with openings for doors and drawers. Face-frame cabinets are widely used in applications where a traditional aesthetic is desired (*Figure 75*).

Figure 74 Frameless "Eurostyle" cabinets.
Source: Pavel Adashkevich/Shutterstock

Figure 75 Face-frame cabinets.
Source: Essential Image Media/Shutterstock

Manufacturer Code Numbers

Cabinet manufacturers assign catalog code numbers to identify each type of cabinet and describe its dimensions. Let's use W361824 as an example. The letter(s) preceding the cabinet nomenclature refer to the type of cabinet, such as W for wall, B for base, SB for sink base, CW for corner wall, DB for drawer base, and PB for peninsula base.

The first number or pair of numbers after the letter(s) refers to the cabinet width. In the example, the number is 36. This means that the cabinet is 36" wide. If there is a second pair of numbers in the cabinet nomenclature, it refers to the cabinet height. In the example, the cabinet height is 18". If a third pair of numbers is included, it indicates the depth of the cabinet. In the example, the cabinet is 24" deep. If the third pair of numbers is omitted, the cabinets are standard depth; in the case of wall cabinets, 12" deep.

If the cabinet nomenclature has letters following the sets of numbers, they generally indicate a special feature. For example, a B24SS is a 24"-wide base cabinet with sliding shelves. The cabinet nomenclature may also have an L or R after it, indicating the hinge side of cabinet doors.

5.1.1 Cabinet Surfaces and Accessories

A wide range of materials is used in manufactured cabinets. Low-priced cabinets are usually made from panels of particleboard with a vinyl film or melamine applied to exposed surfaces. The surface is printed with either a solid color or with a wood-grain pattern to give the appearance of real wood.

The construction of visible exterior cabinet surfaces such as frames, doors, and drawers are made from **veneer** and a wide variety of softwoods and hardwoods including ash, beech, birch, cherry, oak, maple, and pine (*Figure 76*).

Veneer: A thin layer or sheet of wood intended to be overlaid on a surface to provide strength, stability, and/or an attractive finish. Thicknesses range between $1/16$" and $1/8$" for core plies and between $1/128$" and $1/32$" for decorative faces.

Figure 76 Assorted veneer samples.
Source: Visharo/Shutterstock

Hardboard, particleboard, plywood, or medium-density fiberboard (MDF) is commonly used for interior panels and drawer bottoms, and as the base for plastic-laminate countertops. Some points to look for in higher-grade cabinets include:

- Face-frame cabinets should have a $1/2$" to $3/4$" hardwood face frame, $1/2$" plywood or particleboard side frames, and a $1/4$" back panel.
- Frameless cabinets should use $5/8$" to $3/4$" particleboard or plywood for the entire cabinet. Wood veneer or high-pressure laminates are preferable to melamine for the exterior finish.
- Drawers should slide smoothly with little play. They should also close quietly and completely, with some including soft-close slides and accessories to eliminate the risk of loud or damaging slamming. These drawers can be regular or full-extension drawers, based on the specification. Look for a $1/2$" to $3/4$" solid wood or 9-ply solid birch plywood drawer box with integral wood dado, shoulder, or dovetail joints (*Figure 77*). The box is typically attached to the drawer front with screws.
- Door hinges and catch mechanisms should function without binding.
- Rabbeted joints should be used where the top, bottom, back, and side pieces join.

Various accessories may also be used to provide cabinets with a finished appearance. For instance, filler strips matching the material and finish of the base and wall cabinets are used to fill any gaps between the end base or wall cabinet units and the room walls, or between adjacent cabinet units when no combination of standard sizes can fill an existing space. These filler pieces are usually supplied in 3" or 6" widths and then ripped to the required width on the job (*Figure 78*).

Figure 77 Drawer with a dovetail joint.
Source: robcocquyt/Shutterstock

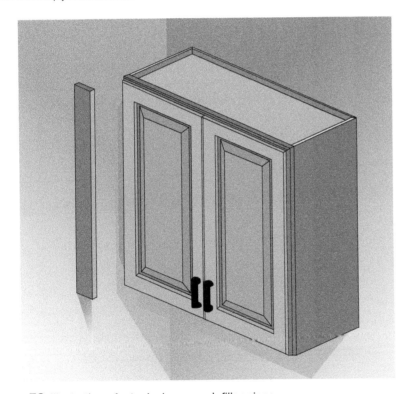

Figure 78 Illustration of a typical casework filler piece.

Other accessories include finished cabinet end panels and face panels for exposed ends and openings at appliances and at the end of cabinet runs. Attractive matching moldings are often used to trim stock or semicustom cabinets to make them more distinctive.

5.1.2 Cabinet Doors

Cabinet doors have a functional purpose but they can also provide decorative elements for the cabinet unit. Cabinet doors are generally swing or slide, with many configurations using a combination of both styles. Most cabinet doors are made of solid plywood or are built with panel-and-frame construction.

Panel-and-frame doors consist of a solid wood frame mounted around a panel of plywood or solid wood. The frame is joined by mortise-and-tenon, mitered spline, or stile joints. Some cabinet doors have a frame covered on each side with thin plywood, like a hollow-core interior door. Many different trim designs are used with cabinet swinging doors; however, the most common are flat-panel doors for a modern style and inset and shaker-style doors for a more traditional aesthetic (*Figure 79*). Shaker-style doors are some of the most common selections for custom projects, but the construction is simple with two vertical stiles on each side, one rail on top and bottom, and a center panel.

Figure 79 Popular cabinet door profiles.
Sources: Inna photographer/Shutterstock (left), Berkay Demirkan/Shutterstock (middle); ML Harris/Alamy (right)

Three main types of hinged swinging doors are used with cabinets: lipped, overlay, and flush (*Figure 80*). These terms refer to the way the door is mounted on the cabinet. A lipped door is rabbeted along all edges so that they overlap and conceal the cabinet opening when the door is mounted. Lipped doors are relatively easy to install because they do not require exact fitting in the opening and the rabbeted edges create a stop against the face frame of the cabinet.

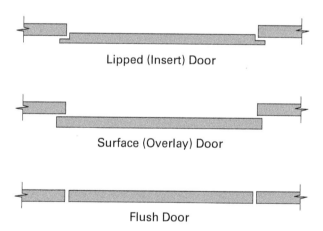

Lipped (Insert) Door

Surface (Overlay) Door

Flush Door

Figure 80 Types of cabinet door installations.

Overlay doors, or surface doors, are mounted on the outside frame and overlay the cabinet opening, concealing it. The overlay door is also easy to install because it does not require exact fitting in the opening and the face frame of the cabinet acts as a doorstop. Similarly, frameless (Eurostyle) cabinet door types completely overlay the front edges of the cabinet.

As the name entails, flush doors fit flush with the face of the cabinet opening and are the most difficult to install because they require an exact fit in the cabinet opening. A clearance of about $\frac{1}{16}$" must be made between the opening and the door edges. Stops must be provided in the cabinet against which the door will close.

5.1.3 Cabinet Drawers

Drawer fronts are generally made of the same material as the cabinet doors. The sides and backs of drawers are typically made of $^1\!/_2$"-thick solid lumber or plywood. The front and sides can be joined using dovetail, lock-shouldered, or rabbeted joints (*Figure 81*).

Cabinet drawers are also classified as overlay, lipped, and flush, depending on the way the drawer front fits the opening. The overlay drawer (*Figure 82*) is

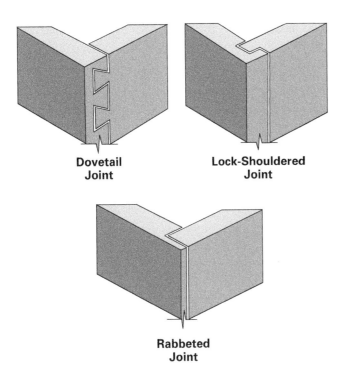

**Dovetail
Joint**

**Lock-Shouldered
Joint**

**Rabbeted
Joint**

Figure 81 Joints used to join drawer pieces.

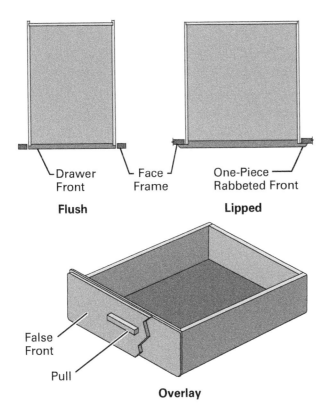

Drawer
Front

Face
Frame

One-Piece
Rabbeted Front

Flush

Lipped

False
Front

Pull

Overlay

Figure 82 Drawer construction.

formed by a self-contained drawer unit comprising a bottom, two sides, a back, and a false front. An overlay front, which conceals the cabinet drawer opening, is fastened to the false front with screws from the inside.

A lipped drawer is made in basically the same way as an overlay drawer, except that it does not have a false front. Both ends of the drawer front are rabbeted to receive the drawer sides. Also, the rabbets on each end are made large enough to overlap and conceal the drawer opening in the cabinet. The flush door is constructed similarly to the lipped drawer, except the front is cut to fit the overall height and width of the drawer opening. For this reason, a drawer stop is provided at the back of the drawer so that the drawer closes flush with the cabinet frame.

Several types of metal drawer guides can be used to support the drawer and limit lateral movement. They also ensure that the drawer does not tilt downward when fully opened. *Figure 83* shows a typical drawer-guide mechanism. Drawer guides are designed to handle loads of various weights and they are available in side-mount, bottom-mount, and undermount options. Generally, side-mounted drawer guides with stainless steel ball-bearing rollers carry more weight than bottom-mounted, single-rail drawer guides.

Figure 83 Typical drawer-guide mechanism.
Source: Bilanol/Shutterstock

Soft-Close and Push-to-Open Accessories

One of the most popular features in modern kitchens and bathrooms are soft-close and push-to-open hardware accessories, which can be added to sliding drawer guides and swinging door hinges. Touch catches and other push-to-open hardware (*Figure 84*) can be installed with screws to the inside of cabinet units or inlaid into the frame itself. With a gentle touch of pressure, the door will push outward, allowing for easy opening with little effort. This mechanism is especially useful for modern flat-panel ("slab") cabinets without face hardware, because the lack of knobs and handles makes opening flush-set frames more challenging.

Push-to-open latches can also be paired with hydraulic hinges (*Figure 85*) that propel cabinet doors into a stable semi-open position. This pairing is best for top-hinged overhead wall cabinets that open upward because you don't have to hold the door open as you retrieve items from storage.

Soft-close hardware (*Figure 86*) adds even more safety to convenience with built-in hydraulic mechanisms that provide resistance as the door or drawer closes. This cushioning effect reduces the risk of slamming your fingers, damaging the cabinet unit, or making loud noises when closing doors and drawers.

Figure 84 Push-to-open hardware.
Source: Alexander Borisenko/Alamy

Figure 85 Hydraulic hinges.
Source: Kutlayev Dmitry/Shutterstock

Figure 86 Soft-close hardware.
Source: Ivelin/Shutterstock

5.1.4 Cabinet Shelves

Most cabinets have shelves or dividers for organizing cookware and various items (*Figure 87*). Cabinet shelves can be made from solid wood, glass, plywood, and particleboard covered with plastic laminate. Most wood shelves are made of $\frac{3}{4}$" stock and should be no longer than 3' without supportive bracing.

Figure 87 Organized shelves.
Source: Paustowski/Shutterstock

Minimalist and farmhouse architectural designs lean toward a more open concept when it comes to shelving so you'll often find floating shelves above kitchen counters in place of larger wall cabinets (*Figure 88*). Floating shelves are supported by heavy-duty cylindrical brackets fastened to studs or backing. They are built hollow with an open back side unless they are designed to hold heavier objects like large dishes, in which case small holes are drilled into solid boards to accommodate the support brackets.

Figure 88 Floating shelves.
Source: David Papazian/Shutterstock

5.2.0   Fasteners and Hardware

Door and drawer hardware are available in a variety of types, shapes, sizes, and finishes. Using the proper hardware for a cabinet is important because these components can dramatically change the aesthetic and functionality of the finished product. In addition to using the correct style of hardware, the proper size is also important. If the hardware is too large for a cabinet, it will make the cabinet appear smaller. If small hardware is used on a larger piece, it will look out of scale.

5.2.1   Hinges

Numerous types of decorative hinges are available to match other detailed components. *Figure 89* shows a small sample of common hinges. Cabinet door hinges can be divided into several categories, including the following:

- *Surface-mount hinges* — Fastened to the exterior surface of the door and frame. The back side of the hinge leaves can be straight for use with flush doors or offset for use with lipped doors.
- *Offset hinges* — Used with lipped doors. A semiconcealed offset hinge has one leaf bent to a $\frac{3}{8}$" offset that is screwed to the back of the door. A concealed-offset type is one in which only the pin is exposed when the door is closed.
- *Butt hinges* — Used on flush doors when it is desired to conceal most of the hardware. The leaves of the hinge are set into the edges of the frame and the door.
- *European hinges* — The most commonly used hinge in modern construction due to its concealment and adjustability. As shown in *Figure 90*, the Euro-hinge's depth, height, and side-to-side positioning can be adjusted to obtain a level, flush match from door to door, and to compensate for any irregularities in the cabinetry.

Surface-Mount

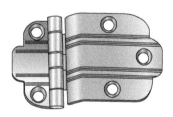

Offset

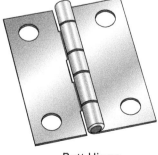

Butt Hinge

Euro-Hinge

Figure 89 Common cabinet door hinges.
Source: Cliff Day/Alamy (Euro-Hinge)

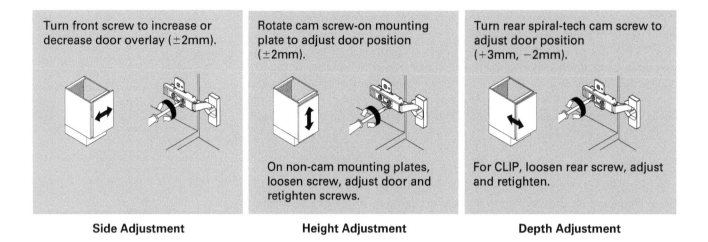

| Side Adjustment | Height Adjustment | Depth Adjustment |

Turn front screw to increase or decrease door overlay (±2mm).

Rotate cam screw-on mounting plate to adjust door position (±2mm).

On non-cam mounting plates, loosen screw, adjust door and retighten screws.

Turn rear spiral-tech cam screw to adjust door position (+3mm, −2mm).

For CLIP, loosen rear screw, adjust and retighten.

Figure 90 Euro-hinge adjustments.

5.2.2 Door Catches

Swinging doors without self-closing hinges (*Figure 91*) may require the use of door catches to hold the door closed. Magnetic, roller, flex (friction), and bullet catches (*Figure 92*) are common.

For magnetic catches, an adjustable magnet plate is fastened to the inside of the case, and the metal plate is fastened to the door at a desired location. Roller and flex (friction) catches are installed in a similar way as magnetic catches. The adjustable section is fastened to the case and the other section to the door. Ball-point (bullet) catches are spring-loaded catches that fit into the edge of the door. When the door is closed, the catch fits into a recessed plate mounted on the frame.

Figure 91 Self-closing hinge.
Source: bilanol/123RF

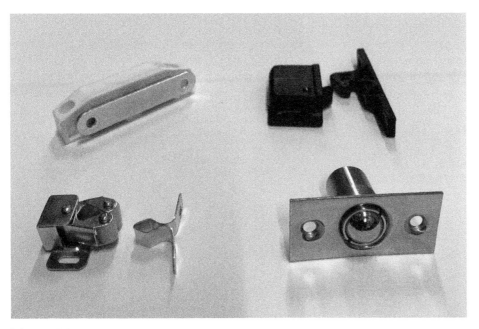

Figure 92 Cabinet door catches.

Full-Extension Drawer Slides

Full-extension drawer slides or guides are a little more expensive than ordinary drawer guides, but they are much more convenient because they permit access to the entire drawer. In contemporary kitchen designs with slab door profiles, you'll often see full-extension drawers hidden behind swinging doors to create a more sleek and seamless appearance than traditional aesthetics that break visual lines of larger cabinet configurations with individual pull-out drawers. The materials shown can also reduce installation time and material costs.

Source: PK-Donovan/Shutterstock

5.2.3 Knobs and Pulls

Many cabinet doors and drawers need pulls or knobs (*Figure 93*) to open and close them. They are available in many styles and designs and are made of decorative metal, plastic, wood, porcelain, or other material. Select matching hardware for both drawers and cabinet doors to improve visual continuity.

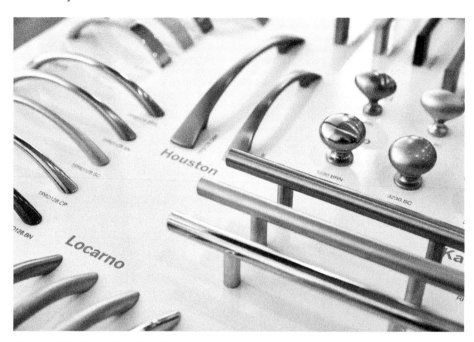

Figure 93 Examples of cabinet door and drawer pulls and knobs.
Source: PhotoMavenStock/Shutterstock

Knobs and pulls are usually installed by boring a hole through the door or drawer front and fastening them using a threaded bolt or screw that is normally supplied with the hardware. Countersink the hole if the supplied bolt or screw is not long enough to engage properly. You may use a job-built jig to expedite this process or purchase a manufactured hardware jig. Kreg® jigs are some of the most popular choices. These tools save time and ensure you drill precise holes for your hardware attachments (*Figure 94*). Knobs and pulls may also have escutcheon plates (*Figure 95*) that can be placed behind them to add a decorative look to the hardware.

Figure 94 Cabinet hardware jig.
Source: kasarp studio/Shutterstock

Figure 95 Ornamental escutcheon plate.
Source: Ronnie McMillan/Alamy

5.2.4 Cabinet Shelf Hardware

Some commercial cabinet shelves are adjustable with the use of slotted metal shelf rails (standards) and clips. Four standards are used, two installed on each side of the cabinet interior. The standards are cut to the required length and can be either surface mounted on the sides of the cabinet or recessed in dadoes. If mounted on the sides of the cabinet, the shelves are cut short enough to slip between the rails.

The more common type of adjustable shelf support used today involves two rows of stopped, drilled holes at the front and rear of each side of the cabinet interior. The number of such holes drilled in each row depends on the number of shelves the unit can hold. Typically, a series of holes spaced about $1\frac{1}{2}$" to 2" apart is drilled to cover an area of about 6" in the vicinity of each shelf's location. Two L-shaped plastic or metal pegs or spade pins are inserted into these holes at identical heights to support each end of the related shelf (*Figure 96*).

Figure 96 Adjustable shelf supports.
Sources: ND700/Shutterstock (left)

5.0.0 Section Review

1. Frameless cabinets are also known as _____.
   a. contemporary cabinets
   b. Eurostyle cabinets
   c. modern-design cabinets
   d. Scandinavian cabinets

2. The four most common categories of hinges are European, offset, butt, and _____.
   a. surface-mounted
   b. lipped
   c. emergency
   d. flex

6.0.0 Installing Cabinets

Performance Tasks

5. Identify and lay out various types of base and wall units following a specified layout scheme.
6. Install a selected type of cabinet and countertop.

Objective

Explain how to lay out and install a basic set of cabinets.

a. Describe common surface preparation techniques required before cabinet installation.

b. Explain how to install wall cabinets.
c. Describe the process for installing base cabinets and countertops

This section provides general procedures for installing stock kitchen cabinets and countertops, because this work requires the most casework components. The installation of other types of cabinets will use the same basic methods and procedures.

Cabinets are installed in accordance with the architectural blueprints or approved casework drawings, illustrating the exact location and configuration of cabinets and countertops within each room (*Figure 97*).

Figure 97 Kitchen casework design sketch.
Source: New Africa/Shutterstock

As with most carpentry procedures, the installation of kitchen cabinets can be accomplished in more than one way. Some carpenters prefer to start by installing the wall-mounted cabinets first, while others prefer to begin with base unit installation.

6.1.0   Surface Preparation

To allow base and wall cabinets to be installed square and plumb, the underlying surface must be prepared properly before wall-finishing applications like drywall begin. Backing boards (for wood frame construction) or 6-inch-wide, 20-gauge sheet-metal straps (for steel frame construction) should be installed to anchor wall cabinets (*Figure 98*). These backing locations should also be illustrated in the architectural drawings of the project blueprints.

Figure 98 Backing boards.
Source: Dennis Axer/Alamy

Check the wall and floor surfaces with a straightedge for unevenness that can cause cabinets to be misaligned, resulting in twisting doors and drawer fronts. You should also shave or sand down excess plaster or drywall to remove high spots along the wall face at this time. Low spots can be shimmed later during cabinet installation.

Thoroughly clean the cabinet installation area and maintain your work area by regularly removing sawdust and storing idle tools. If required, remove any existing baseboard or other trim from the walls where you will be installing casework components (*Figure 99*).

Store cabinets in a temperature- and humidity-controlled area that mimics the environment of the finished building. Move the cabinets into the room and allow them to acclimate to the environment before installing them. These simple steps will reduce the expansion and shrinkage of cabinet materials and minimize the risk of product damage.

Figure 99 Clean and organized work area.
Source: valentyn semenov/Alamy

| 6.2.0 | Installing Wall Cabinets |

To install wall cabinets, proceed as follows:

Step 1   Locate and mark the position of all wall studs in the area where cabinets are installed (*Figure 100*). At each stud location, draw plumb lines on the wall where they can easily be seen after the cabinets are in position.

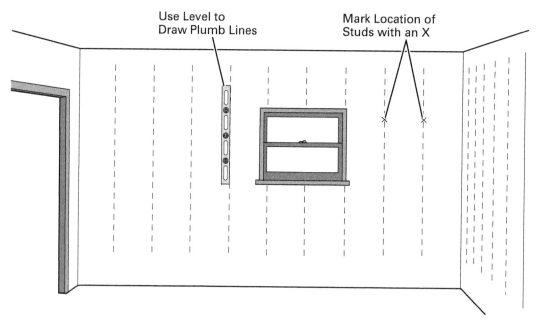

Figure 100 Marking stud locations on a wall.

Step 2   Using a long straightedge and a 6' level, check the floor area for high spots where the base cabinets are to be installed. Find the highest point on the floor, then snap a level chalk line on the wall at this high point around the wall as far as the cabinets will extend (*Figure 101*). Using this line as a reference, measure up and mark the required distances for the tops of the base cabinets, typically $34\frac{1}{2}$" above finish floor (AFF); bottoms of the wall cabinets (typically 54" AFF); and tops of the wall cabinets (typically 84" AFF). Snap level chalk lines around the walls at these heights. Many carpenters use a water level or laser level to lay out the horizontal high-point line.

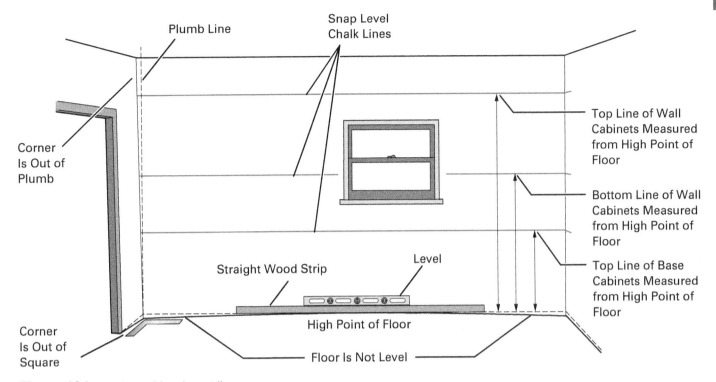

Plumb Line

Snap Level
Chalk Lines

Corner
Is Out of
Plumb

Top Line of Wall
Cabinets Measured
from High Point of
Floor

Bottom Line of Wall
Cabinets Measured
from High Point of
Floor

Top Line of Base
Cabinets Measured
from High Point of
Floor

Straight Wood Strip

Level

Corner
Is Out of
Square

High Point of Floor

Floor Is Not Level

Figure 101 Marking cabinet layout lines.

Step 3 Mark the outline for all cabinets on the wall using a story pole (layout stick). A story pole is a narrow strip of wood that is cut to the exact length or height of a wall and marked with the different widths or heights of the cabinets to be installed on that wall. It is used to lay out and mark the locations of the cabinets on the wall by transferring the location marks from the story pole to the wall.

Step 4 Remove the shipping skids, braces, and other packaging from the cabinets. Label each component before removing all the doors, drawers, and adjustable shelves. Store these items in a safe and secure location to avoid damaging them.

Step 5 Start the wall cabinet installation with a corner unit. First, lay out and mark the back of the cabinet for the stud locations (*Figure 102*), then predrill through the top and bottom cabinet mounting rails at these points. Countersink the drilled holes on the mounting rails.

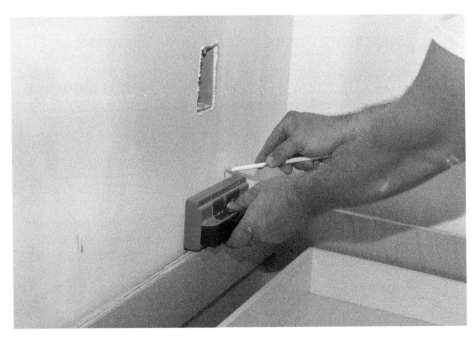

Figure 102 Marking stud locations behind cabinets.
Source: ungvar/Shutterstock

Step 6 Position the corner cabinet in place with its bottom on the chalk line and temporarily hold it in place using a continuous temporary cleat as shown in *Figure 103*. Ensure this cabinet is installed plumb and level, using shims, as necessary. Fasten the cabinet in place with screws of sufficient length to hold the cabinet securely against the wall, but do not fully tighten the screws.

Temporary Cleat

Figure 103 Shimming wall cabinets.

Step 7 Position, brace, and fasten the adjoining wall cabinet so that leveling may be done without removing it. Align the adjoining stiles so that their faces are flush with each other, then clamp them together. Drill through the stiles and screw them together. Continue this procedure for the remaining wall cabinets.

Step 8 Use a level to check the horizontal and vertical cabinet surfaces. Shim between the cabinets and the wall until the cabinets are plumb and level. This is necessary if the doors are to fit, swing, and close properly. When the cabinets are level and plumb, tighten all mounting screws and cut/install any filler strips, when necessary.

6.3.0 Installing Base Cabinets and Countertops

Base cabinets are commonly installed after the wall cabinets to provide better access for the carpenter during wall cabinet installation. The following general procedure for installing base cabinets and countertops assumes that the countertop used with the base cabinets is a manufactured unit that is ready to be installed.

Step 1 Start the installation with a corner base unit (*Figure 104*). Slide it into place, then continue to slide the remaining base cabinets into position. Ensure that the cabinet configuration matches the architectural drawings or cabinet layout sketch.

Step 2 When all base cabinets are in their correct positions, shim the corner cabinet until it is level and plumb, and its top aligns with the level line marked on the wall. Fasten the cabinet to the wall by driving $2\frac{1}{2}$" to 3" screws through the cabinet mounting rail into the studs or blocking.

NOTE

Most frame-style cabinets have a $\frac{1}{4}$" overlay on the stiles at the sides for applying finished sides. When joining two cabinets together, a gap of about $\frac{1}{2}$" between the sides is created. A good practice is to shim and screw this gap to hold the cabinet square.

Figure 104 Install corner cabinets first.
Source: Serghei Starus/Shutterstock

Step 3   Clamp the stiles (frame members) of the corner and the adjoining cabinet together (*Figure 105*). Shim as necessary to ensure that the horizontal frame members form a level and straight line and the frame faces are flush. After drilling countersunk pilot holes, fasten the cabinets together using screws. One screw should be installed close to the top of the cabinet end stile and one near the bottom.

Figure 105 Joining base-cabinet stiles.

Step 4   Continue installing the remaining base cabinets in the same manner. Check the cabinet tops from front to back and across the front edges with a level (*Figure 106*). Shim as necessary between the wall and cabinet backs and the floor and cabinet bottoms until the cabinets are plumb and level and the tops are aligned with the mark on the wall.

Step 5   Fasten the cabinets to the wall by driving $2\frac{1}{2}$" to 3" screws through the cabinet mounting rails into the wall studs. Use a chisel or utility knife to cut off any shims flush with the edges of the cabinet.

Step 6   If required, scribe and cut a filler strip to fit the remaining space between the cabinet and end wall. (Scribing is discussed in the following section.) Clamp, then fasten the filler strip to the cabinet stile with countersunk screws. Use glue with thin filler strips. These strips can be attached at the end of a run or at a corner to provide drawer and door-handle clearance, or to eliminate gaps between

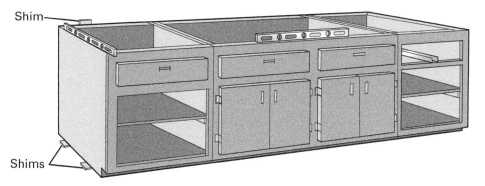

Figure 106 Shimming cabinets plumb and level.

the cabinet and wall (*Figure 107*). While filler strips are usually not needed for custom cabinets, they are a common necessity for stock cabinets. Rip the filler strip to the required width and scribe it to compensate for irregularities in the wall surface or to fit around baseboards.

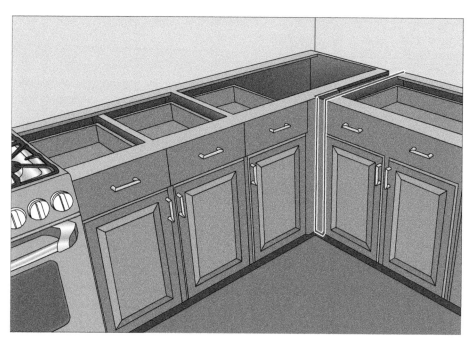

Figure 107 Filler strip.

Step 7   Place the countertop on top of the base cabinets, firmly against the wall (*Figure 108*). If the countertop does not fit tightly against the wall, scribe it to match the irregular wall surface. Next, remove the countertop, place it on sawhorses, and plane or belt-sand it to the scribed line. It is important to ensure that the overhang is uniform before scribing the countertop.

Step 8   Reposition the countertop on top of the base cabinets once more to guarantee a tight fit. Fasten the countertop from underneath by drilling and screwing through the corner blocks and/or diagonal bracing in the top of the base cabinets, without driving the fastener through the countertop surface.

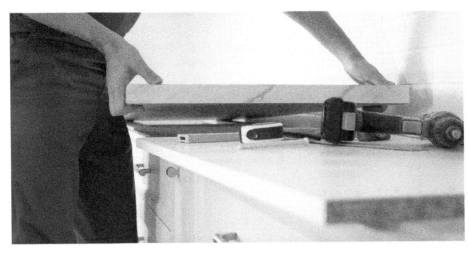

Figure 108 Countertop installation.
Source: New Africa/Shutterstock

6.3.1 Scribing Adjoining Pieces

Scribing is the action of fitting a piece against a surface that is not perfectly straight and smooth. Use the following procedure to scribe one piece to another:

Step 1 Cut a piece of the appropriate material to length for a snug fit. Use a piece that is slightly wider than the space needed.

Step 2 Place the outside edge of the piece against the surface, so that the contoured edge faces away from the surface.

Step 3 Use a piece of material matching the width of the finished piece as a marking block. Slide the block along the surface, flat against the piece. Be sure to hold the point of the pencil firmly against the corner between the filler and the marking block and keep the block firmly against the surface.

Step 4 Once the piece has been marked with the surface contour, use a belt sander, hand plane, portable power plane, band saw, or jigsaw to duplicate the marked contour.

6.3.2 Tight-Joint Fasteners

Tight-joint fasteners (*Figure 109*) are fasteners used at the joint of two mating pieces, such as between long countertop sections, the junction of 45-degree countertop miter joints, or other pieces that cannot be installed in one section. Tight-joint fasteners are concealed from the finished side and placed into a drilled hole in the two pieces to be joined. The fastener has a built-in bolt that is tightened with a wrench to pull the two pieces together to form a tight joint.

To complete the installation, replace the doors, drawers, and shelves in their respective cabinets. Ensure that the doors open and close properly, and that they are installed level and properly aligned with the doors on the other cabinets.

CNC Machines: The Future of Finish Carpentry?

Now that you know the basics of cutting, finishing, and installing trim, moldings, cabinets, and countertops, you can build on these skills and hone your valuable craft. There will likely always be a need for talented craft professionals who have the knowledge, dexterity, and attention to detail required to perform finish carpentry tasks that turn a structure into a comfortable home or working environment.

Figure 109 Tight-joint fasteners.
Source: Sarah Macor/Shutterstock

However, as technology continuously improves, we will see a constant wave of optimized tools and software that will expedite workflows and increase productivity in nearly every phase of construction. One such tool that will impact the future of finish carpentry is the CNC machine (*Figure 110*).

CNC stands for computer numerical control. Described very simply, these machines take input from a human operator and pre-programmed software to control a spindle, lathe, or router head as it cuts and shapes wood and other materials. Although these tools require specialized training and are just recently becoming a fixture of the construction material manufacturing industry, their potential for speeding up work processes with exact precision and near-zero tolerance is exciting.

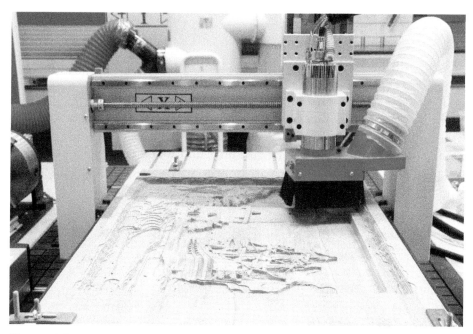

Figure 110 CNC machine.
Source: Marina Demkina/Shutterstock

6.0.0 Section Review

1. Unevenness of the wall surface can result in installed cabinets being _____.
   a. unsafe for use
   b. unusable
   c. misaligned
   d. damaged

2. Carpenters commonly mark the outlines of each of the cabinets on the wall, using a _____.
   a. story pole
   b. builder's transit
   c. laser level
   d. folding rule

3. Which base cabinet unit should you install first?
   a. The unit farthest to the left
   b. The unit on the end, nearest the oven
   c. The unit closest to the fridge
   d. The corner unit

1. Maintaining an organized work area can reduce _____.
   a. trip hazards
   b. productivity
   c. waste
   d. safety planning

2. When changing or tightening router bits, you should always _____.
   a. fill out a hot work permit
   b. wear safety goggles
   c. disconnect the router from the power source
   d. notate the activity in a maintenance log

3. A key difference between a power miter saw and a compound miter saw is _____.
   a. the power miter saw uses smaller blades
   b. the compound miter allows you to tilt the blade at a vertical angle
   c. the compound miter saw has a rotating miter table for angled cuts
   d. a power miter saw only uses fine-toothed blades

4. A type of trim used at the bottom edge of a wall where the wall meets the floor is _____.
   a. base casing
   b. bed base
   c. cap shoe
   d. base shoe

5. Wainscoting is sometimes capped with a type of molding called _____.
   a. crown molding
   b. chair rail
   c. cove
   d. corner guard

6. A typical height above the floor for installation of chair-rail moldings is _____.
   a. 20" to 25"
   b. 26" to 29"
   c. 30" to 35"
   d. 36" to 40"

7. The top third of a crown molding has a profile that is _____.
   a. convex
   b. corrugated
   c. complicated
   d. concave

8. To shape the end piece of molding so that it matches the ornately curved or grooved profile of an adjoining piece, you would use a _____.
   a. dovetail saw
   b. backsaw
   c. coping saw
   d. miter saw

9. A useful attachment that allows more flexibility and control with a portable jigsaw is the _____.
   a. bevel blade
   b. Rawlings dome attachment
   c. miter guard
   d. Collins coping foot

10. The joint between two pieces of baseboard molding in inside corners is usually _____.
    a. mitered
    b. butted
    c. coped
    d. bisected

11. Flat miter joints are most common when cutting _____.
    a. chair rail
    b. base cap
    c. crown molding
    d. door and window casing

12. If using both 15-gauge and 16-gauge pneumatic finish nailers to fasten trim, the 16-gauge nailer typically is used to fasten _____.
    a. thinner trim materials
    b. thicker trim materials
    c. hardwood trim
    d. window stools

13. Interior door trim is applied _____.
    a. after the base trim is installed
    b. to make the door functional
    c. to match the window trim
    d. before the finish floor is laid

14. The space between a door or window casing and the edge of the door or window jamb is called the _____.
    a. setback
    b. backset
    c. interval
    d. reveal

15. If the jamb edges of a window are lower than the adjoining wall surface, it must be brought flush with the wall by installing _____.
    a. jamb adjusters
    b. window spacers
    c. jamb extensions
    d. weephole covers

16. Instead of trying to hide gaps between countertops and walls with silicone, it is typically better to _____.
    a. level the counter with shims
    b. cover the gap with veneer
    c. color match the paint and countertop material
    d. build a backsplash with a revealed edge

17. When choosing which wall to begin installing base or ceiling molding, it is best to start with _____.
    a. the side closest to the door
    b. the long wall opposite the door
    c. any side you prefer
    d. the wall across from the closet

18. Crown molding is installed to cover the wall/ceiling joint at a _____.
    a. 35-degree angle
    b. 15-degree angle
    c. 45-degree angle
    d. 75-degree angle

19. Building codes normally require that there be at least 24" to 30" of clearance between the cooking unit and any overhead wall cabinet, as well as a minimum clearance between the sink and any overhead wall cabinet of _____.
    a. 22"
    b. 24"
    c. 32"
    d. 18"

20. Frameless cabinets are also known as _____.
    a. island cabinets
    b. Eurostyle cabinets
    c. Shaker-style cabinets
    d. RTA cabinets

21. The three main types of hinged swinging cabinet doors include lipped, overlay, and _____.
    a. slab
    b. mitered
    c. flush
    d. coped joint

22. To avoid shrinkage and other damage to cabinet components, you should store them in a temperature- and humidity-controlled environment to _____.
    a. cure
    b. expand
    c. acclimate
    d. contract

23. The action of fitting a cabinetry component against a surface that is not perfectly straight and smooth is called _____.
    a. joining
    b. mitering
    c. scribing
    d. backsawing

Answers to Odd-Numbered Module Review Questions are found in *Appendix A*.

Answers to Section Review Questions

Answer	Section	Objective
Section One		
1. c	1.1.3	1a
2. b	1.2.0	1b
Section Two		
1. d	2.1.1	2a
2. a	2.2.1	2b
Section Three		
1. a	3.1.1	3a
2. b	3.2.1	3b
3. b	3.3.0	3c
Section Four		
1. b	4.1.3	4a
2. c	4.2.2	4b
Section Five		
1. b	5.1.0	5a
2. a	5.2.1	5b
Section Six		
1. c	6.1.0	6a
2. a	6.2.0	6b
3. d	6.3.0	6c

Doors and Door Hardware

Objectives

Successful completion of this module prepares you to do the following:

1. Describe the safety hazards related to working with doors.
   a. Explain the common safety guidelines for handling and installing doors.
2. Identify common door frames and their installation methods.
   a. Discuss common wood door frames and their installation methods.
   b. Discuss common metal door frames and their installation methods.
   c. Explain the importance of door swing.
3. Identify the different types of residential doors, their hardware, and their common installation techniques.
   a. Describe the different types of residential doors.
   b. Identify common residential door hardware and its applications.
   c. Explain the basic steps for installing residential doors and door hardware.
4. Identify commercial doors, their hardware, and their common installation techniques.
   a. Identify common commercial door types.
   b. Identify common commercial door hardware and its applications.
   c. Explain the basic steps for installing commercial doors and door hardware.

Performance Tasks

Under supervision, you should be able to do the following:

1. Install a metal door frame in an instructor-selected rough opening.
2. Lay out and install an instructor-selected set of hinges.
3. Lay out and install an instructor-selected lockset.

Overview

The installation of interior and exterior doors and their hardware is a specialized skill that requires careful measuring and skillful use of tools. Some doors are prehung in a frame and require accurate installation to keep the frame level and plumb. In other cases, a carpenter must learn how to install hinges to hang the door in a finished opening. Both skills will be useful to learn and master as you expand your carpentry skill set.

CODE NOTE

Codes vary among jurisdictions. Because of the variations in code, consult the applicable code whenever regulations are in question. Referring to an incorrect set of codes can cause as much trouble as failing to reference codes altogether. Obtain, review, and familiarize yourself with your local adopted code.

Digital Resources for Carpentry

Scan this code using the camera on your phone or mobile device to view the digital resources related to this craft.

1.0.0 Door Installation Safety

Performance Tasks

There are no Performance Tasks in this section.

Objective

Describe the safety hazards related to working with doors.
a. Explain the common safety guidelines for handling and installing doors.

At first glance, it appears that there would be few safety hazards involved in door installation, but this is not the case. Some of the potential safety hazards include back injuries resulting from lifting heavy doors or overreaching, eye injuries resulting from improper tool use or lack of personal protective equipment (PPE), and injuries from falling off ladders.

1.1.0 General Safety Guidelines

A trained craft professional must always be aware of the safety rules in every phase of construction. A building interior nearing completion may give a worker a false sense of security. However, the element of danger is always present. You can protect yourself from common work-loss injuries by adhering to the following guidelines:

- *Keep work area clean* – Cluttered work areas invite injuries and create tripping hazards.
- *Avoid dangerous equipment* – Avoid using power tools in wet locations. Keep your work area well-lit and avoid chemical or corrosive environments. Store tools in a dry, secure location when they are not in use. Only use tools and extension cords with functional and intact ground fault circuit interrupters (GFCI), and red tag/replace any equipment without grounding components.
- *Secure your work* – Use clamps or vises to hold materials in place, keeping both of your hands free to operate the tool with better control and precision.
- *Dress properly* – Avoid wearing loose clothing or jewelry. Loose clothing, drawstrings, and jewelry can be caught in moving parts. Rubber gloves and nonskid footwear are also recommended.
- *Use safety goggles* – Wear safety glasses or goggles while operating power tools. Wear a respirator or dust mask if the operation creates dust. These precautions are necessary for everyone in the area where power tools are being operated.
- *Do not overreach* – Keep proper footing and balance at all times. Follow the "belt buckle" rule, i.e., never allow your core to extend past the outside rails of a ladder.
- *Use proper lifting form* – Lift with your legs, and keep your back straight when picking up or moving heavy materials and equipment. Engage your core, and avoid twisting when carrying larger items. Ask for help, or use equipment

such as a forklift, pallet jack, or hand truck when possible. These equipment options will allow your team to organize and move your equipment more efficiently, as well as reduce the time and energy expended while carrying everything by hand.

- *Maintain tools with care* – Keep tools sharp and clean for safer performance. Follow instructions for lubricating and changing accessories. Inspect tool cords periodically, and if they are damaged, have them repaired by an authorized service facility. Have all worn, broken, or lost parts replaced immediately. Keep handles dry, clean, and free from oil and grease.

- *Outdoor extension cords* – When using a tool outdoors, only use extension cords marked as suitable for use with outdoor appliances. Store them indoors when not in use.

- *Check damaged parts* – Inspect all tools before use to ensure they will perform their intended functions. Check for alignment of moving parts, binding of moving parts, breakage of parts or mountings, and any other conditions that may affect operation. A guard or other part that is damaged should be properly repaired or replaced by an authorized service center, unless otherwise indicated in the instruction manual. Inspect tool cords periodically, and if they are damaged, have them repaired by an authorized service facility. A service facility can also replace any defective switches, as well as other worn, broken, or lost parts still in production.

- *Stay alert* – Use common sense, and never operate a tool when you are tired or under the influence of medication, alcohol, or drugs.

WARNING!

An extension cord with damaged outer insulation, or a loose or missing ground pin, can cause serious injury or death. Always unplug power tools when changing or adjusting accessories. Plan each task, and always put safety first.

1.0.0 Section Review

1. If you notice a damaged part on a power tool, you should _____.
   a. keep using it until it breaks
   b. take it out of service and send it to an authorized service center for repairs and maintenance
   c. fix it with a job-built replacement part
   d. swap the damaged tool out with a replacement tool and return it to its normal storage area without tags or red tape

2.0.0 Common Door Frames

Objective

Identify common door frames and their installation methods.

a. Discuss common wood door frames and their installation methods.

b. Discuss common metal door frames and their installation methods.

c. Explain the importance of door swing.

Performance Task

1. Install a metal door frame in an instructor-selected rough opening.

Door frames provide the wall opening for doors. Door frames are attached to the jack studs (trimmers), which provide the structural framing for the door. These common trim components also provide a connection point for door hardware and complete the finished appearance of the interior/exterior molding. Proper installation is key since a frame that is installed off plumb or unlevel will need field modifications and potentially costly rework if the finish wall material is completed.

2.1.0 | Wood Door Frames

The majority of door frames delivered to the jobsite from the millwork supply house are $3/4"$ thick. These frames are often finger-jointed when they require a paint finish, but since designers may desire a more natural finish, frames can also be made from the same wood as the rest of the interior trim for accurate color matching. The width of the door frames must match the thickness of the wall or partition where they will be placed. The types that are most often used on prehung door units are shown in *Figure 1*.

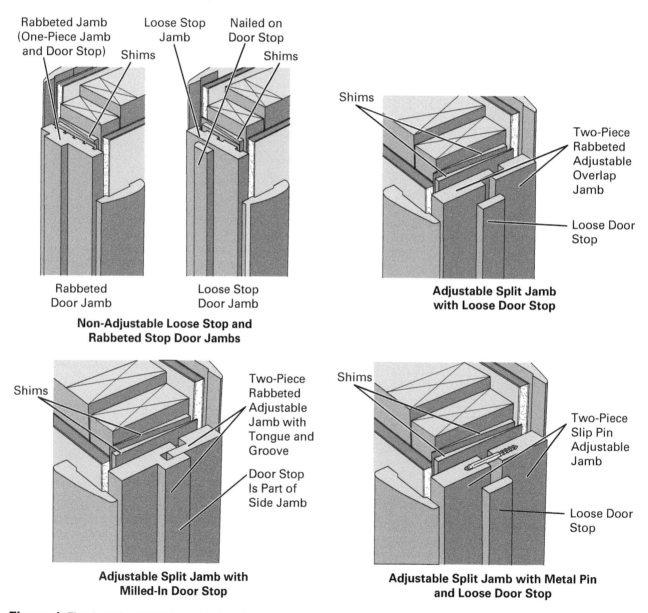

Figure 1 Fixed- and adjustable-width door frames.

When the frames are delivered to the jobsite, they may be assembled or unassembled (knocked down). If they are knocked down, place the head jamb into the dado joint on the side jambs, and fasten it with nails. Saw **kerfs** or grooves help prevent the jambs from warping (*Figure 2*).

Kerf: A slot or cut made with a saw.

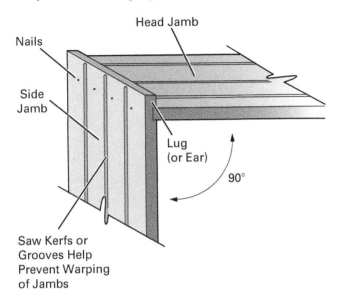

Figure 2 Top-left corner of interior door jamb.

Check the jamb opening for height, width, and squareness. The rough opening for wood jambs should typically be $1^1/_2$" higher and 2" wider than the door. Ensure the face of the wall is plumb. If the jambs are slightly higher than the rough opening, trim the lugs off the top so the jamb will fit. If they are trimmed off flush with the head jamb, shimming of the head jamb at the corners during installation is recommended to prevent separation of the head and side jambs. Place the jambs in the rough opening and check the head jamb to see if it is level. If it is not level, cut off the bottom of the highest side jamb by the amount required to make the head jamb level. Then, adjust the jamb height as needed.

The door opening is built to the exact size specified, and the door is cut in the field, if necessary. As shown in *Figure 3*, the inside width of the jambs should be the width of the door specified for the opening plus $^3/_{16}$" for clearance. For example, for a 2'-6" door, the jamb-to-jamb width would be $2'-6^3/_{16}$" (2'-6" + $^3/_{16}$"). Check the specifications for required clearances. The inside height of the jambs, from the bottom of the finished floor to the bottom of the head jamb, should be the door height plus $^3/_8$" to $^1/_2$" plus the finished floor thickness.

Step 1   Cut a strip of wood the same width as the jambs and the same length as the distance between the jambs at the head. This piece is used to keep the jambs the correct distance apart during installation.

Step 2   Measure in from the edge of the jambs by a distance equal to the thickness of the door, plus one-half the width of the stops. Do this on the side of the jambs where the door will be located. Draw a pencil line from the bottom of both side jambs to the head jamb. This mark outlines the nailing location through the jambs over which the door stop will be placed. The door stop will conceal both the line and the nail heads.

Step 3   Using a long level as a straightedge, ensure the faces of the wall on both sides of the opening are plumb. Do not attempt to install the door jambs if the walls are out of plumb, since the door will likely catch against the floor or close by itself. You can often plumb a stud wall by bumping the bottom plate with a sledgehammer. Use a piece of blocking to protect the wall when attempting this procedure.

Step 4   If the wall faces are plumb, proceed to plumb and level the jambs in the opening. Using shims as wedges, place two on each side between the jamb and the jack stud at the top and bottom. Ensure the edges of the jamb are flush with the finished wall.

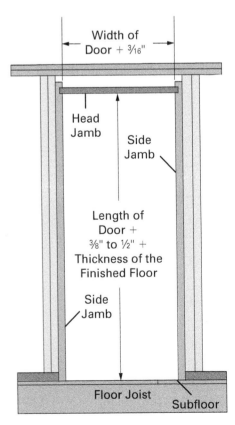

Figure 3 Finding the width and length of the head and side jambs.

Checking and Correcting the Plumb and Square of Wood Door Framing

Check all king and jack studs for plumb in both planes before installing doors. One plane is parallel, and the other is perpendicular to a wall. If the king and jack studs are plumb, check the door header for squareness. If the framing for a door is out of plumb or the king and jack studs are bowed or crowned, the door cannot be hung properly. Even if it is hung, the door would have to be trimmed at an angle on the top and bottom to swing, close, and latch.

Check door framing before finish wall installation is complete. Correcting a major framing problem at that point will require the removal and replacement of some of the finish wall material to reframe the door. However, most minor out-of-plumb conditions can be corrected as illustrated, with minor repairs to the finish wall material. In addition, a minor crown in the king and jack studs can be corrected by cutting both the jack and king stud and closing the saw kerf with screws. Several cuts spaced along the length of both studs may be required to achieve an almost-straight condition. This type of correction can be accomplished with the finish wall material installed; however, the wall material will require some minor repairs.

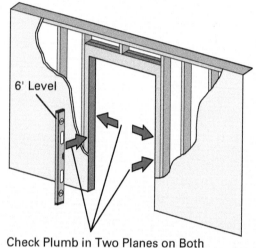

Check Plumb in Two Planes on Both Sides of Door Framing with a Long Level

Checking Plumb and Square of Wood Door Framing

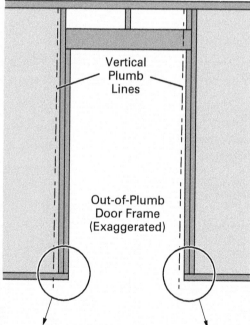

NOTE:
In some construction, the jack studs extend to the floor and do not rest on the bottom plate. If that were true in this example, then the left hand jack studs would have to be cut flush with the bottom plate before being moved and a support block installed under it after it was plumb. The right jack studs would be pried out and the block would be installed between it and the end of the bottom plate.

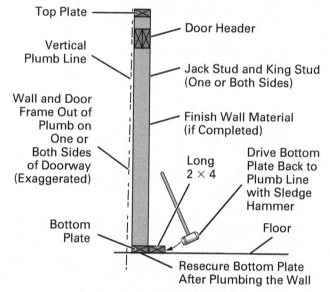

Correcting Minor Out-of-Plumb Walls

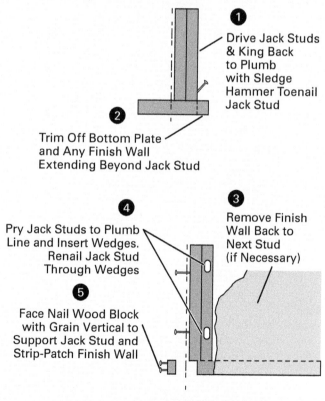

Correcting Minor Out-of-Plumb Door Frame

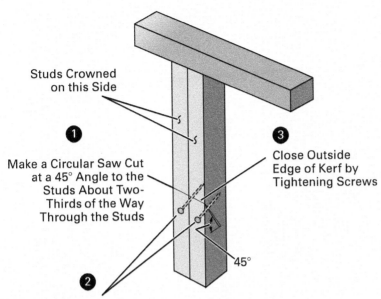

Studs Crowned on this Side

①

Make a Circular Saw Cut at a 45° Angle to the Studs About Two-Thirds of the Way Through the Studs

③

Close Outside Edge of Kerf by Tightening Screws

45°

②

Angle a Screw Up Through Each Stud at a 45° Angle (90° to Saw Kerf) at About One-Third of the Way into the Kerf

Correcting a Crown in the King Stud and Jack Stud

Step 5   Use a long level as a straightedge to check the butt jamb for plumb. Adjust the top and bottom shims as required, and fasten the butt jamb nails at the top and bottom along the nailing line. Don't drive the nails flush. Check the lock-strike side jamb for plumb, and nail it in the same manner. Place three pairs of shims behind the butt jamb; one pair goes behind the location of each butt. If only two butts (one pair) are being used, you'll need three pairs of shims placed at the recommended butt locations shown in *Figure 4*.

Step 6   Use three pairs of shims on the lock side. Use one pair at the lock location, and two pairs halfway between the lock location and the top and bottom. Temporarily fit the door to the opening, making sure the frame is plumb, square, and correctly spaced around the door. Remove the door, and complete the nailing of the jambs (*Figure 5*).

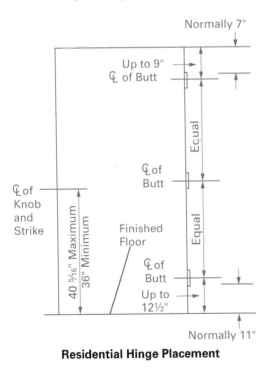

Residential Hinge Placement

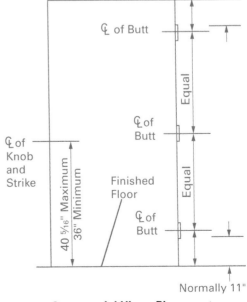

Commercial Hinge Placement

Figure 4 Elevation of door showing hardware locations.

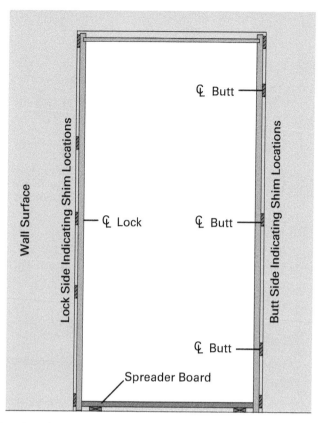

Figure 5 Elevation of door jambs showing shim locations.

Step 7 After the jamb is installed, but before it is nailed into place, check the door opening for square. One method to check for squareness is called cross stringing. To perform this method, tightly stretch a string between the diagonal corners of the jamb, on both sides. If the distances are equal, the jamb is square.

Step 8 If necessary, use casing nails on either side of the stop location and $^1/_2$" from the shim locations to secure the jambs to the jack studs (*Figure 6*).

Step 9 Hang the door, and install the stops as described in the next sections.

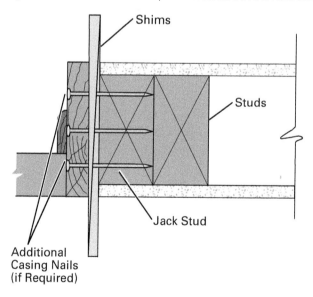

Figure 6 Plan view of additional jamb nailing.

2.1.1 Double Wood Door Frames

Typically composed of two separate wooden frames joined at the center with a mullion, this design enables the installation of two adjacent doors (*Figure 7*). Additionally, double wood door frames have excellent insulation

Figure 7 Double wood entry doors.
Source: David Papazian/Shutterstock

and soundproofing qualities. The installation procedure for a double wood door frame is like the procedure for a single wood door frame, with one exception: Double wood door frame installation requires a cross-story (double-X) method outside of the jamb.

2.1.2 Door-Stop Strips

Install loose door-stop strips after the door is hung for any door jambs that are not rabbeted (grooved). The stop on the hinge-jamb side of the door must have a $^1/_{16}$" clearance between it and the door, as shown in *Figure 8*. The stop on the lock-jamb side of the door must fit flush, so the face of the door is flush with the face edge of the jamb. The stop on the head jamb of the door should line up with both the side jamb stops. Note that by cutting the bottom of the stops on the vertical jambs at a 45-degree angle, a **sanitary stop** can be created. Use finish nails to fasten the door stops to the jambs.

Sanitary stop: A door stop with a 45-degree angle cut at the bottom of the vertical jambs.

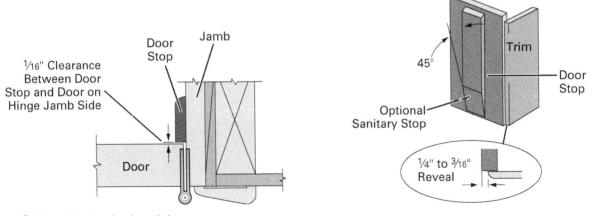

Figure 8 View of a door jamb and door.

2.2.0 Metal Door Frames

Metal door frames are common in commercial and institutional applications, but they are also commonly installed for exterior residential doors. *Figure 9*, *Figure 10A*, and *Figure 10B* show some of the styles and types of typical metal door frames that may be used in both residential and commercial applications.

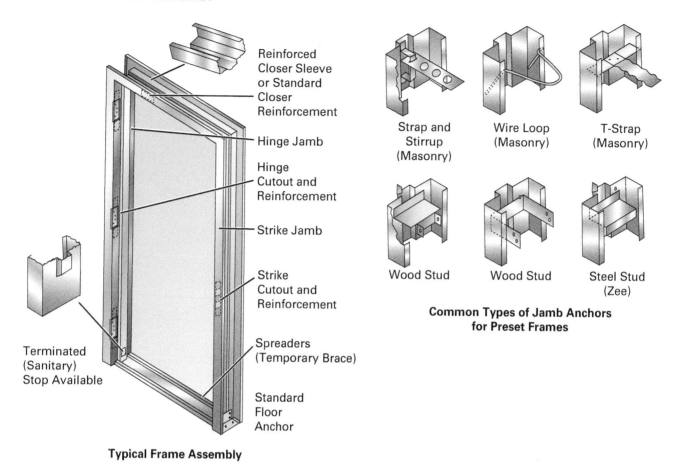

Common Types of Jamb Anchors for Preset Frames

Typical Frame Assembly

Figure 9 Typical fixed-width, single-piece jamb and casing metal frames.

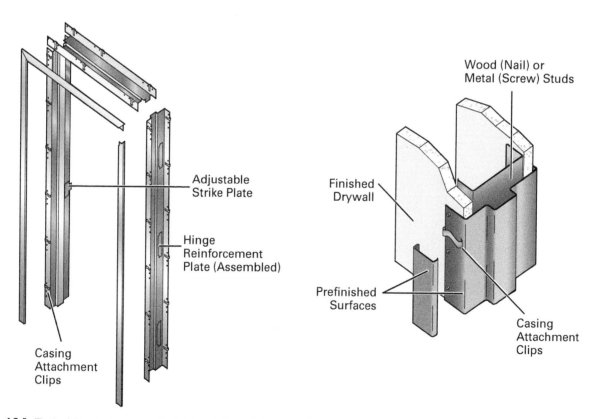

Figure 10A Typical fixed-width or adjustable-width metal frame with snap-on casing (1 of 2).

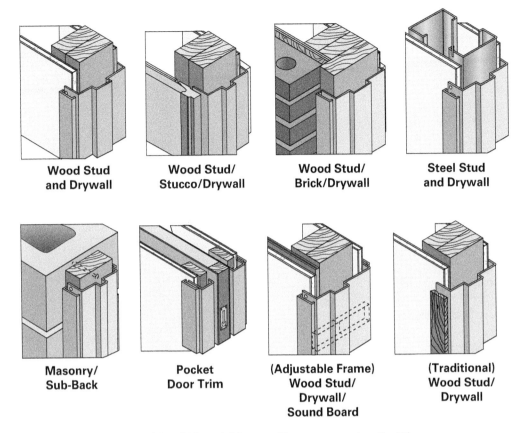

Wood Stud and Drywall

Wood Stud/ Stucco/Drywall

Wood Stud/ Brick/Drywall

Steel Stud and Drywall

Masonry/ Sub-Back

Pocket Door Trim

(Adjustable Frame) Wood Stud/ Drywall/ Sound Board

(Traditional) Wood Stud/ Drywall

Figure 10B Typical fixed-width or adjustable-width metal frame with snap-on casing (2 of 2).

The two basic types of metal door frames are a pre-assembled, welded unit and an unassembled knockdown (KD) unit. Either type is available in fixed or adjustable wall widths (timely frames). Fixed widths typically range from $2^1/_2"$ to $7^1/_2"$ in $^1/_8"$ increments. Adjustable units are available to accommodate wall thicknesses ranging from $3^3/_4"$ to $9^1/_4"$. Some frames have manufactured and finished snap-on casings made of wood, metal, or synthetic materials, while others are available with sanitary door stops.

Any type of wood door, or a matched metal door, can be attached to the frames, and the frames can be fastened to metal structural studs, wood studs, or masonry. This section will cover the typical installation of one type of assembled and unassembled metal frame in wood or masonry construction.

2.2.1 Installing Welded Metal Door Frames in Wood-Framed Construction

Perform the following steps to install a welded metal door frame in wood-framed construction. Usually, a metal frame will typically have wood stud anchors welded to it. Some wood stud anchors will be furnished loose, as shown in *Figure 11*. In some cases, reinforced closer sleeves are required if the reinforcements have not been welded into the frame.

Adjustable Wood Stud Anchor Furnished Loose

Wood Stud Anchors Welded to Frame

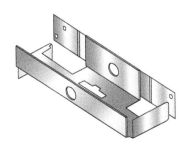

Jamb-Fit Type Combination Stud Anchor Furnished Loose

Figure 11 Wood stud wall anchors.

The Importance of a Plumb and Level Metal Door Frame

Ensure that each metal frame installation is plumb and square because improper frame installation will affect the performance of the door or cause the door to not fit into the frame. A little extra time spent installing the frame will make door installation much easier.

Use a magnetic plumb bob device to guarantee that the metal door frame is plumb. The magnetic bed of the device is attached to the frame. A short arm extending from the base supports the string from the plumb bob. The plumb bob is allowed to hang from the arm. Measure from the frame to the string at the top, and between the frame and tip of the plumb bob at the bottom. If the two measurements are the same, the frame is plumb.

The loose anchors are installed inside this metal frame by forcing or expanding the anchors between the two casing flanges of the frame.

Step 1   Place the anchor into position at an angle, and twist it into a horizontal position, as shown in *Figure 12*. If required, place a closer sleeve inside the head jamb frame at the proper location. It should fit tightly to prevent movement.

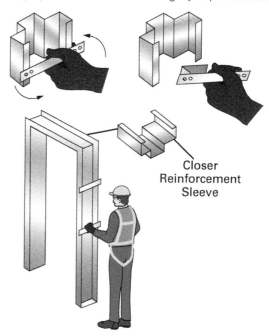

Closer Reinforcement Sleeve

Figure 12 Install anchors and closer reinforcement sleeve.

Step 2   Check the head jamb for level, and shim the bottom of the side jambs as required, subject to the manufacturer's recommendations. Check the location of the metal frame again, and, using an appropriate fastener, attach the floor anchors.

Step 3   The bottom of the door frame must be attached to the floor or the framed opening, per the manufacturer's instructions or local code requirements.

Step 4   Using an accurate level, plumb the jambs, and make sure all studs are square and true with the metal frame. Bend the jamb anchors around the wood studs, nail them in place, and recheck for plumb and squareness.

2.2.2 Sound Attenuation and Grouting for Metal Frames

Grout is installed in frames to provide **sound attenuation** and/or fireproofing. Some frames may be grouted before installation, while others may have to be grouted in place. Any necessary attachment clips must be installed in the door frame before grouting. Cover all butt plates with duct tape at the approximate location of the door closer, so screw holes can be drilled and tapped in that location. If possible, install screws for **weather stripping**, so you can back them out and install the weather stripping later.

2.2.3 Installing Welded Metal Door Frames in Steel-Framed Construction

Installing a welded metal door frame in steel-framed construction is done in much the same manner as with wood studs. Steel stud anchors are furnished

Sound attenuation: The reduction of sound as it passes through a material.

Weather stripping: Strips of metal or plastic used to keep out air or moisture that would otherwise enter through the spaces between the outer edges of doors and windows and the finish frames.

Metal Door Frame Support

Be sure to use at least three spreaders spaced equally within a metal frame when grouting a frame that has been installed. These spreaders must be accurately cut to maintain adequate door clearance.

with a metal door frame. Some anchors are welded to the frame, and some are loose, as shown in *Figure 13*.

Step 1   Loose anchors are installed inside the metal frame by forcing the anchor between the two casing flanges of the frame. Place the anchor into position at an angle, and twist it into a horizontal position. If required, place a closer sleeve inside the head jamb frame at the proper location. It should fit tightly to prevent movement.

Step 2   Move the assembled metal door frame into position, and ensure that the spacing of the jambs at the sill is the same as at the top of the frame.

Step 3   Check the head jamb for level, and shim as required, subject to the frame manufacturer's recommendations. Check the location of the metal frame again, and fasten the floor anchor by local code requirements.

Step 4   Using an accurate level, check that all the studs are square and true with the metal frame. Ensure that the frame is plumb and square because improper frame installation will affect the performance of the door opening, closing, and swinging.

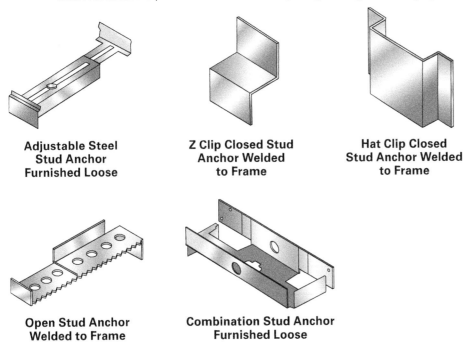

Adjustable Steel Stud Anchor Furnished Loose

Z Clip Closed Stud Anchor Welded to Frame

Hat Clip Closed Stud Anchor Welded to Frame

Open Stud Anchor Welded to Frame

Combination Stud Anchor Furnished Loose

Figure 13 Anchors.

Using a Door Buck

A door buck is a cross-braced open frame that is the same size as the finish door. Use a door buck in the frame opening to check that the door fits during the metal frame installation.

2.2.4 Installing Unassembled Metal Door Frames in Drywall Construction

An unassembled metal door frame is shipped to the jobsite knocked down and must be installed in the opening one piece at a time. The wood stud opening must be 1" higher than the metal door and 2" wider, giving it a 1" clearance on all three sides, as shown in *Figure 14*.

Step 1   Attach the sill anchors (*Figure 15*) to the bottom of each jamb. Some frame designs are furnished with a screw hole at the base of each jamb instead of the standard base anchor.

Step 2   Remove the hinge jamb from the knockdown frame parts, and slip the jamb into position over the wall. Hold the top in place, then push in the bottom to move it toward the wall. Check the plans to be sure that the hinge jamb is on the correct side. If a closer is required, and the head jamb frame does not have reinforcement, install a closer sleeve at the proper location. Position the head jamb frame over the wall. Align the head tabs with the jamb slots. Then, slide the head jamb toward the hinge jamb, and engage the tabs in the slots (*Figure 16*).

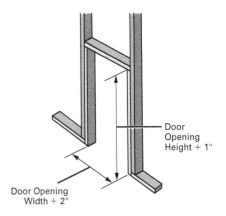

Door Opening Height + 1"

Door Opening Width + 2"

Figure 14 Door opening height and width.

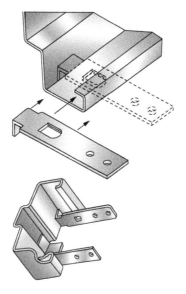
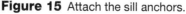

Figure 15 Attach the sill anchors.

Figure 16 Slip the hinge jamb into position.

Step 3 Slip the strike jamb over the wall, as shown in *Figure 17*. Push the top of the remaining jamb over the wall, and mate the jamb slots and head jamb tabs. Push in the bottom of this jamb to move it toward and over the wall. Level the head jamb.

Step 4 At the base of each side jamb is a shimming angle for vertical shimming (*Figure 18*). Use these shimming angles to level the head.

Step 5 Adjust the grip-lock anchors in both jambs by turning the anchor screw counterclockwise until the anchors hit the studs. Continue turning the anchors until the metal door frame is rigid and secure.

Step 6 Plumb and square the jambs with a level. When the hinge jamb is plumb, fasten the sill anchor at the hinge jamb with nails or screws.

Step 7 One of the most common side-jamb-to-head-jamb connections you'll find in the field is a tabbed "punch and dimple frame" that slides together and is fastened with screws, as seen in *Figure 19*.

Figure 17 Slip the strike jamb into position.

Figure 18 Shimming angle.

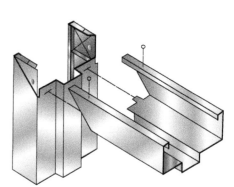

Figure 19 Punch and dimple side-jamb-to-head-jamb connection.

2.2.5 Installing Welded Metal Door Frames in Existing Masonry Construction

The welded metal door frame will be furnished with an appropriate quantity of one type of existing masonry wall anchor. These anchors are furnished loose or welded to the frame, as shown in *Figure 20*.

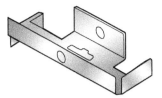

Jamb-Fit
Combination
Existing Masonry
Wall Anchors
Furnished Loose

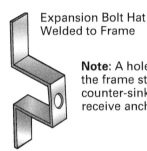

Expansion Bolt Hat
Welded to Frame

Note: A hole is provided through
the frame stop with either a
counter-sink or dimple to
receive anchor bolts.

Figure 20 Masonry wall anchors.

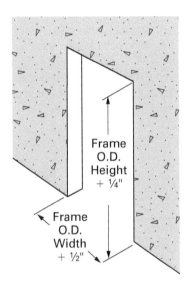

Recommended
Rough Dimensions
of Existing Opening
in Masonry Wall and
Frame with 2" Trim

Frame
O.D.
Height
+ 1/4"

Frame
O.D.
Width
+ 1/2"

Figure 21 Rough opening dimensions.

NOTE

As shown in *Figure 21*, the recommended
rough-opening height in an existing
masonry wall should be the frame height
plus $1/4$". The rough-opening width
should equal the frame width plus $1/2$",
giving the frame a $1/4$" clearance on all
three sides.

Step 1 Install loose anchors in the metal frame by forcing the anchor between the two
casing flanges of the frame. Place the anchor into position at an angle, and twist
it into a horizontal position, making sure the holes through the anchor are aligned
with the hole in the frame stop. If required, insert a closer reinforcement sleeve
inside the head jamb frame at the proper location.

Step 2 Place the metal door frame in the existing rough masonry opening. The frame
may have a spreader bar welded across the sill for shipping purposes. Remove
this spreader bar and discard it. Place a wood spreader bar at the sill to match
the width of the top of the frame. Two middle spreader bars should be installed
if the frame is grouted in place. These spreaders reduce the risk of the frame
warping or bowing from grout pressure (*Figure 22*).

Punch and Dimple Frames

Masonry anchors that are already in place are utilized for openings awaiting
installation of the frame. These types of anchors are commonly known as
"punch and dimple" frame anchors, as seen in *Figure 23*. Typically, a tube and
strap channel are welded to the frame where it has been punched and dimpled
in the soffit. This channel is designed to receive expansion bolts, which are then
installed through the existing block.

Step 1 Plumb the jambs with a level. Level the head and square the corners of the metal
frame. Ensure that the frame is plumb and square. Improper frame installation will
affect the performance of the door opening and closing.

Figure 22 Wood spreader bar and
middle spreader bars.

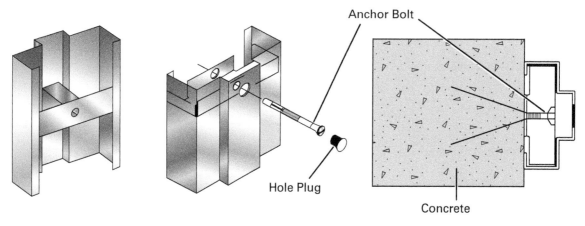

Figure 23 Punch and dimple frame-to-masonry connection.

Step 2 If the frame requires expansion shields and machine bolts, the location of the shields must be marked at this time and the frame removed from the wall. Using the marked locations, drill the depth and diameter hole recommended by the shield manufacturer. Install the shields, reinstall the metal frame, and fasten it into place. If expansion shields are not to be used, place a $3/_8$" drill bit through the $3/_8$" existing holes in the metal frame, and drill no less than $1 3/_8$" into the masonry.

Step 3 Using the sleeve anchor shown in *Figure 24*, fasten the metal door frame in the opening. The installation is now complete. Remove the spreaders, and install a wood door or a matched metal door.

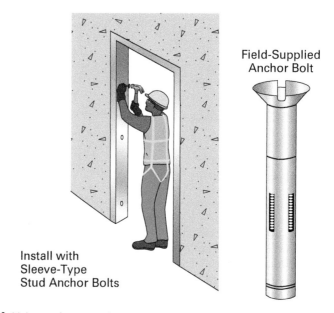

Figure 24 Using a sleeve anchor.

2.2.6 Installing Welded Metal Door Frames in New Masonry Construction

A welded metal door frame will be furnished with a specified masonry wall anchor, furnished loose or welded to the frame, as shown in *Figure 25*. If necessary, install a closer reinforcement sleeve in the head jamb frame.

Step 1 Once you're sure the door frame is in the correct location, check the head jamb for level and shim, as needed. Check the location of the metal frame again, and fasten the floor anchors by code requirements.

Step 2 Place and fasten blocking on the floor. Use telescoping braces to support the frame. Plumb the jambs, and check the level of the head jamb. Square the corners and add two wood spreaders in the center of the frame.

Step 3 Ensure that the frame is plumb, level, and square during masonry installation.

Completing Metal Door Frame Installation in New Masonry

The erection of a block wall around the metal door frame should complete the installation. Prepare the frame for grouting, then install a wood door or a matched metal door.

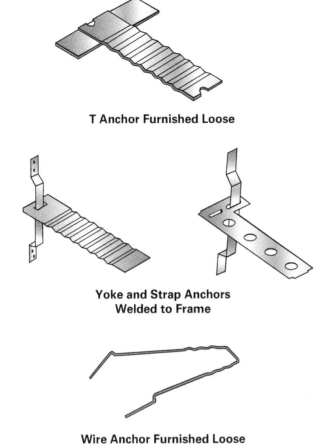

T Anchor Furnished Loose

**Yoke and Strap Anchors
Welded to Frame**

Wire Anchor Furnished Loose

Figure 25 Masonry anchors.

2.3.0   Door Swing

The direction in which a door opens is called the hand or swing of the door. Before hanging a door, a craft professional must have some knowledge about the hand or swing. Ensure that the doors delivered to the jobsite accurately match the project door schedule for size and swing.

The incorrect determination of the door swing is one of the most common mistakes in construction since no universal standard exists. In some cases, the hand can be defined as either the handle location or the hinge location.

One method of identification dictates that a door swing be described in four ways, as shown in the plan-view drawing in *Figure 26*. The door swing is the direction in which a door swings and is determined when facing the door from the outside (public side). This is the most common method for contractors, as well as door and lockset manufacturers.

The most common field technique for determining swing is mentally placing yourself in the floor plan of the building at the doorway with your back against the hinge locations. The side of the jamb on which the hinges (butts) are located determines the swing of the door, as shown in a plan view in *Figure 27 (A)*. This is commonly known as "butt-to-butt." Some door manufacturers use other methods, so always reference manufacturer specifications before installation. Generally, door swing is based on which side the handle or hinge is on when the door is pulled closed, as shown in *Figure 27 (B)*.

To mark the swing of the door for a particular opening, mark the jamb on the hinge side. Mark the location of the hinges on the **hanging stile** of the door.

Delivery Checks

Every interior door must be checked for straightness and damage upon delivery, to validate any returns or claims against the manufacturer or dealer. All parties concerned must be notified immediately upon delivery of any defective or damaged doors.

Hanging stile: The door stile to which the hinges (butts) are fastened.

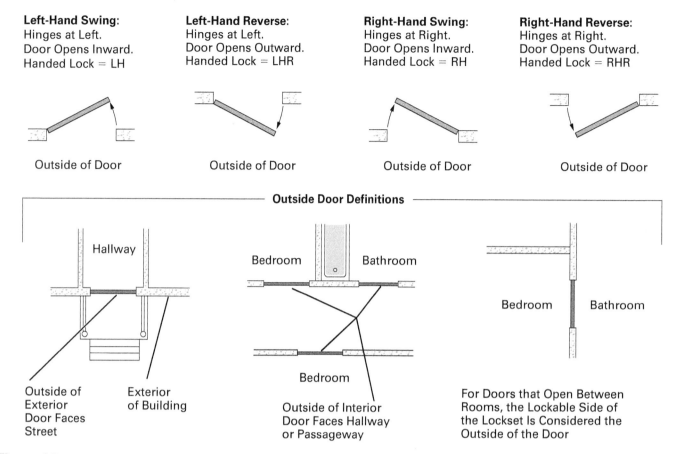

Left-Hand Swing:
Hinges at Left.
Door Opens Inward.
Handed Lock = LH

Left-Hand Reverse:
Hinges at Left.
Door Opens Outward.
Handed Lock = LHR

Right-Hand Swing:
Hinges at Right.
Door Opens Inward.
Handed Lock = RH

Right-Hand Reverse:
Hinges at Right.
Door Opens Outward.
Handed Lock = RHR

Outside of Door

Outside of Door

Outside of Door

Outside of Door

Outside Door Definitions

Hallway

Bedroom   Bathroom

Bedroom   Bathroom

Bedroom

Outside of
Exterior
Door Faces
Street

Exterior
of Building

Outside of Interior
Door Faces Hallway
or Passageway

For Doors that Open Between
Rooms, the Lockable Side of
the Lockset Is Considered the
Outside of the Door

Figure 26 Door swing when facing the outside of a door.

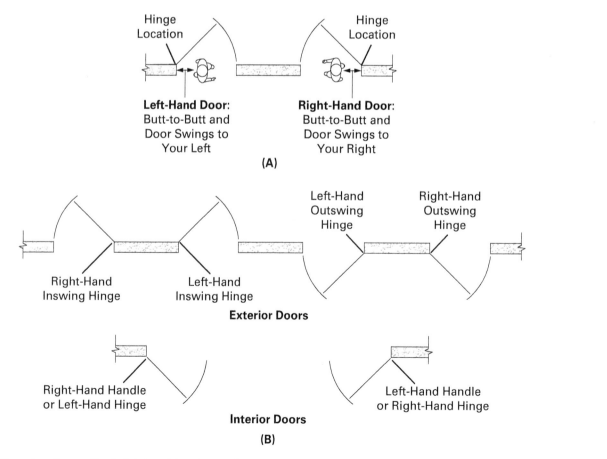

Hinge
Location

Hinge
Location

Left-Hand Door:
Butt-to-Butt and
Door Swings to
Your Left

Right-Hand Door:
Butt-to-Butt and
Door Swings to
Your Right

(A)

Left-Hand
Outswing
Hinge

Right-Hand
Outswing
Hinge

Right-Hand
Inswing Hinge

Left-Hand
Inswing Hinge

Exterior Doors

Right-Hand Handle
or Left-Hand Hinge

Left-Hand Handle
or Right-Hand Hinge

Interior Doors

(B)

Figure 27 Hinge-location method of determining door swing.

2.0.0 Section Review

1. The purpose of the kerfs on the back side of wood jambs is to _____.
   a. allow alignment of the head jamb with the side jambs
   b. space the nails used to fasten the head jamb to the side jambs
   c. mark the cutoff lines to be used when reducing the width of a jamb
   d. help prevent warping of the jambs

2. When grouting a metal door frame that has been installed, be sure to use at least three equally spaced _____.
   a. kerfs
   b. angle brackets
   c. shims
   d. bucks

3. What is the most common method for determining door swing?
   a. Back-to-knob
   b. Butt-to-butt
   c. Back-to-back
   d. Hinge-to-jamb

3.0.0 Residential Door and Hardware Installation

Objective

Identify the different types of residential doors, their hardware, and their common installation techniques.
   a. Describe the different types of residential doors.
   b. Identify common residential door hardware and its applications.
   c. Explain the basic steps for installing residential doors and door hardware.

Performance Tasks

2. Lay out and install an instructor-selected set of hinges.
3. Lay out and install an instructor-selected lockset.

Residential doors come in a wide range of sizes, shapes, materials, and operation types, but they all perform the same function. Residential doors provide security and privacy, while also offering unique aesthetic qualities that make a structure feel like a home. In the following section, we will cover many of the most common residential door types. Although you may find similar door styles in commercial projects, this content will provide you with a basic overview to get you started.

3.1.0 Residential Doors

Common types of residential doors include bypass doors, bifold doors, pocket doors, wood folding doors, and metal doors. Although metal and composite doors are more common in commercial projects due to their durability and ease of installation, many situations in residential construction will call out for these characteristics as well.

3.1.1 Bypass Doors

Bypass doors, as shown in *Figure 28*, are usually hung in pairs, but there may be more doors in an opening. These doors are suspended from an overhead track and move on rollers, like sliding barn doors (*Figure 29*).

A common method for hanging bypass doors is to start by screwing the head track into position. The adjustable rollers are then screwed to the top door edge. Some installation instructions require that doors be shortened on the bottom edge.

Figure 28 Bypass closet doors.
Source: Slavun/Shutterstock

Figure 29 Barn doors.
Source: Neil Podoll/Shutterstock

Preplanning during the framing phase should prevent the need to cut doors. Hang the doors in position by placing the rollers on the track. Next, hang the door closest to the rear first. Mark the location of the doors on the floor and screw the nylon guide strip to the floor. The door pulls are mounted typically 38" above the finished floor (AFF) and on the stiles nearest the jambs. The pulls must be flush with the face of the door.

3.1.2 Bifold Doors

Bifold doors (*Figure 30*) allow full access to a closet, while also allowing furniture to be placed close to the side trim of the closet opening. Usually, this door unit

Figure 30 Bifold doors.
Source: myboys.me/Shutterstock

has two pairs, with each pair operating independently of the other. This four-door unit is used for a 6'-wide opening. For openings of less than 4', only two doors, or one pair, are used.

To hang the unit in a prepared opening, simply screw a slide guide track to the bottom of the head jamb according to the manufacturer's instructions. Then screw an adjustable jamb bracket at the base of the side jambs against the floor on each side.

Both pivot and guide pins are spring-loaded, while a pivot pin and bracket secure the door next to the jamb. The first door and second door are hinged together. If the doors are not plumb or fail to meet properly in the center, adjust the screw height at the jamb bracket on the floor or head jamb until the doors align and function properly. The second door slides along a track on a roller guide and is secured to the head jamb. A complete bifold door assembly should open and close smoothly and display an even reveal (clearance gap) between the doors and jambs, as well as where the doors meet in the center when closed.

3.1.3 Pocket Doors

Pocket-door units are space-saving designs that you will often find covering closet spaces of small bedrooms, hallways, and bathrooms. They use track and rollers that are partially concealed in the head pocket of the unit. The portion of the unit called the "pocket" (*Figure 31*) may be covered with any interior wall finish to match the overall design.

After the door is positioned in the opening and the rollers are adjusted so that the door hangs plumb with the jambs and square with the head, a guide strip is fastened to the bottom edge, as shown in *Figure 32*.

Stops are then nailed on either side of the pocket opening. The stop thickness decreases the width of the opening, preventing the door from being pushed out of the opening. The spacing of the stops also acts as a guide for the guide strip on the door bottom. The purpose of the guide strip is to position the door so the surfaces of the door are not scratched when sliding in and out of the pocket. When the door is completely inside the pocket, the edge of the door facing the opening is flush with the face of the stops.

Fastening Finish Wall Material to a Pocket Door Unit

When fastening the wall finish to the horizontal members of a pocket door unit, ensure the nails, staples, or screws are short enough to not extend into the pocket area, where they might scratch the door or restrict door movement.

Figure 31 Pocket doors.
Source: Chris Haver/Alamy

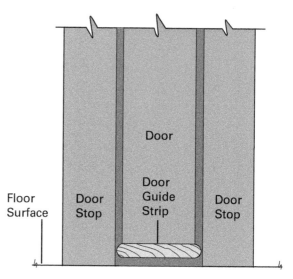

Figure 32 Door guide strip.

3.1.4 Folding Doors

Folding doors are assembled with vertical, prefinished door panels (*Figure 33*). You can install the door to open from either side, like an accordion. The common panel width, sometimes called the stack width, is $4^1/_4$".

Figure 33 Folding doors.
Source: myboys.me/Shutterstock

Molding: Material used for decorative trim.

For traditional wood folding doors, a beveled **molding** at the head conceals a surface-mounted head track. The track and molding are drilled for mounting screws supplied with the door. A spring-action catch at the top of the end post (*Figure 34*) engages the nylon head stop for tight stacking of the panels.

The hangers are attached to alternate panels for quiet, smooth operation. The end post hanger has two sets of rollers to keep it aligned with the track (*Figure 35*). The positioning of the rollers allows operation even in openings that are slightly out of plumb.

The panels have special steel alloy springs that run horizontally through the panels and connect to vertical wood moldings. The spring connectors can be detached to disassemble the door or replace damaged panels. Similar mechanisms can be used with the large patio doors that have replaced traditional sliding glass doors in modern residential layouts (*Figure 36*).

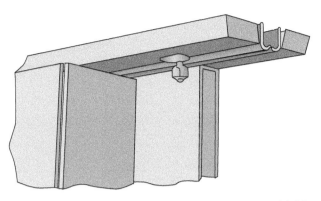

Figure 34 Track molding and head stop of a wood folding door.

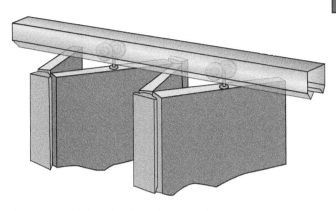

Figure 35 Wood folding-door hangers.

Figure 36 Folding patio doors.
Source: Simon Battensby, Image Source Limited/Alamy

3.1.5 Metal Doors

Metal doors are available in practically any variation and are used mostly in residential exterior and commercial construction (*Figure 37*). They are identified in architectural specifications by a defined coding system.

Figure 37 Metal exterior door.
Source: Maciej Bledowski/Alamy

Mortise lockset: A rectangular metal box that houses a lock. It usually has a latch and deadbolt as part of the unit. Options consist of a locking cylinder and/or thumb turn by which it can be secured. It is used on residential entry doors and commercial doors.

Metal doors are available with butts, cylindrical locksets, **mortise locksets**, unit (mono) locksets, panic devices, flush bolts, and other installed hardware. Door lights and louvers are also available to provide more natural light to the interior.

Metal door cores are designed to meet specification requirements, as shown in *Figure 38*. For example, a metal door with a lead core can protect occupants working in high-radiation areas.

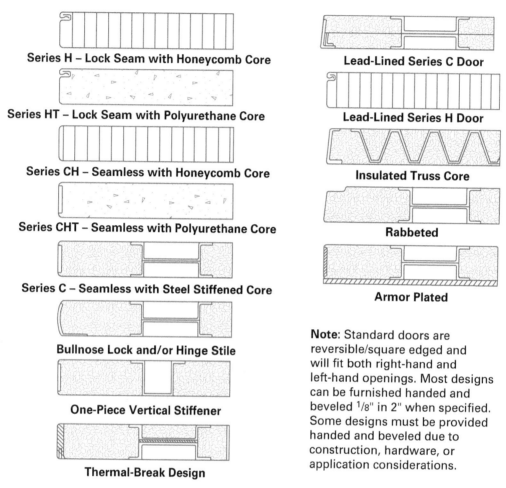

Series H – Lock Seam with Honeycomb Core

Series HT – Lock Seam with Polyurethane Core

Series CH – Seamless with Honeycomb Core

Series CHT – Seamless with Polyurethane Core

Series C – Seamless with Steel Stiffened Core

Bullnose Lock and/or Hinge Stile

One-Piece Vertical Stiffener

Thermal-Break Design

Lead-Lined Series C Door

Lead-Lined Series H Door

Insulated Truss Core

Rabbeted

Armor Plated

Note: Standard doors are reversible/square edged and will fit both right-hand and left-hand openings. Most designs can be furnished handed and beveled $1/8$" in 2" when specified. Some designs must be provided handed and beveled due to construction, hardware, or application considerations.

Figure 38 Standard and engineered door sections.

Steel doors have prepared hinge mortises and lock holes, while hollow metal doors are reinforced with a honeycomb-patterned core. Others may contain steel reinforcement sections to support locks and hinges, or other hardware with drilled and tapped holes, to receive machine screws. *Figure 39* shows a typical metal door construction.

3.1.6 Doors between Dwellings and Garages

As per ICC 2021 IRC Essentials, doors between the dwelling and garage must provide fire resistance but may not require an assembly with a fire-resistance rating. In other words, the frame, hardware, and sealing of the opening are not addressed, only the materials of the door leaf itself. Any one of the following types of doors satisfies this separation requirement:

- $1^3/_8$"-thick solid wood
- $1^3/_8$"-thick solid steel
- $1^3/_8$"-thick honeycomb-core steel
- A listed door with a 20-minute fire rating

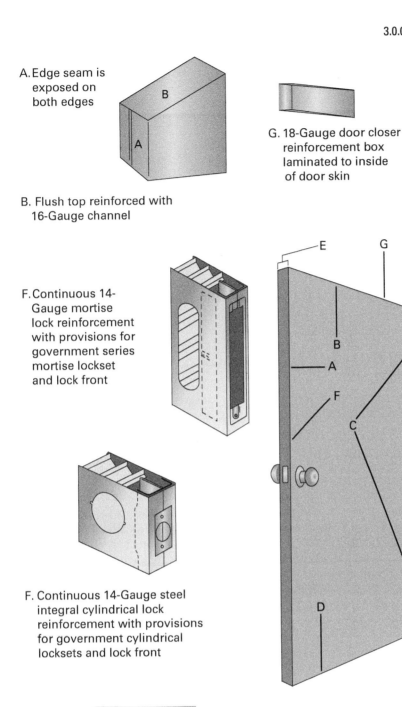
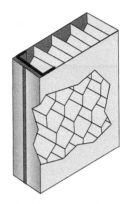
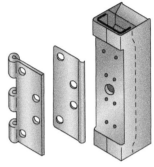
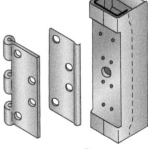

A. Edge seam is exposed on both edges

G. 18-Gauge door closer reinforcement box laminated to inside of door skin

B. Flush top reinforced with 16-Gauge channel

E. Cell honeycomb core

F. Continuous 14-Gauge mortise lock reinforcement with provisions for government series mortise lockset and lock front

C. Continuous 11-Gauge steel integral hinge reinforcement with provision for full mortise template-type hinges

F. Continuous 14-Gauge steel integral cylindrical lock reinforcement with provisions for government cylindrical locksets and lock front

C. Continuous 11-Gauge steel integral hinge reinforcement with provision for full mortise template-type hinges

D. Bottom reinforced with 16-Gauge channel

C. Standard doors are non-handed using hinge fillers

Figure 39 Metal door construction.

3.1.7 Garage Doors

This large door design protects vehicles and garage storage from the outside elements. Garage doors can have a single face that tilts and swivels upwards for access, although the more traditional design has multiple panels that are pulled and released along a curved track by a remote-controlled motor (*Figure 40*). As shown in *Figure 40*, many models have door lights, to allow natural light into the space, and manual release, to operate the door in case of power outages.

Figure 40 Garage doors.
Source: Konstantin L/Shutterstock

Basic residential door hardware includes door hinges and locksets. Door hinges provide a pivot point for a door within the frame, while locksets provide a means for securing a door.

3.2.1 Door Hinges

Door hinges are sometimes referred to as butts or butt hinges. The flat portions of the door hinge that contain the countersunk holes for the screws are called leaves (*Figure 41*). The round portion in the center of the hinge that joins the leaves is called the knuckle. The knuckle is usually divided into five sections, two on one leaf and three on another. A pin running through these five-knuckle sections joins the two leaves, creating a five-knuckle butt hinge.

Common hinges used in residential construction include loose- and fixed-pin models. Loose-pin hinges, commonly used on interior doors, have a pin that can easily be removed, making the door installation and removal much easier. For fixed-pin hinges, the pin is non-removable, providing more security

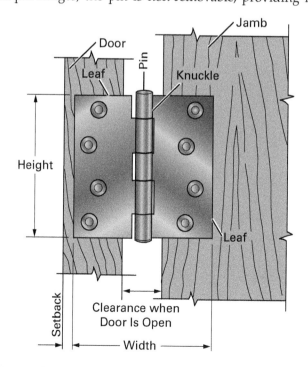

Figure 41 Loose-pin hinge (butt).

than loose-pin hinges. Fixed-pin hinges are also called fast-joint hinges or non-removable pin (NPR) hinges.

The basic dimensions of a hinge are taken with the hinge in the open position and are always stated with the height first and the width last. The height of the hinge is determined by the width and thickness of the door, as shown in *Table 1*.

TABLE 1 Hinge Height

Door Thickness	Door Width	Hinge Height
$1^3/_8$"	To 36"	$3^1/_2$"
$1^3/_4$"	To 36"	4"
	36" to 41"	$4^1/_2$"
2", $2^1/_4$", $2^1/_2$"	42" to 48"	$4^1/_2$" Extra Heavy
	To 42"	5"
	Over 42"	6"

The width of the hinge is determined by the door thickness and the clearance required for the trim (*Figure 42*). This clearance will allow the door to open parallel to the wall without marring the door trim.

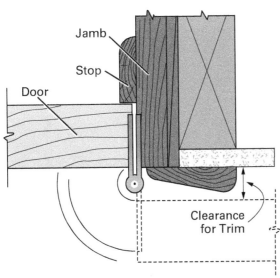

Jamb
Stop
Door
Clearance for Trim

Figure 42 Plan-view section of a door jamb and door.

To determine the width of the hinge, use the door thickness and clearance required, as shown in *Table 2*.

TABLE 2 Hinge Width

Door Thickness	Clearance Required	Hinge Width
$1^3/_8$"	$1^1/_4$"	$3^1/_2$"
	$1^3/_4$"	4"
$1^3/_4$"	1"	4"
	$1^1/_2$"	$4^1/_2$"
	2"	5"
	3"	6"
2"	1"	$4^1/_2$"
	$1^1/_2$"	5"
	$2^1/_2$"	6"
$2^1/_4$"	1"	5"
	2"	6"
$2^1/_2$"	$3/_4$"	5"
	$1^3/_4$"	6"

3.2.2 Locksets

Locksets include the working components of the lock and the trim pieces. Trim pieces include the doorknob or lever, strike plate, and cylinder. Some types of locksets are reversible, while others are designed to be installed in one direction.

The entrance lock (*Figure 43*) is used for entrance doors where locking is required. Turning either the outside or inside knob operates the latch bolt. Both knobs can be locked or unlocked by rotating the turn button on the inside knob or by using the key on the outside knob. However, any exterior lockset for egress doors in residential buildings must allow occupants to exit without keys or special knowledge. IBC codes for ADA egress doors take this concept a step further to ensure that the door handle, pull, latch, or lock must be designed so that it can be operated without tight grasping, tight pinching, or twisting of the wrist. The hardware to open the door must be installed between 34" and 48" above the finished floor.

The patio lock is used for entrance doors with limited entry. Turning either the outside or inside knob operates the latch bolt. Both knobs can be locked and unlocked by rotating the turn button on the inside knob.

The passage latch is typical for doors that do not require locking, such as a closet door. Turning either knob always operates the latch bolt.

A privacy lock is primarily for bedrooms and bathrooms. Turning either knob operates the latch bolt, or rotating the turn button on the inside knob locks the outside knob. The inside knob is always active. An emergency release knob is used as an outside knob to prevent accidental locking. Some of these locksets are furnished with different inside and outside finishes.

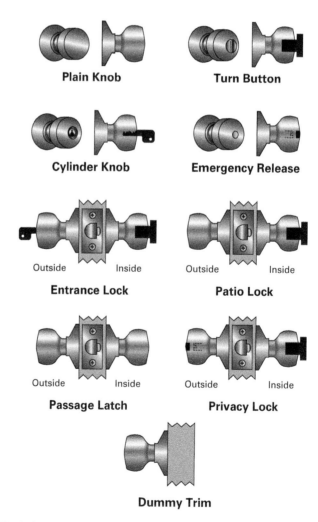

Figure 43 Basic knob-type locksets and knobs.

Understanding knob functions will help in understanding lockset functions. The first knob function is the plain knob. It can be rotated clockwise or counter-clockwise, and the operation of the latch bolt is its only function.

The second knob function is the turn button. The knob function is the same as the plain knob except that by using the turn button, the opposite knob or both knobs can be locked, depending on the lockset function.

The third knob function is the cylinder knob. The knob function is the same as the plain knob except that you can lock it by using the correct key.

The fourth knob function is the emergency release. This knob function is the same as the plain knob except when an accidental locking occurs (such as a child turning the turn button on the room side of the lock), the latch bolt can be released by inserting a special key or pin in the hole of the outside knob.

3.3.0 Residential Door and Door Hardware Installation

Most residential doors for new-home construction and remodeling work are prehung doors, due to their ease of installation. Residential doors are typically lighter weight and do not require some of the complex security hardware used in commercial construction.

3.3.1 Door Jack

A door jack is a piece of equipment that holds a door securely on edge while it is being prepared for installation. Some tool manufacturers make a metal, spring-loaded door jack; however, a simpler door jack (*Figure 44*) can be easily fabricated. The uprights that receive the door are lined with scrap carpeting to protect the door finish and provide a snug fit for the door.

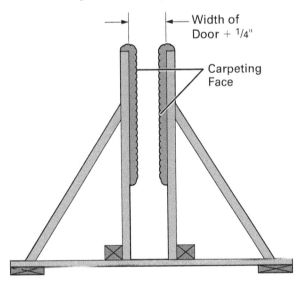

Width of
Door + 1/4"

Carpeting
Face

Figure 44 Door jack.

3.3.2 Manufactured Prehung Door-Unit Installation

Manufactured prehung door units may be supplied assembled or unassembled (*Figure 45*). They are available in fixed jamb widths or with adjustable-width split jambs (*Figure 46*).

Most prehung door units do not have lugs at the top of the side jambs, so the head jamb is not dadoed into the side jambs. As a precaution, the head jamb should be shimmed at both top corners when the door is installed to prevent separation of the head jamb from the side jambs.

Before installing an assembled unit, check that the lockset side of the door is not nailed in place through the back side of the strike-plate jamb, and ensure that any shipping braces or tie straps holding the frame together at the bottom of the door will not interfere with the installation. These braces can be removed or cut away when the door is installed and secured in the rough opening.

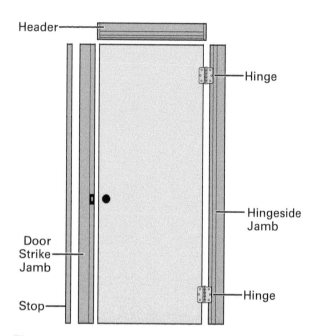

Figure 45 Typical unassembled prehung door unit.

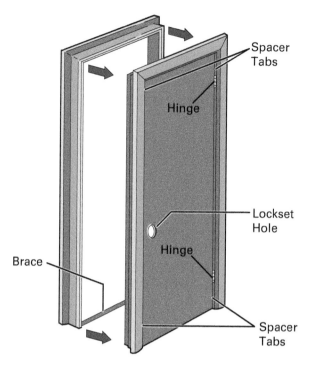

Figure 46 Typical assembled split-jamb, prehung door unit.

The following installation guidelines are for a typical split-jamb, milled-stop, prehung door. However, always refer to the door manufacturer's instructions before installation. Rough openings required for prehung units are typically 2" wider and 1" higher than the door size. Remember to use 4" screws in the jack stud during lockset installation to reduce the door racking, and use 3" screws on deadbolt strike plates for exterior doors.

Step 1 Unpack the split-jamb door unit and separate the two halves. One of the halves will have the door attached to the jamb by the hinges, and the other will contain the assembled top and side jambs (*Figure 47*). Both halves will normally have the trim casing installed as well.

Step 2 Determine the clearance of the door above the floor, and trim the bottom of the jambs and casings to obtain the correct spacing, if necessary.

Step 3 Position the split-jamb half containing the door into the rough opening, and plumb the jamb by placing a level against the side of the casing on the hinge side of the jamb (*Figure 48*). Check that both jambs rest on the floor. If they do not rest on the floor, and a metal door is being hung, insert shims under the jamb and casing. If a wood door is being hung, remove the door unit, and trim the bottom of the appropriate jamb and casing so that both jambs will touch the floor (except on concrete). Then, replace the door unit in the opening, and replumb the unit. Temporarily nail the casing to the jack stud at the top and bottom on the hinge side, to hold the door in place. Then, temporarily nail the strike-side jamb in place after verifying that the head clearance is correct. Do not drive the nails flush.

Step 4 Move to the other side of the wall, and place shims at the top and bottom of both side jambs to hold the unit in place. Shims should span the face of the jack studs. Check that the jambs are plumb, and renail casings and/or adjust the shims, as necessary. Also, place shims behind each hinge and strike-plate location (*Figure 49*).

Step 5 Remove the spacer tabs, and open the door (*Figure 50*). After checking the plumb of the hinge jamb, nail through the jamb behind the edge of the door (not the stop) to the jack stud. Nail above and below each hinge and at the top and bottom of the jamb. Close the door to check for proper and even spacing at the header and strike-plate jamb. Adjust the header and the strike-side jamb. Then, nail the strike-side jamb above and below the strike-plate location, and at the top and bottom of the jamb. If no lugs exist on the side jambs or they were cut off, shim the top corners of the header jamb.

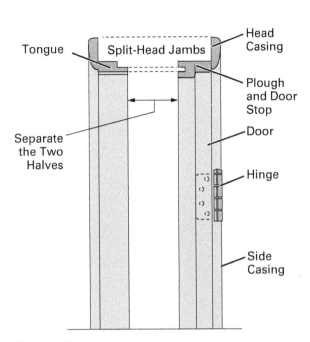

Figure 47 Unpack and separate split jambs.

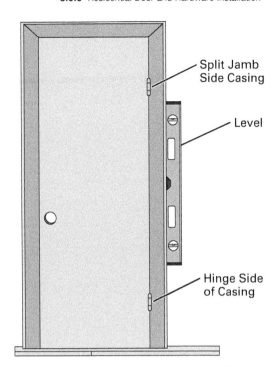

Figure 48 Position and plumb the split-jamb half with the door.

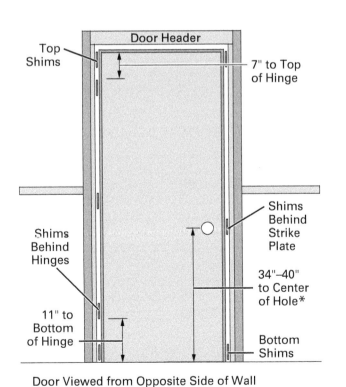

Door Viewed from Opposite Side of Wall

*As Specified by Door Schedule

Figure 49 Shimming the door half of a split-jamb frame.

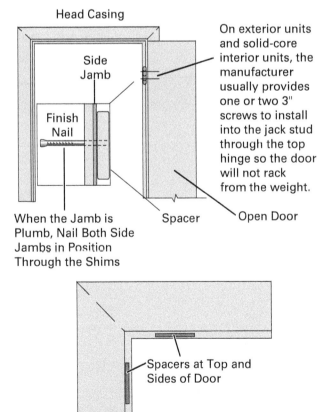

Figure 50 Fastening the jamb to the jack stud and removing the spacers.

Step 6   Remove any tie straps or bracing from the installed half of the split-jamb frame and any bracing from the other half of the split-jamb frame.

Step 7   Slide the tongue of the second half of the split-jamb frame into the plough of the installed half of the frame (*Figure 51*). Nail the casing to the jack studs. Then, nail the jamb to the jack studs at approximately the same locations as the door half of the jamb. Do not nail through the door-stop strip.

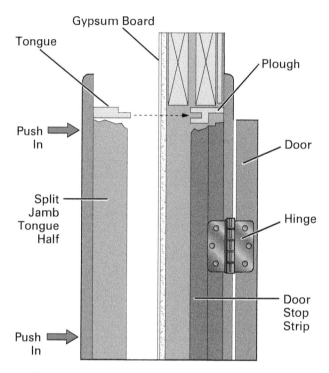

Figure 51 Installing the second half of a split-jamb frame.

Step 8   Complete the nailing of the casing on both sides of the door, and install the appropriate lockset and other hardware, as specified.

3.3.3 Door-Hinge Installation

Manufactured doors are made for simple installation and application of all required hardware. However, you may run into scenarios when you must field-install hinges on solid doors. In these cases, your scope begins with laying out the correct locations of all hardware that is not already installed. Door-hinge locations are typically determined by the architect or contractor. Exact hinge locations vary across different regions and manufacturers. However, most carpenters position the top of the highest door hinge 7" down from the top of the door, the bottom of the lowest hinge 11" above the finished floor, and the third hinge centered between the other two. The clearance at the top of the door may be $1/16$" or $1/8$".

Step 1   Place the door in the opening, and, using a wedge at the bottom, move the door upwards until you achieve a $3/32$" gap at the top. Use a thin scrap piece of wood as a template spacer. Next, wedge the door up against the hinge-side jamb.

Consistent Hinge Location

Hinge locations should be uniform for all doors on a project. If there are both prehung door units and doors that must be hung by hand, make sure that the hinge locations on the prehung units are duplicated throughout the job.

Step 2   Measure and mark hinge locations on the jamb with a knife, and transfer those marks to the edge of the door stile. Avoid marking the door face or the jamb any deeper than the thickness of one leaf of the butts. Remove the door from the opening, and place it in the door jack with the stile to receive the hinges in the up position.

Step 3   The door-hinge backset should be $1/4$" for doors up to $2^{1}/4$" thick and $3/8$" for doors over $2^{1}/4$" thick, as shown in *Figure 52*. Allow $1/16$" clearance for the door when nailing the stop in place.

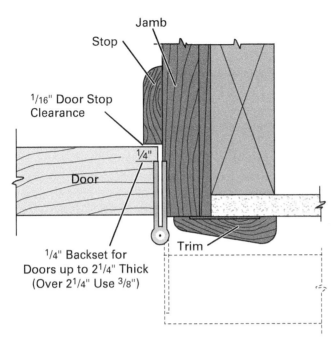

Figure 52 Hinge backset.

Step 4 Place the cutting edge of the chisel across the width of the wood to be removed, and score to the hinge leaf depth about every $1/8$".

Step 5 Place the cutting edge of the chisel along the lower edge of the chisel cuts, and remove all excess wood, as shown in *Figure 53*.

Figure 53 Removing excess wood from the mortise.
Source: Koldunov Alexey/Shutterstock

NOTE

Hinge mortises may be hand-cut and fitted, or they may be cut using a template. There are three measurements to be marked when laying out the mortises:

- Location of the door hinge on the jamb
- Location of the hinge on the door
- Thickness of the hinge leaf on both the door and jamb

Reversible Hinges

Most hinges are reversible and are functional for a left-hand or a right-hand door. If it is an NRP-type hinge, it can be used with either end mounted in an upright position. However, there are hinges specifically manufactured for either a right-hand or left-hand door that are non-reversible.

3.3.4 Installing Hinges and Hanging the Door

Step 1   Remove the loose pin, and place one hinge leaf in the mortise on the door. If the mortise has been properly cut, the hinge leaf may require a few light taps with a soft-faced hammer to seat it flush. Place the other half of the hinge leaf on the mortised jamb, ensuring the hinges on both the door and jamb are in the correct position to allow loose pin insertion from the top.

Step 2   Place a self-centering punch or bit in the countersunk holes of each hinge leaf. If a punch is used, gently tap the pin to mark the location of the pilot hole. The pilot hole will be drilled to receive the hinge leaf screw.

Step 3   Carefully position the door in the opening, engage the door-hinge leaves, insert the hinge pins, and check the door for proper operation and clearance. The door should have a clearance of $^1/_8$" to $^3/_{32}$" at the top and $^3/_{32}$" at both sides. If a slight adjustment at the sides is required, the hinge mortises can be cut deeper, or the hinge can be adjusted using cardboard strips, as shown in *Figure 54*. Ensure that all screws are tightened in the door hinges after adjustments are made.

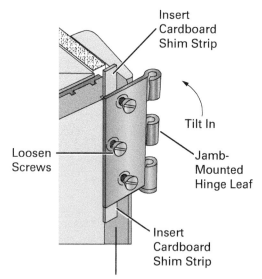 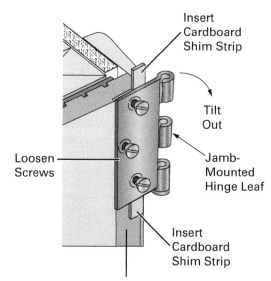

Decrease Hinge Jamb Clearance by Placing the Shim Strip Away from the Edge of the Jamb

Increase Hinge Jamb Clearance by Placing the Shim Strip Toward the Edge of the Jamb

Figure 54 Adjusting the hinge-side jamb clearance using thin cardboard shim strips.

3.3.5 Lockset Installation

All locksets, regardless of the manufacturer, are packaged with installation instructions and templates to mark drill locations. The installation of most exterior and interior locksets is identical, and most locks require workers to drill two holes. The entrance lockset, the passage latch, and the privacy lockset are all shown in *Figure 55*.

 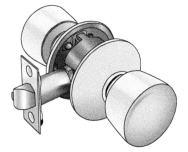 

Entrance Lockset   **Passage Latch**   **Privacy Lockset**

Figure 55 Entrance lockset, passage latch, and privacy lockset.

Figure 56 shows an exploded view of a typical residential tubular lockset. Most locksets consist of three interlocking parts that accommodate rapid installation and correct alignment. The primary part consists of a pre-assembled outside knob, rose, and spindle assembly, while the secondary part consists of a pre-assembled inside knob and rose assembly. The last part is the latch unit that is reversible for doors opening in or out.

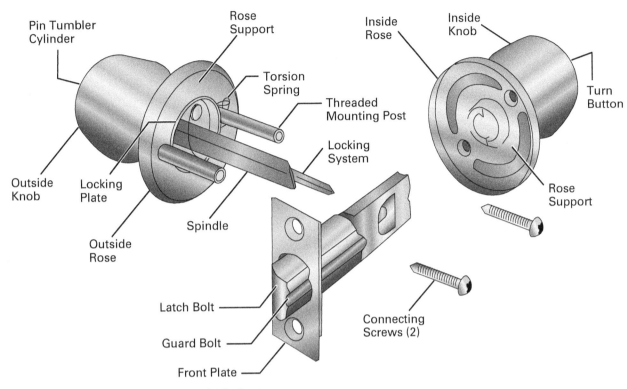

Figure 56 Exploded view of a typical tubular lockset.

Deadbolts

Most residential entrance doors are equipped with single-cylinder key-operated deadbolts in addition to an entrance lockset. In some cases, the primary lockset is not key-operated, and only the deadbolt can be used to secure the door. Also available are electronically locked residential deadbolts that are remote-controlled from a key-fob transmitter and manually unlockable from inside the residence. Entrance doors with large door lights or side lights at the strike allow easy access to the turn-button release of a single-cylinder deadbolt. Such doors can be equipped with a double-cylinder key-operated deadbolt. However, some jurisdictions prohibit their use, because they are a safety hazard and do not permit a quick exit from the building. Always check local codes before installing a double-cylinder deadbolt.

Source: Lost_in_the_Midwest/Shutterstock

Step 1   Remove the lockset from the packing carton, and ensure no parts are missing. Verify that the lockset is the correct one for the target door, and read the installation instructions carefully. Open the door to work comfortably on both sides, and place two wedges under the lock side of the door and at right angles to the face of the door. These wedges will hold the door firmly and allow safe work.

Step 2   Measure up from the floor. The standard height to the center of the doorknob is 38", but this may vary. Place a light pencil mark on the edge of the door. Using a square, draw a horizontal light pencil line on the edge and inside surface of the door, and at the door jamb at 38" above the floor. With the use of the installation template, mark the center of the cross bore and the center of the door edge, as shown in *Figure 57*. The setback of the cross-bore hole is $2^3/_8$" (residential interior) and $2^3/_4$" (commercial and residential exterior) from the edge of the door. Keep in mind that you must adjust the setback if the door edge has a bevel (usually a $1/_{16}$" difference).

Step 3   To determine the location of the vertical center line of the strike, take one-half of the door thickness, as measured from the door stop. Using the strike as a template, mark the outline with a sharp knife or pencil. The location of the attaching screws should also be marked at this time, as shown in *Figure 58*.

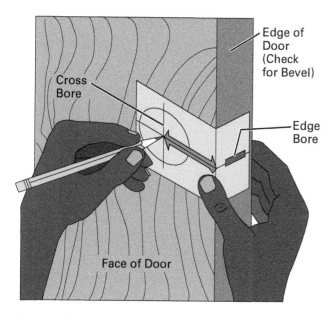

Figure 57 Use of an installation template.

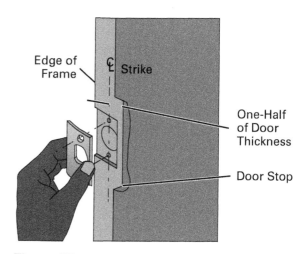

Figure 58 Centerline location of the strike.

Step 4   Drill the holes as directed on the installation template. Drill the cross-bore hole first from both sides of the door to avoid splintering the wood. Each side of the hole should have a clean, sharp edge. A hole saw of the correct size will provide excellent results. Next, using the center mark on the edge of the door, drill the hole for the latch bolt. The hole for the latch bolt should be 3" to $3^1/_2$" deep from the edge of the door and will extend past the cross-bore hole. Insert the latch bolt into the hole in the edge of the door. Use the front trim as a template to mark the outline with a sharp knife or pencil. The location of the attaching screws should also be marked at this time and pilot holes drilled.

Step 5   Chisel out the area marked for the latch bolt to a depth of $5/_{32}$" to allow the front trim to fit flush with the door edge. On the door jamb, drill a 1"-diameter hole $1/_2$" deep at the center point of the strike. Drill the pilot holes for the attaching screws, and chisel out the jamb to a depth of $1/_{16}$", so the face of the strike is flush with the face of the jamb.

Step 6   Insert the latch bolt into the hole in the door edge, with the beveled edge of the latch bolt facing toward the door jamb. Fasten the latch bolt to the door with the two attaching screws.

Step 7   Install the outside knob assembly by inserting the spindle and threaded posts through the holes in the latch bolt. Rotate the locking stem so the knob is unlocked.

Step 8 Install the inside-knob assembly by aligning it with the tip of the spindle of the outside-knob assembly (*Figure 59*). Verify that the turn button is in a horizontal position, then slide the inside-knob assembly up to the door surface. Rotate the rose so the screw holes align with the threaded posts. Insert the connecting screws through the holes in the rose and into the posts. Tighten the connecting screws, but do not overtighten them. Attach the strike to the jamb with the two attaching screws, and you've completed the task.

Figure 59 Assembling a lockset.
Source: Grigvovan/Shutterstock

Step 9 Check the operation of the lockset with the thumb turn, making sure it operates properly. If the lock functions properly, tighten the connecting screws, but do not overtighten them, as this may cause a bind in the latch mechanism.

Cylindrical locksets are used in industrial, commercial, and institutional construction. To simplify architectural specifications and installation, all locks are usually uniform in size, self-aligning, and simple to install. While both knob and lever locksets are essentially the same mechanically, lever locksets are compliant with ADA requirements. Both types of locksets have a pin tumbler cylinder to offer a wide range of keying and control options. A comparison may be made between the internal mechanism of the tubular lockset, as shown earlier, and the internal mechanism of the cylindrical lockset, as shown in *Figure 60*.

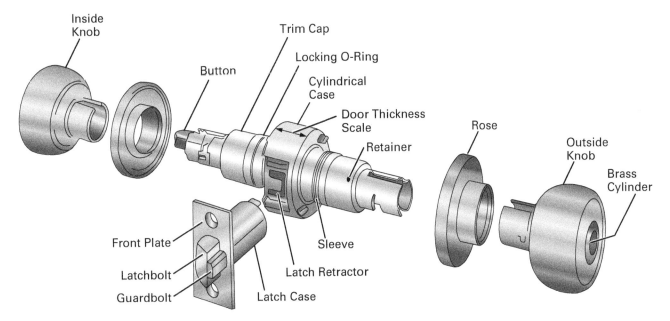

Figure 60 Exploded view of a heavy-duty cylindrical lockset.

3.0.0 Section Review

1. What sliding residential door moves along tracks installed on the exterior of the door opening?
   a. Barn
   b. Pocket
   c. Bypass
   d. Bifold

2. What are the two most common types of residential hinges?
   a. Entrance and exit
   b. Fail-safe and fail-secure
   c. Loose-pin and fixed-pin
   d. Forward-hinge and reverse-hinge

3. What is the standard height to the center of a doorknob?
   a. 24"
   b. 28"
   c. 32"
   d. 38"

4.0.0 Commercial Door and Hardware Installation

Objective

Identify commercial doors, their hardware, and their common installation techniques.

a. Identify common commercial door types.

b. Identify common commercial door hardware and its applications.

c. Explain the basic steps for installing commercial doors and door hardware.

Performance Tasks

There are no Performance Tasks in this section.

As discussed in previous sections, there is a significant overlap between residential and commercial doors. Many commercial projects will use similar doors and door hardware types, but the main difference is that commercial projects prefer function over form in most cases. Commercial projects must select options that will be durable and cost-effective, and that provide features to keep occupants safe and secure.

4.1.0 Commercial Exterior Doors

Commercial exterior doors are designed to provide a secure and convenient means of entering and leaving a building, also known as egress. The type and style selected for an exterior door are based on its intended purpose and function. Doors used exclusively by employees are usually no-frills doors, while those used on loading docks are strictly utilitarian. Public doors are selected for aesthetics and user-friendly function. For example, a trendy retail store will use a different style than a supermarket, and an upscale hotel in New York City will have a different style of door than a budget motel. Manufacturers produce doors in a diverse range of styles and materials to fulfill varying requirements.

Figure 61 shows an example of revolving doors. This beautiful entrance is appropriate for a building that houses offices for major corporations, investment management companies, advertising agencies, marketing firms, a bank, and foreign consulates.

Figure 61 Revolving doors with ornate facade.
Source: Micha Weber/Shutterstock

Figure 62 shows a thermally efficient door on a loading dock; this type of door is used to maintain critical temperature control of the interior of the building.

Figure 62 Thermally efficient loading-dock doors.
Source: hacohob/Shutterstock

CAUTION

When installing a door, you should only perform the work you are trained and qualified to perform. Other work, such as electrical installations, must be completed by a qualified worker.

Installation processes of some commercial doors can be complicated. Due to this complexity, commercial exterior doors are often installed by teams of specially trained manufacturer professionals. In these cases, a carpenter's job is to prepare the installation site according to the building plans and specifications. If you are required to install a commercial exterior door, it is very important that you obtain the door vendor's installation instructions and carefully follow each step.

You can prevent problems during door installation by ensuring that wall construction is completed to specification. Wall angles must be plumb, and door openings must be square and level. Although there are several fixes the installation team can use to compensate for tolerance variations, it's more efficient to ensure the work is correct the first time.

Even if you are not tasked with door installation, you may be asked to accept door deliveries on-site. It is important to carefully inspect each door for damage before accepting it. Note any damage to the door packaging on the delivery invoice, and never accept delivery of severely damaged doors. Instead, report the damage to a supervisor. Don't remove any packaging unless the product is suspected to be damaged, since the packaging material will protect the door during storage.

Once accepted, store doors in a secure location. For example, a shipping container provides an ideal environment for door storage since it has adequate air circulation and substantial protection from the elements.

4.1.1 Fire Doors and Fire Ratings

Fire doors are manufactured to meet specific conditions, and, because of this, the openings are referred to as labeled openings. Labeling agencies use letters A, B, C, D, and E to refer to specific opening locations. Refer to the door schedule for the project, and match each classification door with its appropriate opening. Each location classification requires a specific door rating, depending on the fire hazard involved. An example of these classifications and ratings is shown in *Table 3*.

Average-capacity buildings must allow for at least two points to exit, whereas buildings with larger capacities of 500 to 1,000 people can require three to four exit points. These doors must have a minimum 32" width and 80" height when the

TABLE 3 Labeled Fire-Door Application Chart

Location	Class	Rating
Openings in walls dividing fire areas	A	3 Hour
Elevators, stairwells, 2-hour partitions	B	1.5 Hours
Corridor and room partitions	C	$^3/_4$ Hour
Exterior walls subject to severe fire exposure	D	1.5 Hours
Exterior walls subject to moderate fire exposure	E	$^3/_4$ Hour

door is opened at a 90-degree angle, to avoid bottlenecks as people exit. The landing on each side of these doors must extend at least 44" in the direction of travel, and both landings must match elevations. Other types of exits include interior exit stairways, exit passageways, exterior exit stairways, and horizontal exits that lead occupants directly to the exterior of the building and out of harm's way.

4.2.0   Commercial Hardware

A wide variety of commercial door hardware is available to fit the needs of commercial structures, such as hospitals, schools, and shopping malls. In addition to the basic door hardware, such as hinges and locksets, specialized door hardware, such as touch bars or door **coordinators**, may be required for egress.

Coordinators: Devices used with exit features to hold active doors open until inactive doors are closed.

4.2.1 Hinges

For commercial and institutional doors, three hinges per door are typically specified to accommodate the additional size and weight of the doors. Heavier, high-frequency hinges should be used on doors with high traffic patterns. This type of door is usually specified for commercial buildings, such as an entrance to a department store or mall. Other examples of institutional doors that receive high-frequency usage are school entrances or bathroom doors.

On some jobs, such as a hospital, wider, heavier doors may require a ball-bearing hinge. This hinge type is permanently lubricated and has a fast joint. The tip of the hinge is usually rounded to avoid snagging of clothing or equipment.

4.2.2 Interior Locksets

Other than the basic knob locksets previously discussed, there are lever-handle locksets and deadbolts with, and without, activation devices operated by push buttons, a magnetic pass card, or a remote (*Figure 63*). Many of the activation-device locksets are used in commercial and institutional buildings, such as hotels and hospitals.

Modern technology has transformed interior locksets, providing convenience, security, and ease of use. Some of the latest features include touchless entry through fingerprint or facial recognition, as well as Bluetooth-enabled locks that can be controlled and monitored remotely via smartphones or voice assistants. You can also connect these smart locksets to home automation systems, allowing for automatic locking and unlocking. Some locksets come with advanced security features, such as anti-pick, anti-bump, and anti-drill capabilities, providing an additional layer of protection against intruders.

Keying door locks is necessary for large office buildings, hotels, or motels. This is done so one key will open several doors in a specific location. It eliminates the need for a person to carry multiple keys for each room.

The following terminology should explain most keying methods used in construction today:

- *Simple master key systems* – Each lock has a specific key, which will not operate any other lock in the system. However, all locks in the system can be operated by a master key.

Figure 63 Modern lever-handle lockset with mobile activation device.
Source: Andrew Angelov/Shutterstock

- *Grand master systems* – Each lock has a specific key, as in the simple master key system. The locks are divided into two or more groups, with each group being operated by a master key, and all locks in the system can be operated by a singular grand master key.
- *Great-grand master systems* – Each lock has a specific key, as in the simple master key system. The locks are divided into additional subgroups, as needed, including a master key for each subgroup and a grand master key for each group. All locks in the system can then be operated by a great-grand master key.

4.2.3 External Door Stops, Door Holders, and Door Closers

An external door stop is a rubber-tipped device fastened near the bottom of the door, the wall base, or the floor that prevents the door from striking the wall when the door is fully opened (*Figure 64*). One type of door holder has a plunger device, which is released by foot pressure. The spring holds the rubber tip to the floor. Another type of door holder is screwed into the door and wedges the door open when the lever is dropped to the floor.

Figure 64 Door stop and holder.
Sources: Bowonpat Sakaew/Shutterstock (left); happycreator/Shutterstock (right)

Many building codes require the use of magnetic door holders to control hallway access doors in office and apartment buildings. These devices contain an electromagnet that is wired into the building's fire alarm system. Under normal conditions, the magnets hold the door open. If there is a fire alarm, however, the electrical power to the magnet is automatically turned off, and any open doors will close to inhibit the passage of fire and smoke.

Prior to installing a door closer, the hand or swing of the door must be determined, since closer hardware is only available for regular arm installation on the pull side of the door (*Figure 65*). It is also available for regular or parallel arm installation only on the push side (**transom** bar or top jamb) of the door. Head-frame mounting provides leverage and power to control exceptionally wide doors or doors that are subject to pressure issues.

Transom: A panel above a door that lets light and/or air into a room or is used to fill the space above a door when the ceiling heights on both sides of the door opening allow.

Figure 65 Regular arm installation.
Source: jaojormami/Shutterstock

Manufacturers offer a wide variety of door closers—some of which are non-handed to permit installation on doors of either hand. Some closers can be mounted in different ways and are available in a variety of finishes to complement other door hardware.

4.2.4 Security Hardware

The following sections cover various locking devices and accessories used for entry doors and gates. All electrically operated devices covered in this section are 12VDC (volts direct current) or 24VDC units.

Figure 66 shows an electric deadbolt, along with a typical local power module. Electric strikes and deadbolts provide a remote release of a locked door without requiring the retraction of a latch bolt. They are available as fail-safe or fail-secure. Fail-secure means that the strike remains locked without power. Fail-safe means that the strike opens during power loss. Apart from prisons or mental institutions, most local jurisdictions require a locking device on an exterior exit door to fail-safe upon a fire alarm, sprinkler alarm, or loss of alternating current (AC) power. To prevent the spread of fire and smoke, most codes prohibit the use of fail-safe strikes for stairways and interior fire doors. In these applications, electric locks or latches must be used to allow the doors to remain latched when they are electrically unlocked.

Figure 66 Electric deadbolt.

You can mount most electric strikes on either side of the door frame for a right-hand or left-hand opening door. However, a portion of the outer edge of the door frame must be removed to accommodate the back box that allows the strike lip to swing open. This method can weaken the door frame, allowing easier penetration.

Electric bolt locks (*Figure 67*) are an alternative to electric strikes or magnetic locks because the bolting device that locks a door or gate is mounted on the top and/or sides of the door frame. The door itself has no latch. Multiple electric bolt locks can be used on a door to provide extra security. These locks are available as fail-secure or fail-safe units. In most cases, neither can be used on exterior exit doors due to code restrictions. Some of these devices are designed to fit in narrow door frames and do not require the removal of a portion of the door-frame edge, while others are designed for surface mounts on sliding- or swinging-gate locks.

Figure 67 Electric bolt locks.
Sources: Shutterstock (left); Joe Gough/Shutterstock (right)

The primary use for electric locksets, also called electric latches, is in stairway fire doors on each floor of a building. Building codes generally require that stairway fire doors are never locked on the stair side unless they can be remotely unlocked without unlatching. While providing controlled access and remote unlocking capability, the doors stay latched even when unlocked, maintaining fire door integrity.

Electromagnetic locks (*Figure 68*) are fail-safe and can be used on interior doors and exterior exit doors. However, they cannot be used on interior or stairway fire doors. They have no moving parts and are not subject to wear. Electromagnetic locks are available in direct-hold and shear-hold (concealed) styles. The direct-hold styles are graded for use by the American National Standards Institute (ANSI) as follows:

- *Grade 1* – 1,650 pounds direct holding force, medium security
- *Grade 2* – 1,200 pounds direct holding force, light security
- *Grade 3* – 650 pounds direct holding force, door holding only

There are some electromagnetic locks with 2,000 pounds or more direct holding force. These locks will stay joined even when the door they secure is destroyed. Shear types have holding forces of 2,700 pounds, but they receive a grade 1 rating due to the 90-degree pulling angle. Most electromagnetic locks have integral door position switches to indicate that the door is locked and secure. Shear locks have relocking delay timers activated by the position switch, placing the door at rest before the lock reactivates.

Touch-sensitive bars and handles (*Figure 69*), and switch bars are commonly used in place of exit switches or readers to turn off electromagnetic locks. These touch-sensitive bars and handles are capacitive touch-sensitive switches and

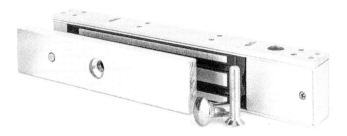

Figure 68 Electromagnetic locks.
Sources: Vladimir Zhupanenko/Shutterstock (top); Surachet Jo/Shutterstock (bottom)

Figure 69 Touch-sensitive bar.
Source: kckate16/Shutterstock

have no moving parts, while switch bars have mechanical switches. Some touch-sensitive bars have electronic timers that delay power shutoff to an emergency exit for a set amount of time. Armored cables are used to connect the bars or handles to the hinge side of the door frame, where the wiring is routed to a controller or to the electromagnetic lock. These touch bars/handles are part of a group of exit devices that are sometimes called request-to-exit (RTE or REX) devices.

Delayed-exit alert locks, sometimes called RTE or REX locks, delay an exit through exterior exit doors in secure facilities. By law, exit doors in other public facilities may not be locked or delayed. When an unauthorized exit is attempted, an alarm sounds, and a signal is sent to guards for closed-circuit television (CCTV) or physical monitoring purposes. After 15 seconds, the exit door is unlocked, permitting an occupant to exit. A fire alarm system signal can also release the lock, allowing unrestricted exits during an emergency. Other buildings may be equipped with an emergency door release button that works in a backup (*Figure 70*).

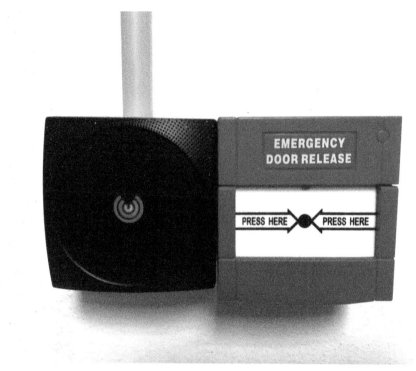

Figure 70 Emergency door release.
Source: axxxelsz/Shutterstock

4.2.5 Weather Stripping

Weather stripping is the application of materials in spaces between the outer edges of doors and windows that keep out air or moisture. For example, *Figure 71* shows a rubber fixed-bottom sweep, which prevents air infiltration into the room.

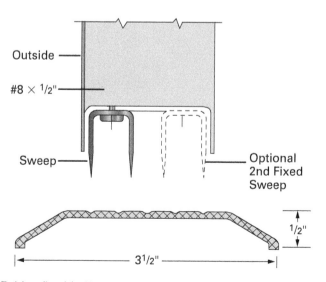

Figure 71 Rubber fixed-bottom sweep.

Other forms of weather stripping are interlocking thresholds (*Figure 72*), a vinyl bulb that compresses when the door is closed (*Figure 73*), and an automatic door bottom. Common forms of weather stripping might also include a spring metal V-strip, a wood-backed foam-rubber strip, or rolled vinyl.

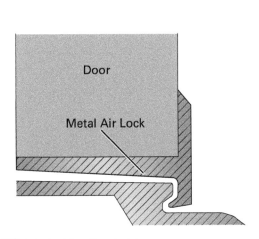

Figure 72 Interlocking threshold.

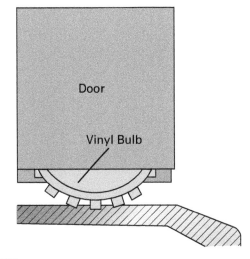

Figure 73 Vinyl bulb.

4.2.6 Thresholds

A threshold is a piece of wood, metal, or stone that is set between the door jamb and the bottom of a door opening. An example of an aluminum threshold is shown in *Figure 74*, along with stop-strip weather seals. The threshold must be set in a bead of caulk to keep air from getting under the door.

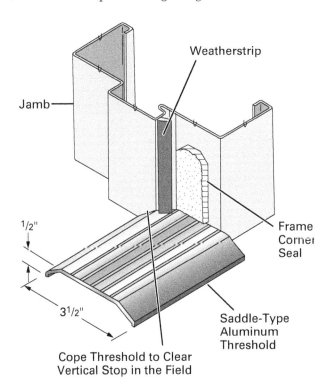

Figure 74 Aluminum threshold.

Ensure that your bevel achieves a 50% slope when the transition increases height by greater than $1/4$". As per ICC 2021 IRC Essentials, thresholds shall not exceed $3/4$" above the finished floor for sliding doors, or $1/2$" above the finished floor for other doors. Furthermore, a landing or floor is generally required on each side of exterior doors, with a maximum threshold height above the landing of 1".

4.2.7 Touch-Bar or Crossbar Hardware

Touch-bar or crossbar hardware, sometimes called panic hardware, provides secure locking of single or double doors from the opposite side of the touch bar but allows easy passage from the touch-bar side. Several panic hardware types are shown in *Figure 75* and *Figure 76*. Touch-bar and crossbar hardware are available with or without an opposite side latch and release lock.

Figure 75 Surface-mounted, double-latch touch-bar device.
Source: Tawatchai45/Shutterstock

Figure 76 Surface-mounted crossbar device.
Source: Florin Burlan/Shutterstock

Touch bars are primarily used on emergency or normal exit doors where local codes prohibit locked doors on the touch-bar side at any time. They are also used on interior building doors that require controlled access and rapid egress. Touch-bar or crossbar hardware is typically installed in commercial and institutional buildings such as stores, hospitals, schools, and government facilities. Closer hardware is typically used with these devices to ensure the door properly closes and latches after an exit.

Alarmed Emergency-Exit Touch Bars

Alarmed emergency-exit touch bars are equipped with a large deadbolt that provides secure, alarmed, code-compliant protection for emergency-only exits. These touch bars can be armed, disarmed, and opened from the inside or outside with a standard rim cylinder key.

Source: Mrs_ya/Shutterstock

4.2.8 Flush Bolts

A flush bolt is a sliding bolt mechanism that is mortised into the door at the top and/or bottom edge. It holds an inactive door in a fixed position on a pair of double doors. A door coordinator is required for automatic and semi-automatic flush bolts on metal, composite, and wood doors. An automatic flush bolt retracts without manual actuation, while a semi-automatic flush bolt engages a door latch when an inactive door closes without the use of a triggering mechanism. The bolt remains extended until it is retracted by a manual release of the bolt-actuating lever.

4.2.9 Door Coordinator

A door coordinator is a device for double doors that holds the active door open until the inactive door is closed. This function allows the normal locking or latching of the lock bolt and the normal overlapping of an **astragal** or rabbeted door (*Figure 77*). Refer to *Figure 78* for a diagram of an astragal set for a pair of doors.

Astragal: A piece of molding attached to the edge of an inactive door on a pair of double doors. It serves as a stop for the active door.

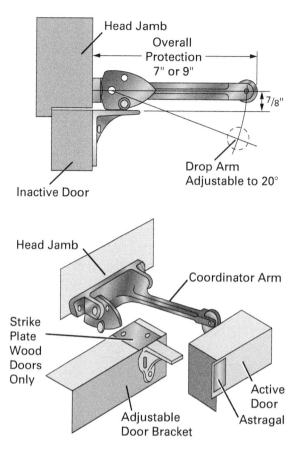

Figure 77 Door coordinator.

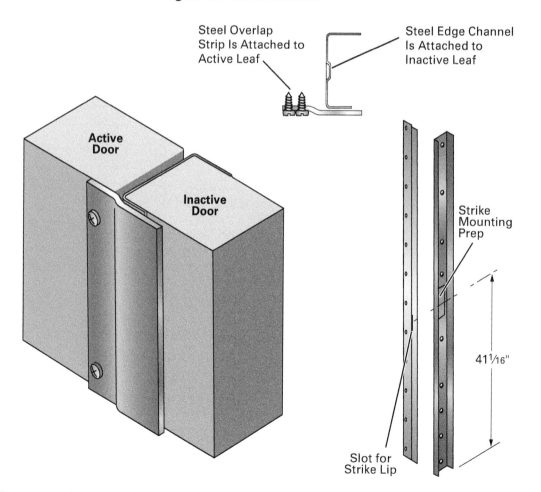

Figure 78 Diagram of an astragal set for a pair of doors.

4.3.0 Commercial Door and Door Hardware Installation

When doors are delivered to the jobsite, they should be checked with the plans of the building to make sure they are the correct thickness, width, and height. Make sure the style and type of wood correspond to the specifications.

4.3.1 Handling, Job Finishing, and Installation Instructions

The following is typical of the handling, job finishing, and installation instructions required to comply with one door manufacturer's warranty policy:

- Store the door flat on a level surface in a dry, well-ventilated building. Cover it to keep it clean, but allow for proper air circulation.
- Wear clean gloves when handling doors, and do not drag doors across one another or across other surfaces to avoid scratches and other damage.
- Deliver doors to the building site once plaster or concrete is completely dry. If doors are stored at the jobsite for more than one week, the top and bottom edges should be sealed.
- Do not subject doors to abnormal heat, dryness, humidity, or sudden changes between environments. They should be conditioned to the average local humidity before hanging.
- Ensure that any door, window, or garage door you install meets the proper wind-load ratings and manufacturer labeling requirements that satisfy the local and regional building codes of the project.
- Use three hinges per door on doors 7'-0" in height or less, and four hinges per door on doors over 7'-0" in height. Hinges should be set flush with the edge surfaces. Be sure that the hinges are set in a straight line to prevent distortion, and allow approximately $3/16$" clearance for swelling of the door or frame during future damp weather periods.
- Immediately after fitting, apply the appropriate weather stripping and/or threshold, and before hanging any interior or exterior door on the job, the top and bottom edges should receive two coats of paint, varnish, or sealer to prevent undue absorption of moisture.

4.3.2 Fitting Doors in Prepared Openings

After selecting the correct door for an opening, measure the opening width and height, and compare it to the size of the door to ensure the sizes match. Having determined the hinge side of the door, fit it to the hinge jamb of the opening. The butt stile is usually left square and will not interfere with the operation of the door. The use of a power plane to fit the door in the opening will save time and energy. If a power plane is not available, you can accomplish this task with a hand plane or fore plane. A fore plane is 18" in length, and its size is somewhere between the larger jointer plane and the smaller jack plane.

The industry standard is $1/8$" clearance between the door and the jamb. The clearance at the bottom of the door is determined by the floor finish or required air circulation. For example, if the floor is a hardwood floor, the clearance over it should be $3/8$" (*Figure 79*).

If the floor finish is a carpet, the clearance should be the carpet clearance plus $3/8$". If there is a threshold under the door, refer to the clearance recommended by the hardware manufacturer. This information is usually included in the hardware manufacturer's installation instructions. Whatever clearance is used between the floor finish and the bottom edge of the door is in addition to the $3/32$" clearance between the head jamb and the top edge of the door.

If you must trim the door at the bottom edge to fit the floor finish, mark your cut line lightly with a pencil. Remove the door and carefully place it on carpet-covered finishing sawhorses.

The bottom edge of a hollow-core or solid-core door with veneer faces must be cut carefully to avoid damaging the veneer. This damage can come in the

Handling Doors

Be careful when handling and installing doors to avoid damaging the finish. Orient the door correctly before installation. Some smooth-finish doors have a specific top and bottom, and should not be installed upside down.

Figure 79 Bottom door clearance.
Source: Lolostock/Shutterstock

form of feathering the surface veneer away from the base veneers when sawing. Avoid this issue by placing a straightedge across the door at the cut line.

Using the straightedge as a guide, run a sharp knife across the grain of the veneer on both sides of the door. The knife cut will prevent the veneer from feathering or tearing. If the veneer is scored only on one side of the door, the scoring should be on the top side, with the door in a flat position when cutting with a power saw. You can also use masking tape to reduce splintering.

Another method of cutting off the bottom edge of a door with veneer faces is with the use of a template and a power saw. A template will reduce friction between the saw and door face to avoid scratches.

Clamp the template to the door in the correct position with the edge up from the bottom of the door by the amount to be trimmed. Run a sharp knife across the bottom of the template, scoring the veneer before sawing.

Reducing Power- and Hand-Plane Surface Drag

Lightly coat the bottom of the plane with a block of paraffin wax to reduce friction between planes and doors. However, use the wax sparingly, because if the doors must be painted or stained, it can interfere with the final finishing process.

Sealing the Bottom and Edges of a Door

Always seal the bottom and edges of the door with varnish or paint, especially after cutting it to length, but avoid spilling any varnish on the door surface. The bottom must be sealed before the door is hung.

After sawing, place the door in a carpet-covered door jack for planing. Plane the edge of the door to the correct width. Put the door in the opening, and with a shim at the bottom, move the door up against the head jamb.

No other steps are required if the door head is square with the head jamb. If the top rail or door head is out of square with the head jamb, scribe, saw, and plane it to fit. When the door has $^3/_{32}$" clearance at the head and both sides, as well as the proper clearance at the bottom edge and the floor, remove it from the opening, and place it in the door jack with the hinge stile down.

Bevel the edge of the lock stile of the door by planing it to an angle of $1/8$" in 2", as shown in *Figure 80*. The door swings on an arc created by a radius that has its center located in the middle of butt hinge pin. This center is well forward of the face of the door stop, so the lock edge of the door must be beveled to clear the face edge of the door jamb.

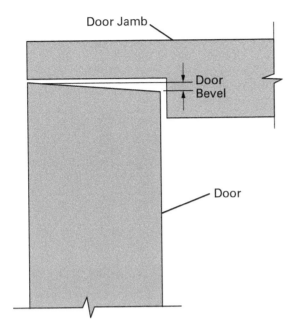

Figure 80 Section of door showing door bevel.

4.3.3 Installation of Door Closers

Generally, the installation of a door-mounted closer will proceed as follows:

Step 1 Using a supplied template, select the desired angle of the opening. Locate and drill holes on the door for the closer body and on the frame for the arm shoe.

Step 2 Install the closer body on the door.

Step 3 Disassemble the secondary arm and shoe assembly from the main arm by removing the elbow screw. Fasten the secondary arm and shoe assembly to the frame face.

Step 4 Place the main arm into the closer pinion shaft, and install and tighten the main arm screw with a $1/2$" wrench.

Step 5 Close the door and adjust the secondary arm assembly so the main arm is perpendicular to the face of the door. Reassemble the secondary arm to the main arm and tighten the screw securely.

Step 6 Adjust the closing tension by using the wrench packed with the door closer on the ratchet. Swing the wrench away from the hinge to wind the spring between 3 and 10 notches, then engage the dog on the ratchet. Increase or decrease the swing power to suit the closing conditions of the door (*Figure 81*).

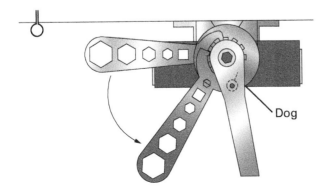

Figure 81 Adjusting closer tension.

4.0.0 Section Review

1. What is the required rating for a Class A fire door?
   a. 1 hour
   b. 1.5 hours
   c. 2 hours
   d. 3 hours

2. What does "fail-safe" mean for a commercial door?
   a. Backup power is available.
   b. It will not fail in an emergency.
   c. The strike opens in the event of a power loss.
   d. The strike stays locked if power is lost.

3. What is the industry-standard clearance between the door and jamb?
   a. $1/16"$
   b. $1/8"$
   c. $1/4"$
   d. $1/2"$

Module 27208 Review Questions

1. What is the term for the safety rule that states you should never reach beyond your center of gravity?
   a. Shoelace
   b. Belt buckle
   c. Ladder rung
   d. Hard hat

2. What is the most common thickness for finger-jointed door frames?
   a. $1/4"$
   b. $1/2"$
   c. $3/4"$
   d. 1"

3. How many pairs of shims should you use on the lock-side jamb?
   a. 1
   b. 2
   c. 3
   d. 4

4. What common metal frame accessory is designed to keep air and moisture from the outer edges of the door?
   a. WRB
   b. Pocket guard
   c. Bypass sealant
   d. Weather stripping

5. What is the proper clearance for framing of a knockdown metal frame being installed over drywall?
   a. 1" on each side
   b. $1/2"$ on the sides, 1" on top and bottom
   c. $3/4"$ on each side
   d. $1/8"$ clearance on the hinge side

6. What type of metal frame is commonly applied to masonry?
   a. Barn
   b. Pocket
   c. Punch and dimple
   d. Bifold

7. What door component is where the hinges (butts) are fastened?
   a. Butt jamb
   b. Hanging stile
   c. Hinge back
   d. Sanitary stop

8. What is the name for the common pair of overlapping residential doors, typically used for closets?
   a. Bypass
   b. Stacking
   c. French
   d. Trifold

9. What would a lead-core metal door protect against?
   a. Fire
   b. Smoke
   c. Radiation
   d. Mold

10. What is the minimum fire rating for doors between dwellings and garages?
    a. 20 minutes
    b. 30 minutes
    c. 1 hours
    d. 2 hours

11. What is the name of the round hinge center that joins the leaves?
    a. Joint
    b. Knee
    c. Toe
    d. Knuckle

12. What equipment would you use to keep doors securely on-edge?
    a. Pallet jacket
    b. Door jack
    c. Door stand
    d. Frame jockey

13. What is the common term for the piece of wood, metal, or stone set between the door jamb and the bottom door opening?
    a. Veneer
    b. Smoke gasket
    c. Threshold
    d. Door guide

Answers to Odd-Numbered Module Review Questions are found in *Appendix A*.

Answers to Section Review Questions

Answer	Section	Objective
Section One		
1. b	1.1.0	1a
Section Two		
1. d	2.1.0	2a
2. d	2.2.2	2b
3. b	2.3.0	2c
Section Three		
1. a	3.1.5	3a
2. c	3.2.1	3b
3. d	3.3.5	3c
Section Four		
1. d	4.1.1	4a
2. c	4.2.4	4b
3. b	4.3.2	4c

Suspended and Acoustical Ceilings

Objectives

Successful completion of this module prepares you to do the following:

1. Identify the components necessary to properly install a suspended ceiling system.
   a. Identify the suspension systems and hardware necessary to properly install a suspended ceiling system.
   b. Identify the system components necessary to properly frame a suspended ceiling system.
   c. Identify the safe material handling and storage procedures required when installing a suspended ceiling system.
2. Interpret a reflected ceiling plan.
   a. Interpret the layout information.
   b. Interpret the mechanical, electrical, and plumbing (MEP) locations.
3. Describe the key considerations, methods, and best practices relating to ceiling installation.
   a. Explain seismic considerations for ceilings.
   b. Identify the layout and takeoff procedures to install a suspended ceiling system.
   c. Identify the tools and equipment to lay out and install a suspended ceiling system.
   d. Identify the installation methods and procedures for a suspended ceiling system.

Performance Tasks

Under supervision, you should be able to do the following:

1. Estimate the quantities of materials needed to install a lay-in suspended ceiling system in a typical room from an instructor-supplied drawing.
2. Establish a level line at ceiling level, such as is required when installing the wall angle for a suspended ceiling.
3. Lay out and install a lay-in suspended ceiling system according to an instructor-supplied drawing.

Overview

Suspended ceilings are found in most commercial buildings. Unlike fixed ceilings, they provide easy access to wiring, cabling, and air conditioning equipment located in the area between the ceiling and the overhead deck. The ceiling tiles suppress sound transmission. Suspended ceilings are built by first installing a grid suspended from the overhead deck, and then installing ceiling tiles in the grid. Proper installation of these ceilings is a skill that takes training and practice.

NOTE

Codes vary among jurisdictions. Because of the variations in code, consult the applicable code whenever regulations are in question. Referring to an incorrect set of codes can cause as much trouble as failing to reference codes altogether. Obtain, review, and familiarize yourself with your local adopted code.

Digital Resources for Carpentry

Scan this code using the camera on your phone or mobile device to view the digital resources related to this craft.

1.0.0 Suspended Ceiling System Components

Performance Tasks

There are no Performance Tasks in this section.

Objective

Identify the components necessary to properly install a suspended ceiling system.

a. Identify the suspension systems and hardware necessary to properly install a suspended ceiling system.

b. Identify the system components necessary to properly frame a suspended ceiling system.

c. Identify the safe material handling and storage procedures required when installing a suspended ceiling system.

Ceiling panels: Acoustical ceiling boards that are suspended by a concealed grid mounting system. The edges are often kerfed and cut back.

Ceiling tiles: Any lay-in acoustical boards designed for use with exposed grid mounting systems. Ceiling tiles normally do not have finished edges or precise dimensional tolerances because the exposed grid mounting system provides the trim-out.

Acoustical materials: Types of ceiling panel, plaster, and other materials that have high absorption characteristics for sound waves.

Suspended ceilings are widely used in commercial construction and to some extent in residential construction. Modern suspended ceilings serve many purposes. They are designed to help keep outside noise from entering the room and to reduce noise levels occurring within the room itself. In some cases, ceilings are integrated with the electrical and HVAC (heating, ventilating, and air conditioning) functions to provide correct lighting and temperature control. Use of attractive **ceiling panels** or **ceiling tiles** helps give a warm, relaxed feeling to a room. Complete ceiling systems offer a wide variety of options, both functional and visual. The type of **acoustical materials**, the plans for the acoustical ceiling, and the method of installing the ceiling depend on the intended use of the room.

Sounds travel through the air in a room as a series of pressure waves. These sound pressure waves travel outward in all directions. When sound waves strike a wall or ceiling, some of the sound-wave energy is absorbed and some is reflected in wave patterns moving in the opposite direction (*Figure 1*). The result

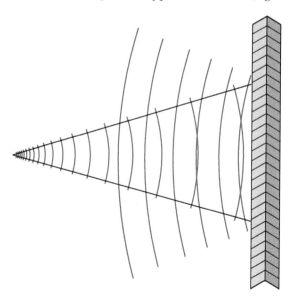

Figure 1 Sound-wave reflection.

is that you will hear both the original sound and its reflected image. The sound is also transmitted through the air in the wall or ceiling cavity to the opposite surface, causing the surface to vibrate and transmit the sound to any adjoining room(s).

Sound waves have a **frequency**. The frequency, or pitch, measured in **hertz (Hz)**, is the number of vibrations or cycles that occur in the wave in one second. The greater the number of cycles per second (Hz), the higher the frequency and the higher the pitch.

The intensity of sound refers to its degree of loudness or softness. An **A-weighted decibel (dBA)** is a unit of measure used for establishing and comparing the intensity of sound sources. It is used to express the value of all sounds in a range from 0 dBA to 140 dBA and higher.

Table 1 shows some typical examples of noise situations and their relative **decibel (dB)** levels. Note that any sounds greater than 120 dBA can produce a physical sensation. Sounds above 130 dBA can cause pain and/or deafness. Continued exposure to sound levels above 85 dBA can cause hearing loss over time.

Some terms you may encounter and should understand when selecting or working with **acoustics** and acoustical ceiling materials include the following:

- *Reflection* — The bouncing back of sound waves after hitting some obstacle or surface such as a ceiling or wall.

- *Reverberation* — The prolonging of a sound through multiple reflections of that sound as it travels back and forth across a room. These multiple reflections of sound occur so fast that they are usually not heard as distinct repetitions of that sound. However, they can cause a higher noise level than the original sound source.

- *Noise reduction coefficient (NRC)* — Used by manufacturers to compare the noise absorbencies of acoustical products. The higher the number, the better the absorbency. The NRC measures the average percentage of noise a material absorbs at four selected frequencies.

- *Sound transmission loss* — The amount of sound lost as a noise travels through a material. Acoustical ceiling assemblies are rated in terms of sound

Frequency: Cycles per unit of time, usually expressed in hertz (Hz).

Hertz (Hz): A unit of frequency equal to one cycle per second.

A-weighted decibel (dBA): A single number measurement based on the decibel but weighted to approximate the response of the human ear with respect to frequencies.

Decibel (dB): An expression of the relative loudness of sounds in air as perceived by the human ear.

Acoustics: A science involving the production, transmission, reception, and effects of sound. In a room or other location, it refers to those characteristics that control reflections of sound waves and thus the sound reception in the area.

TABLE 1 Sound Levels of Some Common Noises

Sound Level (dBA)	Intensity Level	Outdoor Environment	Indoor Environment
140	Deafening	Jet aircraft, artillery fire	Gunshot
130	Threshold of pain	—	Loud rock band
120	Threshold of feeling	Elevated train	Portable stereo headset on high setting
110	Extremely loud	Overhead jet aircraft at 1,000'	Loud nightclub
100	Very loud	Chainsaw, motorcycle at 25', auto horn at 10'	—
90	Loud	Lawn mower, noisy city street	Full symphony band, noisy factory
80	Moderately loud	Diesel truck at 50'	Garbage disposal, dishwasher
70	Average	—	Face-to-face conversation, vacuum cleaner, printers, and copiers
60	Moderately quiet	Air conditioning condenser at 15', auto traffic near an interstate highway	Normal conversation, general office
50	Quiet	Large transformer at 50'	—
40	Very quiet	Bird calls	Private office, soft radio music
30	Extremely quiet	Quiet residential neighborhood	Average residence
20	Nearly silent	Rustling leaves	Quiet theater, whisper
10	Just audible	—	Human breathing
0	Threshold of human hearing	—	One's own heartbeat in a silent room

transmission class (STC). An STC value of 20 to 25 indicates that even normal speech can be easily understood in an adjoining room. On the other hand, an STC value of 50 to 60 indicates that loud sounds will be heard only faintly or not at all. Acceptable STC ratings range from approximately 50 to 65.

- *Articulation class (AC)* — The rating of a ceiling's ability to achieve normal privacy in open office spaces by absorbing noise reflected at an angle off the ceiling into adjacent areas (cubicles). According to ASTM International *E1110* and *E1111* standards, the generally accepted AC ratings for normal privacy in open-plan offices is a minimum of 170, with 190 to 210 preferred.

- *Ceiling attenuation class (CAC)* — The rating of a ceiling's efficiency as a barrier to airborne sound transmission between adjacent work areas, where sound can penetrate **plenum** spaces and travel to other spaces. CAC is stated as a minimum value. Per *ASTM E1264*, CAC minimum 25 is acceptable in an open-plan office, while a rating of CAC minimum 35 to 40 is preferred for closed offices.

- *Absorption* — The energy of sound waves being taken in (entering) and absorbed by a surface of any material rather than being bounced off or reflected.

Plenum: A chamber or container for moving air under a slight pressure. In commercial construction, the area between the suspended ceiling and the floor or roof above is often used as the HVAC return air plenum.

1.1.0 Ceiling Systems

A wide variety of suspended ceiling systems is available, with each system being somewhat different from the others. All use the same basic materials, but their appearances are completely different.

The focus of this module is on the following ceiling systems:

- Exposed grid systems
- Metal-pan systems
- Direct-hung concealed grid systems
- Integrated ceiling systems
- Luminous ceiling systems
- Suspended drywall ceiling system
- Special ceiling systems

Also covered in this module is background information relevant to acoustics and acoustical ceilings, including information on the propagation of sound waves, acoustical ceiling product terminology, and ceiling-related drawings.

1.1.1 Exposed Grid Systems

An exposed grid system is a suspension system for lay-in ceiling tiles (*Figure 2*). The factory-finished supporting members are exposed to view.

Figure 2 Exposed grid system.
Source: Lukassek/Shutterstock

1.1.2 Metal-Pan Systems

The metal-pan system resembles a conventional exposed grid ceiling system except that metal panels or pans are used in place of the conventional sound-absorbing tile. In some cases, the panels or pans are snapped into place from below rather than being laid-in from above the ceiling frame.

1.1.3 Direct-Hung Systems

A direct-hung ceiling system is used if the grid system is to be concealed from view. A mechanical clip or tongue-and-groove joint is used to connect the tiles together. The tiles are then tied to the suspended grid.

1.1.4 Integrated Ceiling Systems

As suggested by its name, the integrated ceiling system incorporates the lighting and/or an air supply **diffuser** as part of the overall ceiling system, as shown in *Figure 3* and *Figure 4*.

Diffuser: An attachment for duct openings in air distribution systems that distributes the air in wide flow patterns. In lighting systems, it is an attachment used to redirect or scatter the light from a light source.

Figure 3 Integrated grid system.
Source: Art Noppawat/Shutterstock

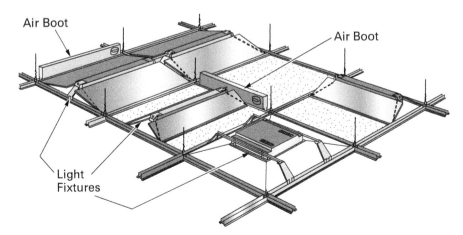

Figure 4 Integrated ceiling schematic.

Some manufacturers of ceiling materials offer specialty ceiling designs. Many integrated ceilings can be used to create distinctive architectural appearances and interior artistic designs. Complete ceiling systems offer dozens of options, both functional and visual.

The functional aspect allows for enhanced lighting and lighting effects. Mechanically, it can incorporate the air supply system through spaced air supply diffusers and return air grilles, all of which have been designed to go beyond function to enhance the artistic appearance of the ceiling.

Integrated ceiling systems are available in units called modules. The common sizes are 30" × 60" and 60" × 60". The dimensions refer to the spacing of the main runners and cross tees.

1.1.5 Luminous Ceiling Systems

Whereas light sources are incorporated into an integrated ceiling, luminous ceiling systems are the light source. Unlike other ceiling systems, in which the light sources interrupt the surface of the ceiling, the surface of luminous ceilings are a continuous light source (only interrupted by the ceiling grid if an exposed grid is used). The choice to make the entire ceiling a light source is often made to provide dispersed light—rather than direct light which tends to create a glare—to an area that needs to be well lit, such as a museum or office. Luminous ceiling systems (*Figure 5*) are available in many styles, such as exposed grid systems with drop-in plastic light diffusers or an aluminum or wood framework with translucent acrylic light diffusers.

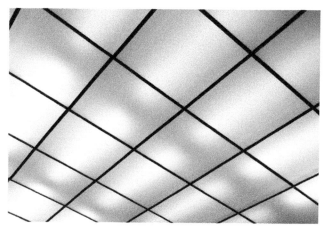

Figure 5 Luminous ceiling system.
Source: Alex Veresovich/Shutterstock

Fluorescent or LED fixtures are generally installed above the translucent diffusers. Standard modules of 2' × 2' up to sizes of 5' × 5' are available. It is also possible to purchase custom sizes for special fit conditions. There are two types of luminous ceilings: standard and nonstandard. Standard systems are, as their name indicates, those that are available in a series of standard sizes and patterns. Nonstandard systems differ in that they deviate from the normal spacing of main supports and may include unusual tile sizes, shapes, and configurations.

All surfaces in the luminous space, including pipes, ductwork, ceilings, and walls, are painted with a 75% to 90% reflective matte white finish. Any surfaces in this area that might tend to flake, such as fireproofing and insulation, should receive an approved hard surface coating prior to painting to prevent flaking onto the ceiling below.

1.1.6 Suspended Drywall Ceiling System

The suspended drywall system is used when it is desirable or specified to use a drywall finish or drywall backing for an acoustical panel ceiling.

1.1.7 Special Ceiling Systems

Ordinary acoustical ceilings with an exposed grid and white, mineral fiber panels are so familiar that they likely go unnoticed by most. On the other hand, some ceilings are indeed special.

This category of eye-catching ceiling systems includes (but is not limited to) those covered by wood or metal tiles or panels, arranged in soothing linear patterns, composed of neatly arranged baffles or blades, spotted with floating acoustical islands, reflecting back at you like a pond, and formed by mesmerizing curves (*Figure 6*).

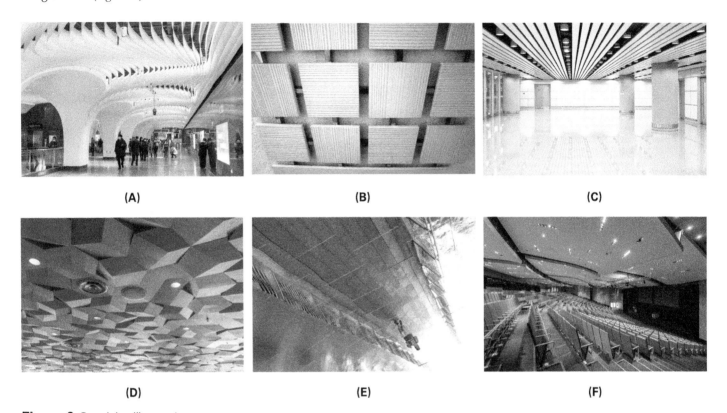

(A)   (B)   (C)

(D)   (E)   (F)

Figure 6 Special ceiling systems.
Sources: Cynthia Lee/Alamy (A); David Ausserhofer/Intro, imageBROKER.com GmbH & Co. KG/Alamy (B); yxm2008/Shutterstock (C); Askar Karimullin/Alamy (D); Benjamin Marcus/Alamy (E); bialasiewicz/123RF (F)

Metallic Ceilings

The malleability of metal allows it to be formed into a wide variety of shapes and styles. Metallic ceilings are also desirable because they reflect rather than absorb heat, making them energy efficient. Additional benefits include ease of maintenance, durability, recyclability, and fire-resistance. Sound-absorbing backing material is installed in metallic ceiling systems to improve acoustical quality.

Pre-manufactured metallic ceiling tiles are normally produced in the standard 2' × 2' and 2' × 4' sizes and may be of the lay-in or surface mount installation type. Custom-made metallic ceiling tiles can be ordered from different manufacturers in a variety of sizes, shapes, and surface patterns.

Linear/Planar Ceilings

Linear (or planar) ceiling systems allow designers to create a sense of visual flow and focus as the lines direct the viewer's eyes. Long planks can be used to achieve a smooth, continuous design or a variety of lengths and dimensions can be used to create interesting patterns.

The uniformity and length of linear tiles speeds up the installation process. The smooth minimalist appearance of the linear tiles makes them compatible with a variety of building exteriors and, therefore, a popular choice for exterior ceilings.

Wood lends itself well as a material for use in linear ceilings because it is relatively easy to cut into uniform, linear sections. Manufactured wood linear ceiling panels are made to lay into or mount to proprietary grid systems. Wood panels can also be surface mounted and given soundproofing, in addition to the natural acoustics of the wood, through the application of insulative backing material.

Reflective Ceilings

When designers want to increase the perceived space within a structure without increasing the height of the ceiling, they might incorporate a reflective ceiling. Reflective ceilings return the existing light within a space and intensify the mood of the lighting.

Metals with high reflective properties and glass (mirrors) have traditionally been used to make reflective ceiling tiles. A glossy coating can be applied to *acoustic stretch ceiling systems* (lightweight fabric stretched between a frame with sound-absorbent membrane) to create a highly reflective result.

Baffle and Blade Ceilings

Baffles are free-hanging panels that obstruct, trap, or redirect the flow of sound. Like other acoustical ceilings, baffle systems are suspended from a support grid. The quantity, size, and perforations in the baffles provide sound absorption.

Blades, slats rotated with the narrow side facing, are often used to create open cell linear ceiling designs by leaving gaps between the blades. Blades provide sound absorption, which is dependent on the size of the blades and their spacing. The spacing of the panels also determines the degree to which the plenum or mechanicals will be visible.

Both baffles and blades are highly customizable in terms of their size, shape, material, and color.

Island/Cloud Ceilings

Similar to baffle/blade systems, acoustical islands and clouds utilize sound-absorbing panels suspended from grid systems, which allow some of the mechanical systems and structure above the panels to show. Islands and clouds are both made from sound absorbing materials that can be produced in a variety of colors and shapes, including curves and waves. The difference between the two is that islands are larger and cover more area than clouds. Islands and clouds are sometimes used in combination with baffles to increase the sound absorption of the system.

Look Up

The next time you are out, look at the ceilings in the stores, theaters, malls, and other buildings you enter. You will see that the available styles, designs, materials, and color schemes are more numerous than you ever imagined. The one thing most of them have in common is that they are some form of suspended ceiling using panels set into or attached to a framework.

Sources: gerenme/Getty Images (left); Pavel L Photo and Video/Shutterstock (right)

1.2.0 Ceiling System Components

A variety of components are combined to create the ceiling grid. Each type of system and each manufacturer's components may be slightly different. Always refer to the manufacturer's instructions before installing a ceiling grid. The grid, composed of main runners, cross runners, and wall angles, is commonly made from light-gauge metal members.

1.2.1 Exposed Grid Components

For an exposed grid suspended ceiling, a light-gauge metal grid is hung by wires attached to the original ceiling or structural members. Tiles that usually measure 2' × 2' or 2' × 4' are then placed in the frames of the metal grid. Exposed grid systems are constructed using the components and materials described as follows and shown in *Figure 7.*

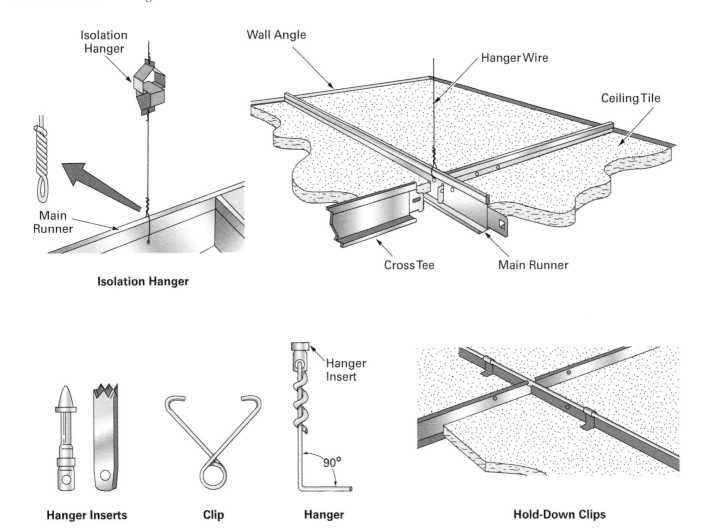

Figure 7 Exposed grid system components.

- *Main runners* — Primary support members of the grid system for all types of suspended ceiling systems. They are 12' in length and are usually constructed in the form of an inverted T. When it is necessary to lengthen the main runners, they are usually spliced together using extension inserts; however, the method of splicing may vary with the type of system being used.

- *Cross runners (cross ties or cross tees)* — Inserted into the main runners at right angles and spaced an equal distance from each other, forming a complete grid system. They are held in place by either clips or automatic locking devices. Typically, they are either 2' or 4' in length and are usually constructed in the form of an inverted T. Note that 2' cross runners are only required for use when using 2' × 2' ceiling tiles.

- *Wall angles* — Installed on the walls to support the exposed grid system at the outer edges.

- *Ceiling tiles* — Tiles that are laid in place between the main runners and cross ties to provide an acoustical treatment. The acoustical tiles used in suspended ceilings stop sound reflection and reverberation by absorbing sound waves. These tiles are typically designed with numerous tiny sound traps consisting

Fissured: A ceiling-panel or ceiling-tile surface design that has the appearance of splits or cracks.

Striated: A ceiling-panel or ceiling-tile surface design that has the appearance of fine parallel grooves.

USDA-Compliant Tiles

The United States Department of Agriculture (USDA) is responsible for ensuring that the nation's commercial supply of meat, poultry, and egg products is safe through safety guidelines and inspections. USDA-compliant ceiling tiles are designed for use in kitchens and central food-preparation areas. These tiles are washable, waterproof, and bacteria resistant.

of drilled or punched holes or fissures, or a combination of both. When sound strikes the tile, it is trapped in the holes or fissures. A wide variety of ceiling tile designs, patterns, colors, facings, and sizes is available, allowing most environmental and appearance demands to be met. Tiles are typically made of glass or mineral fiber. Generally, glass-fiber tiles have a higher sound absorbency than mineral-fiber tiles. Tile facings are typically embossed vinyl in a choice of patterns such as **fissured**, pebbled, or **striated**. The specific ceiling tiles used must be compatible with the ceiling suspension system due to variations in manufacturers' standards.

- *Hanger inserts and clips* — Many types of fastening devices are used to attach the grid-system hangers or wires to the ceiling or structural members located above the suspended ceiling. Screw eyes and star anchors are commonly used and may require a hammer drill for installation. Powder-actuated fasteners are commonly used when fastening to reinforced concrete. Clips are used where beams are available and are typically installed over the beam flanges. Then the hanger wires are inserted through the loops in the clips and secured.

- *Hangers* — The devices attached to the hanger inserts and used to support the main runners. The hangers can be made of No. 12 wire or heavier rod stock. Ceiling isolation hangers are also available to isolate the ceiling from noise traveling through the building structure.

- *Hold-down clips* — Used in some systems to hold the ceiling tiles in place.

- *Nails, screws, expansion anchors, and molly bolts* — Used to secure the wall angle to the wall. The specific fastener used depends on the wall construction and material.

What Is in a Name?

The terms *ceiling panel* and *ceiling tile* have specific meanings in the trade. Ceiling tiles are typically any lay-in acoustical board that is designed for use with an exposed grid system. They do not have finished edges or precise dimensional tolerances because the grid system provides the trim-out. Ceiling panels are acoustical ceiling boards, usually 12" × 12" or 12" × 24", which are nailed, cemented, or suspended by a concealed grid system. The edges are often kerfed and cut back.

1.2.2 Ceiling Panels and Tiles

Ceiling panels and tiles range in size from 12" × 12" up to 60" × 60". Various colors and designs are available. Most are fabricated from mineral fiber and glass fiber. Depending on their design and purpose, mineral-fiber panels and tiles are made with painted, plastic, aluminum, ceramic, or mineral faces. Glass-fiber panels and tiles are made with painted, film, glass cloth, and molded faces.

The three general types of tile/ceiling grid interfaces are lay-in, concealed tee, and profiled edge. Lay-in tiles are widely used and are generally the most cost-effective style. Concealed tee tiles are butt-jointed tiles that provide a monolithic ceiling design with no visible support system. Profiled edge tiles feature a wide selection of edge designs from soft-edged chamfered or curved tiles to highly articulated edges. Beveled and angular reveal-edged tiles provide a three-dimensional look in a suspended ceiling.

Many interior spaces have specific requirements for ceiling tile materials and characteristics. These can include sound control, fire resistance, thermal insulation, light reflectance, and moisture resistance (*Table 2*). Other considerations include maintenance, appearance, and cost considerations.

High-performance acoustical tiles are used to help prevent noise in open-plan and closed types of offices. The following three factors contribute to noise distractions in a workplace:

- *General office noise* — The ability of a ceiling material to absorb general office noise is measured using a value known as the noise reduction coefficient (NRC),

TABLE 2 Ceiling Panel Ratings

Rating	Meaning
NRC (Noise Reduction Coefficient)	Ability of a material to absorb general indoor noise, measured in average percentage of noise absorbed.
AC (Articulation Class)	Ability of material to absorb reflected conversational noise. A rating of 170 is generally the minimum for open plan offices.
CAC (Ceiling Attenuation Class)	Ability of material to absorb sound transmission between adjacent indoor spaces. A minimum of 25 is considered acceptable for open spaces and 35–40 is preferred for closed spaces.
Flame Spread Rating	A measure of the flammability of a material, with an index of 0–200 grouped into classes. Lower is better; a class A or class 1 rating indicates that the material is nearly incombustible (0–25 index).
Fire Resistance Assembly Rating	Degree to which an entire assembly prevents the spread of fire without losing its structural function, measured in time. The rating applies to the entire floor-ceiling or roof-ceiling assembly, not to the ceiling alone.
Thermal Insulation	When required by codes, must be lightweight. Fiberglass is preferred; if insulation is installed on non-fiberglass panels it must be R-19 with vapor barrier facing down and perpendicular to cross tees. Insulation should not be installed above fire-resistant ceiling panels.
High Humidity Resistance	Resists sagging caused by moisture and prevents growth of mold or mildew. Used in humid environments (indoor pools, bathrooms, etc.) and to control moisture from cycling HVAC systems.
High Light Reflectance (LR)	Percentage of visible light reflected. Used to maximize natural lighting in an area and reduce artificial lighting costs and environmental impacts. A rating of .83 (83% of visible light reflected) is desirable.

which is used by tile manufacturers to compare the noise absorbency of their ceiling tile products. The higher the number, the better the absorbency. The NRC measures the average percentage of noise a tile absorbs at various frequencies.

- *Reflected conversational noise that angles off ceilings into adjacent cubicles* — The ability of a ceiling material to absorb reflected conversational noise is measured using a value known as the articulation class (AC), which rates a ceiling's ability to achieve normal privacy in open office spaces by absorbing noise reflected at an angle off the ceiling into adjacent areas (cubicles). The *ASTM E1110* and *E1111* standards recommend a ceiling with an AC rating above 170 for adequate privacy in an open office.

- *Sound transmission through cubicles, partitions, walls, and ceilings* — The ability of a ceiling material to absorb sound transmission is measured using a value known as the ceiling attenuation class (CAC), which rates a ceiling tile's efficiency as a barrier to airborne sound transmission between adjacent work areas, where sound can penetrate plenum spaces and carry to other spaces. The CAC is stated as a minimum value. Per *ASTM E1264*, CAC minimum 25 is acceptable in an open-plan office, while a rating of CAC minimum 35 to 40 is preferred for closed offices.

Fire-resistant ceiling tiles and support systems are specially made of materials that provide increased resistance against flame spread, smoke generation, and/or structural failure in the event of a fire. Two ratings based on ASTM International, ANSI (American National Standards Institute), and NFPA (National Fire Protection Association) standards are used to evaluate fire-resistant tiles: the flame-spread rating of the material *(ASTM E84)* and the fire-resistance rating of a ceiling assembly *(ANSI/UL 263, ASTM E119,* and *NFPA 251)*.

Basically, the flame-spread rating of a ceiling material is the relative rate at which a flame will spread over the surface of the material. This rate is compared against a rating of 0 (highest rating) for fiber-cement board and a rating of 100 for red oak. Class A ceilings have flame-spread ratings of 25 or less, the required standard for most commercial applications. The fire-resistance rating of a ceiling assembly represents the degree to which the entire assembly, not the individual components, withstands fire and high temperatures (measured in hours). Specifically, it is an assembly's ability to prevent the spread of fire between spaces while retaining structural integrity.

Fire-Resistance-Rated Applications

Suspended ceiling systems can be used in fire-resistance-rated applications. Ratings consider factors such as resistance to fire and flame spread for both the tiles and the suspension system. However, the ceiling materials alone do not determine the fire-resistance rating. Rather, the materials and construction methods used in the entire system, including the floor/ceiling or ceiling/roof assembly, all factor into the rating determination. If a suspended ceiling is to be part of a fire-resistance-rated system, you must consult the manufacturer's product literature to determine the ceiling tile and grid that must be used to meet the rating.

Most ceiling tiles provide little thermal insulation between the space above the suspended ceiling and the work area below. When the building design or local codes require insulation above ceiling panels, special care must be taken to avoid placing too much weight on the ceiling system. When insulation is required, fiberglass ceiling panels are recommended. Most manufacturers discourage the use of insulation on mineral ceiling panels because the panels may sag as the humidity increases.

High-humidity-resistant tiles are tiles that have superior resistance to sagging caused by highly humid conditions. Many also inhibit the growth of mold or mildew that may appear in highly humid conditions. Sagging not only diminishes the attractiveness of a ceiling, but also causes ceilings to chip and soil more easily, reducing the light reflectivity of the tiles. High-humidity-resistant tiles are typically installed in humid climates or areas such as kitchens, locker rooms, shower areas, and indoor pools; buildings where the HVAC systems may be shut down for extended periods; or where the ceiling might be installed early in the construction process before the building is fully enclosed.

Ceiling tiles with a high light reflectance (LR) value of 0.83 or greater per *ASTM E1477* help increase effective lighting levels and reduce light fixture costs and energy consumption, especially with indirect lighting systems. Their use also helps to reduce eyestrain. These tiles typically have soil-resistant surfaces that stay cleaner longer than standard ceilings, resulting in a much lower loss in light reflectance over time.

Detailed information regarding the various tile characteristics described in this section can normally be found in ceiling manufacturers' product catalogs and literature. When selecting tiles for a particular application, it is best to follow the manufacturer's recommendations.

1.2.3 Metal-Pan System Components

The metal-pan system is similar to the conventional suspended acoustical ceiling system except that metal panels, or pans, are used in place of the conventional sound-absorbing tile (*Figure 8*). In some cases, the metal pans are snapped in from the bottom of the grid.

The pans are made of steel or aluminum and are generally painted white; however, other colors are available by special order. Pans are also available in a variety of surface patterns. Metal-pan ceiling systems are effective for sound absorption. They are durable and easily cleaned and disinfected. In addition, the finished ceiling has little or no tendency to have sagging joint lines or drooping corners. The metal pans are die-stamped and have crimped edges, which snap into the spring-locking main runner and provide a flush ceiling.

The tools, room layout, and installation of hanger inserts, hangers, and wall angle for the metal-pan system are basically the same as for the conventional exposed grid suspended ceiling.

1.2.4 Direct-Hung System Components

A concealed grid system is advantageous if the support runners need to be hidden from view, resulting in a ceiling that is not broken by the pattern of the runners (*Figure 9*).

The panels used for this system are similar in composition to conventional panels but are manufactured with a kerf on all four edges. Kerfed and rabbeted

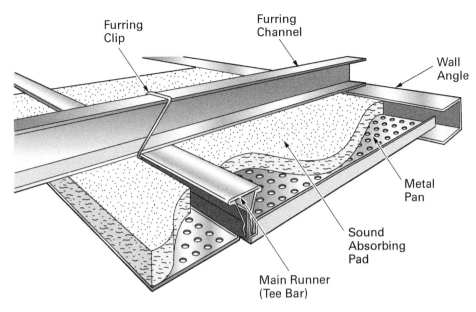

Figure 8 Metal-pan ceiling components.

Figure 9 Concealed grid system.
Source: Ansario/Shutterstock

12" × 12" and 12" × 24" panels are used with this system. Splines are inserted in the kerfs to tie the panels together. Panels of various colors and finishes are available. Refer to *Figure 10* for a diagram of the components in a typical concealed grid system.

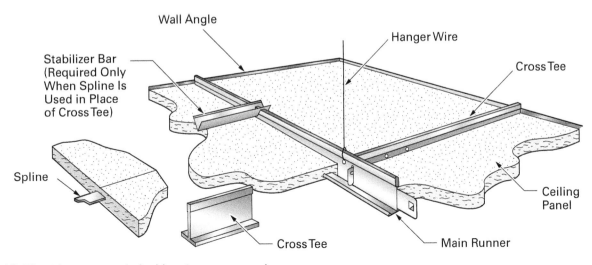

Figure 10 Direct-hung concealed grid system components.

1.3.0   Material Handling and Storage

Ceiling panels or tiles, and in some cases the tracks and runners, are exposed to view. To prevent damage, they must be handled and stored carefully. In addition, working with ceiling materials often presents safety hazards.

1.3.1 Handling and Storing Ceiling Materials

Finish ceiling materials should be stored in their original unopened packages and be protected from damage and exposure to the elements. Materials in their unopened packages can be easily moved to their installation location. The conditions where the materials are stored should be as close as possible to the place where they will be installed. Ensure there is proper support for the ceiling materials being placed. Materials should be stored at the jobsite for a minimal amount of time before being installed. Long-term storage should be avoided. Other considerations for proper handling and storage are as follows:

- Excess humidity during storage can cause expansion of material and possible warp, sag, or poor fit after installation.
- Chemical changes in the mat and/or coatings can be aggravated by excess humidity and cause discoloration during storage, even in unopened cartons.
- Cartons should be removed from pallets and stringers to prevent distortion of material.

1.3.2 Safely Working with Ceiling Materials

When installing grid support members or the panels or tiles themselves, carpenters typically work from ladders, movable scaffold, or lifts, depending on the height of the ceiling. Refer to the NCCER Module 00101, *Basic Safety (Construction Site Safety Orientation)* for specific safety information relating to ladders and lifts. A common type of movable scaffold used for ceiling projects is called a baker's scaffold. Baker's scaffolds are short and lightweight scaffolds with wheels that allow scaffolds to roll easily. Safety guidelines for using a baker's scaffold include the following:

- Inspect the scaffold before each use for defects or damage.
- Do not stand on or attach any equipment to cross braces or diagonal braces.
- Do not place boxes or ladders on a scaffold to increase your reach or height.
- Do not sit or stand on guardrails. Ensure that all guardrails are secured in place on all four sides.
- Never ride on a moving scaffold.
- Do not attempt to move a scaffold by applying a pushing or pulling force at or near the top of the scaffold.
- When hoisting material up to a scaffold platform, ensure the scaffold is attached to a permanent structure to keep the scaffold from tipping.
- Workloads on the scaffold must not exceed the capacity of the lowest-rated scaffold component.
- When working around a scaffold, always wear a hard hat, safety glasses, work gloves, and steel-toe boots.

In addition to standard hand-tool and power-tool safety guidelines that should be followed, eye protection is especially important when installing hanger wire. The loose ends of hanger wire are commonly at eye level when working from a baker's scaffold. Carefully handle the cut ends of wall angle and other support members. The cut ends of these items are very sharp and can cause serious injury.

Using Stilts to Install Ceilings

While stilts are permitted in certain areas, they are not permitted in all jurisdictions. Always refer to the specific safety regulations in effect in the area in which you are working before using stilts.

1.0.0 Section Review

1. Integrated ceiling systems incorporate _____.
   a. both ceiling panels and ceiling tiles
   b. fire-rated panels and sprinkler systems
   c. lighting and/or air supply diffusers
   d. both exposed and concealed grid systems

2. True or False: A typical ceiling grid consists of main runners, cross runners, and wall angles.
   a. True
   b. False

3. Finished ceiling materials should be stored _____.
   a. on pallets
   b. in their original unopened packages
   c. at a constant 60°F
   d. on edge to prevent warping

2.0.0 Reflected Ceiling Plan

Objective

Interpret a reflected ceiling plan.
   a. Interpret the layout information.
   b. Interpret the mechanical, electrical, and plumbing (MEP) locations.

Performance Task

1. Estimate the quantities of materials needed to install a lay-in suspended ceiling system in a typical room from an instructor-supplied drawing.

Some large construction jobs may have a set of reflected ceiling plans. These plans show the details of the ceiling as though it were reflected onto the floor (*Figure 11*). This view shows features of the ceiling while keeping those features in proper relation to the floor plan. For example, if a pipe runs from floor to ceiling in a room and is drawn in the upper left corner of the floor plan, it is also shown on the upper left corner of the reflected ceiling plan of that same room. Reflected ceiling plans show in detail how the ceiling will be constructed. The plans will indicate the following:

- Layout (direction) of the ceiling panels or tiles
- Location of the center (starting) line
- Size of the borders
- Position of the light fixtures
- Location of the air diffusers
- Location of life safety devices
- Position of strobes
- Location of exit devices
- Location of fire calls

As a rule, reflected ceiling plans also show all items that penetrate the ceiling, including the following:

- Return grilles
- Diffusers
- Sprinkler heads
- Light fixtures
- Recessed speakers for sound systems

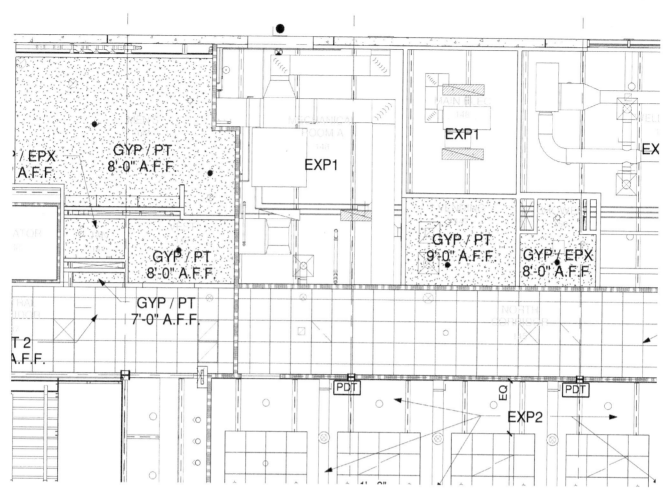

Figure 11 Example of a reflected ceiling plan.

Other information is also used when constructing a ceiling. Your employer may prepare shop drawings that show in detail just how the ceiling should be installed and also indicate the finished appearance. These drawings provide insurance against errors in the details of installation.

2.1.0 Interpreting Ceiling Plans

Pinpointing the locations of lighting fixtures, sprinkler systems, air diffusers, and other features that protrude through ceilings is essential when installing a ceiling. Other professionals working behind you depend on these items to be located accurately. In addition to the location of ceiling components found in the plans, specifications contain pertinent information including construction techniques, specifics about materials, and dimensions.

To avoid mistakes and/or omissions when installing ceilings, an organized and systematic approach should be used for reading the related construction drawings. The following is a general procedure for reading construction drawings:

Step 1   Check the room schedule on the construction drawings.

- Identify the type of material to be used.
- Locate the protrusions into the ceiling.

Step 2   Locate the room on the floor plan.

- Find the room dimensions. If none are found, locate the drawing scale.
- Using the given scale, determine the dimensions of the room.

Step 3   Check to see if there is a reflected ceiling plan. If no reflected ceiling plan is found, check to see if a shop drawing is included.

Step 4   Be sure the construction drawings are the final revised set.

- Construction drawings are often revised several times before the ceiling is ready to be installed. To be sure the construction drawings are the final revised set, check the revision block date against the work order to see if they are the same.
- If work that has already been done is not reflected on the construction drawings, chances are the construction drawings are not the final revised set. This could have a significant impact on the ceiling installation.

Step 5   Read the specifications and general notes.

- Be sure the ceiling to be installed is the same as that listed in the specifications.
- If the job conditions do not agree with what is shown in the specifications and/or plans, call your supervisor and ask for instructions on how to proceed.

Step 6   Check the mechanical and electrical plans prior to the layout of the ceiling.

- On the mechanical plans, locate the air diffusers (HVAC air supply outlets), return grilles, ducts, and sprinkler heads.
- On the electrical plans, locate the light fixtures, fans, and other ceiling protrusions.
- Make sure to cross-reference both plans for items that may be in the same location.

2.2.0 Interpreting the Mechanical, Electrical, and Plumbing (MEP) Drawings

Along with reading and interpreting the reflected ceiling plan, reference the mechanical, electrical, and plumbing (MEP) drawings to ensure that all ceiling penetrations are accounted for. The MEP drawings will indicate where plumbing and electrical risers penetrate the ceiling, and will show the routing of mechanical equipment. *Figure 12A* and *Figure 12B* show the mechanical, electrical, and plumbing drawings for similar areas of the same set of construction drawings.

Plenum Ceilings

The systems that provide heating and cooling for most commercial buildings are forced-air systems. Blower fans are used to circulate the air. The blower draws air from the space to be conditioned and then forces the air over a heat exchanger, which cools or heats the air. In a cooling system, for example, the air is forced over an evaporator coil that has very cold refrigerant flowing through it. The heat in the air is transferred to the refrigerant, so the air that comes out the other side of the evaporator coil is cold. In homes, the air is delivered to the conditioned space and returned to the air conditioning/heating system through ductwork that is usually made of sheet metal. In commercial buildings with suspended ceilings, the space between the ceiling and the overhead decking is often used as the return air plenum. It is often called an open plenum. (A plenum is a sealed chamber at the inlet or outlet of an air handler.) This approach saves money by eliminating about half the cost of materials and labor associated with ductwork.

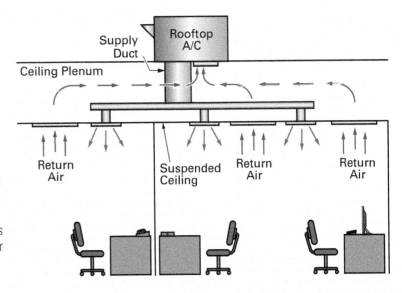

   One thing to keep in mind is that anything in the plenum space (electrical or telecommunications cable, for example) must be specifically rated for plenum use in order to meet fire-resistance ratings. Plastic sheathing used on standard cables gives off toxic fumes when burned. Plenum-rated cable uses nontoxic sheathing.

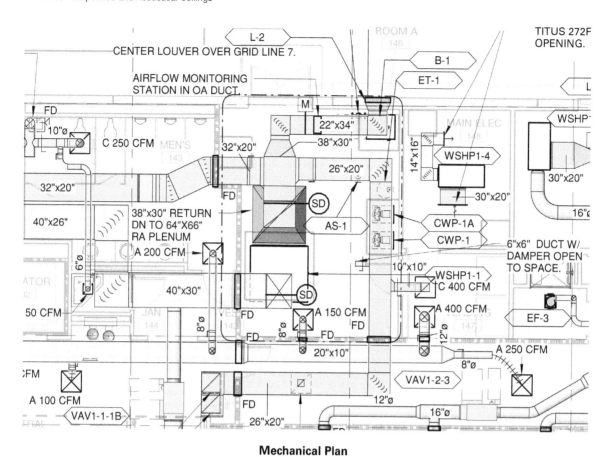

Mechanical Plan

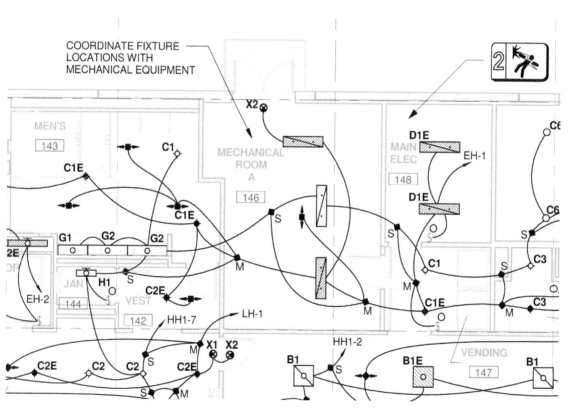

Electrical Plan

Figure 12A Mechanical, electrical, and plumbing drawings. (1 of 2)

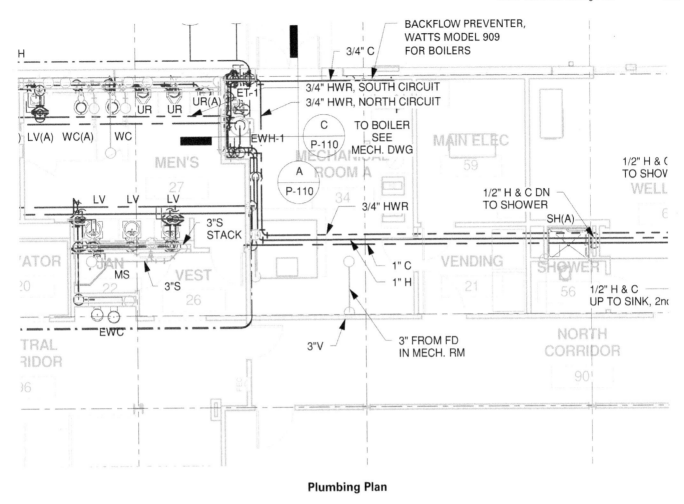

Plumbing Plan

Figure 12B Mechanical, electrical, and plumbing drawings. (2 of 2)

2.0.0 Section Review

1. To be sure that the construction drawings you are using are the final revised set, you should _____.
   a. verify the revision block date
   b. ask your supervisor
   c. call the architect's office
   d. compare them to the specifications

2. MEP drawings indicate _____.
   a. the direction and location of main runners
   b. where open plenums should be installed
   c. datum lines for ceiling elevations
   d. where plumbing and electrical risers penetrate the ceiling

Performance Tasks

2. Establish a level line at ceiling level, such as is required when installing the wall angle for a suspended ceiling.
3. Lay out and install a lay-in suspended ceiling system according to an instructor-supplied drawing.

3.0.0 Laying Out and Installing Suspended Ceiling Systems

Objective

Describe the key considerations, methods, and best practices relating to ceiling installation.

a. Explain seismic considerations for ceilings.
b. Identify the layout and takeoff procedures to install a suspended ceiling system.

c. Identify the tools and equipment to lay out and install a suspended ceiling system.
d. Identify the installation methods and procedures for a suspended ceiling system.

A professional installer must understand a variety of ceiling systems, including their components, acoustical properties, and installation techniques. In addition, the installer should be familiar with the following aspects of suspended ceilings:

- The risk of structural failure due to seismic activity is a concern, especially in earthquake-prone areas. Special building codes apply to the materials and design of ceiling systems depending on the seismic risk category.
- Before a ceiling installation job begins, the cost of materials must be accurately estimated for bidding and/or budgeting purposes.
- To maintain an attractive appearance and retain acoustical properties, ceilings need to be cleaned periodically. Tiles and panels may have specific cleaning instructions.
- Installers should follow plan and manufacturer's guidelines to ensure a superior installation.

3.1.0 Seismic Considerations for Ceilings

Although important in their function, modern ceilings are generally not part of a building's structural system. Ceilings today are used to make rooms and buildings more livable. The most desirable features of a ceiling are those that improve the acoustic properties of an area and those that make it easier to heat, cool, and light. Although ceilings are not structurally significant, the *International Building Code®* (*IBC®*) recognizes that ceiling failures due to seismic activity can make an area unusable.

Although seismic activity is of special importance in earthquake-prone areas, such as the Pacific Northwest region of the United States, it is also an issue in other areas. The *IBC®* uses three factors that place more than half of the US in areas at risk for seismic activity. The factors the *IBC®* uses for determining a building's seismic risk are as follows:

- *Ground movement* — All geographic areas have the potential for ground movement, even if the movement is unlikely to be felt. The primary factor is the amount of potential ground movement. In the US, the state with the potential for the most ground movement is California, while the ones with the least potential are Florida and Texas.
- *Soil classification* — Soil classifications are based on the soil type at the building site to a point 100" below the surface. For seismic purposes, there are six soil classifications, labeled A through F. Class A is solid rock, which is considered the most stable. Class F soil has no positive construction traits and is unusable for building. It is unlikely that a site would be classified at the extremes of A or F. Most sites will fall somewhere in between.

- *Building use/occupancy category* — Risk categories based on occupancy of a structure are divided into four groups. Category I buildings have low occupancy, such as small agricultural buildings, and they represent a low risk of loss of human life in the event of a failure. Category II consists of all buildings and other structures not listed in the other three categories. Structures that represent substantial hazard to human life in the event of structural failure fall into Category III. Category IV structures are designated as essential facilities, meaning they are required to maintain the functionality of structures in the other three categories.

3.1.1 Seismic Design Categories

Based on factors of potential ground movement, soil type, and building use (risk categories), a seismic design category (SDC) has been established. The categories range from A (lowest seismic risk) to F (highest risk). Based on the SDC assigned to a geographical region in which a structure is located, certain building codes apply to the design and installation of ceiling systems. Buildings in categories A and B use the standard suspended ceiling requirements described in this module. Their ceilings are anchored to walls and suspended from the upper decks using standard hardware. Those buildings in category C should have unrestrained ceilings that are free-floating to allow for movement of up to 12" and 45° without damage. Buildings in categories D, E, and F should have restrained ceilings that are reinforced with stabilizing bars, heavier gauge wire, and multiple strands of wire. *Figure 13* is an earthquake hazard map showing seismic risk across the United States.

The ceilings described in categories C through F are more expensive than those in categories A and B. Unrestrained ceilings (used in category C buildings) are not anchored to walls, so they are extremely difficult to install without

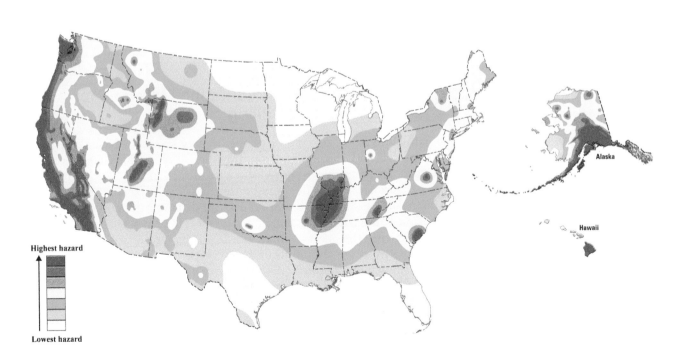

Figure 13 Map of seismic hazard in the US.
Source: US Geological Survey

a fixed reference. Unrestrained ceilings cost about 50% more than standard suspended ceilings—mostly in installation costs. Unrestrained ceilings have the following features:

- The ceiling system grid is not attached to the wall in any way.
- There are no perimeter wires.
- The wall molding must be at least $\frac{7}{8}$" in width.
- The clearance between the ceiling and walls is at least $\frac{3}{8}$" at all points.
- Grids joints must overlap at least $\frac{3}{8}$".
- The ends of the main and cross tees, which are free-floating, must be tied together to prevent spreading.

Restrained ceilings (used in categories D, E, and F) are anchored to at least two adjacent walls and require the use of heavy gauge hardware, as well as a grid system. These ceilings usually cost twice as much as a standard suspension ceiling, with materials accounting for most of the increased cost. Restrained ceilings have the following characteristics:

- A heavy-duty ceiling grid system is required.
- The grid must be attached to two adjacent walls, and the clearance between the ceiling and the opposite wall must be $\frac{3}{4}$".
- The wall molding must be at least 2" wide.
- The ceiling perimeter must have support wires that are spaced not more than 8' from the wall.
- Ceilings with an area of less than 1,000 ft^2 must use heavy-duty hardware for equipment mounted through the ceiling, such as light fixtures and sprinklers.
- Ceilings with an area of more than 1,000 ft^2 must have additional bracing.
- Ceilings with an area of more than 2,500 ft^2 must have seismic separation joints.
- Cable trays and electrical conduits must have their own supports and braces.

Requirements for acoustical tile and lay-in panel ceilings based on seismic design categories are more complex than the overview provided here. For detailed codes and requirements, see the *Additional Resources* section at the end of the book.

3.2.0 Laying Out and Estimating Materials for a Suspended Ceiling

The estimate of materials for a suspended ceiling should be based on the ceiling plan provided with the construction drawings or on a scaled sketch of the ceiling layout. These drawings should show the direction and location of the main runners, cross tees, light tiles, and border tiles. In a typical suspended ceiling, the main runners are spaced 4' apart and are usually run parallel with the long dimension of the room. For a standard 2' × 4' pattern, 4' cross tees are spaced 2' apart between the main runners. If a 2' × 2' pattern is used, 2' cross tees are installed between the midpoints of the 4' cross tees.

If no ceiling plan or sketch is available, a craft professional may need to make one to determine the required quantity of materials. A sketch can be made following the steps shown in Section 3.4.1, *Exposed Grid Systems*. *Figure 14* shows an example of a sketched ceiling plan.

From the ceiling plan or sketch, determine the number of pieces required for the wall angle, main runners, cross tees, and ceiling tiles. Normally, main runners come in 12' lengths, cross tees in 4' lengths, and wall angle in 10' lengths. The following calculations are based on the example room shown in *Figure 14*:

- Using the ceiling plan or sketch, find the number of main-runner sections needed. For our example room, six main-runner sections are required. This

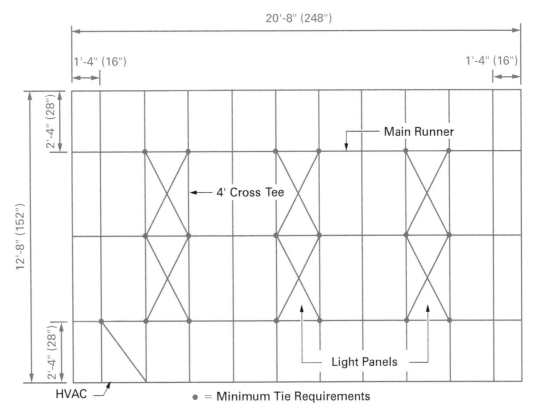

Figure 14 Completed sketch of a suspended ceiling layout.

is because main runners are made 12' in length and no more than two pieces can be cut from any one 12' runner.

$$3 \text{ main runners} \times 20'8'' = 62.001'$$
$$62.001' \div 12' = 5.167 \text{ rounded to 6 lengths}$$

- Find the number of 4' cross tees. For our example room, 40 cross tees are required. Note that the border cross tees must be cut from full-length cross tees.

$$10 \text{ cross tees per row} \times 4 \text{ rows} = 40 \text{ cross tees}$$

  ○ If using 2' × 2' panels or tiles, find the number of 2' cross tees. A 2' × 2' grid is made by installing 2' cross tees between the midpoints of the 4' cross tees. The number of 2' cross tees required for our example room is 44 (11 per row × 4 rows).

- Find the number of sections of wall angle needed. Divide the perimeter of the room by 10' [perimeter = (2 × length) + (2 × width)]. For our example room, the perimeter is 66'-8" (66.667'). Therefore, seven sections are needed (66.667' ÷ 10' = 6.667, or 7 when rounded off).

$$\text{Perimeter} = (2 \times 152'') + (2 \times 248'')$$
$$\text{Perimeter} = 66'8'' = 66.667'$$
$$66.667' \div 10' = 6.667 = 7 \text{ sections}$$

- Find the number of ceiling panels or tiles. One method is to count the total number of ceiling tiles shown on the ceiling plan or sketch. Note that each border tile requires a full-size ceiling tile. Also, subtract one tile for each lighting fixture installed in the ceiling. Assuming the use of 2' × 4' ceiling tiles and six light fixtures, our example ceiling would require 38 tiles.

- Find the approximate number of hanger wires and hangers needed. Assume one hanger device and wire for about every 4' of main runner. For our example ceiling, approximately 16 hangers are required (62.001' ÷ 4' = 15.5, or 16 when rounded off). Multiply the number of hangers needed by the required length of each hanger wire to find the total linear feet of hanger wire needed.

NOTE

Another method for estimating the number of ceiling tiles is to determine the total area (in square feet) of the ceiling by multiplying the length by the width (area = length × width). Then, divide the total ceiling area by the coverage (in square feet) printed on the carton of the tiles intended for use. If using tiles that cover 64 ft² per carton, our example room ceiling would require 4.09 cartons of ceiling tiles (12.667' × 20.667' = 261.789' ÷ 64 = 4.09, or four cartons plus one extra tile).

3.2.1 Alternate Method for Laying Out a Suspended Ceiling Grid System

Another common method for laying out the grid system for a suspended ceiling is given here:

Step 1 Locate the room centerline parallel to the long dimension of the room and draw it on the ceiling sketch.

Step 2 Beginning at the centerline and going toward each side wall, mark off 4' intervals on the sketch. If more than a 2' space remains between the last mark and the side wall, locate the main runners at these marks. If less than a 2' space remains between the last mark and the side wall, locate the main runners at 4' intervals beginning 2' on either side of the centerline. This procedure provides for symmetrical border tiles of the largest possible size. Remember to consider the locations of light fixtures and air diffusers in the room.

Step 3 Locate the 4' cross tees by drawing lines 2' on center at right angles to the main runners. To obtain border tiles of equal size, begin at the center of the room using the same procedure as in Step 2.

Step 4 If using a 2' × 2' grid pattern, locate the 2' cross tees by bisecting each 2' × 4' module.

Step 5 Estimate the materials for the grid system, using the information shown on the ceiling sketch in the same manner as described previously.

3.2.2 Establishing Room Centerlines

For a grid system to be installed square within a room, it is necessary to lay out two centerlines (north-south and east-west) for the room. When correctly laid out, these centerlines will intersect at right angles (90 degrees) at the exact center of the room. If the ceiling of the room located above the proposed suspended ceiling is solid and flat, such as with a plaster or drywall ceiling, then the centerlines can be laid out on the ceiling. If the ceiling is not flat, such as an open ceiling with I-beams or joists, the centerlines can be laid out on the floor. In either case, the centerlines that are laid out on the ceiling or floor can be transferred down from the ceiling, or up from the floor, to the level of the suspended ceiling by use of a plumb bob and **dry lines**. The following procedure describes one common method for establishing the centerlines in a rectangular room:

Dry lines: A string line suspended from two points and used as a guideline when installing a suspended ceiling.

Step 1 Measure and mark the exact center of one of the short walls in the room. Repeat the procedure at the other short wall.

Step 2 Snap a chalk line on the ceiling (or floor) between these two marks. (See *Figure 15*, chalk line A-B). This is the first centerline.

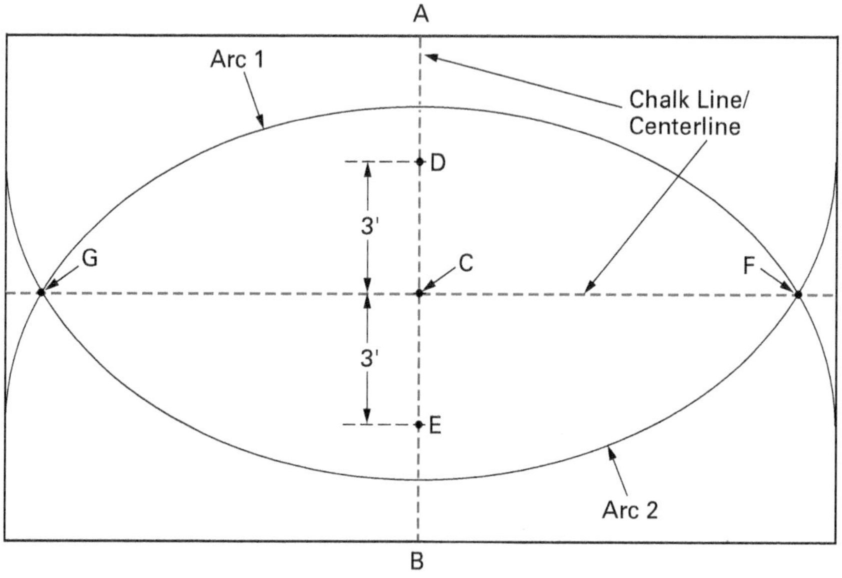

Figure 15 Method for laying out the centerlines of a room.

Step 3   Measure the length of the room along the chalk line. Find its center, and then place a mark on the chalk line at this point (point C).

Step 4   From point C, measure a minimum of 3' in both directions along the chalk line, then place a mark on the chalk line at these points (points D and E).

Step 5   Drive a nail at point D and attach a string to the nail. Extend the string to the side wall so that it is perpendicular to the chalk line, then attach a pencil to the line.

Step 6   Making sure to keep the string taut, draw an arc on the ceiling (floor) from the wall, across the chalk line, to the opposite wall (arc 1).

Step 7   Repeat Steps 5 and 6, starting at point E on the chalk line and draw another arc (arc 2).

Step 8   Mark the intersecting points of arcs 1 and 2 on both sides of the chalk line (points F and G).

Step 9   Snap a chalk line from wall to wall on the ceiling (or floor) that passes through points F and G. If done correctly, you should now have two centerlines perpendicular to each other that cross in the exact center of the ceiling (floor).

3.3.0 Suspended Ceiling Tools and Equipment

Depending on the type of suspended ceiling being installed, a variety of tools and leveling equipment is needed. Reviewing the NCCER Module 00101, *Basic Safety (Construction Site Safety Orientation)* will refresh your knowledge of safety guidelines and considerations for working on a construction project.

3.3.1 General Tools

Preparation for installing a suspended ceiling includes gathering the necessary tools and equipment. Having the right equipment ready before installation begins will set up the rest of the process for success. The following tools are required when installing an exposed grid ceiling system:

- Aviation snips (tin snips)
- Clamping pliers or vise grips with plastic or rubber corners
- Clamps and brackets
- Chalk line
- Dry line
- 50' or 100' tape measure
- Hammer
- Awl
- Keyhole saw
- Lath nippers
- Magnetic punch
- Scribe or compass
- Plumb bob
- 9" lineman's pliers
- Straightedge for cutting
- Board
- Tile knife
- Ladders
- Laser level
- Bubble level
- Pop-rivet gun
- Powder-actuated tool
- Scaffold
- Special dies for cutting suspension members
- Whitney punch

Layout Equipment

Most of the work of installing a ceiling will be done at ceiling level. Ladders, portable scaffolding, scissor lifts, or drywall stilts will all get the job done. Several factors such as the height of the ceiling, the complexity of the job, the projected timeline for completion, and the budget will impact your choice of access equipment.

Proper installation of suspended ceilings begins with taking accurate measurements. You will need a tape measure to find the center of the walls in the room, and to measure spacing of the main runners and cross runners.

Cutting and Fastening Equipment

Lineman's (electrician's) pliers (*Figure 16*) are designed to cut, bend, straighten, strip, twist, grip, and otherwise manipulate electrical wire. They also happen to be a handy tool for cutting and twisting hanger wire when installing ceiling grid systems. When cross runners, also known as cross ties or cross tees, need to be cut to length to fit the main runners, either tin snips (*Figure 17*) or aviation snips can be used.

NOTE

Powder-actuated fasteners are not permitted to be loaded in tension (such as supporting a ceiling) in high seismic risk areas.

Figure 16 Lineman's (electrician's) pliers.
Source: Paul Nichol/Alamy

Figure 17 Tin snips.
Source: v_zaitsev/Getty Images

Rivets are fastened where the wall angle (frame that attaches all the way around the room and supports the edge of the grid) and cross runners meet as well as at the intersection of every other cross runner and main runner to stabilize the grid. A rivet tool (*Figure 18*), also referred to as a rivet gun, is used to pop the rivets through the members being attached.

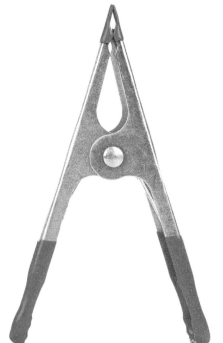

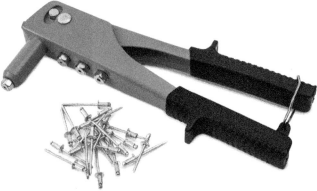

Figure 18 Rivet tool.
Source: ohotnik/123RF

Spring clamps (*Figure 19*) or ceiling grid clamps are used at the intersections of the main runners and cross runners to temporarily hold them in place as adjustments are being made to the grid. Once the grid is completely assembled and squareness has been checked, the clamps can be replaced with rivets.

A different set of fastening tools is needed when the framework is attached directly to a ceiling deck instead of being suspended. To fasten a ceiling grid to a concrete surface, either a powder-actuated fastener or hammer drill can be

Figure 19 Spring clamp.
Source: Tpopova/Getty Images

used. When a furring system is being installed to support drywall on a ceiling, the $\frac{1}{2}$" cold-rolled channel is attached to the ceiling deck with an impact driver.

3.3.2 Ceiling Leveling Equipment

To install suspended ceilings, carpenters use various types of leveling devices to find the level plane of a ceiling. These devices include the carpenter's level and laser level.

In the application of acoustical ceilings, a level, as shown in *Figure 20*, is a tool that comes in handy to check the ceiling-grid main runner and cross runner installation. When using a level to install a suspended ceiling, place it at right angles to the runners as you install them. If the tool is perfectly level, the bubble will appear centered between the crosshatches. The leveling should be checked every 6'.

Figure 20 Level.

Many types of laser levels are available that can be used to aid in the installation of suspended ceilings. To use a laser level when installing a ceiling, follow the manufacturer's instructions for the laser level being used. Generally, the procedure involves mounting the laser either on a wall/ceiling mount (*Figure 21*) or on a tall tripod (*Figure 22*). The laser beam is rotated either at the finished ceiling height or at a reference point. A special target is snapped to a grid, and then the grid is moved up or down until the laser beam crosses the target's offset mark. The grid is then secured in place.

Figure 21 Laser level—wall or ceiling mount.
Source: DEWALT Industrial Tool Co.

Figure 22 Laser level—tripod mounted.
Source: vivooo/Shutterstock

WARNING!

OSHA (Occupational Safety and Health Administration) *CFR 1926.54* covers safety regulations for the use of lasers. Some guidelines are as follows:

- Avoid direct eye exposure to the laser beam.
- Only qualified and trained personnel are permitted to operate laser equipment.
- Place a standard laser warning sign conspicuously at major approaches to the instrument use area.
- Always turn off the laser when transmission of the beam is not required.

NOTE

Make sure all inspections for areas and/or equipment above the grid line have been completed before beginning the ceiling installation.

CAUTION

Be careful when removing ceiling material from its packaging and when handling it. Keep all ceiling material and the grid system clean and undamaged.

3.4.0 Installing Suspended Ceiling Systems

Observing the following general guidelines will help to achieve the desired level of professionalism when installing ceilings:

- Ceiling tiles should be arranged so that units less than one-half width do not occur unless otherwise directed by the reflected ceiling plans or job conditions.
- All tiles, tile joints, and exposed suspension systems must be straight and aligned.
- All acoustical ceiling systems must be level to $\frac{1}{8}$" in 12'.
- Tile must be neatly scribed against butting surfaces and to all penetrations or protrusions where moldings are not required.
- Tile surrounding recessed lights and similar openings must be installed with a positive method to prevent movement or displacement of the tiles.
- Tiles must be installed in a uniform manner with neat hairline-fitted joints between adjoining tiles.
- Wall moldings must be firmly secured, the corners neatly mitered, or corner caps used, if preferred.
- The completed ceiling must be clean and in undamaged condition.

Doing overhead work often requires the use of scaffolding or ladders to reach the work area. Working from an elevated platform adds an element of danger to the job. It is important to always place safety first. Use equipment that is the correct height for the job and be sure that all equipment is in good working order. Any defective items must be tagged right away so they will not be used by mistake. Scaffolding must be inspected by a competent person before it is used on each shift, whenever it is moved, and whenever it is changed in any way. Some scaffolding must be designed by an engineer, so know the practices at your worksite. If you are unsure of the site requirements, ask your supervisor or foreman before you begin work.

When working aboveground, always wear the appropriate personal protective equipment. Maintain control of materials and stay alert for personnel working on the ground below you. Resist the urge to over-reach for your equipment. If working on the ground when overhead work is being done, stay alert for falling debris or tools.

WARNING!

Scaffolds must be used and assembled in accordance with all local, state, and federal/OSHA regulations. OSHA requires the building, moving, or dismantling of all scaffolding to be supervised by a competent person who has the training, knowledge, and experience to identify hazards on the jobsite and the authority to eliminate them. Mobile scaffolds, such as baker's scaffolds, should only be used on level, smooth surfaces that are free of obstructions and openings. OSHA regulations also require that mobile scaffold casters have positive locking devices to hold the scaffold in place. When moving a mobile scaffold, apply the moving force as close to the scaffold base as possible to avoid tipping it over. Never move a scaffold when someone is on it.

3.4.1 Exposed Grid Systems

The following sections describe the general procedure for installing an exposed grid ceiling system, also called a direct-hung system.

Install an exposed grid suspended ceiling system according to the following guidelines:

Step 1 Check the room number and location.

- Ensure the correct ceiling is going into the correct room.
- Refer to the construction drawings, reflected ceiling plan, or shop drawing to determine the correct height for the ceiling.
- Check the electrical drawings to ensure the lights will be placed as indicated on the reflected ceiling plan or shop drawing.
- Check the specification sheets or shop orders for special instructions to ensure the proper hangers and fasteners are provided.

Step 2 If needed, install the scaffold.

- Set the scaffold at the correct height to permit the driving of inserts or other fasteners into the overhead ceiling and to permit the connection of hangers.
- Set the lower portion of the scaffold to permit installation of the grid members and ceiling tiles.

Step 3 Establish benchmarks. In some situations, the floor may not be level in all areas of the room you are working in. Make sure to establish benchmarks that are exactly the same height throughout the room and are consistent in all rooms. These benchmarks can be located with the use of a laser level. The benchmarks will provide an accurate starting point for locating the desired height of the finished ceiling.

- Using the level of choice, locate benchmarks at each end of the room near the corners. If the room is extremely large and long, locate the benchmarks at 15' to 20' intervals.
- It is best to locate the benchmarks at eye level (about 5' above the floor).
- Always put benchmarks on each wall and also on protruding walls. It is better to have too many than not enough.

Step 4 Establish the height for the top of the wall angle.

- Once all benchmarks are located and marked, establish a measurement to the bottom of the wall angle.
- To that measurement, add the height of the wall angle.
- Measure from each benchmark and establish a mark at the top of the wall angle. Place a mark at various intervals on the wall.
- Snap a chalk line on those marks. This will establish the top of the wall angle. Be sure to snap the chalk line at the top of the wall angle so that the chalk line will not be visible when the wall angle is installed.

Step 5 Install the wall angle.

- Be sure that the top of the wall angle is even with the chalk line at all points.
- Nail or screw the wall angle to the wall. The wall angle should fit securely to prevent sound leaks.
- Miter the wall angle to fit at the corners or use corner caps to cover the joints.

Install the wall angle according to the following procedure:

Step 1 Square the room by first establishing length and width centerlines on the floor or ceiling, as presented in the Section 3.2.2, *Establishing Room Centerlines*.

NOTE

The procedures given in this module are general in nature and are provided as examples only. Because of the differences in components made by different manufacturers, it is important to always follow the manufacturer's installation instructions for the specific system being used.

NOTE

Omit Steps 3 and 4 if a laser level is being used to establish the correct and level ceiling height, as described in Section 3.3.2, *Ceiling Leveling Equipment*.

NOTE

The following procedure assumes that the centerlines have been laid out on the floor. The procedure would be done in a similar manner if the centerlines were laid out on the ceiling. The only difference is that the centerlines would be transferred down from the ceiling, instead of up from the floor, to the level of the suspended ceiling grid.

Step 2　Transfer the positions of the centerlines from the floor/ceiling to the level required for the suspended ceiling grid as follows:

- At either one of the long walls, use a plumb bob to locate the center of the wall just above the wall angle. Do this by moving the plumb bob until it is directly over the centerline marked on the floor (*Figure 23*). Then, mark the wall just above the wall-angle flange and insert a nail behind the wall angle at this point. Repeat the procedure at the opposite long wall. Run and secure a taut dry line between the nails in the two long walls.
- Repeat the preceding procedure for the two short walls and run a second dry line perpendicular to the first.
- Make a final check using the plumb bob at all four points.

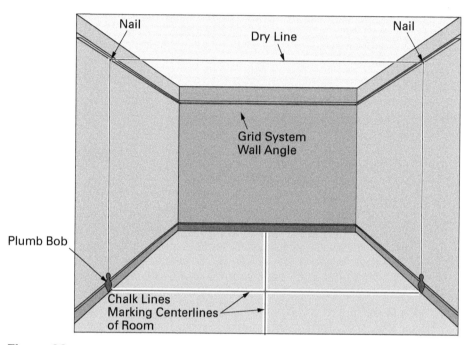

Figure 23　Transferring room centerlines to the height of the suspended ceiling grid.

When completed, you will have two dry lines intersecting at right angles at the center of the ceiling. Use the 3-4-5 method to ensure that the intersection of the dry lines is square. This is important because these intersecting dry lines will be used as the basis for all subsequent measurements used to lay out the ceiling grid system.

In some cases, reflected ceiling plans or shop drawings may indicate a centerline and specify border width. In these instances, follow the suggested layout. If you have neither to follow, you will need to establish the centerline and plan the layout of the ceiling so that the border tiles adjacent to the facing wall are the same width and are not less than one-half the width of a full tile.

"Seeing" through the Ceiling

It is a good idea to keep a package of colored thumbtacks in your toolbox when installing a suspended ceiling. When other trades may have to get back in to complete their work, insert different colored thumbtacks in ceiling tiles to mark the locations of electrical and mechanical services. The marking will allow the other trades to locate their components without having to raise a lot of tiles and will therefore help minimize finger smudges and damage to the tiles.

One method for determining the width of the border tiles is to convert the wall measurement from feet to inches, then divide this amount by the width of the ceiling tiles. For example, assume the room measurements are 42'-6" × 30'-6".

What will be the width of the border tiles running parallel to the two 42'-6" walls if 2' × 2' tiles are being used? To find the answer, proceed as follows:

Step 1   Take the measurement of the short wall (30'-6") and convert feet to inches:

$$30' \times 12" = 360" + 6" = 366"$$

Step 2   Divide that amount by 24" (the width of a tile):

$$366" \div 24" = 15 \text{ tiles with a remainder of } 6"$$

Step 3   If the division does not result in a whole number, add the width (in inches) of a full board to the remainder (in this case, it is 6"):

$$6" + 24" = 30"$$

Step 4   Divide this by 2; the result is 15. This would be the width of a border tile on each side. When adding the width of a tile to the remainder in Step 3, you must delete one full tile from the total. This means that there would be 14 full tiles plus a 15" border on each end (*Figure 24*):

$$14 \times 2' + 30" = 30'\text{-}6"$$

Step 5   Determine the width of the border tiles for the other two remaining walls in the same manner.

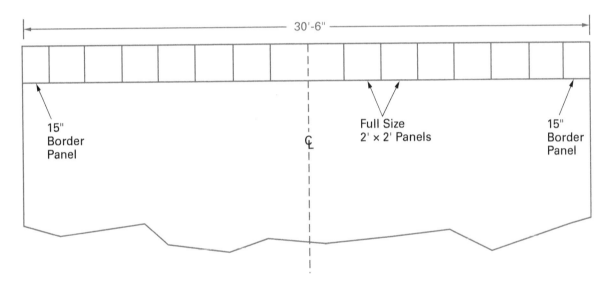

Figure 24 Fitting the tiles to the room.

Once all border unit measurements have been established, it is necessary to install eye pins, also known as ceiling clips or hanger clips, to secure the hangers. Prior to installing the eye pins, you must locate the position of the main runners and cross runners. When the direction of the main tees and cross tees is not indicated on the plans, the installer will decide the best direction. The stronger main runners are more expensive, so installers usually choose to run the main tees in the shortest direction to minimize use of materials. In *Figure 25*, the plans indicated the main runner should run in the long direction. Line A in *Figure 25* represents the main runner, and Line B represents the first cross runner line.

Step 1   Locate the first dry line (A) as shown in *Figure 25*. The dry line should be installed approximately $1\frac{1}{8}"$ above the flange of the wall angle. The dry line will also be used to indicate the bend in the hanger wire.

Step 2   Fasten a second dry line (B) so that it intersects line A at right angles with the proper measurement for the border tiles.

Step 3   The intersection point will be the location of the first hanger wire. Set the first eye pin directly over the intersection of dry lines A and B. Install the other eye pins every 4' on center. Be sure to follow dry line A as a guide for the locations.

NOTE

Most main runners measure $1\frac{1}{8}"$ from the bottom of the runner to the hole for the hanger wire.

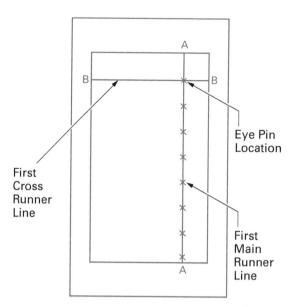

Figure 25 Locating the positions of the first main runner and cross runner.

WARNING!

Powder-actuated tools (PATs) are to be used only by trained operators in accordance with the operator's manual. Operators must take the following precautions to protect both themselves and others when using powder-actuated tools:

- Operate the tool as directed by the manufacturer's instructions and use it only for the fastening jobs for which it was designed.
- To prevent injury or death, make sure that the drive pin cannot penetrate completely through the material into which it is being driven.
- To prevent a ricochet hazard, make sure the recommended shield is in place on the nose of the tool.

As you install each eye pin, suspend the hanger wire from the eye pin. Use pretied wire whenever possible. If not using pretied wire, as you install each eye pin and wire, allow enough wire to ensure proper tying to the eye pin and the main runner. These two ties will require from 10" to 24" of wire. More hanger wire may be required if the ceiling is hung in part from structural members along with the eye pins. All hanger wire is precut and can be obtained in various lengths and gauges. Refer to the specifications for the gauge of hanger wire to use. There are cases in which there may be cast-in-place inserts for hanger wires. In such cases, the insertion of eye pins and installation of wire will not be necessary.

The eye pins and hanger wires should be located on 4' centers in both directions. In most cases, it is best to hang both at the same time, as it requires less movement of the scaffold. To mark and bend the wires, proceed as follows:

Step 1   Mark the hangers where they touch the dry line.

Step 2   Twist the wire using side cutters and bend the other end to the mark (*Figure 26*).

Install the main runners as follows:

Step 1   Measure and cut the main runner so that the cross runner will be at the proper distance for the border tile.

Step 2   Suspend the first length of main runner from the first row of hangers. It is important that the first few hanger wires be perpendicular to the main runner. If they are not, use a Whitney punch and make new holes.

NOTE

At least three turns should be made when twisting the wire. The wire should also be tight to the runner.

NOTE

As indicated previously, in most exposed grid systems, the main runners are available with pre-punched holes. The upper edges of the holes generally measure $1\frac{1}{8}$" from the bottom of the runner; the dry line will have to be adjusted prior to bending the hanger wire.

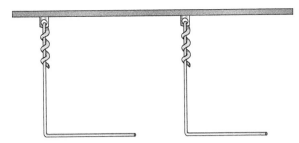

Figure 26 Bending hanger wire.

Step 3   Continue to install the balance of the first row of main runners; splices will be needed. Note that various types of splices are used with various grid systems and are supplied by the grid system manufacturer.

Step 4   After installing the last full length of main runner, measure, cut, and install an end piece from a full length of main runner to complete the first run.

Step 5   After each end of the main runner is resting on the wall angle, insert all hanger wires and twist the wires to secure them in place (*Figure 27*).

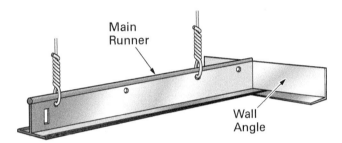

Figure 27 Inserting hanger wires in the main runner.

Use the following procedure to install the cross runners:

Step 1   Measure the distance from the main runner to the wall angle. This width was determined earlier when completing the room layout.

Step 2   Using the snips, cut the cross runner to the correct border width. Be sure to cut and save the correct end or it will not slide into the main runner.

Step 3   Insert the factory end into the main runner and let the other end rest on the wall angle (*Figure 28*).

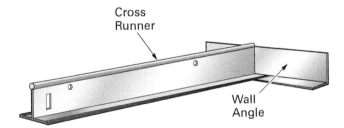

Figure 28 End of cross runner resting on wall angle.

Step 4   When the main runner and the cross runner are perpendicular to each other, lock the cross runner in place. Install the remaining cross runners between the main runners.

Step 5   To stabilize the grid system, install pop rivets at every other cross runner.

Step 6   Double-check the squareness of the entire grid.

Hold-Down Clips

Some ceiling tiles require hold-down clips to secure the ceiling tiles to the grid. For example, clips are used with lightweight tiles to prevent them from reacting to drafts. One manufacturer specifies that clips be used if the tiles weigh less than 1 pound. Hold-down clips are not necessarily required for ceilings used in fire-resistance-rated applications. Check the manufacturer's instructions.

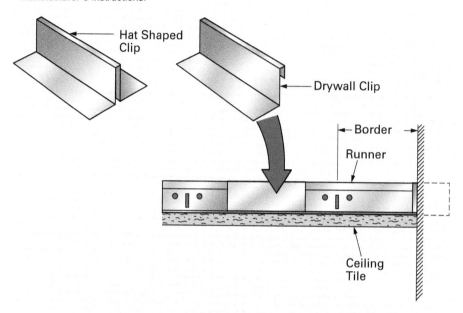

Did You Know

Squaring It Up

If the main runners and cross runners in a ceiling grid are out of square, the tiles will not fit properly into the grid, which may result in gaps between the grid and the tiles. Some quick measurements and a little math can be used to prevent this problem.

You have likely heard of the Pythagorean theorem:

$$a^2 + b^2 = c^2$$

The letters in the formula each represent one side of a triangle. The letters a and b are the straight sides and the letter c is the diagonal side. In a ceiling grid, the measurements of side a and side b would be sections of cross runners and main runners. Square (multiply them by themselves in mathematical terms) these two measurements and add them together. Measure the diagonal side of the imaginary triangle, side c, and square this measurement. If that section of the ceiling grid is perfectly perpendicular (squared in construction terms), the square of side c should equal the added squares of sides a and b. If these measurements are not equal, make slight adjustments to the main runners and cross runners to achieve squareness.

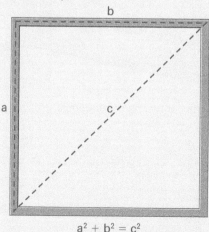

NOTE

Some grid systems use automatic locking devices to secure the cross runners into the main runners. Other grid systems use clips. Once all cross runners have been installed into the main runners, the grid system is ready to receive the ceiling panel or tile.

Prior to installing the ceiling tiles, wash your hands to keep the tiles clean. It is good practice to wear white gloves when handling ceiling tiles. Start by cutting the border tiles and inserting them into position. After the border tiles are installed, proceed to install the full tiles. As the full tiles are being placed, install hold-down clips, if required. General installation guidelines include the following:

- Install the ceiling tiles in place according to the job progression. Do not jump or scatter tiles in the grid system.

- Exercise care in installing tiles so as not to damage or mar the surface.

- Always handle tiles at the edges and keep your thumbs from touching the finished side of the tiles. If gloves are not worn, use cornstarch, powder, or white chalk on your hands.

- If the ceiling tile has a deeply textured pattern, insert it into the grid system so that the directional pattern will flow in the same direction. Such matching of ceiling tiles adds beauty to the finished suspended ceiling.

Upon completion of the ceiling, clean up the work area as follows:

Step 1   Dismantle the scaffold.

Step 2   Pick up all tools and equipment.

Step 3   Secure all equipment in a safe place.

Step 4   Pick up all trash.

Step 5   Sweep up any dust or debris.

3.4.2 Metal-Pan Systems

Metal-pan systems use $1^1/_2$" U-shaped, cold-rolled steel furring channel members for pan support. They are normally installed 4' on center. However, it may be necessary to install them at lesser distances depending on the location of the light fixtures. The furring channels are installed by looping the hanger wires around them. Twist and secure using saddle wire. It is important that the furring channels are properly leveled to ensure a level ceiling.

Once the furring channels have been installed, the main runners (tee bars) must be placed. These bars are manufactured, cold-rolled or zinc-coated steel or aluminum. They are fastened to the furring channels using special clips. The tee bars run at right angles to the furring channels (*Figure 29*).

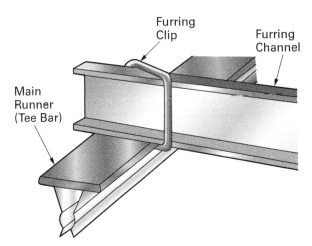

Figure 29 Main runner (tee bar) clipped to a furring channel.

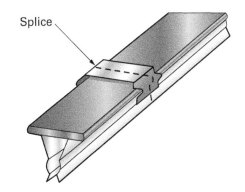

Figure 30 Tee bar splice.

The tee bar has a spring-locking feature, which grips the metal pan. This feature allows the pan to be removed for access to the area above the pan. If the room plan dimensions are longer than the length of the tee bar, use a tee bar splice to couple the tee bars together, extending them to the required length (*Figure 30*).

High-Durability Ceilings

Highly durable ceiling tiles provide for long life and easy maintenance wherever ceilings are subjected to improper use, vandalism, or frequent removal for plenum access. They are used in applications where resistance to impact, scratches, and soil are major considerations. Ceilings in areas such as school corridors or gymnasiums need to withstand abuse, including surface impact. In any areas where lay-in ceiling tiles frequently need to be removed for plenum access, scratch-resistant tiles are highly desirable. Otherwise, the tile surface can be scratched, scuffed, or chipped as it is slid across the metal suspension system components. Ceilings installed in laboratories, clean rooms, and food preparation areas are normally required to meet special standards to ensure they can withstand repeated cleaning.

Use the following procedure to install the tee bars:

Step 1   Install the first tee bar at right angles to the $1\frac{1}{2}$" furring channel. Be sure it is correctly aligned.

Step 2   When in position, place tee bar clips over the furring channel and insert the ends under the flange of the tee bar. Hang additional clips in the same manner along the length of the bar.

Step 3   Install the second tee bar parallel to the first one. The spacing should be from 1' to 4', depending on the size of the metal pans.

After installing all the tee bars in one section of the room (depending on the working area of your scaffold), begin installing the metal pans. Take care in handling the pans. Use white gloves or rub your hands with cornstarch to prevent any perspiration or grease marks from marring the surface of the pans. If care is not taken, fingerprints will be plainly visible when the units are installed. Use the following procedure to remove the pans from their container:

Step 1   Place the pan's finished surface face down on some type of raised platform, such as a table, which has been covered with a pad to protect the surface of the pan.

Step 2   Insert the wire grid into the back side of the metal pan. The wire grid is installed between the metal pan and backing pad to provide an air cushion between the two surfaces.

Step 3   Over the grid, place the paper- or vinyl-wrapped mineral wool or fiberglass batt or pad.

When installing the metal pans, begin at the perimeter of the ceiling, next to the walls, because the units must fit into the channel wall angle. Install the pans as follows:

Step 1   From the room layout or reflected ceiling plan, obtain the width of the border units. Measure the pan to this width.

Step 2   Using a band saw, cut the pan $\frac{1}{8}$" short of the desired width along the edge to be inserted into the wall angle.

Step 3   Slide the pan into the wall angle. Do not force the pan all the way into the molding. Leave $\frac{1}{8}$" for expansion.

Step 4   Insert two crimped edges of the metal pan into the spring-locking tee bars.

Step 5   When the pan is in position, insert the spacer clip into the channel-molding cavity and over the cut edge of the pan. This will prevent the pan from buckling along this cut edge.

Step 6   Cut the wire grid and backing pad to fit the border unit. Place the unit into position.

Step 7   At the corners of the room, install the pans in the order shown in *Figure 31*.

NOTE

In some cases, the metal pans come with the wire grid and pad already assembled.

Indoor Swimming Pools

All-aluminum grid systems are not recommended for use above indoor swimming pools because chlorine gases cause aluminum to corrode.

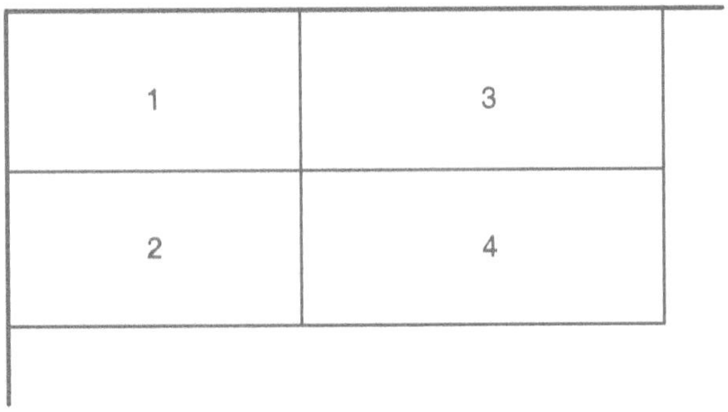

1	3
2	4

Figure 31 Tile layout.

Once the perimeter pans are installed in each row, the full-size pans can be put into place as follows:

Step 1 Grasping a pan at its edges, force its crimped edges into the tee-bar slots. Use the palms of your hands to seat the pan.

Step 2 After installing several of the pans as noted in Step 1, slide them along the tee bars into their final position. Use the side of your closed fist to bump the pan into level position if it does not seat readily.

Step 3 If metal pan hoods are required, slip them into position over the pans as they are installed. The purpose of the hood is to reduce the travel of sound through the ceiling into the room.

If a metal pan must be removed, a pan-removal tool is available (*Figure 32*). To pull out a pan, insert the free ends of the device into two of the perforations at one corner of the pan and pull down sharply. Repeat this at each corner of the pan. Following this removal procedure eliminates the danger of bending the pan out of shape. *Figure 33* shows a finished metal-pan ceiling.

Figure 32 Pan-removal tool.

Figure 33 Metal-pan ceiling.
Source: David R./Alamy

Placing Main Runners

One way to locate the position of the first main runner is to convert the width of the room to inches and divide it by 48 (assuming you are installing 48" tiles). Then add 48" to any remainder and divide that result by 2 to obtain the distance from the wall to the first main runner. The rest of the main runners are then placed at 4' intervals. Try an example assuming the width of the room is 22'.

Converting 22' into inches yields 264". Dividing that result by 48 gives 5' with a remainder of 6". Adding 48" yields 54", which is then divided by 2. The first main runner is placed 27" from the wall.

3.4.3 Direct-Hung Concealed Grid Systems

The installation of the concealed grid system begins in the same way as the conventional exposed grid system previously discussed.

Step 1 Once chalk lines have been established on the walls at the required height above the floor, the next step is to install the wall angle. This molding provides support for the grid and panel or tile at the wall, as it does for the other ceiling systems. The molding should be fastened with nails, screws, or masonry anchors. At corners, miter the inside corner and use a cap to cover the outside corner.

Step 2 Lay out the grid and install the hanger inserts and hangers according to the reflected ceiling plan or lighting layout. Once this has been completed, install the concealed main runners as is done for the exposed grid system (see Section 1.1.1, *Exposed Grid Systems*). The main runners are the primary support members. The method of coupling them to attain a specific length will vary with the manufacturer of the system, but they can all be spliced to the desired length.

Step 3 Install the cross-stabilizer bars and concealed cross tees at right angles to the main runners. They rest on the flange of the main tee runners.

Step 4 Place the ceiling panel into position. Cut in the first row of panel at a line perpendicular to the main runners, as previously established. Attach the panel to the wall using spring clips.

Flat splines (*Figure 34*) are metal or fiber units that are inserted by the installer into the unfilled panel kerfs between the concealed main runners. Splines are used to prevent dust from seeping through the kerfs in adjacent panels.

If access is needed to the area above the ceiling, special systems are available that can be incorporated into the ceiling (*Figure 35*).

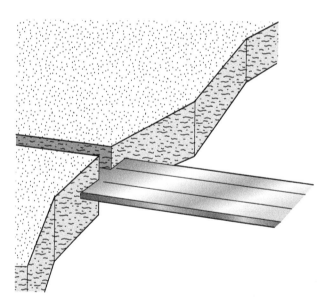

Figure 34 Ceiling-panel spline.

Insulation

Insulation is usually not recommended over mineral ceiling tiles because the additional weight could cause the tiles to sag. If the use of insulation cannot be avoided because of occupancy codes, limit the insulation to R-19 (0.26 pounds per square foot). Use only roll insulation and lay it perpendicular to the cross tees, so that the grid supports the weight of the insulation. If batts are required by code, 24" × 24" ceiling tiles should be used. Check codes carefully before applying insulation in a fire-resistance-rated system.

Facts about Ceiling Tiles

- Ceiling tiles are often called pads.
- Certain lighting fixtures are manufactured in the same sizes as tiles to drop directly into suspended ceilings.
- Some tiles are treated to inhibit the growth of mold, mildew, fungi, and certain bacteria.
- Special tiles are designed for use below sprinkler systems. These tiles will shrink and fall out of the grid as the room temperature rises, allowing the water to reach the fire.
- Most ceiling tiles can be recycled.

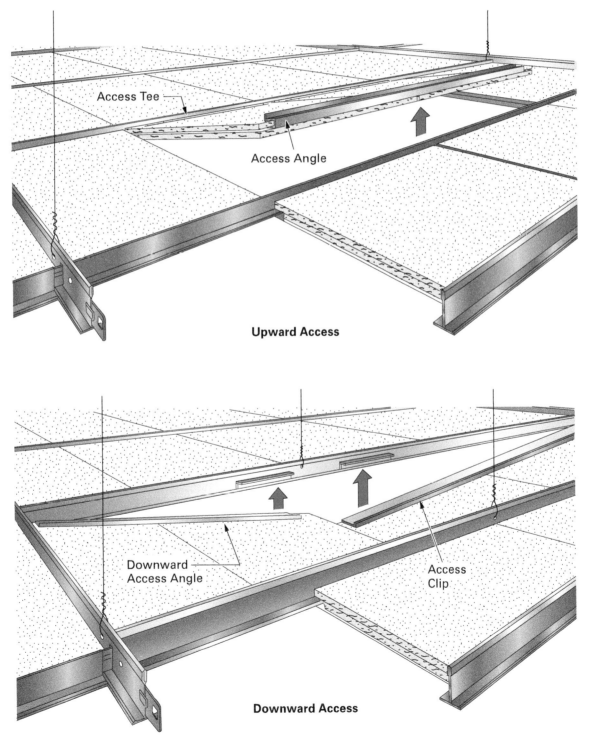

Figure 35 Access for concealed grid ceilings.

3.4.4 Luminous Ceiling Systems

A standard luminous system is installed in the same way as an exposed grid suspension system except for the border cuts. Luminous ceilings are placed into the grid members in full modules. Any remaining modules are filled in with acoustical material that has been cut to size.

Luminous panels are used with a 2' × 2' or 2' × 4' standard exposed grid to provide the light diffusing element in the system (*Figure 36*). These panels are laid in between the runners. Many panel sizes and shapes are available.

Figure 36 Luminous office ceiling.
Source: August0802/Getty Images

A nonstandard luminous ceiling is installed in the same way as a standard system with regard to room layout, hanger insert, installation of hanger wire, main supports, secondary supports, and attachment of some type of wall angle. Because nonstandard ceilings can differ so much in terms of size, complexity of the system, and the exactness of the installation, shop drawings are required. As always, when installing a ceiling system, consult the manufacturer's information for specific details.

3.4.5 Suspended Drywall Ceiling Systems

When drywall is used instead of panels or tiles to cover the surface of a ceiling, it is either fastened directly to the ceiling deck or a drywall suspension grid is installed. Drywall suspension systems are advantageous because they are quick and easy to install in comparison to other ceiling systems, and they can be shaped into the framework for more intricate designs, such as coffered, step soffit (*Figure 37*), and tray ceilings. Suspended drywall ceiling can use either the drywall furring system or the drywall grid system.

Figure 37 Step soffit ceiling.
Source: Dennis Axer/Alamy

A suspended drywall furring system is used when it is desirable or specified to have a drywall finish or drywall backing for an acoustical tile ceiling. *Figure 38* is an example of a suspended drywall furring system. The installation of a suspended furring system requires the use of special carrying channels.

Figure 38 Suspended drywall furring system.
Source: Don Wheeler

Installing Carrying Channels

Carrying channels, also known as $1\frac{1}{2}$" cold-rolled channel or black iron, are used for drywall furring systems to compensate for the additional weight of the systems. To install carrying channels, proceed as follows:

Step 1 Use a laser level to establish the reference line.

Step 2 Install the wall angle, if used.

Step 3 Fasten the hanger inserts into the ceiling structure above. Install a row of hanger inserts 4' on center, or as specified by the manufacturer, into the ceiling structure.

Step 4 Install the hangers. Use No. 8 gauge wire or as specified. Hang the wires from the hanger inserts, then twist and secure them. Allow enough wire to tie the hanger to the insert and the carrying channel. In general, 24" to 28" of wire is adequate.

Step 5 Mark the places where the laser target touches the hanger wires. Mark each hanger wire to the spot where it is to be bent. Bend the wire to a 90° angle.

Step 6 When all the hanger wires in one row have been marked and bent, fasten the carrying channel to the hanger wires (*Figure 39*). Loop the wires around the carrying channel at the point where the wires have been bent. Twist and fasten the wires using a saddle tie. It is important that the carrying channels are installed level.

Step 7 If it is necessary to extend the length of a carrying channel because of the room size, face the open U of the second channel toward the U of the one already installed. Overlap by 8" to 12" and tie together with wire (*Figure 40*).

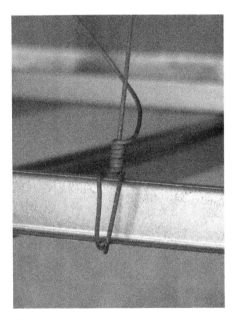

Figure 39 Carrying channel supported by hanger wire.
Source: Don Wheeler

Figure 40 Splicing carrying channels.

Step 8 Continue to install the hanger inserts, wires, and carrying channels on 4' centers over the entire ceiling.

Installing Furring Members and Drywall

After the carrying-channel support members are installed, proceed with the installation of the cross-furring members. There are different kinds of furring members that can be used at right angles to the carrying channels at 16" on center (maximum). For example, a hat furring channel (*Figure 41*) is used when it is desirable to screw the drywall board to the furring channels with self-tapping screws. Install the furring channels as follows:

Step 1   Install the furring channels perpendicular to the carrying channels. Attach the furring channel to the carrying channel using tie wire (*Figure 42*).

Step 2   Continue to install the furring channels 16" on center. If it is necessary to extend the furring channels beyond their normal length, lap one end of the channel over the next piece a minimum of 8" and wrap the splice well with tie wire (*Figure 43*). Avoid installing furring channels over positions where any fixtures are to penetrate the ceiling.

Step 3   Screw the drywall to the furring channels (*Figure 44*). Be sure to use screws of sufficient length. The screw length should be long enough to go through the board and extend through the furring channel about $\frac{5}{8}$" with three exposed threads.

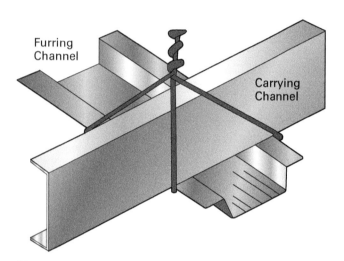

Figure 41 Hat furring channel.

Figure 42 Furring channel attached to carrying channel with clips.
Source: Don Wheeler

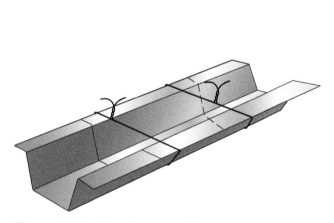

Figure 43 Splicing (lapping) furring channels.

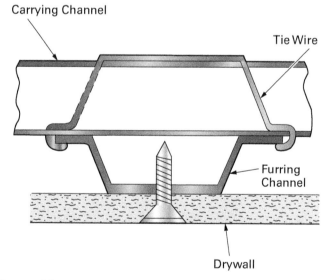

Figure 44 Drywall screwed to a furring channel.

NOTE

Keep acoustic panels clean during installation.

3.4.6 Installing Acoustical Panel

If acoustical panel is to be used as the finish over the drywall, proceed as follows:

Step 1   Lay out the room. Measure the length of one of the two shorter walls. Divide this measurement in half. Mark this halfway point on the drywall next to this point on the wall. Next, measure the length of the facing wall at the opposite end of the room. Divide this wall length in half and mark this point on the drywall.

Step 2   Set up a laser level and align the horizontal beam with the two marks you made on the drywall.

Step 3   Align the vertical laser beam at the midpoint of the horizontal line. If your laser level does not generate two beams simultaneously, use tape to mark the horizontal beam's location and then align a vertical beam to the tape line.

Step 4   Install the ceiling panel per the manufacturer's instructions beginning at the junction of the two centerlines.

3.4.7 Installing Furring Channels Directly to Structural Members

Sometimes furring channels are installed directly to structural members such as open-web steel joists or wood I-joists instead of suspended carrying channels. In these instances, install the furring channels as follows:

Step 1   Mark the walls to indicate the height to be used to level the ceiling. Set the ceiling-height control marks to indicate the position of the bottom edge of the furring channels. Check the level of the structural joists. If they are not level, use the low point as the common level for installing the furring channels.

Step 2   Measure up from the benchmarks to the height of the lowest point on the joists. Determine the distance of the low joist point minus the height of the furring channel and mark it on all walls to indicate a uniform level for the lower edge of all furring channels. Be sure to measure up from the benchmarks to establish this common level on all walls.

Step 3   Establish the level lines using the ceiling control marks.

Step 4   Install the wall angles (if used). New level lines may have to be used to guide the installation process.

Step 5   Install the furring channels in the same manner as described in *Installing Furring Members and Drywall* in Section 3.4.5, with the following exceptions:

- When attaching furring channels to steel joists, tie them together with the appropriate-gauge tie wire. Wrap the wire around the two so that it bridges the joist and supports the furring channel on either side of the joists. Tie the wire ends together at the side of the union. Furring channel can also be attached to steel joists using an 8" overlap and four #8 screws (two screws through each of the small $\frac{1}{2}$" flanges).
- Shimming may be needed between the joists and the furring channel. Be sure to check the level of the furring channel along its entire length. Shim where needed to correct any deviation from level.

3.4.8 Drywall Suspension Systems

The drywall grid system (DGS) is a suspension ceiling system for drywall ceilings. In this system, a grid is suspended from the ceiling, and the drywall panels are installed on the grid (*Figure 45*). Suspending the drywall below the ceiling deck, as opposed to attaching it directly, allows ductwork and other items to remain in place while still achieving the finished drywall look.

This system is ideal for large ceilings, long flat ceilings, and floating ceilings. It can also be used in small areas, called short-span areas, such as hallways. In addition, DGS can be used to create dramatic, curved ceilings and distinctive designs.

Figure 45 Suspended ceiling drywall grid system (DGS).
Sources: Dennis Axer/Alamy (left); Sever180/Shutterstock (right)

Special Furring Systems

Some manufacturers make furring-system cross tees in 14", 26", and 50" lengths. These sizes can reduce the time it takes to install an F-type lighting fixture from as long as 30 minutes to less than 1 minute.

Drywall grid systems are installed much like the standard exposed grid system and use the same size main tees, cross tees, and wall angle. However, the components for this system are more particular and are normally purchased from specialized suppliers. As with all proprietary systems, consult the approved construction documents and manufacturer's information for specific installation instructions.

An extruded aluminum drywall perimeter trim used in conjunction with the DGS provides a great finished edge for floating ceilings with straight or curved edges. Curved perimeter trim is factory pre-curved to the desired radius/radii. This trim will not only create a solution at the perimeter edge but is also a labor cost reducer and time saver that will ensure a better-looking perimeter edge than drywall. The perimeter trim is secured to the drywall suspension with factory clips, which are screwed into the drywall for a seamless appearance. Clean and crisp perimeter edges are a guarantee if extruded trims are used. These trims are generally factory primed and field-finished along with the drywall.

3.4.9 Ceiling Cleaning

Immediately after installation and after extended periods of use, suspended ceilings may require cleaning to maintain their performance characteristics and attractiveness. Dust and loose dirt can easily be removed by brushing or using a vacuum cleaner. Vacuum cleaner attachments such as those designed for cleaning upholstery or walls do the best job. Make sure to clean in one direction only to prevent rubbing the dust or dirt into the surface of the ceiling.

After loose dirt has been removed, pencil marks, smudges, or clinging dirt may often be removed using an ordinary art gum eraser. Most mineral-fiber ceilings can be cleaned with a moist cloth or sponge. The sponge should contain as little water as possible. After washing, the soapy film should be wiped off with a cloth or sponge slightly dampened in clean water. Vinyl-faced fiberglass ceilings and Mylar™-faced ceilings can be cleaned with mild detergents or germicidal cleaners.

Cleaning Ceiling Tiles

The materials and methods used for cleaning ceiling tiles vary widely, depending on the finish and texture of the tiles. One manufacturer prescribes eight different cleaning methods for its line of ceiling tiles. It is very important to check the manufacturer's instructions before proceeding. When cleaning grids, the ceiling tiles should be removed to prevent cleaning solution or dirt from getting on the tiles. Always wear gloves when handling or cleaning tiles.

3.0.0 Section Review

1. Which soil classification is considered the most stable?
   a. Class A
   b. Class B
   c. Class D
   d. Class F

2. If a room does not have a flat ceiling, the centerlines can be laid out on the floor and then transferred upward using _____.
   a. a laser level
   b. a plumb bob and dry lines
   c. a story pole
   d. triangulation (the 3-4-5 rule)

3. When installing the grid for a suspended ceiling, you should use a level to check the installation at intervals of _____.
   a. 3'   c. 6'
   b. 4'   d. 8'

4. After the first eye pin, where should other eye pins be installed?
   a. 2' on center
   b. 4' on center
   c. 6' on center
   d. 8' on center

Module 27209 Review Questions

1. From their source, sound waves travel _____.
   a. on a line of sight
   b. in all directions
   c. vertically
   d. at right angles

2. The intensity of a sound wave refers to _____.
   a. the number of wave cycles it completes in a second
   b. its loudness
   c. its frequency
   d. the degree of loudness or softness

3. A sound level of 100 dBA is considered _____.
   a. very loud
   b. moderately loud
   c. loud
   d. quiet

4. The term used by manufacturers to compare the noise absorbency of their materials is the _____.
   a. ceiling attenuation class (CAC)
   b. articulation class (AC)
   c. noise reduction coefficient (NRC)
   d. sound transmission classification (STC)

5. A(n) _____ is used if the grid system is to be concealed from view.
   a. metal-pan system
   b. direct-hung ceiling system
   c. integrated ceiling system
   d. luminous ceiling system

6. All ceilings, walls, pipe, and ductwork in the space above a luminous ceiling will normally be painted with a _____.
   a. light-refracting finish
   b. 75% to 90% reflective matte white finish
   c. 90% to 100% reflective matte finish
   d. luminous matte finish

7. _____ are inserted into the main runners at right angles and spaced an equal distance from each other, forming a complete grid system.
   a. Hangers
   b. Wall angles
   c. Hold-down clips
   d. Cross runners

8. In an exposed grid ceiling system, hangers are used to support the _____.
   a. 4' cross tees
   b. 2' cross tees
   c. wall angle
   d. main runners

9. Acoustical panels and tiles used in suspended ceilings stop the transmission of unwanted sounds by the process of _____.
   a. reflection
   b. reverberation
   c. absorption
   d. refraction

10. Butt-jointed tiles that produce a monolithic ceiling design with no visible support system are known as _____.
    a. beveled edge tiles
    b. concealed tee tiles
    c. lay-in tiles
    d. profiled edge tiles

11. Which of the following is a safety guideline for using a baker's scaffold?
    a. Never ride on a moving scaffold.
    b. Inspect the scaffold once per week.
    c. Apply pushing and pulling force at or near the top of the scaffold.
    d. Erect a ladder on top of the scaffold to increase your reach or height.

12. A(n) _____ shows the features of the ceiling while keeping those features in proper relation to the floor plan.
    a. elevation plan
    b. mechanical plan
    c. reflected ceiling plan
    d. orthographic ceiling plan

13. If job conditions do not agree with what is shown in the specifications, you should _____.
    a. change the job conditions to match the specifications
    b. match the specifications as closely as you can
    c. change the specifications to match the job conditions
    d. contact your supervisor and ask for instructions on how to proceed

14. The points at which electrical and plumbing risers penetrate a ceiling are shown on _____.
    a. MEP drawings
    b. floor plans
    c. section views
    d. elevation drawings

15. For a suspended ceiling that is 10' × 16', using ceiling tiles that are 2' × 4', with the main runners parallel with the ceiling's long dimension and the 4' length of the tiles parallel with the ceiling's short dimension, the first main runner should be installed at a distance from the wall of _____.
    a. 18"
    b. 24"
    c. 36"
    d. 48"

16. For a suspended ceiling that is 10' × 16', using ceiling tiles that are 2' × 4', with the main runners parallel with the ceiling's long dimension and the 4' length of the tiles parallel with the ceiling's short dimension, the dimensions (width and length) of the border ceiling tiles to be installed along the long dimension of the ceiling will be _____.

    a. 18" × 24"
    b. 24" × 30"
    c. 24" × 36"
    d. 36" × 48"

17. When correctly laid out, the centerlines for a grid system will intersect at a _____ at the exact midpoint of the room.

    a. 25-degree angle
    b. 45-degree angle
    c. 90-degree angle
    d. 120-degree angle

18. When laying out eye pins for hanger inserts, two dry lines are used to determine the _____.

    a. exact center of the room
    b. positions of all cross ties
    c. centerline of the first ceiling tile
    d. location of the first wall angle

19. When installing the metal pans in a metal-pan ceiling system, begin by installing them _____.

    a. in the middle of the ceiling
    b. at the perimeter of the ceiling next to the walls
    c. at the corners of the room
    d. at the most convenient location

20. Splines installed into the unfilled panel kerfs in a direct-hung concealed grid ceiling system _____.

    a. prevent buckling of the ceiling
    b. are used in place of stabilizer bars
    c. prevent shifting of the panels
    d. prevent dust from seeping through the kerfs in adjacent panels

Answers to Odd-Numbered Module Review Questions are found in *Appendix A*.

Answers to Section Review Questions

Answer	Section	Objective
Section One		
1. c	1.1.4	1a
2. a	1.2.0	1b
3. b	1.3.1	1c
Section Two		
1. a	2.1.0	2a
2. d	2.2.0	2b
Section Three		
1. a	3.1.0	3a
2. b	3.2.2	3b
3. c	3.3.2	3c
4. b	3.4.1	3d

MODULE 27405

Finished Stairs

Objectives

Successful completion of this module prepares you to do the following:

1. Examine the procedure for installing various stairway systems.
   a. Explain the procedure for cutting and installing treads and risers for closed stairways.
   b. Identify the procedure for installing open, combination, L-shaped, and U-shaped stairways.
   c. Describe the procedure for installing handrails, guards, and balustrades.
2. Describe the procedure for installing custom stairway systems.
   a. Outline the procedure for installing shop-built stairways.
   b. Describe the procedure for installing winder treads.
3. Summarize the process of building exterior stairways.
   a. Describe the procedure for building exterior wood stairways.
   b. Explain the procedure for building cast-in-place forms for concrete stairways.

Performance Tasks

Under supervision, you should be able to do the following:

1. Install treads and risers on open, closed, and/or combination open/closed main stairways.
2. Miter a finished stringer and risers.
3. Install a return nosing.
4. Install a post-to-post balustrade system.
5. Build a cast-in-place form for a concrete stairway.

Overview

Stairways represent an important part of the interior design of a structure. A great deal of finish work needs to be done to transform a basic set of stairs into an attractive stairway.

Observe the following guidelines when installing any stairway components or a balustrade system:

- Always cut and fit all parts before gluing or permanently installing them. This will significantly reduce the chance of error and the amount of waste material.
- To provide a solid and durable balustrade installation and to avoid squeaks, use wood screws, lag bolts, rail bolts, and a quality wood glue or construction adhesive to glue all joints.
- Newel posts are the primary support of any balustrade system. They are recommended at the start, at the end, and at all changes of direction on a stairway. In addition, they are recommended at intervals of 5' to 6' on a balcony or a landing that is 10' or more in length.
- Always consult local building codes prior to installation for information concerning handrail heights, baluster spacing, or any other regulations that may apply.

NOTE

Building code requirements vary from jurisdiction to jurisdiction. Because of the variations in code, check the approved construction documents and/or the applicable code whenever regulations are in question. The code requirements for residential and commercial construction can vary significantly.

Digital Resources for Carpentry

Scan this code using the camera on your phone or mobile device to view the digital resources related to this craft.

1.0.0 Installing Stairway Systems

Performance Tasks

1. Install treads and risers on open, closed, and/or combination open/closed main stairways.
2. Miter a finished stringer and risers.
3. Install a return nosing.
4. Install a post-to-post balustrade system.

Objective

Examine the procedure for installing various stairway systems.

a. Explain the procedure for cutting and installing treads and risers for closed stairways.

b. Identify the procedure for installing open, combination, L-shaped, and U-shaped stairways.

c. Describe the procedure for installing handrails, guards, and balustrades.

CAUTION

Be careful and plan ahead. Always think about safety. Create a job hazard analysis before undertaking any task. Accidents do not just happen; they are generally caused by carelessness and unsafe practices.

A variety of stairway designs are available for residential and commercial applications. Stairways can be classified as open, closed, combination, L-shaped, and U-shaped. In inhabited areas of a structure, stairways can range from simple to ornate. This type of stairway is usually finished with birch, oak, maple, beech, cherry, poplar, or ash. If carpeting is being used, the treads and risers may be partially finished with pine, fir, or plywood. Uninhabited areas of a structure, such as basements and attics, usually have simple stairways. Southern pine, yellow pine, white pine, fir, or plywood is commonly used for the treads and risers for this type of stairway.

1.1.0 Cutting and Installing Treads and Risers

A closed stairway may have skirt boards to finish into, or it may have gypsum board as a finish completely to the rough stairway. In either case, the risers and treads must be installed without damaging the gypsum board or skirt board.

When finishing stairways, care and accuracy are needed on the part of the carpenter. A power compound miter saw is best for cutting stair stock.

This section presents the steps for finishing a closed stairway. The sample stairway has an 11" run and a 7" rise, with a nosing equal to the riser thickness. If for some reason the rise and run are not known, you must determine the rise and run of the stairway, as well as the required nosing dimension. See *Figure 1* and *Figure 2*.

Computerized Main Stairway Design and Takeoff Systems

Some stair parts manufacturers have computerized stairway design and takeoff software programs. These software programs are usually available for use at a distributor, or they can be purchased separately. Most of these programs provide automatic balustrade and tread/riser takeoff lists along with a 3-D rendering of a main stairway based on the total run, rise, and width of the stairway. Typically, they also price the stair parts and provide price comparisons between different stair-tread and balustrade styles. They can usually produce printouts of the stairway layout plan view with key parts called out, along with building code selections.

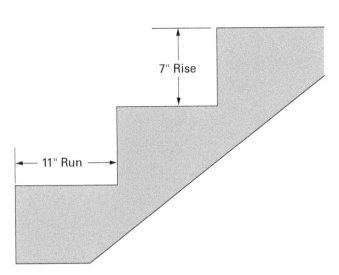

Figure 1 Rise and run of a stairway.

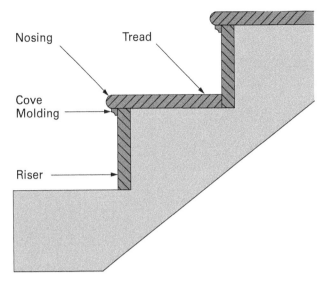

Figure 2 Stair treads, risers, and nosing.

1.1.1 Riser and Tread Templates

Templates may be used when determining the correct length and angle of the risers and treads. These templates can be made from $\frac{1}{4}$" tempered hardboard or plywood and will help to minimize layout errors. The template shown in *Figure 3* is used for finding the correct length and angle (if any) of the riser. The other template, as shown in *Figure 4*, is used to establish the correct length and angle (if any) of the tread.

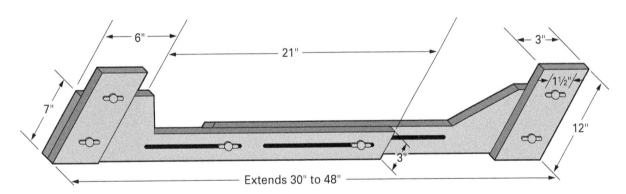

Figure 3 Riser template.

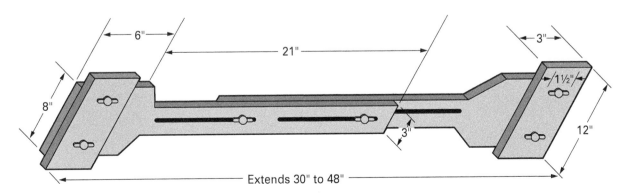

Figure 4 Tread template.

A riser template will extend from 30" to 48". The end pieces adjust in or out a distance of $\frac{3}{4}$". All moving pieces are secured with a bolt, washer, and wing nut. The bolt fits into the tempered hardboard or plywood through a slot in the other piece and is secured with the washer and wing nut. To use the template, extend the pieces to fit snugly against the walls, and tighten the wing nuts. This allows for transfer to the stock and for marking the pattern layout. The use of the riser template will allow the riser, when cut, to fit snugly, and it will not gouge the gypsum board.

As shown in *Figure 4*, the tread template is made in almost the same fashion as the riser template. Both templates are helpful for finishing stairs.

NOTE

A power compound miter saw is recommended for cutting stair materials.

WARNING!

When preparing to finish a stairway, always barricade the stairway from above to prevent others from falling into the floor opening.

1.1.2 Cutting and Fitting Risers

Start the stairway by removing the first four or five rough treads at the bottom of the stairway. Clean the area around the stringers to prevent injury to yourself and to ensure that the riser will fit properly against the stringers. The first riser will be shorter than the other risers by the thickness of a tread because the stringers are set on the finish floor. This maintains equal riser heights for the stairway.

Use the following procedure when cutting and fitting risers:

Step 1   Cut a piece of riser stock 1" longer than required.

Step 2   Rip the riser stock to the proper height.

Step 3   Using the riser template, loosen all wing nuts and place it against the stringer.

Step 4   Extend the template so it is snug against the wall or skirt boards. Tighten all wing nuts and remove.

Step 5   Place the template on the ripped riser stock. Make sure the bottom of the template is even with the bottom of the riser. Mark both ends and remove the template.

Step 6   Using a power compound miter saw, cut the riser but leave the lines marked in the previous step. This will allow the riser to fit snugly.

Step 7   Put the cut riser in place. If it is too tight, remove the riser and use a wood rasp on the back side of the cuts to create a slight end bevel. Avoid rasping the front edge of the riser, as this could make it too loose.

Step 8   Replace the cut riser and secure it, using three finish nails for each stringer. Leave the nails $\frac{1}{8}$" away from the riser face and set the nails with a nail set. Allow the nail heads to penetrate $\frac{1}{16}$" to $\frac{1}{8}$" into the stock. Repeat Steps 1 through 8 for the rest of the risers.

NOTE

Always refer to the approved construction document for the appropriate size finish nails to use when installing treads and risers.

1.1.3 Cutting and Fitting Treads

To find the total tread width, add the tread depth and nosing depth, as shown in *Figure 5*. Depending on code requirements, the nosing depth may be the same as the riser thickness.

As an example (see *Figure 5*), assume that dimension A is 11" and dimension B is $\frac{3}{4}$". The total tread width of dimension C will be the sum of dimensions A and B, or $11\frac{3}{4}$".

Tread Material Hardness

Depending on the type of material used for the treads and the material's hardness, you may need to drill a pilot hole through the tread for the nail before fastening the tread to the stringers. This prevents splitting treads or bending nails when securing the treads to the stringers.

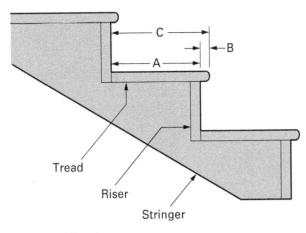

A – Tread Depth
B – Nosing Depth
C – Total Tread Width

Tread Depth + Nosing Depth = Total Tread Width
(From Riser)

Figure 5 Tread dimensions.

After installing four risers, lay out and install three treads using the following procedure:

Step 1 Loosen all wing nuts on the tread template and place it on top of the stringers and against the riser.

Step 2 Extend the template so it is snug against the walls or skirt boards. Tighten all wing nuts.

Step 3 Place the template on the tread stock, ensuring that the back of the template is flush with the rear of the tread stock. Hold the template in place and mark the ends.

Step 4 Remove the template. Cut the tread using a power compound miter saw, leaving the line.

Step 5 After cutting, place the tread on the stringers. If it is too tight, file and fit the tread the same way as the risers were fitted.

Step 6 Replace the tread and drill three pilot holes in the tread at each stringer, and two holes at the front of the tread into the riser below.

Step 7 After drilling, secure with three finish nails into the stringer, and two finish nails into the edge of the riser. Use a nail set to countersink the nails. Repeat Steps 1 through 7 for the rest of the treads.

The top of the stairway may be finished in one of two ways depending on how the stringers were installed. *Figure 6* illustrates the last tread set below floor level. *Figure 7* illustrates the last tread nearly even with the floor level.

If the stringers are installed as shown in *Figure 6*, a landing tread (*Figure 8*) must be used. The lip of the tread overlaps the subfloor, and the finish floor is installed around it. If finished as in *Figure 7*, a full tread is installed at the top of the stringer and is butted flush to the finish floor. When a landing tread is used, it must be installed properly to prevent rocking. *Figure 9* shows a completed open stairway.

NOTE

Glue or construction adhesive applied to the tread surfaces where it meets the stringers and risers will prevent the stairway from squeaking. The most common application is to use glue and screws to connect the riser to the tread.

Securing a Landing Tread

If a landing tread is not secured properly, it becomes a weak point. After a time, it will begin to rock from the constant foot pressure it receives. The landing tread should be drilled and secured with screws or threaded nails.

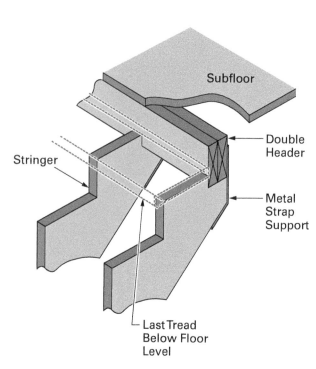

Figure 6 Last step set below the floor level.

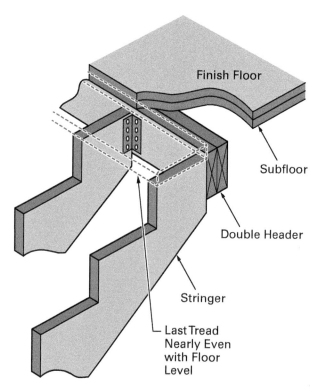

Figure 7 Last step set even with the floor level.

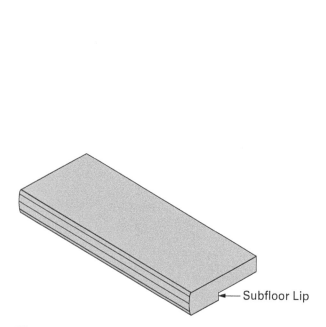

Figure 8 Landing tread.

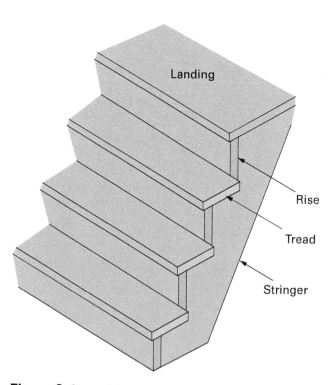

Figure 9 Open stairway.

1.1.4 Alternate Stairway Construction

The two basic methods for connecting stair treads to the risers are shown in *Figure 10*. The most common method is the butt joint. When a butt joint is used, as is typical when the stairway is site built, the riser butts against the tread and is generally covered with a cove or other type of shaped molding. If a rabbet joint is used, a groove is cut along the top portion of the riser and a groove is cut in the

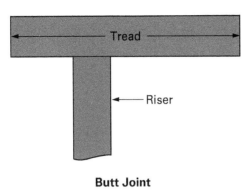

Butt Joint

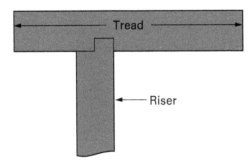

Rabbet Joint

Figure 10 Joints used to connect the tread to the riser.

bottom of the tread at a predetermined measurement. This type of joint is used in manufactured or custom-built stairs, including housed-tread and riser stair systems.

After installing the risers and treads, a length of **cove molding** must be installed under the nosing of the tread to seal the joint between the riser and tread (except in a rabbeted tread). It must also be installed along the wall and skirt board. Install the cove molding under the nosing of each tread using the following procedure:

Cove molding: A decorative strip that, when attached to the underside of the tread nosing, covers the joint between the tread and riser.

Step 1 Measure the distance between the skirt boards at a point directly under the nosing.

Step 2 Using a miter box, cut the molding to the proper length.

Step 3 Place the molding under the nosing against the riser.

Step 4 Nail in place using three or four finish nails. Use a nail set to countersink nails.

Step 5 Repeat Steps 1 through 4 for each step.

A finished closed stairway is shown in *Figure 11*.

Figure 11 Finished closed stairway.

Eliminating a Bow or Sag in Center Stringers

If center stringers are used, be sure no bow or sag exists at the center stringer. If a bow exists, scribe the tread, and plane the bottom until the tread fits. Other methods to eliminate a bow or sag at the center stringer require forcing a small shim between the stringer face and the riser to eliminate a bow or shimming the tread to eliminate a sag.

1.2.0 Installing Open, Combination, L-Shaped, and U-Shaped Stairways

Open, combination, L-shaped, and U-shaped are common stairway designs. In many ways, the installation of these types of stairways resembles that of a closed stairway. However, these designs have unique characteristics that set them apart. The design and layout of the structure, the preference of the owner, and safety considerations typically determine the type of stairway used.

1.2.1 Open Stairways

Open stairways come in a variety of designs from simple to elaborate as shown in *Figure 12*. Some stairways are open on both sides the entire length of the stairway (*Figure 12B*), while others are open at the bottom, but then have a closed (*Figure 12C*) or combination (*Figure 12A*) stairway near the top. This section will focus on the open parts of the stairway.

(A)

(B)

(C)

Figure 12 Open Stairways.
Sources: Stockernumber2/Getty Images (A); Cormac Byrne/Alamy (B); EricVega/Getty Images (C)

Bullnose Starting Step

A completed bullnose starting step is shown here. Most starting steps are furnished with a riser height that varies according to code. The riser must be cut to the required height for the first riser of the stairway. Make sure to account for the finish floor height when cutting the riser.

Source: yevgeniy11/Shutterstock

If the stairway will have a bullnose as the starting step, follow this procedure for installing and placing the starting tread and riser:

Step 1   Set the rip fence on a table saw to the required starting riser height.

Step 2   With the starting step tread removed, trim the bottom edge of the starting riser on the table saw.

• If a doweled newel post (*Figure 13*) or some other newel mounting system will be used that requires access from the bottom of the starting step to mount the newel and no access will be possible from the back side of the stairway, do not permanently attach the starting step until the balustrade is laid out and fitted. Cut and fit the starting step riser and the starting tread, and temporarily fasten them in place until the newel post is measured and is ready to be installed. If one of the other newel post mounting systems shown in *Figure 13* is used, the step can be secured in place. Section 1.3.2 *Installing Guards* and Section 1.3.3 *Installing Balustrade Systems* provide more information on the requirements for newels.

Step 3   Trim the bullnose riser length to fit across the stairway, as shown in *Figure 14*.

Step 4   Secure the bullnose riser to the stairway stringers, as shown in *Figure 14A* (preferred) or *Figure 14B*. A **glue block** is usually required along the inside bottom edge of the riser to anchor the front of the riser.

Step 5   Install the second riser as described in Section 1.1.2 *Cutting and Fitting Risers*, then see the following instructions for installing open treads.

Glue block: A block of wood that is attached to the hidden side of a tread and riser joint. This is done for structural soundness and to eliminate squeaking.

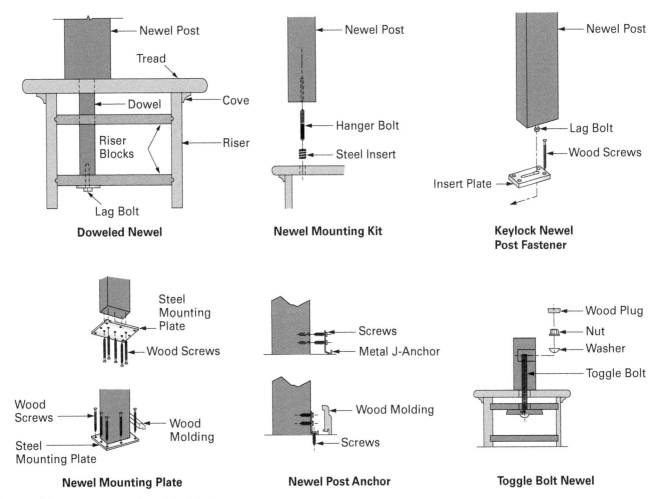

Figure 13 Common newel-post installations.

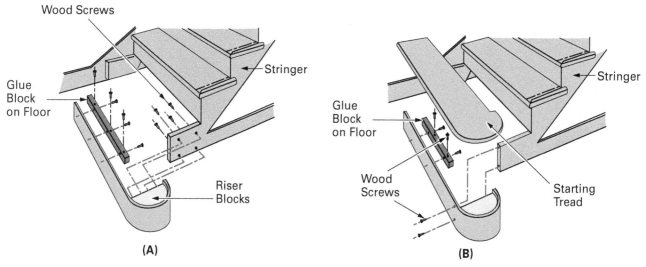

Figure 14 Typical starting-step installation.

NOTE

If treads with preinstalled return nosing are used, the treads are installed as previously described, and the following steps can be omitted.

After installing several risers, treads must be installed up to each riser. When treads are used on an open stairway, a tread return nosing is used on the end to seal the end grain of the tread (except on a bullnose starting step). The tread return nosing usually measures $1\frac{1}{4}$" wide; however, be sure to check the manufacturer's specifications.

The procedure to cut and install tread and tread return nosing is as follows:

Step 1   Measure the length required for the tread. Allow an extra 3". Place the tread on the stringer, tight against the skirt board. Scribe to fit, if necessary.

Step 2   After cutting the scribed end, place the tread on the stringers.

Step 3   Where the tread and outer finish stringer meet, place a mark on the underside of the tread.

Step 4   Place the tread on a solid surface, and, using a square, extend the mark the full width of the tread.

Step 5   From this line (Cut No. 1 in *Figure 15*), extend the tread length by the width of the tread return nosing width (usually $1\frac{1}{4}$"), and scribe a second line parallel to the first line.

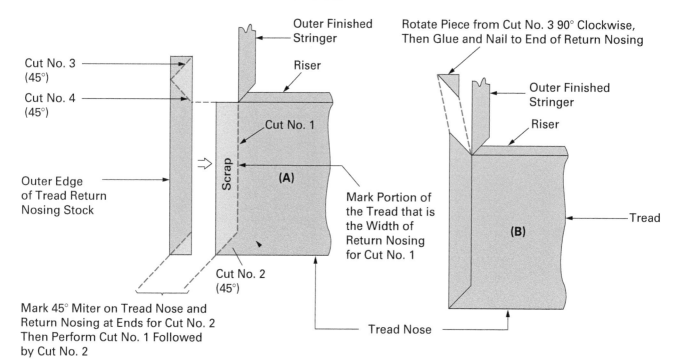

Figure 15 Fitting tread return nosing and installing tread.

Step 6 Using a combination square, mark a 45° angle from the second line at the tread nose back to the first line (Cut No. 2 in *Figure 15*).

Step 7 Using a power compound miter saw and a handsaw, cut away the portion of the tread bounded by Cut No. 1 and Cut 2.

Step 8 Measure and cut off a length of tread-return nosing stock that is equal to the width of the tread plus more than two times the width of the return nosing stock.

Step 9 On a power compound miter saw, cut parallel 45° angles at both ends of the piece of the tread-return nosing stock (Cut No. 2 and Cut No. 3). Retain the small triangular piece obtained from Cut No. 3.

Step 10 Place the tread return nosing against the tread so the miters meet. Using a combination square, set it on the tread return nosing so the 45° edge of the blade is even with the rear of the tread. Mark and cut the tread nosing. Cutting the end of the tread return nosing seals the end grain by the insertion of the mitered piece of tread return nosing.

Step 11 Glue and nail the mitered tread return nosing to the tread using finish nails. Drill, glue, and nail the mitered piece of tread return nosing from Cut No. 3 to the rear end of the tread return, as shown in *Figure 15B*.

Step 12 Place the completed tread in place against the skirt board and stringers. When this is done, the tread is secured in place. The tread return nosing should overlap the outer finish stringer, as shown in *Figure 15B*.

Step 13 Fit, cut, glue, and nail cove molding under the exposed edges of the tread. The finished tread should look like *Figure 16*.

Step 14 Repeat Steps 1 through 13 to finish installing the other treads.

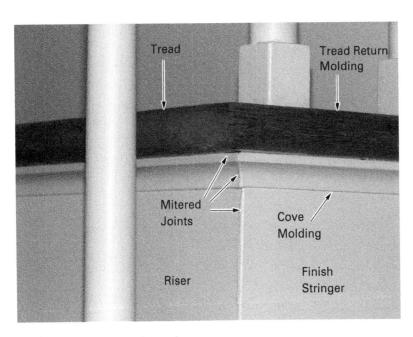

Figure 16 Finished open stair tread.

1.2.2 Combination Stairways

Like the open stairway, a combination stairway can have variations as shown in *Figure 17*. A combination stairway has one open side and one closed side. The open side can have treads like an open stairway (*Figure 17A*), or it can have a skirt board on both sides (*Figure 17B*) to attach the treads like on a closed stairway.

In either case, the instructions for cutting and fitting risers still apply. If the open side has treads like an open stairway, follow the instructions for installing treads for an open stairway. The only difference will be that, in this case, the treads will only have tread nosing on one side. If the open side has a skirtboard, follow the instructions for installing treads on a closed stairway.

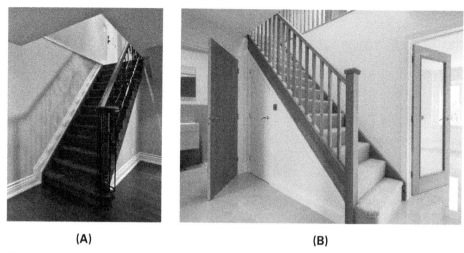

(A)   **(B)**

Figure 17 Combination stairways.
Sources: trishz/123RF (A); Michael Higginson/Alamy (B)

The stairway shown in *Figure 18* is an illustration of a combination stairway that is open on one side and has a wall on the other.

In this example, the outer wall under the stairway is flush with the outer rough stringer. This will require an outer finish stringer to fit against the outer rough stringer. The rise of each step on the outer finish stringer must be mitered to receive a mitered riser (*Figure 19*). On the inside wall, a straight skirt board is inserted between the gypsum board and the rough stringer.

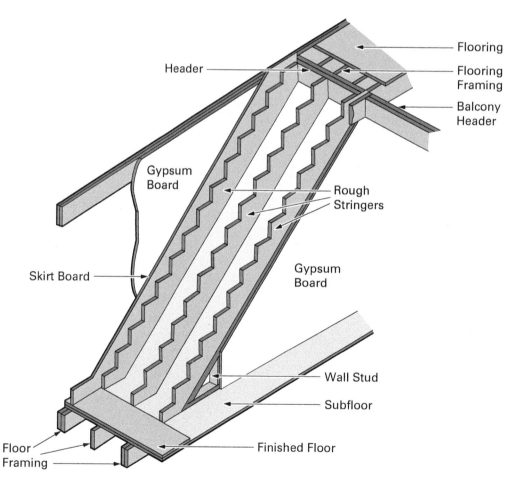

Figure 18 Framed stairway opening.

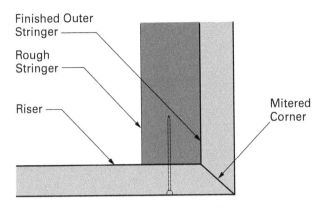

Figure 19 Riser mitered to outer finish stringer (top view).

The outer finish stringer may be fabricated using one of the two following methods:

- The tread and riser of the finish stringer can be laid out with a framing square or **pitch block** in the same manner as a rough stringer.
- The finish stringer is placed against the rough stringer and tacked. Then the tread width and riser height are transferred to the finish stringer with a pencil mark.

If the finish stringer is marked by the second method, use a pitch block to establish the floor cut. This is where the finish stringer will sit on the floor. Make sure the bottom edge of the outer stringer is positioned flush with the bottom of the rough stringer. Then, tack the finish stringer to the rough stringer and use the framing square to mark the finish stringer, as shown in *Figure 20*.

Pitch block: A triangular piece of wood used for laying out the rise and run on stringers, and when trimming easement fittings.

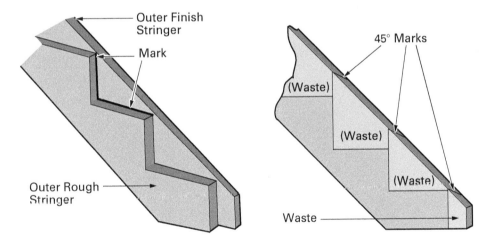

Figure 20 Marking the outer finish stringer.

After the finish stringer is marked with the tread and riser, set one end of the finish riser on the floor and the other end on a sawhorse or workbench. With a combination square, mark each riser at a 45° angle. Using a power compound miter saw, cut each riser at a 45° angle up to the tread mark. The tread mark is cut square.

After the finish stringer has been cut, secure it to the rough stringer using finishing nails. Use a nail set to countersink the nails. The stringer is now ready for risers and treads to be installed.

Pre-Engineered and Pre-Assembled Stairways

Schedules and labor costs dictate the use of some form of pre-engineered and (in many cases) pre-assembled stairway systems in most new residential and commercial construction. For residential main stairways, a temporary stairway may be built at the jobsite and then removed when a pre-engineered or assembled stairway is installed during or after the interior wall-covering application.

In most cases, a residential finish balustrade system may be the only part of the stairs that is site built. In certain very high-end construction or renovation applications, stairways may be completely custom fabricated on-site.

1.2.3 L-Shaped Stairways

An L-shaped stairway includes one 90° turn with a landing between two straight-run stairways. The following are types of L-shaped stairways:

- Closed
- Open
- Combination

Figure 21 shows one type of L-shaped stairway. This stairway will be finished in the same manner as the straight-run stairway. The platform will accept either a landing nosing or a stair tread, depending on the method used for placing the stair stringers.

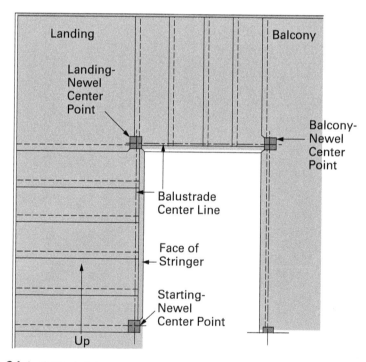

Figure 21 L-shaped stairway.

As shown in *Figure 22*, this type of stairway is typically open on one side and closed on the other. Depending on the architectural design, the stairway could be finished in one of two ways:

- The risers and skirt board are mitered, and the treads overhang the skirt board.
- The treads and risers are installed in the same fashion as the closed stairway.

Figure 22 Combination L-shaped stairway.

Sources: PORNTHIP ALOUNTHONG/Getty Images (left); Wavebreak Media ltd/Alamy Images (right)

1.2.4 U-Shaped Stairways

A U-shaped stairway (*Figure 23*) can have the following variations:

Figure 23 U-shaped stairway.

Sources: alexandrumagurean/Getty Images (left); Slobo/Getty Images (right)

- Open on both sides, lower landing
- Combination open/closed, lower landing
- Closed on both sides
- Closed on one side
- Open to second landing

This section describes a narrow U-shaped stairway that is open on both sides (lower portion), with a spacing of 2' between the end of the stairway and the wall. Refer to *Figure 24* for a sectional view of a U-shaped stairway.

Figure 25 is a plan view of the stairway. The lower portion of this stairway is open on both sides, and a balustrade will be used on both sides of the lower portion. The upper portion is partially open. Therefore, the **balustrade system** will make a U-turn and continue to the second-floor landing without interruption. The balustrade on the right side of the lower portion, as viewed from the first floor, will continue across the landing and to the wall.

Balustrade system: A collective term that refers to the newels, balusters, and handrail(s) on a stairway.

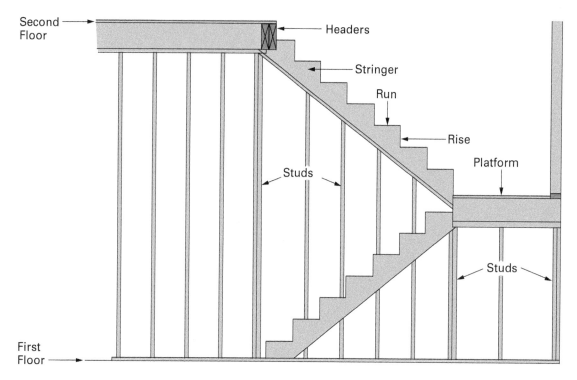

Figure 24 Sectional view of a U-shaped stairway.

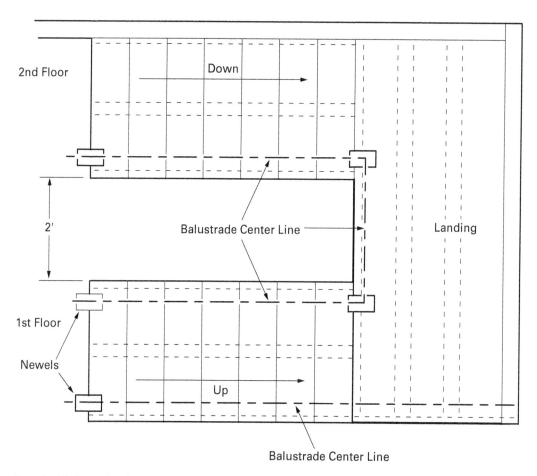

Figure 25 Plan view of a U-shaped stairway.

1.3.0 Installing Handrails, Guards, and Balustrades

To complete a stairway, the last task is the installation of the handrail, guards, and/or balustrades. The type of staircase built will determine which of the following components are needed:

- Handrails are used on all closed sections of stairways, including fully closed stairways, and closed sections of combination, L-shaped and U-shaped stairways.

- Balustrades are used on all open sections of stairways, including fully open stairways, and open sections of combination, L-shaped, and U-shaped stairways.

- Guards are used on elevated walkways. The *IBC*® requires guards along open-sided walking surfaces, including mezzanines, equipment platforms, aisles, stairs, ramps, and landings located more than 30" measured vertically to the floor or grade below at any point within 36" to the edge of the open side.

Guards and Handrails

OSHA provides the following definitions [OSHA 1910.21(b)]:

"*Guardrail system* means a barrier erected along an unprotected or exposed side, edge, or other area of a walking-working surface to prevent employees from falling to a lower level."

"*Handrail* means a rail used to provide employees with a handhold for support."

OSHA Part 1926, Subpart M requires that "the top edge height of top rails, or equivalent guardrail system members, shall be 42 inches plus or minus 3 inches above the walking/working level. When conditions warrant, the height of the top edge may exceed the 45-inch height." [OSHA 1926.501(b)(1)]

IBC® and *IRC*® define a *guard* as "a building component or a system of building components located at or near the open sides of elevated walking surfaces that minimizes the possibility of a fall from the walking surface to a lower level." The *IRC*® requires a guard to be a minimum of 36" tall measured from the walking surface, and the *IBC*® requires 42" tall guards.

The *IBC*® and *IRC*® define a *handrail* as "a horizontal or sloping rail intended for grasping by the hand for guidance or support."

1.3.1 Installing Handrails

The *International Residential Code*® (*IRC*®) notes standard handrail heights of 34" to 38", measured from the top of the handrail to the sloped plane adjoining the nosing on the rake of the stairway. Handrails must be provided on at least one side of each continuous run of treads or flights of stairs of four or more risers.

The *International Building Code*® (*IBC*®) states that stairways used for commercial applications must have handrails on each side except for stairs for dwelling units and spiral stairways. Handrails must be located so that all portions of the stairway width are within 30" of a handrail.

Handrails for residential applications are usually supported by brackets along their length. For commercial applications, more support brackets may be required over the length of the rail and the ends must usually be returned to a wall or terminated in a post or safety terminal.

Handrails are simple in profile and usually have a circular cross section. Handrails with circular cross sections must have an outside diameter of at least $1\frac{1}{4}$" and no more than 2" (*IRC*® Type I handrails). If a handrail is not circular, the perimeter dimension should be 4" minimum and $6\frac{1}{4}$" maximum with a maximum cross-sectional dimension of $2\frac{1}{4}$" (*IRC*® Type II handrails). The handgrip portion of handrails should have a smooth surface with no sharp edges or corners.

The *IBC*® and *IRC*® require that a handrail should not be less than $1\frac{1}{2}$" from the wall or other surface or project more than $4\frac{1}{2}$" from the wall or other surface

Handrail Extensions

The *Americans with Disabilities Act* (*ADA*) and the *International Building Code*® (*IBC*®) require that the upper end of wall rails or handrails that are not continuous have a minimum horizontal extension of 12" beginning at the face of each top riser.

Source: Kopyoc Oleg/Alamy Images

Rake Rail and Height

Because a handrail is slanted (raked) to be parallel with the stairway rise and run, the handrail is often called a rake rail and its height above the stairway is sometimes referred to as the rake height.

on either side of the stairway. A handrail, wall, or other surface adjacent to the handrail shall be free of any sharp or abrasive elements.

To install a handrail, establish the proper rail height by checking the approved construction drawings and local building code (*Figure 26*). Then establish the location of the wall brackets.

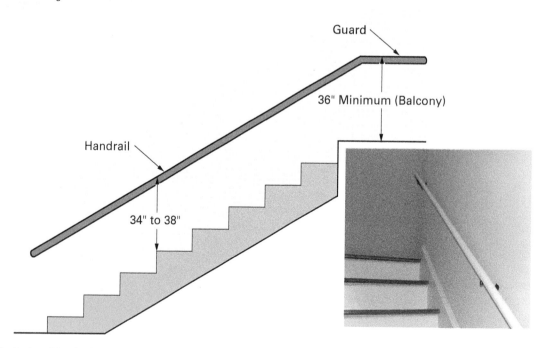

Figure 26 Typical residential handrail height.

Install **wall rail brackets** (*Figure 27*) at the bottom and top of the stairway. Secure the brackets by screwing them into a stud. Where no stud exists, backing may be installed prior to covering with gypsum board, or the gypsum board should be cut and blocking inserted.

1.3.2 Installing Guards

Any stairway that includes an open walkway at an elevated height, such as a landing or balcony, requires a guard. Some stairways, including L-shaped and U-shaped stairways, may include guards as part of the stairway. Other stairways may be connected to a landing or balcony at the top of the stairway. In stairway systems, guards are often included as part of the balustrade installation on stairways that require them.

Use the following procedure for the assembly and installation of balcony guards and balusters:

Step 1 If a half-newel against a wall will be used (*Figure 28*), trim the half-newel to the height of the balcony newel and place it against the wall. Then, plumb the newel and fasten it with lag bolts and adhesive. Toggle bolts may be used if a stud is not available. Drive a finish nail through the newel at the top to help stabilize the newel. If a rosette (*Figure 29*) is used to secure the handrail to the wall, secure the rosette to the handrail with two wood screws and adhesive.

Step 2 If a half-newel is used, mark and trim an **opening cap** to fit the half-newel, as shown in *Figure 30*. Fasten the trimmed cap to the balcony guard with a rail bolt and adhesive.

Step 3 Place the guard against the wall and mark the length to the end of the rake handrail quarter turn (*Figure 31*). Square-cut the guard using a compound power miter saw, and temporarily fasten the guard to the quarter turn with a rail bolt. For a half-newel wall mount, temporarily rest the other end on the half-newel post. For a rosette wall mount, locate and mark the bottom of the guard on the wall at the same height as the balcony newel. Then, temporarily secure the rosette to the wall with screws (*Figure 32*).

Wall rail brackets: Metal supports for a wall rail.

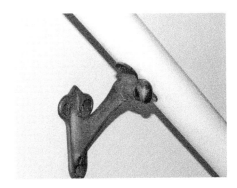

Figure 27 Wall rail bracket.

Opening cap: A handrail fitting at the start of a level balustrade system.

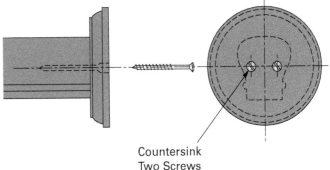

Figure 29 Fastening a rosette to a handrail.

Countersink Two Screws

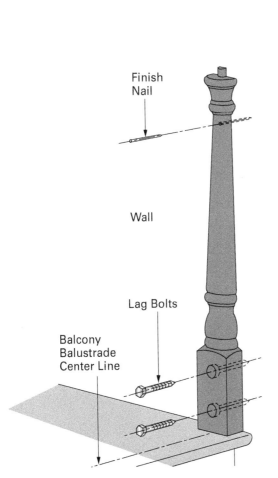

Figure 28 Installing a half-newel.

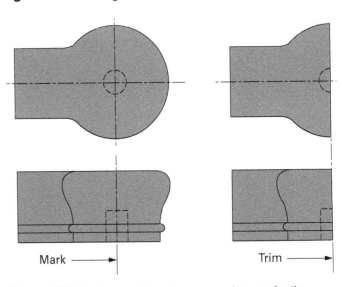

Mark   Trim

Figure 30 Marking and trimming an opening cap for the half-newel.

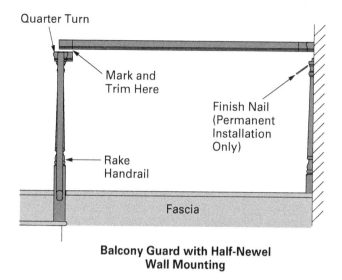

**Balcony Guard with Half-Newel
Wall Mounting**

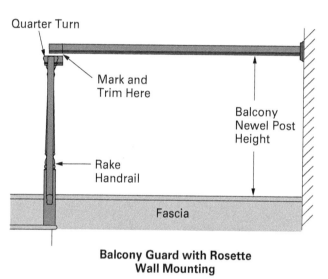

**Balcony Guard with Rosette
Wall Mounting**

Figure 31 Marking the balcony guard at the quarter turn.

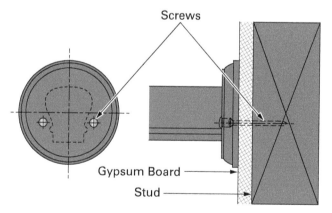

Figure 32 Temporarily fasten the rosette to the wall.

Step 4 Determine the balcony baluster spacing by referring to *Figure 33* as a reference and performing the following calculations to obtain the spacing.

Equation 1:

Horizontal distance (A) + Width of one baluster square = Total horizontal distance (B)

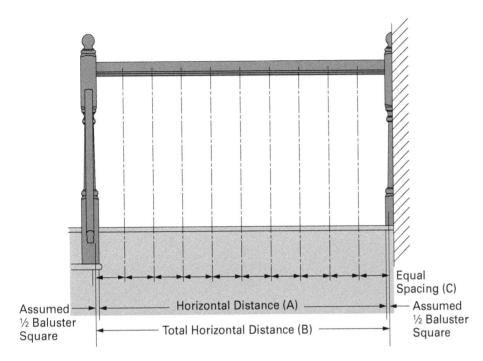

Figure 33 Measuring balcony baluster spacing.

Equation 2:

Total horizontal distance (B) ÷ spacing per code = number of spaces (rounded up)

Equation 3:

Total horizontal distance (B) ÷ number of spaces (rounded up) = spacing (C)

Step 5 Mark the points on the floor, starting from the assumed one-half baluster spacing and using the baluster spacing (C) obtained from Equations 1–3. Check that the total number of spaces, including the two assumed one-half baluster spaces, add up to the total horizontal distance (B).

Step 6 Remove the guard and lay it down next to the balcony balustrade centerline with the end against the wall (not a baseboard). With a framing square, transfer the baluster spacing marks to the center of the guard. If pin balusters are being used, drill the appropriately sized holes in the guard and balcony floor to accept the balusters.

Step 7 Place the guard in its proper position, and temporarily secure it to the quarter turn and the wall, as applicable.

Step 8 Measure and trim the balusters at their base. Install baluster pins or screws and secure the balusters to the balcony floor using glue or adhesive.

Step 9 Install the top of the balusters in the guard and secure with glue or adhesive. Permanently secure the guard to the quarter turn and wall with glue or adhesive. Tighten the rail bolts or screws holding the rosette (if used) or apply glue or adhesive to the half cap and half-newel to secure the guard. Drive a finishing nail through the half cap into the half-newel. Drive a finishing nail at the top of the balusters into the guard to prevent the baluster from rotating and to secure the guard to the baluster. Install fillet trim if required.

Step 10 Install wood plugs in the holes for the lag bolts holding the half-newel (if used). Sand the plugs flush.

Guard Post Connection

The *IRC*® requires guards to resist a 200 lb concentrated load. Where the top of a guard system is not required to serve as a handrail, the single concentrated load shall be applied at any point along the top, in the vertical downward direction and in the horizontal direction away from the walking surface. Where the top of a guard is also serving as the handrail, a single concentrated load shall be applied in any direction at any point along the top. Concentrated loads shall not be applied concurrently.

Guard in-fill components (all those except the handrail), balusters and panel fillers shall be designed to withstand a horizontally applied normal load of 50 pounds on an area equal to 1 ft^2. This load need not be assumed to act concurrently with any other live load requirement.

IRC Section R507.10.2, Wood Posts at Deck Guards, states that where 4" by 4" wood posts support guard loads applied to the top of the guard, such posts shall not be notched at the connection to the supporting structure.

1.3.3 Installing Balustrade Systems

Before beginning an installation, refer to applicable codes to determine the required rake handrail height, balcony height (if applicable), baluster spacing, and handrail clearance as illustrated in *Figure 34*. For this balustrade system, the balcony height, rake height, and baluster spacing requirements would be applicable.

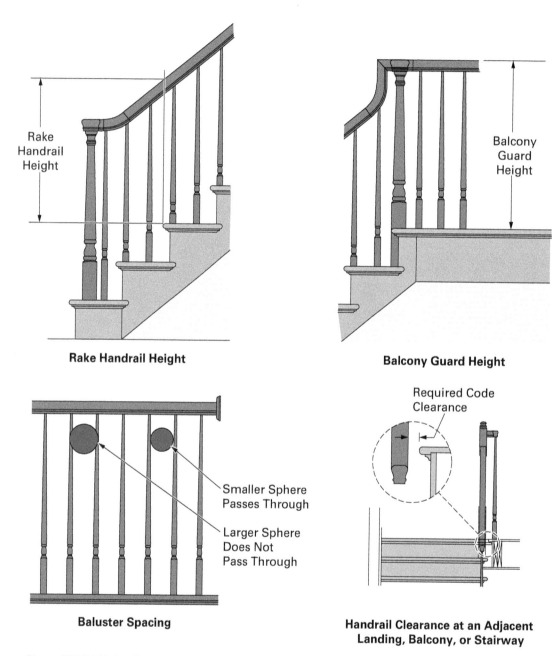

Rake Handrail Height

Balcony Guard Height

Baluster Spacing

Handrail Clearance at an Adjacent Landing, Balcony, or Stairway

Note: IRS R312.1.3 Opening Limitations. Required *guards* shall not have openings from the walking surface to the required *guard* height that allow passage of a sphere 4" in diameter.

Exceptions:
1. The triangular openings at the open side of *stair*, formed by the *riser*, and bottom rail of a *guard*, shall not allow passage of a sphere 6" in diameter.
2. *Guards* on the open side of stairs shall not have openings that allow passage of a sphere 4³/₈" in diameter.

Figure 34 Critical balustrade system code requirements.

Preparation for Balustrade Installation

Use the following procedure to prepare for the installation of a balustrade system:

Step 1 Mark the balustrade centerline, as shown in *Figure 35*. This example stairway is an open tread style stairway. The centerline should be spaced in from the face of the finish stringer by a distance equal to one-half of the width of the base of the selected style of baluster (for example, $\frac{5}{8}$" for a $1\frac{1}{4}$" square-base baluster).

Step 2 From the baluster spacing required by the applicable code, determine the number of balusters required per tread by dividing the run (tread width) by the maximum spacing allowed by the code. Then divide the run by the number of required balusters to determine the center-to-center distance between the balusters.

- For example, if the run is 10" and the maximum spacing required by code is 4", then 10 divided by 4 equals $2\frac{1}{2}$ balusters per tread. See *Figure 36B*. Rounding up to the next whole number yields 3 balusters per tread. Dividing the 10" run by 3 equals a baluster spacing of 3.333" (or $3\frac{5}{16}$") center-to-center. The required baluster spacing is then marked on the balustrade centerline of each tread, starting with the first baluster on each tread. The first baluster is spaced in from the face of the riser under the tread by a distance equal to one-half the width of the base of the selected style of baluster (for example, $\frac{5}{8}$" for a $1\frac{1}{4}$" square-base baluster, as shown in *Figure 36*).

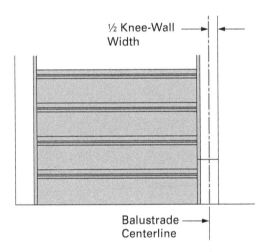

Knee-Wall Stairway (Front View)

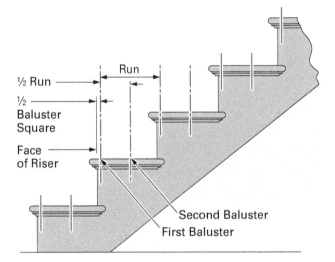

(A) Two Balusters Required per Tread

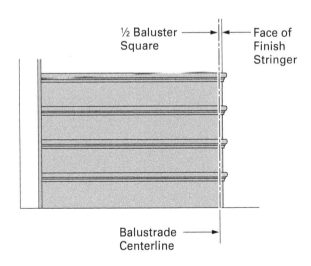

Open-Tread Stairway (Front View)

Figure 35 Marking the balustrade centerline.

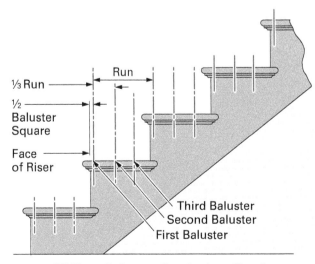

(B) Three Balusters Required per Tread

Figure 36 Marking balustrade center points.

Step 3 Make a pitch block by using a piece of off-fall, as shown in *Figure 37A*. Off-fall is the piece of scrap from one of the rise-run cutouts of the stair stringers. If this scrap is not available, clamp a piece of handrail or similar wood against the nosing of the treads, as shown in *Figure 37B*. Then, set the squared corner of a piece of ½" plywood on one tread and against the nose of the next tread. Mark the plywood on the underside of the clamped handrail or other wood and cut off the resulting triangular pitch block. The pitch block will be used in the marking, cutting, and installation of balustrade fittings and handrails. Make sure it accurately reflects the rise and run of the stair treads. Mark the appropriate sides of the pitch block as RAKE, RISE, and RUN.

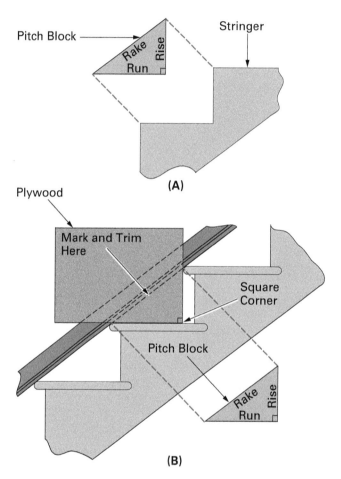

Figure 37 Making a pitch block.

Rail-Bolt Installation Method

Handrails, newel posts, and fittings are secured by means of rail bolts that are installed as described in the following steps:

Plan Ahead

When nailing the treads to the stringers, keep the nails out of the path of the baluster. You don't want to find a nail under your drill bit when you drill the treads to receive the balusters.

NOTE

When predrilled quick-connect fittings are used, all the drilling, except for bolt hole drilling in the handrail, is eliminated.

Angled-Rule Method of Baluster Spacing

Shown here is a nonmathematical method of determining baluster spacing. A ruled straightedge is marked off with the code-required spacing until the spaces marked exceed the total horizontal distance (B), which is the distance between newels plus one baluster square. With one end of the straightedge at one of the assumed one-half baluster-square marks, swing (angle) the other end of the straightedge away from the other newel post until the last space marked on the straightedge lines up with the assumed one-half baluster-square mark at that post. This can be determined using a framing square set parallel with the centerline between the newel posts and at the assumed one-half baluster-square mark. After the angle of the ruled straightedge is set, mark off the centers of the remaining spaces at the newel-post centerline using the framing square, as shown.

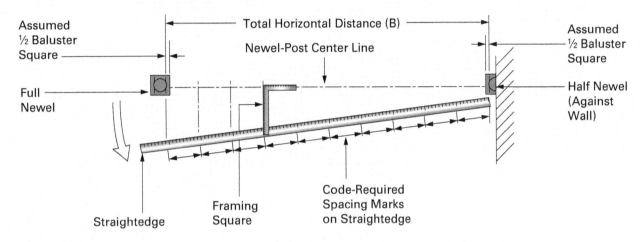

Step 1 Make a drilling template to mark rail-bolt hole centers for fittings and handrails by square-cutting a $\frac{1}{8}$"-thick piece of the handrail. Drill a $\frac{1}{16}$" hole through the template, as shown in *Figure 38*. Mark one side RAIL and the other side FITTING.

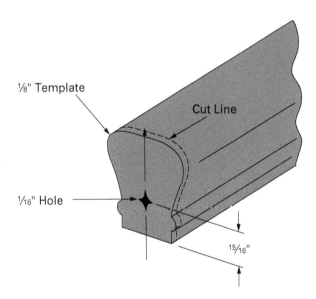

Figure 38 Rail-bolt drilling template.

Step 2 When using the template, align the template on the fitting or rail with the corresponding marked side visible. If two fittings are being joined, treat one fitting as a rail. Carefully mark the hole position on the fitting or rail through the $\frac{1}{16}$" hole. Mark both mating surfaces of the mating pieces. To obtain the closest possible fit, extreme care in marking and drilling the mating pieces is required. If desired, the marks on the mating surfaces can be lightly center punched to aid in accurately positioning the drill bit prior to drilling. Make sure all holes are drilled exactly perpendicular to the surface being drilled even though the piece may be curved.

Step 3   Drill all holes to the depth and diameter shown in *Figure 39*. Use a Forstner bit for the 1" hole.

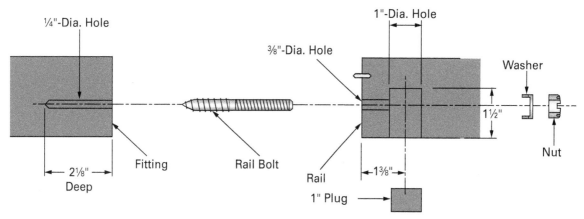

Figure 39 Holes required for a rail bolt.

Step 4   On one of the mating surfaces, drive two small finish nails into the surface, leaving ⅛" exposed (*Figure 40*). Cut the heads off. The resulting two pins will keep the parts from rotating when they are assembled.

Step 5   Double nut the rail bolt to drive it into the rail, leaving 1⁷⁄₁₆" of the bolt exposed (*Figure 41*). Remove the nuts and temporarily assemble the parts by using a finger to position the washer and nut on the end of the bolt. Align the parts and tighten the nut with a ½" box-end wrench. Use glue or construction adhesive only on final assembly after the entire balustrade is cut and fitted.

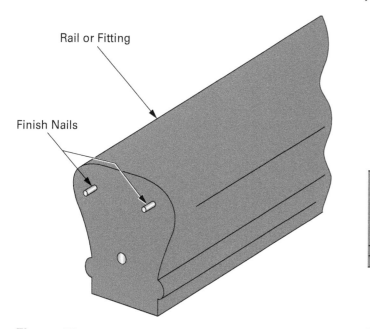

Figure 40 Anti-rotation pins.

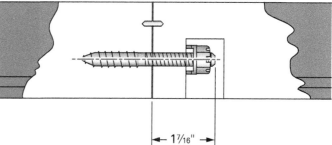

Figure 41 Fitting and/or rail assembly.

Finished Rake Rail

The finished rake rail shown here used glued pin-top balusters that were secured with a finishing nail. The hole for the finishing nail was filled and then the balusters and underside of the handrail were painted, while the rest of the rail and newel post was finished in a mahogany stain and varnished.

Newel Notching

Part of balustrade installation usually involves notching newel posts so they can be installed on the balustrade centerline. Newel posts must be notched and installed at a point prior to baluster installation and after the proper newel heights have been established.

To notch a newel, an outline of the material to be removed from the newel is drawn with the aid of a square. The exact dimensions of the notch(es) will depend on the newel and stair dimensions. Rough cuts are made $\frac{1}{8}$" to $\frac{1}{4}$" from the outline. Further cuts are made to carefully pare away material up to the line. Any remaining material is hand-chiseled away.

Step 1   If required, notch a starting newel, as shown in *Figure 42*. Once notched, the starting newel will sit on the floor and the first tread. It will also drape over the first riser and the stringer.

Step 2   If required, notch a landing newel as shown in *Figure 43*. Once notched, the landing newel will sit on the first tread on the second flight of stairs, the landing, and the top tread of the first flight of stairs. It will also drape over the risers and stringers of both flights of stairs.

Step 3   If required, notch a balcony newel, as shown in *Figure 44*. Once notched, the balcony newel will sit on the floor of the balcony and the last tread of the stairway. It will also drape over the last riser of the stairway.

NOTE

The notch will reduce the capacity of the newel and needs to be accounted for when ensuring that the guard system can still meet the 200 lb concentrated load requirement.

Newel Notching

Unless you are experienced at performing newel notching operations, practice marking and notching on a 4 × 4 or other scrap material. This is especially critical when performing a compound notch, such as that used on a landing newel.

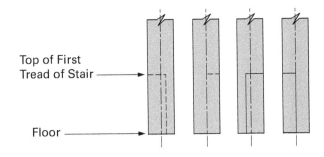

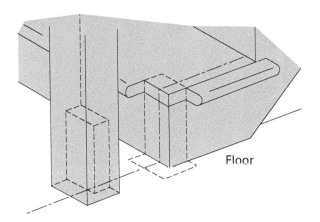

Figure 42 Notching a starting newel.

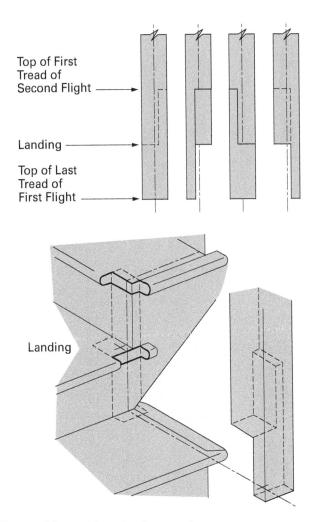

Figure 43 Notching a landing newel.

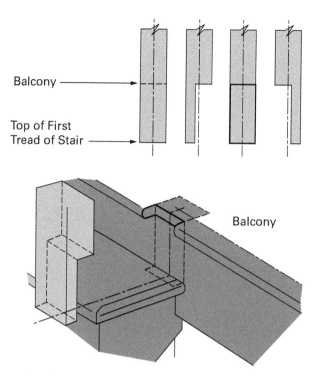

Figure 44 Notching a balcony newel.

Post-to-Post Connection

Post-to-post: Balustrade system where the handrail is not continuous. The handrail is lagged into the face of a square-top newel.

An open stairway can also use the **post-to-post** connection system. In constructing this type of stairway, a starting step may or may not be used. If a starting step is not used, the starting newel is installed by drilling and attaching the newel to the first riser with lag bolts. See *Figure 45*.

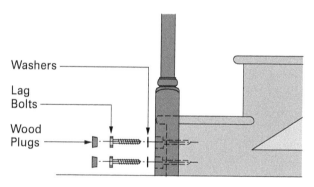

Figure 45 Newel post connection.

The newel height is determined by adding 1" for the block reveal (*Figure 46*) to the rake or balcony handrail height plus the necessary extension to allow for fastening of the newel. Landing or balcony newels with an extended top block may be used, or gooseneck fittings may be used instead of the extended block.

After marking the location of the balusters, lay the handrail on the rake of the stairs, and transfer the marks with the combination square to the handrail. At the same time, mark the handrail to fit between the newels.

After cutting and drilling the handrail (unless a plowed handrail is used), drill the holes in the treads for the balusters. Set the handrail at the proper rail height (check local codes) and measure the length of the balusters. Install the balusters and handrail as previously described.

Secure the handrail to a newel using a rail bolt at an angle, or drill through the starting newel and screw the handrail to the newel with a lag screw. Use wood plugs to fill the holes.

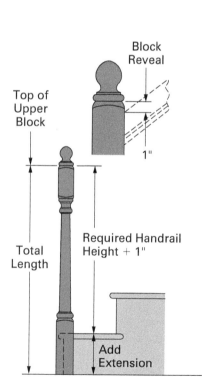

Figure 46 Post-to-post newel height.

Combination Open/Closed Stairway Balustrade

The open section of a combination stairway often has a knee wall, which is a short, non-structural section of wall used to conceal the open space beneath the stairway. Depending on the type of balusters being installed, a shoe rail, or finishing strip, may be fastened to the rake of the knee wall as a mounting base for the balusters. Check the approved construction documents to determine the appropriate construction for the stairway and balustrade system.

The stairway in this example has a knee wall. Construct the balustrade system as follows:

Step 1 After cutting the newel post (check local codes), put it against the knee wall; then plumb the post and drill three pilot holes through it and into the knee wall.

Step 2 Drill three 1" holes $\frac{3}{4}$" deep at the pilot hole.

Step 3 Secure the newel post to the knee wall using lag bolts and washers.

Step 4 Using 1" wood plugs (obtained through the manufacturer), match the wood grain in the plugs with the grain of the newel post, add glue to the plugs, and drive them into the hole. Use a block of wood and hammer to drive the plugs.

Step 5 If it is a post-to-post balustrade, cut a half-newel post to the correct height and install it at the upper end of the knee wall. Plug any fastener holes as necessary.

Step 6 Use a wood chisel to trim excess wood from the plugs. Sand with a power sander.

This procedure used a starting newel that made a post-to-post connection; that is, the handrail butts into the newels. The other option is an **over-the-post** connection, which is outlined next.

Over-the-post: A balustrade system that utilizes fittings to go over newel posts for an unbroken, continuous handrail.

The installation of the starting newel post is the same as previously described. When mounting the handrail, the railing is mounted against the wall with a rosette instead of a half-newel post.

Trim and mount the handrail for either type of balustrade system, but do not secure it. Then determine baluster spacing as follows:

Step 1 Use the same method as used for balcony balustrades. Lay out the dimensions on the floor under the knee wall using a plumb bob to establish and mark the initial dimensions that are required (*Figure 47*).

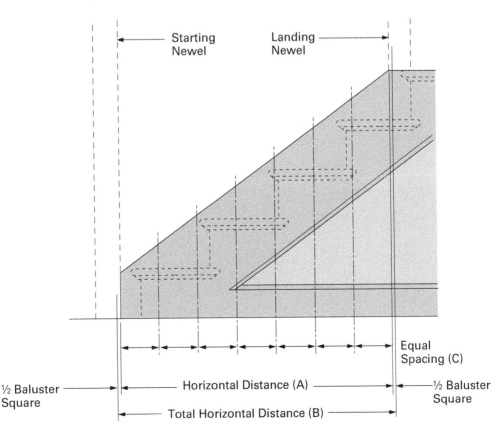

Figure 47 Baluster spacing schematic for knee wall.

Step 2   After the spacing is determined, transfer the spacing marks back to the top of the knee wall using a plumb bob and square. Then, remove the handrail and clamp it to the knee wall in the proper relationship to the newels.

Step 3   Using a combination square, transfer the marks across the bottom of the handrail and the plow of the shoe rail.

Step 4   If pin-top balusters are used, drill the railing using a $\frac{3}{4}$" bit.

Step 5   The bottom of the balusters must be cut to be used on the knee wall. Use a pitch block to mark the angle on the balusters. If a bottom pin is used on the balusters, drill vertical holes in the shoe rail.

Step 6   Measure the length of each baluster and cut.

Step 7   Put a small amount of slow-setting glue on the end of each baluster and insert the balusters into the handrail. If bottom pins are used, insert the balusters into the shoe rail. While the glue for the balusters is still wet, fasten the handrail to the newel posts.

Step 8   After securing the handrail and while the baluster glue is still wet, make sure that the balusters are firmly seated in the correct position on the shoe rail. If bottom pins are not used, or if desired, drill pilot holes at the base of the balusters and secure them to the shoe rail with finish nails.

Step 9   If the baluster is loose at the rail, drill a pilot hole through the baluster into the rail, and secure it with a finish nail.

Step 10   Nail or glue fillet inserts between the balusters on the shoe rail.

Quick-Connect Balustrade Parts

Some manufacturers offer lines of quick-connect parts for most over-the-post and some post-to-post balustrade systems. These parts have various predrilled bolt holes and assembly pockets that eliminate much of the close-tolerance, time-consuming drilling required for assembly of standard balustrade parts. The parts have multiple pockets to allow cutting the parts for various mating angles with handrails, upeasings, and other fittings.

After finishing the balustrade system, install a wall rail to the top of the stairs.

Another version of the open/closed stairway is the partially open stairway (*Figure 48*), which is finished in the same fashion as the open stairway.

Figure 48 Partially open stairway.

1.0.0 Section Review

1. Which tool should be used for cutting stair stock?
    a. Table saw
    b. Saber saw
    c. Power compound miter saw
    d. Circular saw

2. Which of the following is *true* of combination stairways?
    a. They must have a skirt board on both sides.
    b. They must have a landing.
    c. They have one side open, and one side closed.
    d. They can only use a skirt board on one side.

3. What is required on any stairway that includes an open walkway at an elevated height, such as a landing or balcony?
    a. Baluster
    b. Half-newel
    c. Wall-rail bracket
    d. Guard

2.0.0 Custom Stair Systems

Objective	Performance Tasks
Describe the procedure for installing custom stairway systems. a. Outline the procedure for installing shop-built stairways. b. Describe the procedure for installing winder treads.	There are no Performance Tasks in this section.

Custom stairway systems include shop-built and winder stairways. The layout and construction of custom stairways offer greater challenges to carpenters.

2.1.0 Shop-Built Stairways

Some millwork companies now fabricate stairways. A shop-built stairway consists of two housed stringers. The housed stringers are usually made from $5/4$ stock. These members are routed on specialized equipment or by using a template and router. After the stringer stock is routed, it will resemble *Figure 49*. Wedges are made to secure the treads and risers. The wedges start at $5/8$" and taper down to a fine edge.

Wedges: Triangular-shaped wood pieces that are coated with adhesive and used to drive treads and risers tightly into a routed stringer.

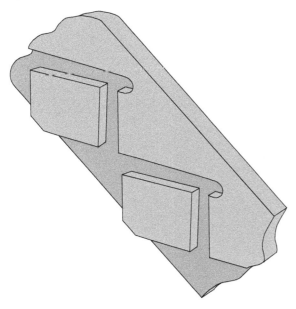

Figure 49 Housed stringer.

Shop-built stairways are usually built on an assembly table. Glue is placed in the tread dadoes and the treads are slid into place. The risers are installed in the same manner. Small blocks are then glued and tack-nailed or stapled to the back of the risers and the bottom of the treads, as shown in *Figure 50*.

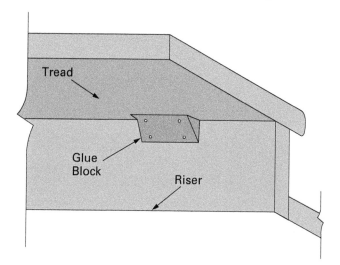

Figure 50 Placement of a glue block.

Manual Stair Jig

A simple jig that can be used with a handheld power router is available. It can be used to create housed stringers, especially in the restoration of old stairways where the stringer may be of special dimensions or made of a wood not commonly available in millwork stairs.

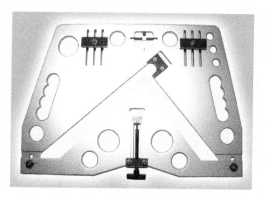

To ensure a tight fit and a squeak-free stairway, some manufacturers mortise the riser and tread together, as shown in *Figure 51*. These risers and treads are installed in the same manner, except the risers are stapled or screwed to the treads. Wedges and glue blocks are also used (*Figure 52*).

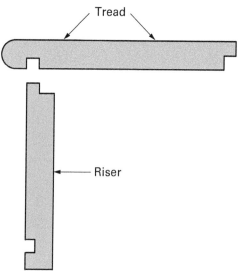

Figure 51 Mortised riser and tread.

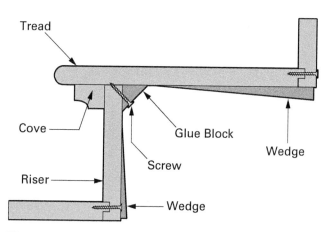

Figure 52 Housed tread and riser connections.

Figure 53, Figure 54, and *Figure 55* show the construction details of a portion of a typical shop-built, partially open, straight stairway made of oak stock that was machine-cut and then glued and stapled together. The assembly of this type of stairway is usually made ½" smaller in width than the finished opening of the stairwell. Once the stairway is slid into place, cove molding will be used to cover any space between the housed stringers and the finished walls.

Figure 53 Bottom view of a partially open shop-built stairway.

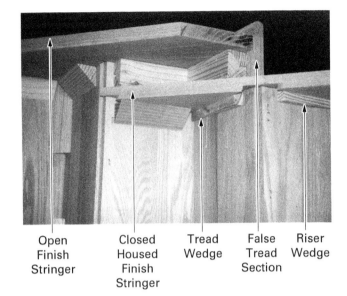

Open	Closed	Tread	False	Riser
Finish	Housed	Wedge	Tread	Wedge
Stringer	Finish		Section	
	Stringer			

Figure 54 Closed-to-open stringer transition.

Figure 55 Completed open/closed stairway.

When shop-built stairs are used, the construction of the stairwell and surrounding area must be precisely and accurately done; there is no room for error. Because shop-built stairways are installed in the finish (trim out) portion of construction, any error in the framing (rough portion) will be obvious and could cause a major problem in trying to install the shop-built stairway. If a job is going to use shop-built stairways, the carpenter should take the following steps prior to gypsum-board installation:

- Check the walls of the stairwell with a straightedge and level.
- Check the finish floor-to-finish floor measurements.
- Specify in writing to the stairway manufacturer the rough dimensions and finished dimensions of the opening.
- Specify in writing what the total run of the stairway must be.
- If possible, send detail drawings, elevations, and sectional drawings of the stairway to the stair manufacturer.

Specifications for the shop-built stairs should not be sent to the manufacturer until they are verified with field measurements from the construction site. Provide as much information as possible about the stairway.

In today's construction industry, where quick production is a major factor, some carpenters may inadvertently make an error in framing that could go unnoticed until the final stages of construction. An example of this is the installation of a double header at the stair opening that is not set properly. Another example is the walls of the stairwell being out-of-plumb and out-of-square. Either error would cause problems and prevent proper installation of the shop-built stairway. The errors would have to be corrected prior to installation of the stairway.

2.2.0 Winder Stairways

NOTE

A winder stairway should be used only when the construction of other types of stairways is impractical.

Most stairway construction incorporates a straight stairway, or L-shaped or U-shaped stairways with landings. Where stair space is further restricted, risers and treads are built on a landing. This is called a *winder stairway*. Winder stairways turn from 90° to 180° from one level to another. Winder stairways can be built as open, closed, or combination stairways (*Figure 56*).

Figure 56 Winder stairway.
Sources: EricVega/Getty Images (left); TerryJ/Getty Images (right)

Winder Treads

The *IBC*® and *IRC*® have different requirements for winder treads.

IBC Section 1011.5.2: Winder treads shall have a minimum tread depth of 11" between the vertical planes of the foremost projection of adjacent treads at the intersection with the walkline and a minimum tread depth of 10" within the clear width of the stair.

IRC Section R311.7.5.2.1: Winder treads shall have a tread depth of not less than 10" measured between the vertical planes of the foremost projection of adjacent treads at the intersection with the walkline. Winder treads shall have a tread depth of not less than 6" at any point within the clear width of the stair. Within any flight of stairs, the largest winder tread depth at the walkline shall not exceed the smallest winder tread by more than $\frac{3}{8}$" (9.5 mm). Consistently shaped winders at the walkline shall be allowed within the same flight of stairs as rectangular treads and shall not be required to be within $\frac{3}{8}$" (9.5 mm) of the rectangular tread depth.

The best method of establishing the correct dimension and shape of the winder is to lay it out full size on sheets of heavy paper or cardboard as shown in *Figure 57*; this design is used to make the winder treads shown in *Figure 58*.

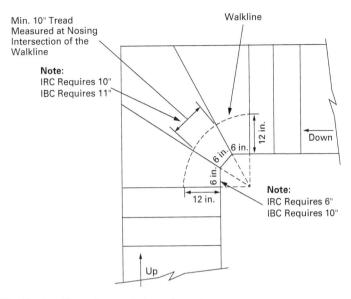

Figure 57 Winder dimension and shape layout.

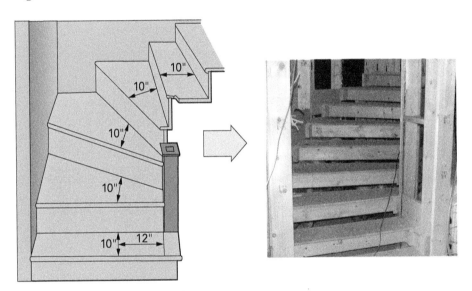

Figure 58 Actual winder treads.

Figure 59 is an example of the arrangement of winders in a U-shaped stairway. The diameter of the U is given, but the construction at this point could also be square, as shown by the dashed lines. The circular line illustrates the method of

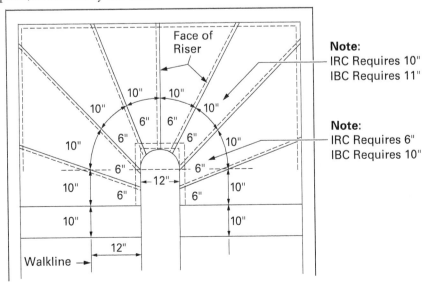

Figure 59 Width of winder treads.

spacing the risers at the inside of the turn. In this type of stairway construction, only a certain number of risers may be included in the half-turn. The method of spacing the risers is similar to the L-shaped stairway. It is important to establish a walkline and maintain an equal tread width through the winding area as in a straight flight.

2.2.1 Constructing Stairways with Winder Treads

When building a stairway with winder treads, the landing is built first. The stairway construction is as follows:

Step 1	Establish the exact distance from the top of the lower floor level to the top of the floor above (total rise). Lay out this distance on a story pole.
Step 2	Using dividers, space the number of risers equally over the distance marked on the pole.
Step 3	Locate the stringer position for the flight above the winders. Cut and install the stringers.
Step 4	Establish the height of the landing. Be sure that the top of the landing is at the height of the first winder tread. Build the landing and secure it in place.
Step 5	Once the landing is securely nailed, lay out the location of the winders on the landing. Allow for any subsequent wall finish and/or skirt board.
Step 6	Build a framework to represent the outline of the winders. Stay back 1" from the winder layout line when installing this framework. This 1" will be the space required for the finished stringer.
Step 7	Measure up on the studs the height of one riser minus the thickness of a winder tread.
Step 8	Level the framework at the marks established on the studs and nail securely.
Step 9	Follow the same procedure for the other winder treads. Be sure to conform to the winder layout on the landing.
Step 10	Brace the winder landings by running cross bracing where the spaces are more than 16" in the framework of the riser landings.
Step 11	Locate the stringer position for the flight below the winders. Cut and install the stringers.
Step 12	Install the risers and treads for all steps.

2.0.0   Section Review

1. During which portion of construction are shop-built stairs installed?
   a. Rough framing
   b. Interior work
   c. Trim out
   d. Dry in

2. What type of stairway is used when stair space is restricted?
   a. U-shaped
   b. L-shaped
   c. Combination
   d. Winder

3.0.0 Exterior Stairways

Objective

Summarize the process of building exterior stairways.
  a. Describe the procedure for building exterior wood stairways.
  b. Explain the procedure for constructing cast-in-place forms for concrete stairways.

Performance Task

5. Build a cast-in-place form for a concrete stairway.

Many structures require one or more exterior stairways. Some stairways may have only one or two steps, while others may have enough steps to connect different levels of a structure. There are a variety of ways to construct exterior stairways depending on the design of the building and the purpose of the stairway. This section will focus on exterior wood stairways and cast-in-place forms for concrete stairways.

3.1.0 Exterior Wood Stairways

The construction of exterior wood stairs is similar to that of interior stairways. However, when building exterior wood stairs, moisture and weather are two problems that must be taken into consideration. All material used on exterior wood stairs should be preservative treated or naturally durable wood. Many wood exterior stairs have no risers; therefore, they are considered open riser stairs. The tread material is usually made of 2" nominal stock, such as 2 × 6, 2 × 10, or 2 × 12 lumber.

Most exterior wood stairs are built on either a concrete footing or concrete base with a clearance of at least 2" from the finished grade to the top of the concrete base. When preparing to build a concrete base to receive exterior steps, the footing must be at least 12" deep or below the frost line.

On exterior stairs, treads should have a slight slope to provide for water run-off and drainage. The landing or deck should also have a slight slope for the same reason. *IRC Section R311.7.7* requires the walking surface of treads and landings to have a slope not steeper than 2% (1 unit vertical in 48 units horizontal). The exception to this is when code requires the surface of a landing to drain surface water. In this case the slope of the walking surface of the landing should not be more than 5% (1 unit vertical in 20 units horizontal) in the direction of travel. A simple way to build a slope into exterior stairs is to cut the slope into the bottom of the stringer.

Exterior wood stairs that are constructed out of stair stock and have no slope in either the tread or landing must be spaced to allow rain to pass through. The basic procedure used to cut a stringer generally applies to exterior wood steps from decks or porches to the ground level.

A landing may be required for wood stairs at an exterior door. Exterior stairs going to a basement may also need a landing outside of the exterior door. Handrails shall be provided on at least one side of each flight of stairs with four or more risers. Always consult the approved construction documents and local building codes before constructing exterior stairways.

The bottom of the stair stringers should be connected to a footing. One example of this connection is a preservative-treated 2 × 4 fastened across the area where the stringer will be located. The stringer is notched to fit over the 2 × 4 to prevent movement. A typical set of exterior stairs is shown in *Figure 60*.

NOTE

Be sure to check the safety data sheet (SDS) and follow applicable safety precautions when handling, cutting, and disposing of preservative-treated lumber.

Open Riser Stairs

According to *IBC Section 1011.5.3* and *IRC Section R311.7.5.1*, the distance between treads on an open riser stairway generally cannot be more than 4" (102 mm). There are exceptions in the *IBC®* that allow for unrestricted riser height openings.

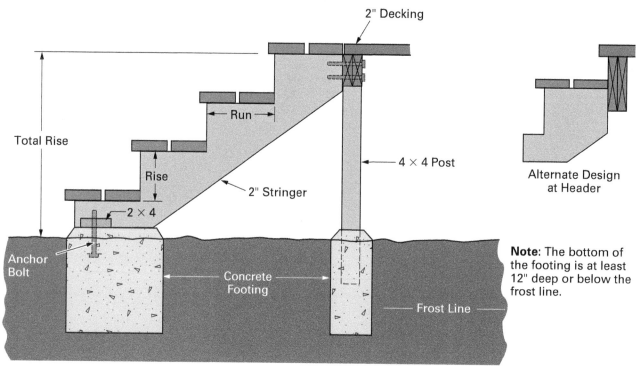

Figure 60 Exterior stairs.

3.2.0   Concrete Stairways

Most contractors will subcontract for any stair work that is not related to the carpentry trade. A carpenter will normally construct cast-in-place forms (*Figure 61*) for concrete stairways and may even pour and finish the concrete on some projects.

NOTE

Check applicable codes for the requirements for concrete stairs in your jurisdiction.

NOTE

Stringers for cast-in-place forms for concrete stairs may also be called *side forms*. They are located where cut stringer would be in a wood stairway, but the "side forms" will be removed once the concrete sets.

Figure 61 Cast-in-place form.
Source: Aleksei Ignatov/Shutterstock

The procedure used to calculate the tread and riser dimensions based on the total rise and run of the stairway is also used for concrete stairways. To construct a cast-in-place form for a concrete stairway, follow these steps:

Step 1 Determine the total rise, total run, and width of your stairway.

Step 2 Calculate the number of risers you will need. Landing steps at the top and bottom of the stairway may have a larger tread depth than the rest of the stairs. Take this into account in your calculations.

Step 3 Determine the length of your stringers.

Step 4 Stake out the base of your steps by hammering a wooden stake into the ground at each corner of the base of your stairs. Run a length of string around the perimeter of the stakes to make sure they are aligned and square.

Step 5 Excavate the area for the base of your steps. The approved construction documents will specify the depth required for the base.

Step 6 Lay a subbase and tamp it down so it is very tightly packed and level. The approved construction documents will specify the material required for the subbase.

Step 7 Cut the stringers and risers for your form using 2 × 6 or 2 × 8 boards, or even wood structural panels. Scrap wood or low-grade lumber is commonly used for this.

Step 8 Set the stringers in place. They can be attached to the structure and the wooden stakes. Angled braces can also be used to prevent the form from bowing. If stringers are not to be used and the concrete stairs will be connected to a wall, draw the stringer outline onto the wall and fasten risers to the wall.

Step 9 Attach the risers to the stringers. Depending on the width of the stairway, braces may be attached to the risers to prevent bowing. For a closed stairway, risers may be attached to the walls instead of stringers.

Step 10 Check the approved construction documents to determine details on guards, handrails, or any other attachments. If the stairway requires a guard and handrail that is intended to be set in the concrete, place the guard and handrail assembly inside the formwork perimeter and stabilize it prior to the concrete pour. Make sure the guard assembly is plumb and that the handrail is at the required height and includes the proper extensions at the top and bottom of the stairs. (See previous information in this module regarding guard and handrail regulations.)

Step 11 Check the form for square, level, and plumb. It may have a slight slope to allow water to run off the steps.

Step 12 If the project calls for concrete reinforced with rebar, place the rebar in the form.

Step 13 Once the concrete is poured and has set long enough to hold its own weight, the forms can be removed.

Another option is to use preformed concrete stairs (*Figure 62*). These stairs come in many different dimensions and are delivered to the jobsite ready to be installed.

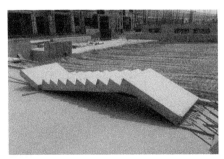

Figure 62 Preformed concrete stairs.
Sources: Pictures-and-Pixels/Getty Images (left); Another77/Shutterstock (right)

3.2.1 Stair-Tread Facing and Nosing

A great variety of safety and anti-wear tread facings and nosings are available from many manufacturers for use on wood or concrete exterior stairways. *Figure 63* provides a small sample of the available treads and nosings. They are available in materials that resist extreme wear and corrosive conditions. They are also available with several mounting options. Some are removable and replaceable from anchors that are embedded in concrete. Others have integral anchors and are embedded when the concrete is placed and finished. Still others are fastened to the stair surface using concrete or wood screws.

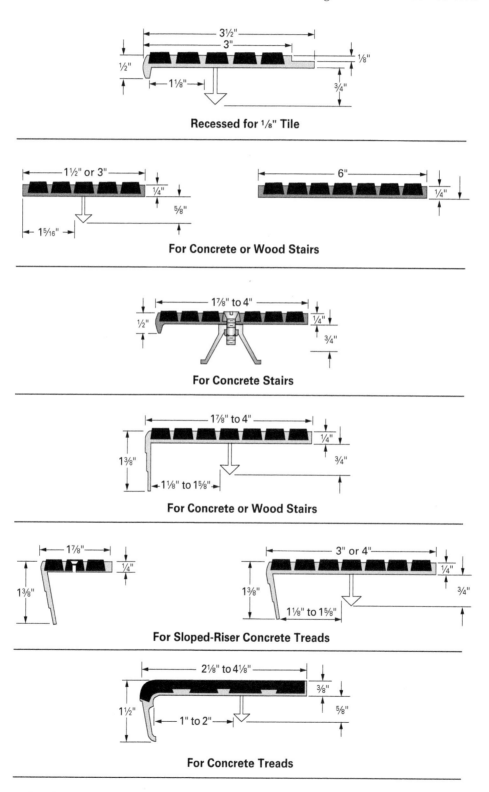

Figure 63 Tread facing and nosing.

3.0.0 Section Review

1. Which of the following is a major problem to consider when building exterior wood stairways?
   a. Rise and run
   b. Handrail installation
   c. Fasteners
   d. Moisture

2. What measurement does a carpenter need, along with total rise and total run, to build a cast-in-place form for a concrete stairway?
   a. Volume
   b. Width
   c. Depth
   d. Area

Module 27405 Review Questions

1. What is the distance a riser or tread template will extend to?
   a. 12" to 22"
   b. 24" to 36"
   c. 30" to 48"
   d. 36" to 50"

2. How much shorter than other risers should the first riser be?
   a. The thickness of the finish floor
   b. The thickness of a tread
   c. Twice the thickness of a tread
   d. The thickness of the nosing

3. How far from the face of the riser should the finish nails be?
   a. $\frac{1}{8}$"
   b. $\frac{1}{4}$"
   c. $\frac{3}{8}$"
   d. $\frac{1}{2}$"

4. When must a landing tread be used?
   a. On all stairways
   b. When the last step is set even with floor level
   c. On L- and U-shaped stairways
   d. When the last step is set below floor level

5. Which method of connecting stair treads to risers has a groove cut along the top portion of the riser and a groove cut in the bottom of the tread?
   a. Wedge
   b. Butt joint
   c. Pitch block
   d. Rabbet joint

6. At what angle should the miter cuts be made for treads and tread nosings for an open stairway?
   a. 15°
   b. 30°
   c. 45°
   d. 60°

7. What is required along the inside bottom edge of the riser to anchor the front of the riser on a bullnose starting step?
   a. Pitch block
   b. Glue block
   c. Wedge
   d. Nosing

8. Which of the following is true of L- shaped stairways?
   a. They have a 90° turn with a landing.
   b. They are closed stairways.
   c. They can have a maximum of five stairs to the landing.
   d. They are open stairways.

9. What is the minimum distance from the wall the *IBC*® and *IRC*® require for a handrail?
   a. 1"
   b. $1\frac{1}{2}$"
   c. $1\frac{3}{4}$"
   d. 2"

10. How is the number of balusters required for a stairway determined?
    a. Divide the rise by the maximum spacing allowed by code.
    b. Multiply the number of treads by 2.
    c. Divide the run by the maximum spacing allowed by code.
    d. Multiply the run by 0.5.

11. What are used to secure the treads and risers on shop-built stairways?
    a. Glue blocks
    b. Wedges
    c. Pitch blocks
    d. Newel posts

12. What type of fastener is used when installing treads and risers on shop-built stairs?
    a. Glue
    b. 3d finish nails
    c. Rail bolts
    d. 6d finish nails

13. What type of stringers do shop-built stairways have?
    a. Housed
    b. Rough
    c. Finish
    d. Skirt board

14. What is used to cover any space between the housed stringer and the finished wall when a shop-built stairway is installed?
    a. Baluster
    b. Cove molding
    c. Skirt board
    d. Gypsum board

15. What degree of turn can winder stairs have from one level to another?

    a. 45° to 60°

    b. 75° to 80°

    c. 90° to 180°

    d. 200° to 360°

16. What minimum tread depth does the *IRC*® require at the walkline on a winder stairway?

    a. 6"

    b. 8"

    c. 10"

    d. 11"

17. How deep must the footing be for a concrete base that will receive exterior steps?

    a. At least 8" deep or below the frost line

    b. At least 10" deep or below the frost line

    c. At least 12" deep or below the frost line

    d. At least 15" deep or below the frost line

18. What is the maximum slope allowed by the *IRC*® for the walking surface of treads and landings on exterior stairs?

    a. 2%

    b. 4%

    c. 6%

    d. 8%

19. What is the minimum required clearance from the finished grade to the top of the concrete base for exterior wood stairs?

    a. 1"

    b. 2"

    c. 3"

    d. 5"

20. When can a cast-in-place form be removed from concrete stairs?

    a. After the concrete has cured

    b. Within 24 hours of the concrete being poured

    c. When the concrete pulls away from the form

    d. Once the concrete can hold its own weight

Answers to Odd-Numbered Module Review Questions are found in *Appendix A.*

Answers to Section Review Questions

Answer	Section	Objective
Section One		
1. c	1.1.0	1a
2. c	1.2.2	1b
3. d	1.3.2	1c
Section Two		
1. c	2.1.0	2a
2. d	2.2.0	2b
Section Three		
1. d	3.1.0	3a
2. b	3.2.0	3b

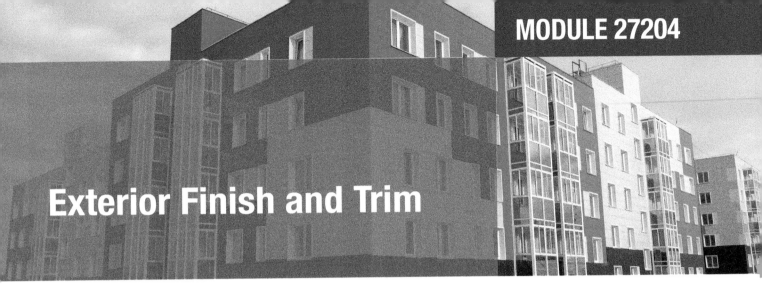

MODULE 27204

Exterior Finish and Trim

Objectives

Successful completion of this module prepares you to do the following:

1. Explain the common hazards you must identify and mitigate in your site safety plan when installing exterior trim and finish materials.
    a. Describe common safety hazards of working from ladders, elevated platforms, and aerial lifts.
    b. Discuss common safety hazards of exterior finishing tools and materials.
2. Explain the steps to prepare the exterior for siding installation.
    a. Discuss which construction documents will provide useful information for the exterior.
    b. Explain the basic concepts of performing a material takeoff for siding materials.
    c. Discuss common types of insulation, flashing, furring, and building wrap that might be applied to exterior walls.
3. Discuss common types of exterior siding materials and their installation processes.
    a. Describe the basic principles of exterior siding layout.
    b. Discuss common siding styles and their installation methods.
    c. Identify exterior trim components including soffits, fascia, and cornices.

Performance Tasks

Under supervision, you should be able to do the following:

1. Perform a wall inspection, and apply flashing, furring, insulation, and building wrap to prepare the wall surface for exterior siding.
2. Lay out and install three of the most common siding types in your area.

Overview

There are a wide variety of finishing materials used to cover the exteriors of homes and commercial buildings, including wood, brick, vinyl, metal, and fiber-cement. Each type of material has its own preparation requirements and installation practices. These materials produce an aesthetic quality, but they also protect the building from the elements and seal the building at roof overhangs with soffit and fascia. Always follow the engineer's and manufacturer's instructions to maintain warranties and improve the longevity of your project.

Digital Resources for Carpentry

Scan this code using the camera on your phone or mobile device to view the digital resources related to this craft.

SCAN ME

1.0.0 Exterior Finishing Safety

Performance Tasks

There are no Performance Tasks in this section.

Objective

Explain the common hazards you must identify and mitigate in your site safety plan when installing exterior trim and finish materials.

a. Describe common safety hazards of working from ladders, elevated platforms, and aerial lifts.

b. Discuss common safety hazards of exterior finishing tools and materials.

Safety should be everyone's top priority on a construction site, and it is every craft professional's duty to ensure they and their team use common sense and are constantly aware of their surroundings. Always wear appropriate high-vis work attire and all required personal protective equipment (PPE) (*Figure 1*). Follow all applicable Occupational Safety and Health Administration (OSHA) standards, as well as local and national building codes. A job hazard analysis and fall protection work plan are necessary to perform exterior finishing, but you should immediately address new hazards as they arise on the jobsite.

Figure 1 Workers wearing personal protective equipment (PPE).
Source: MNBB Studio/Shutterstock

Since the installation of exterior finishing materials takes place in the elements, your team should always be aware of rapidly changing weather conditions. Extreme wind, sun, snow, ice, and rain conditions can all increase the risk of potentially fatal injuries. When carrying large pieces of exterior siding, consider wind direction. Wind striking the flat surface of these materials could cause you to lose your footing on an elevated work platform, leading to serious injuries or death. Strong winds can also carry away unsecured ladders and scaffolding, causing severe equipment damage and injury to people working below.

1.1.0 Working at Elevations

Falls from elevated surfaces are one of the leading causes of death among craft professionals. When installing exterior finishing materials, craft professionals are commonly required to work above ground level on ladders, scaffolds, aerial lifts, or other elevated work platforms. OSHA *Subpart M* requires fall protection for platforms or other work surfaces with unprotected sides or edges that are 6' or higher above the ground or the level below it.

However, some state OSHA regulations and company policies may require fall protection and fall arrest systems (*Figure 2*) for heights less than 6'. Identify the task-specific fall protection guidelines while you develop your daily safety plan, and ensure that all team members are aware of potential hazards. Every worker on-site is a safety professional, and it is everyone's duty to identify and proactively address safety concerns before an accident occurs.

Figure 2 Worker wearing lanyard-style fall arrest system with full-body harness.
Source: King Ropes Access/Shutterstock

1.1.1 Ladders

Ladders are used to install exterior finishing materials and accessories at elevations. Reduce the risk of falls or equipment failure by carefully inspecting a ladder before use and instructing all workers on proper ladder safety. Check the rungs and rails for cracks or other damage, and take the ladder out of service if they fail inspection. OSHA requires regular inspections of ladders before each use. General guidelines for ladder safety follow:

- All ladders must be inspected and approved by a competent person, trained in the proper use and maintenance of a ladder.
- All safety labels must be present and legible.
- Do not stand on or above the second uppermost of a stepladder (*Figure 3*) or above the fourth rung from the top of an extension ladder. This is one of the most common ladder safety problems you will see in the field, and it can often lead to dangerous falls. Take the extra time to get a taller ladder instead of putting yourself at risk.
- Ensure the ladder is on a firm and stable footing before climbing. Never elevate the ladder base with unsecured waste (or "donage") and avoid setting a ladder up near slopes or excavation.
- Keep your body's center of gravity between the ladder rails when climbing. You may hear the term "belt-buckle rule" in the field to describe this practice. This helps workers remember that your center of gravity, or the area where you would place a belt buckle, should never extend beyond the outer rails of a ladder.

Figure 3 Improper use of step ladder.
Source: Rachid Jalayanadeja/Shutterstock

- Secure extension ladders at the base level and top platform of the structure for permanent entry or emergency exit and egress from elevated levels. The side rails of these permanent ladders should not extend more than 3' above the walking surface of the top platform and fall protection systems and safety gates should be installed properly around the entire perimeter of the top level before a competent person approves a permanent ladder for general use.

Duty rating: Load capacity of a ladder.

- Do not exceed the **duty rating** of the ladder. See *Table 1*.

TABLE 1 Ladder Duty Ratings

Duty Ratings	Load Capacities
Type IAA	375 lb, extra heavy duty/professional use
Type IA	300 lb, extra heavy duty/professional use
Type I	250 lb, heavy duty/industrial use
Type II	medium duty/commercial use
Type III	200 lb, light duty/household use

- No more than one person is permitted on a ladder at a given time.
- Place an extension ladder at a 75-degree angle. The bottom of an extension ladder should be set back 1' for every 4' of ladder length. Tie off the ladder when it is in the proper position. A simple trick to ensure your ladder is at the proper angle is to stand straight with your toes touching the bottom cleats. If you cannot touch the side rails when reaching forward, your ladder is set up at an unsafe, shallow angle.
- Always look up for overhead hazards before climbing a ladder, and avoid setting up your ladder near power lines.
- Climb a ladder while facing it, and maintain three points of contact (*Figure 4*). Use ropes to lift tools and handheld materials from the ground to upper levels to keep both hands free to maintain the three points of contact as you climb the ladder.

A complete discussion of ladders and ladder safety is included in NCCER Module 00101, *Basic Safety (Construction Site Safety Orientation)*.

Figure 4 Worker using three points of contact while climbing a ladder.
Source: Glasshouse Images/Alamy

Ladder Jacks and Walk Boards

Ladder jacks are triangular brackets that attach to portable ladders and support each side of a ladder jack scaffold—creating a stable, low-elevation, horizontal working platform for craft professionals. These devices are available in single-, double-, and triple-rung bracket variations for accommodating OSHA-approved planking and walk boards.

Secure both vertical extension ladders at the top and bottom at a 4-to-1 angle, as mentioned in the previous ladder guidelines. Inspect the ladders for damage, and investigate the area for trip hazards and overhead power lines before beginning installation. Note that OSHA regulations forbid ladder jack scaffolding to support a platform higher than 20', and ladder jack scaffolding is not intended for working loads higher than 25 lbs per square foot. No more than two workers may use the platform at the same time, and the span between vertical support ladders must not exceed 8'. Any craft professionals working above 6' must also wear fall arrest systems unless an authorized guardrail system is in place.

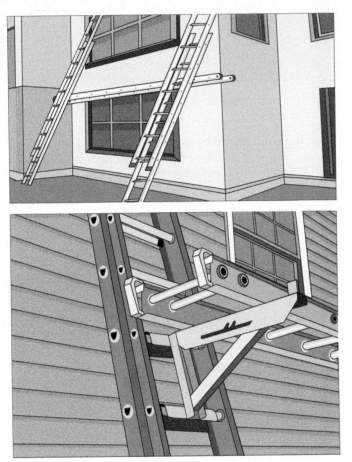

1.1.2 Scaffolds

Scaffolds provide safe elevated work platforms for craft professionals and materials. Scaffolds are designed and constructed to support a specified load, but heavy materials, frequent use, and outdoor weather conditions can weaken them over time, making them less stable and unsafe. That's why it is always important to have a competent person inspect and clear all scaffolds before using them each day (*Figure 5*). Even after the scaffold is green-tagged and ready for use, you should always be vigilant and check for bent, broken, or rusted tubes, or loose connections that may have been missed during inspection. If you find any of these issues or have a question, immediately bring it to the attention of the competent person and your supervisor.

Figure 5 Competent person inspecting scaffold.
Source: Lynne Sutherland/Alamy

Adjustable aluminum scaffold systems (*Figure 6*) are common for providing a working platform when installing exterior finishing materials on large buildings. Exact specifications on spacing dimensions, planking, permissible heights and loads, and other details are provided in OSHA regulations. All scaffolds must be placed on a firm and level footing. Per OSHA, if a scaffold is more than 6' high, it must be equipped with top rails, midrails, and toeboards, unless a personal fall arrest system is utilized. However, some company-specific safety rules may require fall protection systems at heights above 4'.

Figure 6 Adjustable aluminum scaffold system.
Source: imageBROKER.com GmbH & Co. KG/Alamy

Aside from sectional, freestanding scaffolds that can be assembled in various configurations, continuously adjustable aluminum scaffolding systems are available for cornice and siding working heights of up to 50'. Typically, this type of OSHA-recognized scaffolding system consists of aluminum poles and a standing platform assembly with a safety railing or safety net. The standing platform can be raised and lowered as an assembly on the supporting poles to achieve an optimum working height. Most systems may be joined both vertically and horizontally for improved security and portability.

For more information, a complete discussion of scaffolds and scaffold safety is included in NCCER Module 00101, *Basic Safety (Construction Site Safety Orientation)*.

1.1.3 Aerial Lifts

Aerial lifts are common equipment that raise and lower workers to and from elevated locations. There are two main types of lifts: boom lifts and scissor lifts (*Figure 7*).

Figure 7 Boom lift and scissor lift.
Sources: RJH_IMAGES/Alamy (left); Zigmunds Dizgalvis/Shutterstock (right)

Boom lifts have a single arm that extends a work platform/basket to higher elevations. Some models have a jointed (articulated) arm that allows the work platform to be positioned both horizontally and vertically. Scissor lifts raise a work enclosure vertically by means of crisscrossed supports that can move personnel, tools, and materials up and down from a parked position. These lifts are powered by electric and hydraulic motors and fueled by gasoline, propane, or diesel engines.

OSHA regulations require all boom lift operators to wear a full-body harness and lanyard attached to the boom or basket. As with any other form of specialized equipment, all operators must be trained and certified on a boom or scissor lift, and they should carry a copy of certification documentation on their person whenever operating the equipment.

Remember that the other safety precautions previously discussed in this module also apply to aerial lifts. Each manufacturer provides specific safety precautions in the operator's manual. Specifically, OSHA CFR (*Code of Federal Regulations*) *1926.453* defines and governs the use of aerial lifts as expressed in the following precautions:

- Avoid using the lift outdoors in stormy weather or in strong winds.
- Cordon off areas around and below elevated work platforms with safety rope, caution tape, and clear signage to prevent people from walking beneath the platform.
- Use a personal fall arrest system with approved anchorage points as required for the type of lift you're using.

- Never use an aerial lift on uneven ground.
- Lower the lift before moving the equipment; shut off the engine, set the parking brake, and remove the key before leaving the lift unattended.
- Stand firmly with both feet on the floor of the basket or platform. Do not lean over the guardrails of the platform, and never stand on the guardrails. Do not sit or climb on the edge of the basket or use planks, ladders, or other devices to gain additional height.
- Never climb out of an aerial lift or use it to reach an elevated platform, especially one without proper fall protection in place (*Figure 8*).

Figure 8 Unsafe use of a boom lift.
Source: Roy LANGSTAFF/Alamy

1.2.0 Tools, Equipment, Materials, and Related Safety

Always use the most efficient tool for each exterior siding and finishing task, and follow manufacturer guidelines to ensure you are using the tool as it was intended. Maintain all power tools, and keep necessary safety guards in place. The following section describes a few common tools that craft professionals use to install exterior finishing materials.

1.2.1 Hand and Power Tools

Power staplers, power saws, tin snips, utility knives, claw hammers, and pneumatic- and battery-operated nailers are a few of the most common tools you will use to cut, fasten, and install siding materials in the field (*Figure 9*).

| Finish Nailer | Finish Stapler | Circular Saw | Tin Snips | Utility Knife |

Figure 9 Siding installation tools.
Sources: Courtesy of Stanley Black & Decker, Inc. (left to right labels: finish nailer, finish stapler, circular saw, utility knife); Klein Tools (tin snips)

Carbide-tipped swivel head shears (*Figure 10*) are a common tool for cutting fiber-cement siding. Using power shears also minimizes the amount of toxic silica dust compared to that produced from cutting the material with a power saw. A power saw using a fine-toothed, carbide-tipped, or dry-diamond circular saw blade, tin snips, or a utility knife with a tungsten-carbide tip may also be used as alternatives. A circular saw is the most common power tool for cutting wood and fiber-cement boards and panels, while individual vinyl and metal panels are cut with tin snips and utility knives.

Figure 10 Swivel head shears.
Source: Courtesy of Stanley Black & Decker, Inc.

Treat all cutting tools with caution, and never remove manufacturer safety guards. Wear cut-resistant gloves, eye protection, and hearing protection when using power tools to cut potentially sharp-edged siding materials, and follow all other general tool safety guidelines found in NCCER Module 00101, *Basic Safety (Construction Site Safety Orientation)*.

1.2.2 Equipment

The following equipment is required when applying metal or vinyl siding:

- *Portable brake* – A portable brake is a useful tool for jobsite bending custom trim sections, such as **fascia** trim, window casing, and sill trim. These machines are lightweight and portable, with various sizes and brake styles available. As shown in *Figure 11*, some are equipped with a lengthwise rolling cutter to allow workers to size trim stock to the desired width.

Fascia: A horizontal board that encloses and protects the front face of projecting eaves.

Figure 11 Portable brake.
Source: valentyn semenov/Alamy

WARNING!

Exercise extreme care when working with a portable brake. A portable brake has many pinch points, which can cause injury.

- *Cutting table* – A cutting table (*Figure 12*) allows a standard circular saw to be mounted in a carrier and held away from the work to avoid damaging the siding. This table can be used for measuring and crosscutting, as well as for making miters and bevels. The table is constructed of lightweight aluminum and can be easily set up on the jobsite by one worker.

Figure 12 Cutting table.
Source: Southern Tool

1.2.3 Material Safety

Some of the most common siding materials are wood (natural and engineered), metal, fiber cement, stone, brick, and stucco. Masons and plasterers install stone, brick, and stucco, while exterior finish carpenters are typically tasked with prepping, cutting, and installing wood, metal, and fiber-cement boards or panels. Each material has its own specific safety issues that should be considered prior to working with it. Always follow general safety precautions, like those discussed in NCCER Module 00101, *Basic Safety (Construction Site Safety Orientation)*, but be aware of site-specific hazards (*Figure 13*) as your team completes safety plans for each phase of work.

Figure 13 Site safety signage.
Source: Justin Kase z12z/Alamy

Observe all safety precautions when cutting certain siding materials, including western red cedar, masonry coatings, treated lumber, and fiber-cement siding. Cutting or mixing such products can produce toxic or allergenic dust. Consult the manufacturer **safety data sheet (SDS)** for any applicable hazards before cutting siding products.

An SDS must accompany every shipment of hazardous substances and be available at the jobsite. These documents are also available online for the project team's records. SDSs are used to manage, use, and dispose of hazardous materials safely. Information found on an SDS includes the following:

- Exposure limits
- Physical and chemical characteristics of the substance
- Types of hazards the substance presents
- Precautions for safe use and handling
- Emergency first-aid procedures
- Manufacturer contact information

Additional information on SDSs is found in NCCER Module 00101, *Basic Safety (Construction Site Safety Orientation)*.

Safety data sheet (SDS): A document with essential information about a hazardous substance.

Western Red Cedar (WRC)

Dust from western red cedar (WRC) is a common allergen that can cause respiratory ailments, including asthma and rhinitis. WRC dust can also cause eye irritation and skin disorders, including dermatitis, itching, and rashes. Wear the required PPE when handling western red cedar siding materials to avoid inhaling dust or exposing your skin and eyes.

Fiber-Cement Board

Proper respiratory protection must be used when working with fiber-cement siding to avoid inhalation of toxic silica dust, which can cause a fatal lung disease called **silicosis**. Many masonry and siding products contain silica, which is one of the most dangerous byproducts in modern construction. Use wet-cutting techniques while wearing an approved respirator, and equip tools with high-efficiency particulate air (HEPA) filtered accessories to minimize the risk of inhaling silica dust and developing silicosis and other life-threatening respiratory conditions later in life (*Figure 14*). Even while wearing an approved respirator during cutting processes, you may also consider making a dust-control plan to minimize exposure to nearby workers and the general public if you are working near high-traffic areas.

Silicosis: A life-threatening lung disease caused by inhaling large amounts of crystalline silica dust.

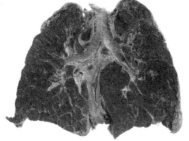

Figure 14 Worker wearing proper PPE, to protect against silica dust exposure, and a human lung with silicosis.

Sources: bigwa11/Shutterstock (top); MicroScape/Science Source (bottom)

1.0.0 Section Review

1. OSHA regulations require fall protection for platforms or work surfaces higher than _____.
   a. 2'
   b. 36"
   c. 6'
   d. 5'

2. Information on the use, management, and disposal of a hazardous substance is contained in the _____.
   a. SDS
   b. owner's manual
   c. HSSS
   d. hazard warning document

2.0.0 Exterior Siding Preparation

Performance Task

1. Perform a wall inspection, and apply flashing, furring, insulation, and building wrap to prepare the wall surface for exterior siding.

Objective

Explain the steps to prepare the exterior for siding installation.
   a. Discuss which construction documents will provide useful information for the exterior.
   b. Explain the basic concepts of performing a material takeoff for siding materials.
   c. Discuss common types of insulation, flashing, furring, and building wrap that might be applied to exterior walls.

The first step of exterior siding and trim installation is familiarizing yourself with the scope of the project and ensuring your team has the proper tools, materials, and information needed to complete all essential tasks. Although many of these preliminary planning tasks, such as wall inspections and ordering materials, will likely fall to your supervisors and project managers, it will be helpful to understand an overview of these basic concepts as you continue your career as a craft professional.

2.1.0 Construction Documents

Looking through the construction documents is one of the first steps for familiarizing yourself with a new project. As you learn how the exterior finishing systems work with one another, you may find specific details that need clarification. Asking the right questions early in the project-planning stage, and proactively following up those questions with requests-for-information (RFIs), will reduce the risk of budget and schedule conflicts later in the project.

With the continuous evolution of construction software and the widespread adoption of laptops, cell phones, and tablets, the construction field is progressing into more convenient and speedier document management and team-to-team communication. Although construction documents still have the same titles and basic appearances, the use of technology improves the construction process from traditional paper plans and redlined RFIs (*Figure 15*).

Every project will have its setbacks, and you will need to work with your team, and likely other trades, to solve these issues. However, the more knowledgeable you are about your scope and how it fits into the big picture, the more equipped you will be to make simple, informed solutions to complex problems. You can find the most essential information you will need to complete an exterior siding installation by studying the manufacturer instructions for selected products and the following approved construction documents:

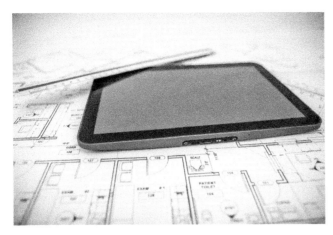

Figure 15 Construction documents.
Source: Zoonar GmbH/Alamy

Site Plan

Since exterior finish and trim work focuses on the outside of the building, it's a good strategy to begin your project planning with a thorough overview of the project site plan (*Figure 16*). A site plan acts as a bird's-eye map describing all site development phases from earthwork to final landscaping. Knowing the workflow of the excavators and other trades will help you work with the project-planning team to optimize your time on-site and reduce trade stacking. A clear, detailed site plan can also improve material logistics and project efficiency by ensuring your team will have proper entry and exit routes for material deliveries and aerial lifts.

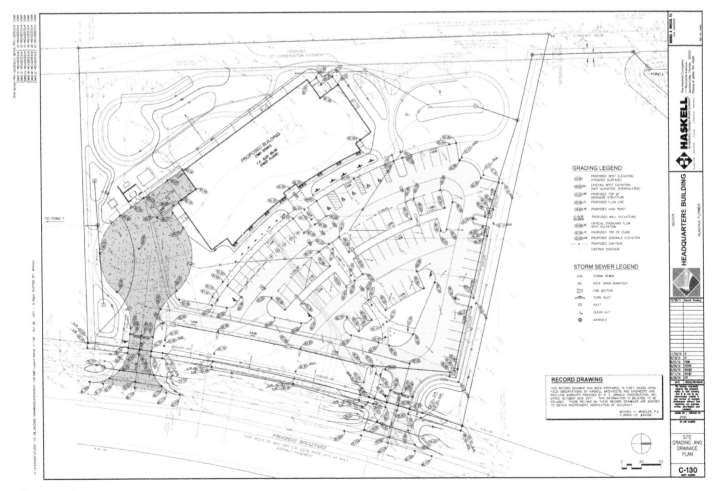

Figure 16 Site plan.

Elevation Drawings

Elevation drawings (*Figure 17*) are representations of the final project as if you were looking straight on toward the various exterior faces, or facades, of the structure. Exterior elevations provide the height of each floor on multistory buildings, approximate door and window locations and configurations, as well as detail callouts—all of which will inform the material takeoff and task preparation for complex transitions between various siding materials and manufacturer-recommended installation products.

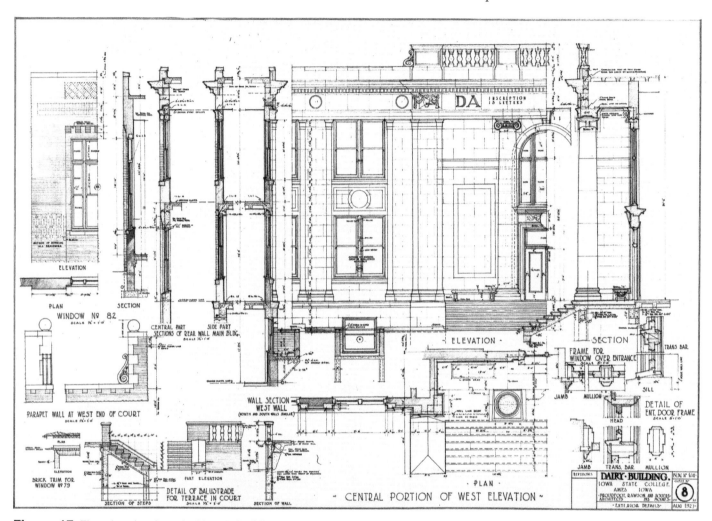

Figure 17 Elevation drawing of a historic building.
Source: Tango Images/Alamy

Detail Drawings

Detail drawings provide in-depth visual illustrations of a single component or system of assemblies. One of the most helpful details for a trade professional tasked with installing exterior siding and trim is the vertical cut view that shows the building's interior and exterior envelope systems and how they all fit together. This detailed view will show what type of insulation, sheathing, building paper, siding, and any other accessories are needed to protect the structure.

In commercial construction, it is a common practice to build a field mockup (*Figure 18*) to troubleshoot and test some of these more challenging transitions. This practice structure also provides your team members with an opportunity to train on any installation processes that they are unfamiliar with.

Finishing Drawings

Finishing drawings will provide detailed information on material colors, finishes, and layout requirements for exterior items that require structural reinforcement

Figure 18 Exterior siding mockup.
Source: Yalcin Sonat/Shutterstock

connections, like signs and large architectural elements. This drawing set may also contain reflected ceiling plans (RCPs), which show exterior **soffit** configurations, with locations for lighting, vents, and other exterior accessories.

Soffit: A nonloadbearing material beneath eaves that encloses and protects the space between a roofline and exterior wall.

2.2.0 Material Takeoff

Before beginning any exterior finishing project, check local codes and manufacturer's instructions for important information that can affect the installation of materials. Such information may include important considerations like whether metal siding must be grounded, whether metal drip caps must be installed on cornices, and the size of rain gutters.

Knowing this information will help you take stock of various components that need to be installed with typical exterior siding and trim materials. Although the estimator and supervisor will typically perform the material takeoff (*Figure 19*), knowing the basic principles of this phase will allow you to double-check material deliveries and installation processes to guarantee you have everything you need to complete your scope.

Figure 19 Workers performing a material takeoff.
Source: nd3000/Shutterstock

To estimate the amount of siding material required for a project, proceed as follows:

Step 1   Determine the total area of the structure to be covered by adding up the areas of all walls and gables.

Step 2   Subtract the total area of all openings.

Step 3   For board-type siding, add the waste percentages for the size and lap, as listed in *Table 2*. For metal or vinyl siding, add 10 percent.

TABLE 2 Waste Allowances

Siding	Size and Lap	Percent to Add
Beveled	$1 \times 4 - {}^3/_4"$	38
	$1 \times 5 - {}^7/_8"$	45
	$1 \times 6 - 1"$	33
	$1 \times 8 - 1^1/_4"$	33
	$1 \times 10 - 1^1/_2"$	29
	$1 \times 12 - 1^1/_2"$	23
Shiplapped	1×4	28
	1×5	21
	1×6	19
	1×8	16
	1×4	23
	1×5	18
	1×6	16
	1×8	14
Triangular areas or diagonal installation*		10

*The 10 percent is in addition to other allowances.

Watch Your Rating!

Although you may not have much say in the products you install for commercial and residential projects, it's always best practice to stay up to date on the most recent codes and product information. Knowing a product's rating and developing an experienced understanding of how various exterior systems perform over time will help you make informed suggestions to protect your company's reputation and save your client considerable replacement costs for using subpar materials.

For example, between 1990 and 1996, Louisiana-Pacific (LP), a manufacturer of the composite wood siding TrimBoard®, came under fire and several class-action lawsuits because it marketed its product as an exterior siding option when, in reality, it retained moisture, leading to decay and pest infestation. The most noticeable signs of these issues were visible veneer deterioration and severe warping as siding panels swelled with moisture absorption. LP was forced to recall TrimBoard® and similar products, and many builders and consumers switched to manufactured moisture-resistant alternatives.

2.3.0   Wall Surface Preparation

Once you have familiarized yourself with the scope of the project and have acquired the necessary tools and materials to get started, the next phase is wall surface preparation. This step begins with an exterior inspection where you will be looking for any signs of damaged sheathing or improper framing. New construction will have fewer issues to mitigate than repairing siding on an existing structure, but finding these issues early in the preparation phase will give you the time and information you need to identify and address these concerns before they negatively affect the project schedule and budget. Focus on the following common concerns when you perform an exterior inspection (*Figure 20*):

- Inspect the sheathing installation for gaps at the seams, overnailing, and underdriven nails. The building should've passed the framing inspection at this point in the project; however, if you find something that has been missed, it will be much less expensive and time-consuming for the framers to fix these issues before you install building wrap and siding.

Figure 20 Exterior sheathing of new construction.
Source: Darryl Brooks/Alamy

- Look for signs of moisture damage. Trapped moisture can deteriorate exterior sheathing, affecting structural integrity and inviting the potential of mold, mildew, and pests. Closely inspect areas where water may accumulate, such as exterior openings or floor-to-exterior connections, for stains and discoloration (*Figure 21*).

Figure 21 Water-damaged structural sheathing.
Source: Michael Vi/Alamy

- Ensure the walls are level, plumb, and match the dimensions illustrated in the approved construction set. Drastic design changes may require supplementary materials, manufacturer accessories, or alternative installation methods to complete the project.

If you are tasked with updating siding on an existing structure, one of the most common practices in residential is overlayment, where you would install lighter vinyl siding materials over existing siding by using the original siding material as the substrate. This is a risky, cost-saving technique that will likely lead to several durability issues and a possibly voided warranty for many products. Wood siding may only need to be replaced in some areas; however, the safest and most thorough method for siding replacement is taking the structure down to structural sheathing and replacing everything all at the same time. Keep the following points in mind when preparing a surface for replacing exterior finishing materials on a remodeling project:

- Perform the same steps previously mentioned for inspecting new construction wall surfaces.
- Check for uneven spots in wall planes and corners, and build out (shim) if required.
- Remove downspouts and other items that will interfere with the installation of new siding.
- Tie shrubbery and trees back from the base of the building to avoid damaging them during siding installation.
- Cut off windowsill extensions to allow the J-channel to be installed flush with the window casing. However, if the building owner wishes to maintain the original window design, coil stock can be custom-formed around the sill instead of cutting away the sill extensions.

2.3.1 Building Wrap

Building wrap or weather-resistant barrier (WRB) protects structural sheathing from any moisture that penetrates exterior cladding, regulates airflow between the two materials, and enables ventilation for water vapor to exit the building.

Some of the most popular building wraps are a spun-bonded olefin material such as HardieWrap™ from James Hardie™ or Tyvek® wraps (*Figure 22*). These materials protect the exterior sheathing but breathe to allow water vapor to pass from inside a structure to the exterior.

Figure 22 Building wrap.
Source: B Christopher/Alamy Images (left)

Some local codes may not require building wrap, but it is necessary for a wind- and weather-resistant seal, and it is a vital step in maintaining manufacturer warranties for other components of the building envelope system.

All horizontal and vertical joints must be sealed with manufacturer-approved tape or adhesive. Proper installation and taping are especially important for preventing damage from high winds before the exterior finishing material is completed.

Aluminum Foil Underlayment

Although aluminum reflector foil is more commonly installed with structural sheathing for roofs, as is the case with LP® TechShield® (*Figure 23*), it can also make an excellent insulator for siding in regions where structures require higher insulation standards. Aluminum underlayment may also be stapled directly to the existing wall or over $^3/_4$"-furring strips to provide additional air space and improved insulation. Use perforated or breather-type reflector foil to allow for better moisture ventilation.

Figure 23 Aluminum foil over structural sheathing product.
Source: breakermaximus/Shutterstock

Install foil with the shiny side facing the air space (outward with no furring; inward if applied over furring). Foil is generally available in 36"- and 48"-wide rolls. Nail or staple the foil just before applying the siding.

When applying foil over furring, be careful not to let the foil collapse into the air space. Place the foil as close as possible to openings and around corners where air leaks are likely to occur. Overlap the side and end joints by 1" to 2". For more information on envelope systems, reference NCCER Module 27203 *Thermal and Moisture Protection*.

2.3.2 Furring

Furring strips provide a smooth, even backing on which to attach siding boards and panels. Furring may not be necessary for new construction, but older homes often have uneven walls, and furring or shimming low spots can prevent the siding from appearing wavy. Furring also creates a vital air gap between the siding and weather-resistant barrier, acting as a drainage plane for moisture to run downward and dry before penetrating the exterior sheathing.

For horizontal siding (*Figure 24*), the furring should be installed vertically at a specified spacing (typically 16" OC). Any air space at the base of the siding should be closed off with horizontal strips. Window, door, **gable**, and **eave** trim may have to be built out to match the thickness of the wall furring. The furring for vertical siding is essentially the same process as that used for horizontal siding, except the wood strips are securely nailed horizontally into structural lumber at a specified spacing between 16" and 24" centers (*Figure 25*).

Gable: The triangular section of a wall that is enclosed by the edges of a pitched roof.

Eave: The projecting edge of a roof, extending horizontally beyond the walls of a structure.

Figure 24 Furring strips installed over weather-resistant barrier.
Source: pashapixel/123RF

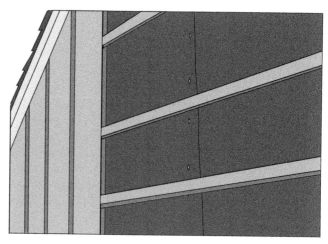

Figure 25 Furring strips installed for vertical siding application.

Undersill Furring

Building out below the windowsill may be required to maintain the correct slope angle if a siding panel needs to be cut to less than full height. The exact thickness required will be apparent when the siding courses have progressed up the wall and reached this point (*Figure 26*). This concept is also true for undereave furring (*Figure 27*).

Figure 26 Undersill furring.

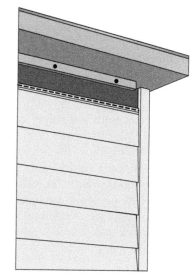

Figure 27 Undereave furring.

Window and Door Buildout

Some trim buildouts at windows and doors may be required to maintain the original appearance of the house when using furring strips or underlayment boards. This is particularly true when the strips or underlayment board are more than 1/2" thick. Thicker furring and underlayment generally provide added insulation value and reduced energy costs for the homeowner.

2.3.3 Flashing

Flashing must be installed around all openings before the exterior finishing materials are applied. The primary purpose of flashing is to prevent water from penetrating the finishing materials and entering the exterior walls, protecting interior surfaces from water damage, mold, mildew, and other forms of decay (*Figure 28*).

IRC® codes stipulate that corrosion-resistant flashings shall be installed in the following locations:

- Exterior window and door openings (in accordance with R703.4.1)
- Intersections of chimneys and other masonry elements and framed walls

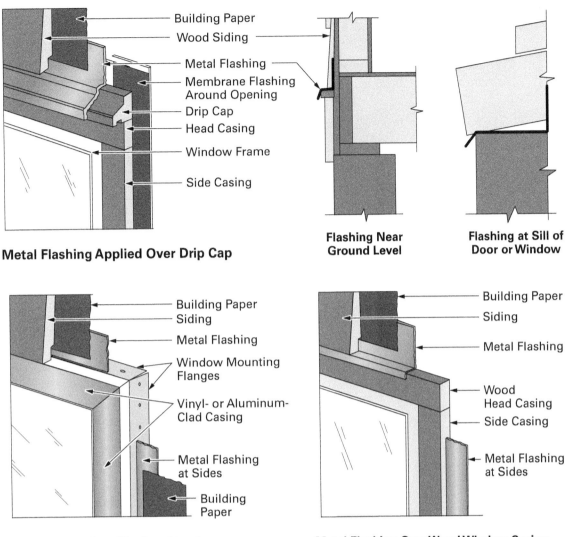

Metal Flashing Applied Over Drip Cap

- Building Paper
- Wood Siding
- Metal Flashing
- Membrane Flashing Around Opening
- Drip Cap
- Head Casing
- Window Frame
- Side Casing

Flashing Near Ground Level

Flashing at Sill of Door or Window

Metal Flashing Over Vinyl- or Aluminum-Clad Window with Mounting Flange

- Building Paper
- Siding
- Metal Flashing
- Window Mounting Flanges
- Vinyl- or Aluminum-Clad Casing
- Metal Flashing at Sides
- Building Paper

Metal Flashing Over Wood Window Casing

- Building Paper
- Siding
- Metal Flashing
- Wood Head Casing
- Side Casing
- Metal Flashing at Sides

Figure 28 Typical flashing installation.

- Under (and at the ends of) masonry, wood, or metal copings and sills
- Continuous application above all projecting wood trim
- Anywhere that exterior porches, decks, or stairs attach to a wall or floor assembly of wood-framed construction
- Wall and roof intersections
- Built-in gutters

Flashing typically consists of galvanized sheet metal, aluminum, copper, stainless steel, or a synthetic material. Aluminum is not used for flashing masonry due to its susceptibility to rust and corrosion. Modularized flashing (galvanized sheet metal) is one of the most common products for directing water away from critical roof areas such as vents, chimneys, and skylights. Bonderized options, commonly known as "paint grip," are common for visible flashing components because they hold paint well to color-match roofing and siding elements. Stainless steel and copper flashing are typically reserved for structures near coastal regions that require extra corrosion-resistance due to the need to protect against the salty air, or on high-end projects and historic restorations (*Figure 29*).

Figure 29 Copper flashing installed on the roof of a historic building.
Source: Istvan Balogh/Alamy

Corner and Joint Flashing

Sealant will be adequate for sealing between siding materials in most regions, but as an added precaution for areas that experience wind-driven rain, vertical flashing is sometimes placed around all outside corners for mitered siding or siding that uses corner boards. This corner flashing should be wide enough to extend 3" to 4" on each side of the corner and installed per the manufacturer's instructions.

Similar considerations should be made for butt or lap joints where horizontal boards and panels meet in the wall field (*Figure 30*). In these situations, *IRC*® R703.10.2 stipulates that lap siding must have a minimum $1^1/_4$" vertical overlap, and be treated with either sealant, joint flashing, or (less common) H-jointer covers. Each system will have its own installation guidelines with manufacturer-approved products. It is important to install these systems in the correct order with these preferred products to guarantee a durable installation that meets the manufacturer's warranty requirements.

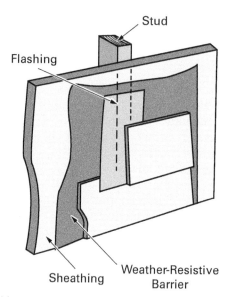

Figure 30 Joint flashing.

2.0.0 Section Review

1. The construction document that provides a straight-on, flat view of an exterior face of a building is known as a(n) _____.
   a. site plan
   b. exterior elevation drawing
   c. reflected ceiling plan (RCP)
   d. cut detail

2. To perform a material takeoff for exterior siding, you will first find the total area of walls and gables, then subtract the _____.
   a. cost of shipping
   b. total roof area
   c. amount of fasteners needed to complete the project
   d. exterior openings for doors and windows

3. The main purpose of flashing is to protect exterior openings, corners, and siding joints from _____.
   a. moisture penetration
   b. fire damage
   c. high winds
   d. impact damage

3.0.0 Exterior Siding Layout, Materials, Styles, and Installation Methods

Objective	Performance Task

Discuss common types of exterior siding materials and their installation processes.
  a. Describe the basic principles of exterior siding layout.
  b. Discuss common siding styles and their installation methods.
  c. Identify exterior trim components including soffits, fascia, and cornices.
	2. Lay out and install three of the most common siding types in your area.

Establishing a Straight Reference Line

A key element of successful siding installation is establishing a straight reference line to start the first course of siding. Use the tops of windows and doorways as another reference break line to ensure the shadow lines match with exterior openings (*Figure 31*).

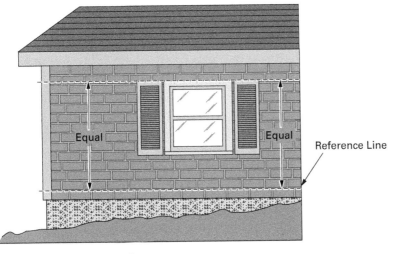

Figure 31 Straight reference line.

Identify the lowest corner of the house and partly drive a nail high enough to clear the height of a full siding panel. Stretch a taut chalk line from this corner to a similar nail installed at another corner. Repeat this procedure on all sides of the house until the chalk lines meet at all corners and until you have established an accurate reference line.

Establishing an initial reference line is even more important for metal siding and architectural panels than it is for wood lap siding because there is less room for adjustment while spacing the courses up to the soffits. Double-check the position of the reference line in relation to the top of exterior openings to ensure that your installation layout will establish even laps as you install each course toward the roofline.

3.1.0 Water Levels and Laser Levels

You can use a water level or a laser level to set the chalk line the width of the starter strip from the lowest point and locate the top of the starter strip at that line. Take the level reading at the corners and centers of the chalk line for best results. The water level can be used for measurements up to 100' and is accurate to $\pm\frac{1}{16}$" at 50'.

One of the most common methods for establishing reference lines is accomplished using story poles and laser levels (*Figure 32*). Simply set up your laser level on a tripod with legs locked tightly in place, in an area where it is unlikely to be disturbed. Ensure your laser location has a clear line of sight to the target wall. You can use a grade rod or fashion a job-built story pole from straight, square lumber to attach your laser detector. You can then use your tape measure to lay out the following courses on the story pole as a template. Just ensure that you hold the pole straight and plumb to get the most accurate measurements.

Figure 32 Laser level.
Source: sir270/123RF

Many newer models of lasers produce bright, highly visible red and green vertical and horizontal lines, but the story pole method is an excellent technique for performing an accurate layout on a sunny day, when lasers may be hard to see.

3.1.1 Layout for Exterior Openings

When approaching an installed window or door, temporarily tack a full-width board to the wall just before the window (or door), as shown in (*Figure 33*). Using a 6" or 7" length of scrap siding of the same width, mark the top, bottom, and side of the opening on the temporarily installed board. Remove the board, and cut out the marked opening. In the same location as the temporary board, permanently install and nail another board with the same width. Then, install the board at the opening edge.

When leaving an installed window or door, tack two scrap siding pieces (one piece for a door) to the wall as temporary spacers (*Figure 34*). For

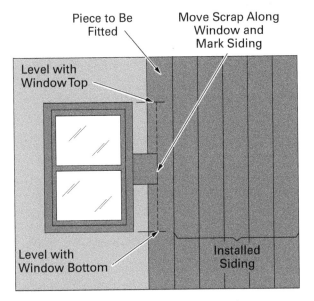

Figure 33 Approaching an installed window.

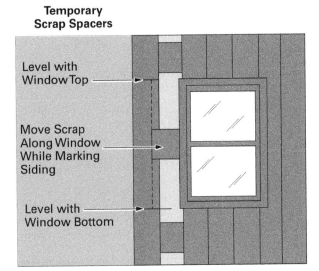

Figure 34 Leaving an installed window.

tongue-and-groove installation, these pieces must be wide enough to extend beyond the opening at positions above and below the opening (above for a door) and installed with the following steps:

Step 1   Temporarily install and tack a board of the same width and length against the scrap pieces, with the tongue of the scrap pieces inserted. For a door, plumb the tongue edge before tacking.

Step 2   Using a 6" or 7" length of scrap siding of the same width with the tongue cut off, mark the top, bottom, and side of the opening on the temporary board.

Step 3   Remove the board and cut out the marked opening. Remove the scrap spacers. Then, install and face-nail the cutout board at the opening edge before blind-nailing the tongue edge.

Step 4   Plan your layout properly to avoid using a narrow sliver of siding at the end of the wall. Stop several feet short of the end of the wall and space off the remaining distance to determine the width of the last board. Use random material widths to provide a reasonably wide ending board; otherwise, rip and regroove several boards to achieve the same effect.

Step 5   Install the boards up to the last board. Then, install the last board temporarily and mark the back side of the board flush with the corner. Rip the board, and permanently install it.

3.2.0   Wood Siding

The most common wood siding materials are western red cedar (WRC), bald cypress, Douglas fir, western hemlock, pine, and redwood. These woods can be manufactured in various style profiles to produce different aesthetics and performance, including lap (clapboard), bungalow, Dolly Varden, tongue-and-groove (T&G), drop, and shakes (*Figure 35*).

Although many designers have moved away from wood siding for more weather-resistant fiber-cement or engineered wood options that mimic the appearance of real wood, natural wood siding is still a popular material for high-end residential exteriors. We will discuss a few of the most popular siding profiles and their installation methods in the next few sections.

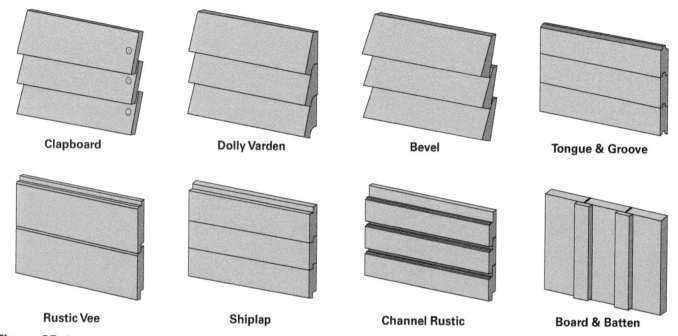

Clapboard   **Dolly Varden**   **Bevel**   **Tongue & Groove**

Rustic Vee   **Shiplap**   **Channel Rustic**   **Board & Batten**

Figure 35 Common wood siding profiles.

3.2.1 Lap Siding

Lap siding (*Figure 36*), sometimes called clapboard siding, is a category of overlapping siding materials comprised of several profile variations, including beaded lap, Dutch lap, and traditional. Beaded lap has a thicker bottom edge that creates a dynamic shadow line; Dutch lap is slightly concave, producing

Figure 36 Residential home with horizontal lap siding.
Source: Graham Prentice/Alamy

a unique reveal; and clapboard is more wedge-shaped, with a wider bottom overlapping a thinner top to create a uniform final product. Beveled and Dolly Varden siding often fall into the lap siding category due to their similar installation process; however, their beveled bottom edges create a more drastic shadow line between boards than flatter, traditional lap siding.

Most of these profiles are available in fiber-cement siding materials as well, with JamesHardie™ being one of the most popular manufacturers of a wide range of exterior siding boards, shingles, and panels that mimic wood, stone, and other natural surfaces. These slight variations provide designers and architects with an array of options to customize a structure's exterior into a cohesive, eye-catching aesthetic. Each installation process will be slightly different (*Figure 37*), so follow the manufacturer's instructions closely to ensure that the selected cladding system performs well and the product warranty is maintained.

3.2.2 Board-and-Batten Siding

Board-and-batten siding is an attractive, versatile, squared-edge siding that is widely used throughout the building industry, especially for residential facades. There are many variations of board-and-batten siding, but the most widely used is the vertical placement of wide boards, with the joints covered by narrow battens (*Figure 38*).

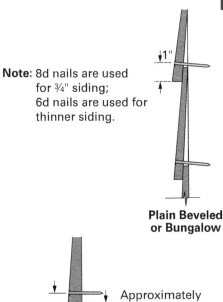

Note: 8d nails are used for ¾" siding; 6d nails are used for thinner siding.

Plain Beveled or Bungalow

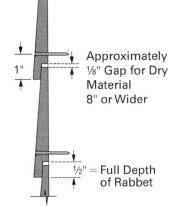

Approximately ⅛" Gap for Dry Material 8" or Wider

½" = Full Depth of Rabbet

Rabbeted Beveled (Dolly Varden) 8" and Wider

Figure 37 Typical nailing for beveled siding.

Figure 38 Residential home with board-and-batten siding.
Source: Hendrickson Photography/Shutterstock

Installing Board-and-Batten Siding

When framing an exterior wall that receives vertical siding, it is necessary to install horizontal blocking between the studs, from top to bottom at a specified spacing (typically 24" OC) to provide nailing locations. When applying the siding, space the underboards $^{1}/_{2}$" apart and drive the nails midway between the edges at each bearing. A major advantage of board-and-batten construction is that with proper nailing, the boards are free to move slightly with changes in moisture content while maintaining a sturdy hold on the battens.

Drive a single nail through the center of a board at each bearing to allow for this movement. Nails should penetrate $1^{1}/_{2}$" into the studs, the studs and wood sheathing combined, or the blocking. If you cannot achieve this depth, use ring- or spiral-shank nails for increased holding power. Drive a nail mid-way between the edges of the underboard at each bearing.

With the underboards in place, fasten the battens with nails. These battens should overlap each edge of the board underneath by at least 1". Drive the nails directly through the center of the batten so that the shank passes between the underboards. The finished appearance should resemble (*Figure 39*).

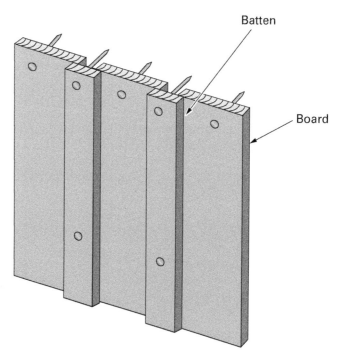

Figure 39 Typical nailing for board-and-batten siding.

3.2.3 Shake or Shingle Siding

Shake or shingle siding is an attractive, rustic, and architecturally interesting siding option. Red cedar and cypress are the most common woods for manufacturing shingles or shakes. These woods are durable and resist decay but must be refinished to maintain their durability over time (*Figure 40*).

However, in today's housing market, there are several fiber-cement options like James Hardie™ HardieShingle™ or Nichicha™ NichiShake™ that are more weather-resistant and easier to install. Some of these options are sold as singular shingles, while others come in overlapped panels with parallel or staggered shadow lines. Since these products have proprietary installation methods with manufacturer-approved accessories, we will cover some of the basic concepts of wood shake and shingle installation in the following sections.

Straightedge Shake/Shingle Application

When single coursing, rest shingles on a straightedge tacked at the butt line for easier spacing selection. For double coursing, use straightedges with a rabbeted edge so that the outer course is about $1/4$" below the inner course. Sort the shakes/shingles for proper seam overlap and lay them butt down on the straightedge before nailing them to the wall. For a ribbon-style double coursing, a reversed straightedge with a deeper rabbet can be used to shift the outer shake/shingle up to expose 1" to $1^1/_2$" of the lower part of the inner shake/shingle. Use a shingling hatchet to trim and fit the edges, if necessary. Butt ends are not trimmed. If rebutted and rejoined shakes/shingles are used, no trimming should be necessary (*Figure 41*).

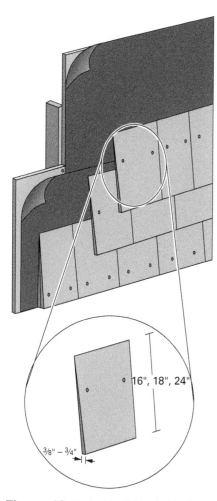

Figure 40 Typical individual shingle application.

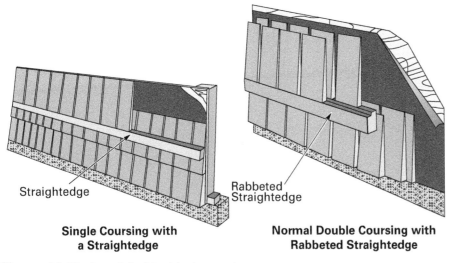

Single Coursing with
a Straightedge

Normal Double Coursing with
Rabbeted Straightedge

Figure 41 Single and double shingle coursing.

Installing Shakes and Shingles

The two basic methods of shingle exterior wall application are single course and double course.

In single-course application, the shingles are applied as in roof construction, but greater weather exposures are permitted. The maximum recommended weather exposures for single-course wall construction are $8^1/_2$" for 18" lengths and $11^1/_2$" for 24" lengths. Shingle walls should have two plies of shingles at every point, whereas shingle roofs will have three-ply construction.

The first course of shingles may be doubled at the bottom. After the first course is applied, lay out the story pole with all the courses indicated on it. Ensure you have the courses arranged to match the top elevations of all door and window openings. It may be necessary to change the exposure slightly on the course at the door and window heads, and windowsills. If such an adjustment is necessary, make sure it is slight so that it is not noticeable when viewing the entire exterior.

Shake/Shingle Architectural Accents

Like board-and-batten siding, special styles of shakes and shingles, known as fancy-butt shingles or shakes, are sometimes used on gables as accents. These specialty siding options are like those used on Victorian-style homes. In other cases, uniformly or randomly spaced staggered-length shakes/shingles are used to enhance a rustic appearance.

Sources: Roaminegg/Alamy (left); Images-USA/Alamy (right)

3.2.4 Tongue-and-Groove (T&G) Siding

Tongue-and-groove (T&G) siding (*Figure 42*) is a popular siding style that features interlocking boards, mated with a tongue on one side fitting into the groove of an adjacent board. This profile style is versatile in its installation as well, creating flush or subtle shadow lines in horizontal, vertical, and

Figure 42 Exterior with tongue-and-groove siding.
Source: Astronaut Images/Getty Images

diagonal orientation. Boards are beveled with a "V-edge" and one face is smooth, while the other (typically the side facing the structure) is rough. The tight fit between each board also makes this siding profile an excellent option for weather resistance in regions that experience high rainfall and humidity.

Installing Tongue-and-Groove and Shiplap Siding

Vertical T&G and shiplap siding are installed in similar manners. Use the following installation techniques when installing T&G vertical siding:

- Install a starting board. Plumb the tongue edge of the board with the grooved edge beyond an outside corner (*Figure 43*), or against an inside corner. Temporarily tack the board in place.

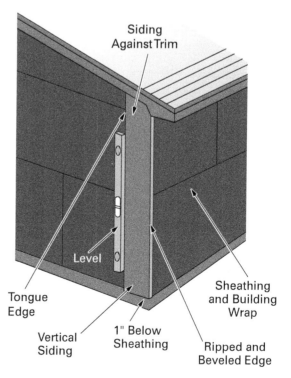

Figure 43 Wall starting board.

- Mark the board along the length of the corner. For an outside corner, mark along the back side of the board, flush with the corner. For an inside corner, mark along the face of the board using a 6" or 7" long spacer that is wide enough to extend beyond the groove depth of the board.

- Rip off the grooved edge on an outside corner and the tongue edge on an inside corner to the mark and slightly back-bevel the edge.

- Position the board at the corner with the ripped edge flush with the corner.

- Install the board and recheck the plumb of the tongue edge before face-nailing the board at the corner and blind-nail the tongue edge. *Figure 44* shows a side profile illustrating how T&G boards fit together and fasten to the exterior.

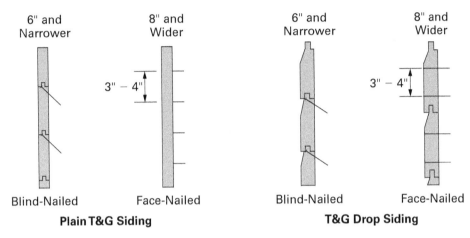

Figure 44 Typical tongue-and-groove nailing.

Mating T&G Siding

In some cases, you may have to force the tongues and grooves (*Figure 45*) of T&G siding together to obtain a uniform mating of each course of siding. If necessary, use a hammer and a scrap block of siding on the course being installed to force it onto the tongue of the preceding course.

If a board is slightly warped and does not mate evenly along its entire length, secure the board with nails at one or both ends up to the point that the warp begins. Set additional nails in the siding beyond the point of warp. Then, drive a flat, broad chisel into the underlayment-nailing surface with its beveled edge against the siding. Use the chisel as a lever to force the siding into position and then nail the siding into place. Repeat along the length of the board until the board is seated and secured, and maintain any required inside-groove spacing.

Figure 45 Tongue-and-groove material profiles.
Source: Joe Ferrer/Alamy

Common Siding Fasteners

The siding nail is considered the best fastener for wood siding except in high-wind areas where spiral-shank nails are often required. Nail sizes vary from 6d to 10d and medium-crown 16- to 18-gauge staples are also common for applying vinyl siding and building wrap. *Figure 46* shows some of the most common nails and staples for fastening siding to the exterior.

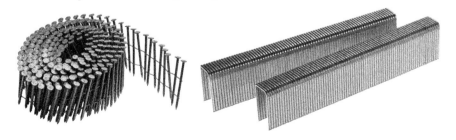

Figure 46 Common siding fasteners.
Sources: Courtesy of Stanley Black & Decker, Inc. (nails) and KYOCERA SENCO Industrial Tools (staples)

You can install exterior siding with pneumatic- or battery-operated nailers and staplers. However, always use a flush-mount attachment or set the driving pressure on the nailer to prevent overdriving fasteners, which will cause the siding to split.

3.3.0   Vinyl and Metal Siding

Vinyl and metal siding are applied in new construction as well as over existing finishes for remodeling work. Both vinyl and metal siding are manufactured in various profiles, colors, and finishes. Inside corner posts, door and window trim, individual corner pieces, starter strips, and butt supports are also available. Metal siding and trim are usually supplied with a baked-on or plastic finish and insulating backing board, which makes the siding less susceptible to exterior damage, while also improving its rigidity.

Metal siding has become an extremely popular choice in recent years for creating a sleek, industrial exterior for tiny homes, commercial residential buildings, and "barndominiums" built as accessory dwelling units (ADUs). These designs often utilize corrugated, bevel, and flat sheet metal, or a combination of these profiles, to produce an eye-catching distinction between different sections of the structure (*Figure 47*).

Figure 47 Metal siding.
Source: Susan Vineyard/Alamy

3.3.1   Metal and Vinyl Siding Materials

Horizontal metal siding is usually limited to single- or double-lap styles for residential structures and large architectural panels for commercial buildings (*Figure 48*). Metal siding is much more durable than vinyl when exposed to extreme temperatures and is also decay-resistant and termite-proof. However, unlike vinyl, it is susceptible to corrosion from salty air near coastal regions and impact damage. Metal siding can also be more difficult to handle and install due to its rigid structure and sharper edges.

The major advantages of vinyl siding are low cost and simple installation. It is also waterproof, decay-resistant, and termite-proof. Its major disadvantage is that cold temperatures significantly reduce its resistance to damage from impacts. Vinyl siding may also break during expansion or contraction when outdoor temperatures quickly change. Another downside of vinyl, which has led many designers and craft professionals to move toward fiber-cement alternatives, is vinyl's low insulation value. Some high-end options like BASF Neopor® GPS can be installed underneath to insulate and contour to vinyl panel profiles, providing better impact resistance and insulation value; however, these extra measures will come with additional installation challenges (*Figure 49*).

Figure 48 Metal siding panels on a commercial building.
Source: Michael Frank/Alamy

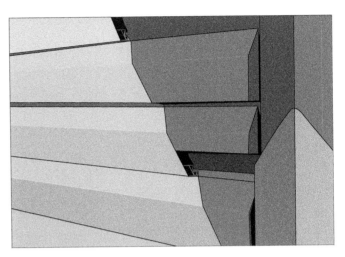

Figure 49 Insulated vinyl siding.

The following materials may be required when installing vinyl or metal siding:

- Building wrap or aluminum breather foil
- Manufacturer-approved sealant
- Aluminum, plain-shank, or spiral-shank nails ($1^1/_2$" for general use; 2" for re-siding; $2^1/_2$" or more to nail insulated siding into soffit sheathing; 1" to $1^1/_2$" trim nails colored to match siding). Plain-shank nails require a minimum penetration of $^3/_4$" into solid lumber. However, you can use screw-shank nails for $^1/_2$" structural sheathing to achieve similar effects.

3.3.2 Installing Vinyl and Metal Siding

Trim accessories are installed to accept the ends of the siding prior to the installation of vinyl or metal siding. These trim accessories include inside corner posts, outside corner posts, starter strips, window and door trim, and gable-end trim pieces (*Figure 50*).

Installing Inside Corner Posts

Inside corner posts are installed before the siding is hung. Depending on the type of siding (insulated or noninsulated), deeper or narrower posts may be required. The post is set in the corner full length, from $^1/_4$" below the bottom of the starter strip to $^1/_4$" from the eave or gable trim. Nail the upper slot at the top, then nail approximately 8" to 12" on both flanges with aluminum nails in the center of the slots. Ensure the post is set straight and true. Nail the flanges securely to the adjoining wall, but do not overdrive the nails, since doing so may cause distortion. If a short section is required, use a hacksaw to cut it. If a long section is required, overlap the posts with the upper piece outside.

Figure 50 Vinyl siding installation at gable roof.
Source: Scott Nodine/Alamy

The siding is later butted into the corner and nailed into place, allowing approximately a $^1/_{16}$" to $^1/_4$" space between the post and the siding to allow for expansion (*Figure 51*).

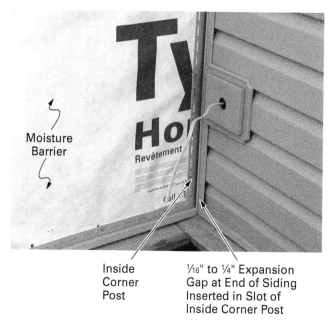

Inside Corner Post

$^1/_{16}$" to $^1/_4$" Expansion Gap at End of Siding Inserted in Slot of Inside Corner Post

Moisture Barrier

Figure 51 Inside corner post.

Installing Outside Corner Posts

The outside corner post produces a trim appearance and will accommodate the greatest variety of siding types. Most outside corner posts are designed to be installed before the siding is hung, like the installation of the inside corner post. Set a full-length piece over the existing corner, running from $^1/_4$" below the bottom of the starter strip, to $^1/_4$" from the underside of the eave. If a long corner post is needed, overlap the corner post sections, with the upper piece outside.

Nail the uppermost slot at the top of the slot, then fasten with aluminum nails on both flanges in the center of the slots every 8" to 12". Ensure the flanges are securely nailed (*Figure 52*) but not overdriven, to avoid distortion. Use a saw to cut material as required, and use deeper corner posts to accommodate insulated siding.

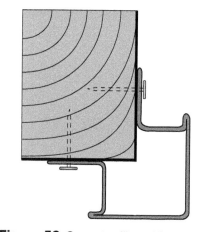

Figure 52 Correct nailing of flanges.

Hanging Vinyl and Metal Siding

Vinyl and metal siding installation is commonly referred to as "hanging siding," since nails should not be driven tightly against the siding surface. Rather, the siding is hung from the nails to allow expansion and contraction of the siding.

Siding will expand when heated and contract when cooled. This expansion will amount to roughly $^1/_8$" per 10' length for a 100°F temperature change. An allowance for this expansion or contraction should be made when installing siding.

Installing a Starter Strip

Using the chalk line previously established as a reference line, take equal-distance measurements, as shown in *Figure 53*, and install the starter strip, or J-channel strip, all the way around the bottom of the building (depending on the material at the base of the building). When using insulated siding, fur the starter strip out to a distance equal to the thickness of the backer. Ensure that the starter strip is straight and connects at all corners since it will determine the line of all siding panels installed. Shim out behind the starter strip at any hollow points in the old wall surface to prevent a wavy appearance in the finished siding.

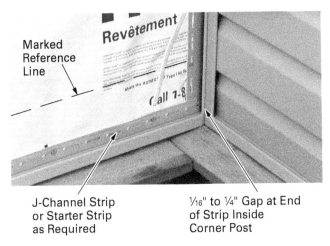

Marked
Reference
Line

J-Channel Strip
or Starter Strip
as Required

¹⁄₁₆" to ¼" Gap at End
of Strip Inside
Corner Post

Figure 53 Installing a starter or J-channel strip.

Install the starter strip up to the edge of the house corner when using individual corner caps. Use aluminum nails to fasten the starter strip, spaced not more than 8" apart. Nail the starter strip as low as possible and avoid bending or distorting the strip by overdriving the nails. Cutting lengths of starter strip is best accomplished with tin snips before butting the sections together.

Starter strips may not work in all situations. For example, other accessory items, such as J-channels or all-purpose trim, may be more efficient for starting the siding course over garage doors and porches, or above masonry. However, these situations must be handled on an individual basis as they occur.

Installing Window and Door Trim

You may need to build out windows and door casings to retain or improve a house's original appearance. To complete this task, nail the appropriate lengths and thicknesses of lumber securely to the existing window casings. During remodeling work, old windowsills and casings can be covered with aluminum coil stock, which is bent to fit on-site. *Figure 54* shows the installation of aluminum window trim.

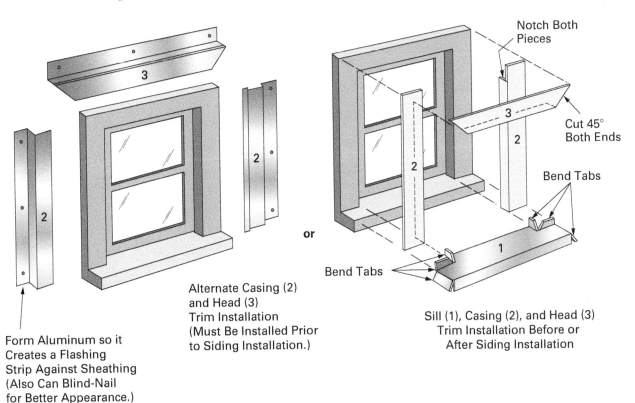

Form Aluminum so it
Creates a Flashing
Strip Against Sheathing
(Also Can Blind-Nail
for Better Appearance.)

Alternate Casing (2)
and Head (3)
Trim Installation
(Must Be Installed Prior
to Siding Installation.)

Notch Both
Pieces

Cut 45°
Both Ends

Bend Tabs

Bend Tabs

Sill (1), Casing (2), and Head (3)
Trim Installation Before or
After Siding Installation

Figure 54 Installing aluminum window trim.

Cover any step in the windowsill by bending two separate sill cover pieces with interlocking flanges, as shown in *Figure 55*. Use tin snips and bend flanges on-site. Box in the sill ends to provide a neat appearance and prevent water penetration.

J-channel is used around windows and doors to receive the ends of siding. Side J-channel members are cut longer than the height of the window or door and notched at the top. Notch the top J-channel member at a 45-degree angle and bend the tab down to provide flashing over the side members (*Figure 56*). Apply sealant behind J-channel members to prevent water infiltration between the window and the channel, as seen in *Figure 57*.

To provide protection against water infiltration, slip a flashing piece cut from coil stock or a precut piece of step flashing under the base of the side J-channel members. Overlap this piece over the top lock of the panel below, as shown in *Figure 58*.

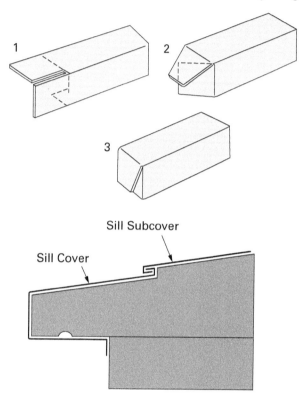

Figure 55 Boxing in sill ends.

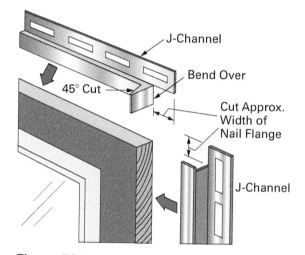

Figure 56 Cutting J-channel.

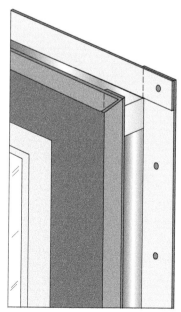

Figure 57 J-channel.

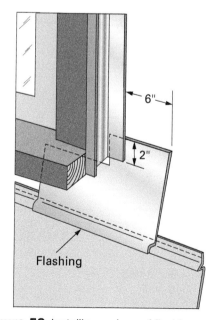

Figure 58 Installing a piece of flashing.

On the sides of the building, start at the rear corner and work toward the front to create a more aesthetically pleasing appearance. On the front of the building, start at the corners and work toward the entrance door for this same reason. Factory-cut ends of the panels should cover the field-cut ends for a more uniform final product.

Metal panels should overlap each other between $^5/_8$" and a minimum of $^3/_8$" as a rule of thumb. Vinyl manufacturers typically recommend a 1" overlap with a double-size nailing flange cut. You must also consider thermal expansion requirements when planning your panel overlaps. Cut away the top lock strip on the overlapped panel by twice the amount of the intended overlap (*Figure 59*).

Avoid panel lengths of 24" or less and overlap the factory-cut ends over field-cut ends. The job should start at the rear of the house and work toward the front, as shown in *Figure 60*.

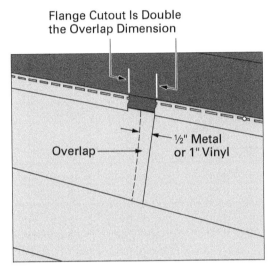

Figure 59 Overlapping panels.

Figure 60 Sequence of installation.

For the best appearance, plan how you will stagger the joints before the installation, and avoid installing siding in a set pattern (*Figure 61*). A set pattern may be more labor- and cost-effective, but typically results in a poor overall appearance. It is best to plan the job so that any two joints aligning vertically are separated by at least two courses and separate panel overlaps on the next course by at least two feet. Joints

Figure 61 Improper staggering of joints.
Source: Nick Cronin/Alamy

should be avoided on panels directly above and below windows. Shorter pieces that develop as work proceeds can be used for smaller areas around windows and doors.

Backer tabs are used with 8" horizontal noninsulated aluminum siding only. They ensure rigidity, even installation, and tight endlaps. They are used at all panel overlaps and behind panels entering corners. After the panel has been locked into place, slip the backer tab behind the panel with the flat side facing out (*Figure 62*). The backer tab should be directly behind (and even with) the edge of the first panel of the overlap before you nail the backer tab into place.

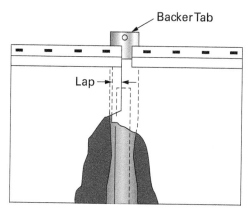

Figure 62 Inserting a backer tab for metal siding.

Corner Caps and Corner Boards

Individual corner caps (*Figure 63*) may be used for horizontal lap siding instead of outside corner posts. The siding courses on adjoining walls must meet evenly at the corners. To allow room for the cap, install the siding with $^3/_4$" clearance from the corner ($^1/_4$" clearance for insulated siding). Complete one wall first. On the adjacent wall, install one course of siding, line the course, and install the corner cap. Each corner cap must be fitted and installed before the next course of siding is installed.

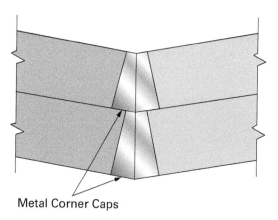

Metal Corner Caps

Figure 63 Corner caps.

Install the cap by slipping the bottom flanges of the corner cap under the butt of each siding panel. Use steady pressure when pressing the cap into place, and never force it. If necessary, insert a putty knife between the panel locks, prying slightly outward to allow room for the flanges to slip in. Gentle tapping with a rubber mallet and wood block can also be helpful.

When the cap is in position, secure it with nails that are long enough for $^3/_4$" penetration into solid wood or sheathing. Nail through at least one of the prepunched nail holes in the top of the corner cap.

Corner boards (*Figure 64*) are another common method of trimming, capping, and protecting outside edges where two courses meet at exterior angles. These boards are thicker than the siding projection, and exposures extend equal widths on either side of the corner.

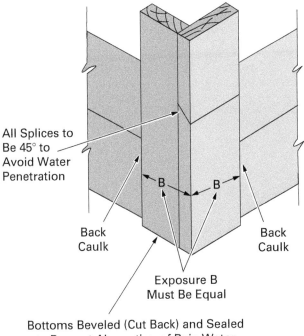

All Splices to Be 45° to Avoid Water Penetration

Back Caulk

B ← → B

Back Caulk

Exposure B Must Be Equal

Bottoms Beveled (Cut Back) and Sealed to Prevent Absorption of Rain Water

Figure 64 Corner boards.

Vinyl Siding Fasteners

Always follow guidance from the Engineer of Record (EOR) and the manufacturer's instructions when applying fasteners to vinyl siding. The *IRC®* also provides the following general requirements:

- Vinyl siding fasteners shall be 0.120" shank diameter nail with a 0.313" head or 16-gauge staple with a $^3/_8$" to $^1/_2$" crown.
- The total penetration into sheathing, furring framing, or other nailable substrates shall be at least $1^1/_4$". Wherever the fastener penetrates fully through the sheathing, the fastener end shall extend a minimum of $^1/_4$" beyond the opposite face of the sheathing or nailable substrate (*IRC®* R703.11.1.2).
- The maximum spacing allowed between fasteners is 16" for horizontal siding and 12" for vertical.

3.4.0 Fiber-Cement Siding

Fiber-cement siding (*Figure 65*) is the most common siding material used in new construction and remodeling work today. The material is comprised of Portland cement, sand, fiberglass, additives, and water that are pressure-formed and heat-cured into planks or panels. The major advantages of fiber-cement siding over wood siding are that it is decay-resistant, noncombustible, and termite-proof. It is highly resistant to impact damage. In some cases, fiber-cement siding can withstand hurricane-force winds of 130 mph or more. It is also an excellent alternative for metal siding, since it resists permanent damage from water and salt spray. This siding is especially suited for use in fire-prone or high-wind areas.

3.4.1 Installing Fiber-Cement Siding

Fiber-cement siding is manufactured either as a panel or lap siding. It can be installed over sheathed or unsheathed walls. Fastener spacing is per the code or manufacturer's recommendations. However, general installation guidelines are as follows:

- Fiber-cement siding may be applied over structural sheathing or insulation board up to 1" thick and with studs spaced no more than 24" OC.

Figure 65 Residential home with fiber-cement siding.
Source: tokar/Shutterstock

Fiber-Cement Siding Styles

Like wood siding, fiber-cement options are available in single-lap siding, vertical panels, and shakes, with several profile variations. These diverse architectural styles of fiber-cement siding are available to achieve different lighting effects and match your preferred aesthetic. Planks and panels are both available with smooth and wood-grained finishes. JamesHardie™ Hardie®Plank fiber-cement siding options are some of the most popular in both the residential and commercial construction industry for their diverse aesthetic options and excellent long-term performance.

Source: ND700/Shutterstock

- Fiber-cement siding may be cut with a power saw using a fine-toothed, carbide-tipped circular saw blade or a circular fiber-cement blade crafted specifically for this purpose. You can also use electric or pneumatic carbide-tipped power shears, or a utility knife with a tungsten-carbide tip.
- Only galvanized steel, copper, or stainless-steel flashing and fasteners may be used when installing fiber-cement siding.

- Never use aluminum trim components or fasteners, because they will corrode when they contact fiber-cement siding.

Details of typical lap and panel siding installation are shown in *Figure 66A*, *Figure 66B*, and *Figure 67.*

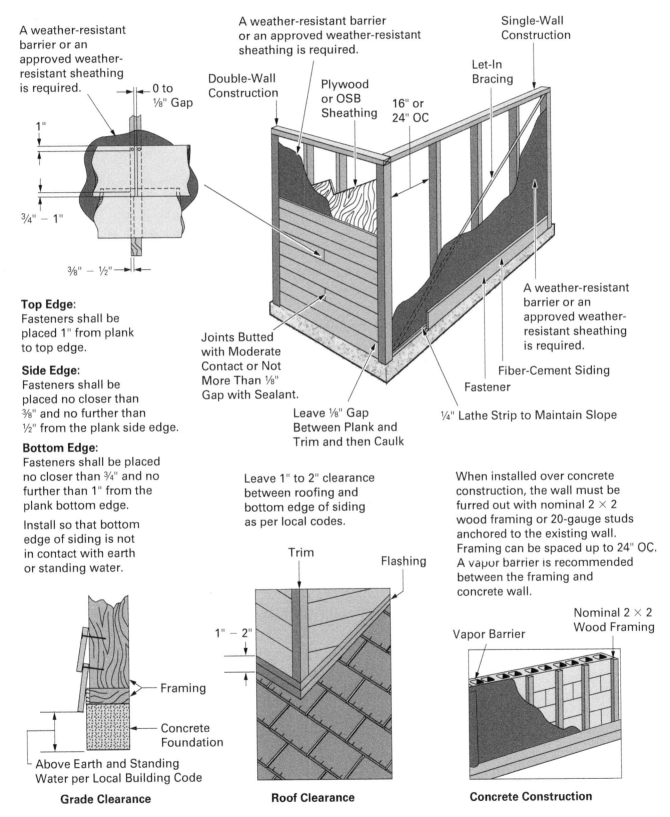

A weather-resistant barrier or an approved weather-resistant sheathing is required.

0 to 1/8" Gap

1"

3/4" – 1"

3/8" – 1/2"

A weather-resistant barrier or an approved weather-resistant sheathing is required.

Double-Wall Construction

Plywood or OSB Sheathing

16" or 24" OC

Single-Wall Construction

Let-In Bracing

A weather-resistant barrier or an approved weather-resistant sheathing is required.

Fiber-Cement Siding

Fastener

1/4" Lathe Strip to Maintain Slope

Top Edge:
Fasteners shall be placed 1" from plank to top edge.

Side Edge:
Fasteners shall be placed no closer than 3/8" and no further than 1/2" from the plank side edge.

Bottom Edge:
Fasteners shall be placed no closer than 3/4" and no further than 1" from the plank bottom edge.

Install so that bottom edge of siding is not in contact with earth or standing water.

Joints Butted with Moderate Contact or Not More Than 1/8" Gap with Sealant.

Leave 1/8" Gap Between Plank and Trim and then Caulk

Leave 1" to 2" clearance between roofing and bottom edge of siding as per local codes.

When installed over concrete construction, the wall must be furred out with nominal 2 × 2 wood framing or 20-gauge studs anchored to the existing wall. Framing can be spaced up to 24" OC. A vapor barrier is recommended between the framing and concrete wall.

Trim

Flashing

1" – 2"

Framing

Concrete Foundation

Above Earth and Standing Water per Local Building Code

Vapor Barrier

Nominal 2 × 2 Wood Framing

Grade Clearance

Roof Clearance

Concrete Construction

Figure 66A Typical fiber-cement lap siding installation details (1 of 2).

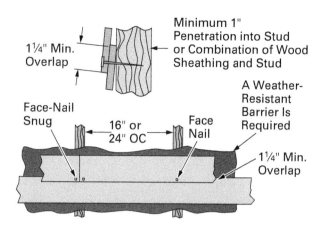

Corrosion-Resistant Nails
(Galvanized‡ or Stainless Steel)

Corrosion-Resistant Screws

Corrosion-Resistant Nails
(Galvanized‡ or Stainless Steel)

Corrosion-Resistant Screws

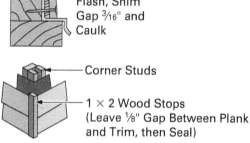

Face-Nailed

**Blind-Nailed
(Not Applicable for 12" Wide Siding)**

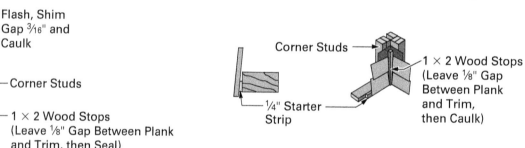

Notes: **Trim Details**

* For face-nail application of 9½" wide or less siding to OSB, fasteners are spaced a maximum of 12" OC.
† The use of a siding nail or roofing nail may not be applicable to all installations where greater wind loads or higher exposure categories of wind resistance are required by the local building code. Consult the applicable building code compliance report.
‡ Hot-dipped galvanized nails are recommended.

Fastening Requirements:
• Drive fasteners perpendicular to siding and framing.
• Fastener heads should fit snug against siding (no air space). (Examples 1 & 2)
• Do not underdrive nail heads or drive nails at an angle. (Example 4)
• If nail is countersunk, caulk nail hole and add a nail. (Example 3)

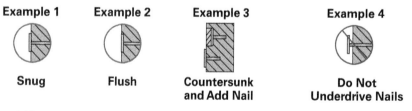

Figure 66B Typical fiber-cement lap siding installation details (2 of 2).

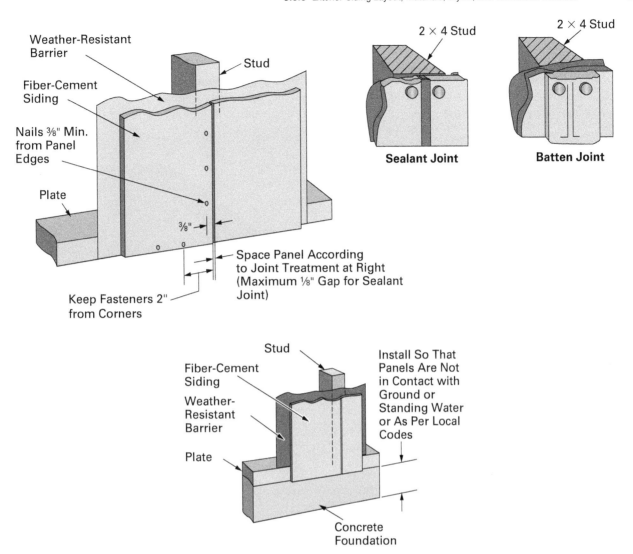

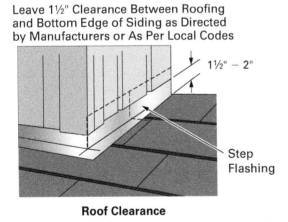

Roof Clearance

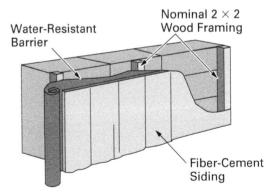

Concrete Construction

Figure 67 Typical fiber-cement panel siding installation details.

3.5.0 Exterior Finishing Components

Now that you have successfully reached the final courses of your exterior siding installation, the final step is to tie the exterior wall siding into the roof structure to create a weather-tight and aesthetically pleasing final product. This final section will discuss the basic principles of installing siding materials at soffits, fascia, and cornices to complete your exterior siding scope.

NOTE

Always check the cutting pattern after each course for accuracy. This is necessary because roof slopes are not always straight.

Installing Siding at Gable Ends

When installing siding on gables, make diagonal pattern cuts to fit the gable slope by using two short pieces of siding as templates (*Figure 68*). Interlock one of these pieces into the panel below. Hold the second piece against the trim on the gable slope. Along the edge of this second piece, scribe a line diagonally across the interlock end panel, and cut along this line with tin snips or a power saw.

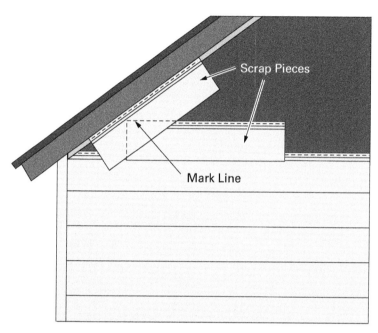

Figure 68 Installing siding at gable ends.

This cut panel can be used to transfer cutting marks to each successive course along the gable slope. You can use this same technique to complete siding for other roof slopes as well. The following steps describe gable-end siding installation for metal or vinyl siding:

Step 1   Slip the angled end of the panel into the J-channel previously installed along the gable end.

Step 2   Lock the butt into the interlock of the panel below, and allow for expansion or contraction where required. If necessary, face-nail with $1^1/_4$" (or longer) painted-head aluminum nails at the highest point of the last panel at the gable peak.

Step 3   Finally, use color-matching touch-up enamel to hide exposed nail heads.

Installing Siding Under Eaves

The last panel course under the eaves will almost always have to be cut lengthwise to fit in the remaining space. Usually, furring will be needed under this last panel to maintain the correct slope angle. Determine the proper furring thickness, and install it. Nail all-purpose trim or J-channel to the furring with aluminum nails. The trim should be cut long enough to extend the length of the wall (*Figure 69*).

To determine the width of the cut required, measure from the bottom of the top lock to the eave, subtract $^1/_4$", and mark the panel for cutting. Take measurements at several points along the eaves to ensure accuracy. Score the panel with the utility knife, and bend back and forth until it snaps. For some vinyl panels, you may need to use a snaplock punch (*Figure 70*) to place raised ears (16" or 24" apart) along the top cut edge so that it will lock into the J-channel.

For aluminum siding, apply gutter seal to the nail flange of the all-purpose trim. Slide the final panel into the trim before engaging the interlock of the panel below.

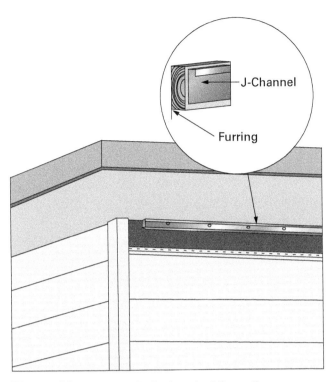

Figure 69 Trim extends the length of the wall.

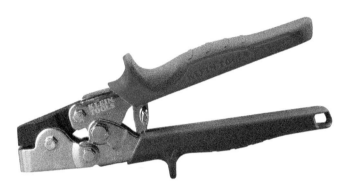

Figure 70 Snaplock punch.
Source: Klein Tools

On metal panels, the lock may be flattened slightly using a hammer and a scrap piece of lumber before the final panel is installed, to allow the panel to grip more securely. Press the panel into the gutter seal adhesive. Refer to *Figure 71*.

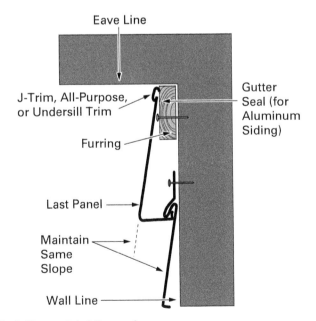

Figure 71 Installing metal siding under eaves.

Installing Cornices

Cornices (*Figure 72*) are constructed of wood, aluminum, vinyl, and other manufactured products, and they are most often found as elaborate eave details on Victorian, colonial, and cape architectural styles. The type of

Figure 72 Ornamental cornices.
Source: HOOB29/Shuuterstock

Cornice: A functional and decorative form of trim or molding located where the exterior walls and roof meet.

cornice required for a particular structure is shown on the wall sections of the construction drawings. The two general types of cornices are the closed cornice and the box cornice.

If a ceiling is installed, the roof area above any ceiling cannot be ventilated unless vertical vents through the soffit trim furring are provided. A closed cornice is the least desirable type of cornice because it does not allow roof ventilation and provides little protection to the side of the building.

With a box cornice, the rafter overhang is entirely boxed in by the roof covering, fascia, and soffit. *Figure 73* shows examples of one type of box cornice.

A roof with no rafter overhang normally has a closed cornice (*Figure 74*). This cornice consists of a single strip called a frieze board. The frieze board is beveled on its upper edge to fit under the overhang of the eaves and rabbeted on its lower edge to overlap the upper edge of the top siding course.

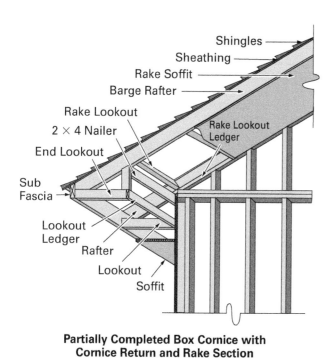

Partially Completed Box Cornice with Cornice Return and Rake Section

Completed Box Cornice with Cornice Return and Rake Section

Figure 73 Box cornice.

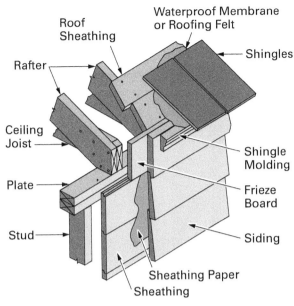

Figure 74 Closed cornice.

Metal Drip Caps and Edges

It is common to use a metal drip edge (sometimes called a drip cap) at the edges of all roof components, including the cornices. Others may permit the use of a row of wood shingles as a drip edge at the cornice edge of the roof or a row of wood shingles at the cornice in combination with a perimeter metal drip cap. The purpose of a drip edge is to direct rainwater away from the cornice fascia.

Source: Ruta Saulyte-Laurinaviciene/Shutterstock

3.5.1 Aluminum or Vinyl Fascia and Soffits

Wood and vented fiber-cement panels are common fascia and soffit materials, but refinished aluminum and vinyl (*Figure 75*) have been used extensively in cornice construction for decades. Much of this section will cover these materials because they are so prevalent in both commercial and residential construction due to their lighter weight and ease of installation.

Figure 75 Aluminum or vinyl soffit.
Source: tokar/Shutterstock

Each manufactured trim system will come with detailed instructions, but *Figure 76* shows a typical installation detail of an aluminum or vinyl fascia eave trim and a soffit. Specific installation instructions will vary according to the manufacturer or engineering instructions. *Figure 77* shows typical soffit and trim components. Each soffit panel piece is cut to the desired width and then slid into place from the end of the cornice, interlocking with adjoining pieces. Depending on the materials used and the amount of overhang, **lookouts** and ledgers may be required to support the soffit, and prevent sag and wind damage.

Lookouts: Joists that extend past the exterior wall of a structure to support roof loads and provide nailing surfaces for fascia.

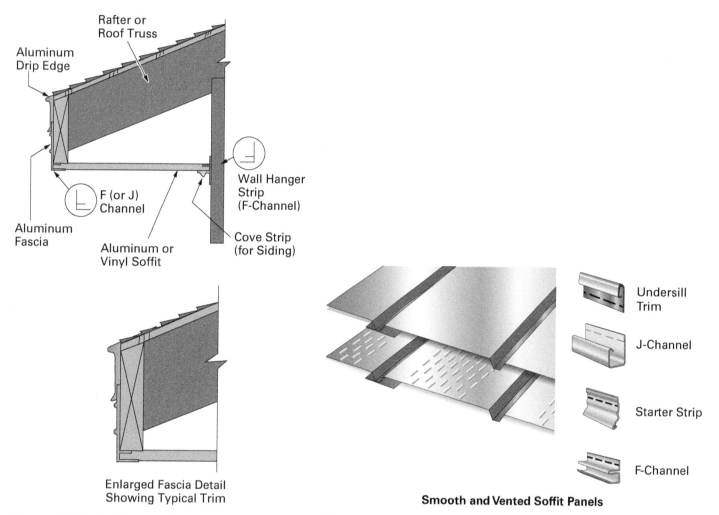

Figure 76 Typical aluminum/vinyl soffit and trim installation.

Figure 77 Typical aluminum/vinyl soffit and trim materials.

IRC® soffit provisions are applied based on an assumed 30 lbs per square foot (psf) design wind pressure. At or below this limit, vinyl soffit panels are required to be fastened at each end, and an unsupported span cannot exceed 16" unless permitted by the manufacturer's product approval. For higher-wind regions, enhanced soffit construction will be required where the design wind pressure exceeds 30 psf. Vinyl soffit panels are required to be fastened at each end, and an unsupported span cannot exceed 12" unless permitted by the manufacturer's product approval.

Guidelines for constructing a box cornice are as follows:

Step 1   Start by using a 6' level to mark the top-of-ledger location, even with the bottom edge of the rafter at intermediate locations on the exterior wall. Then, snap a chalk line between these points to ensure the ledger will be level. Secure the ledger to the wall to accommodate your installation of the lookouts.

Step 2   Use a straight board to mark the lookout positions on the ledger (*Figure 78*).

Step 3   Measure the distance from the face of the ledger to the ends of the rafters to determine the length of the lookouts.

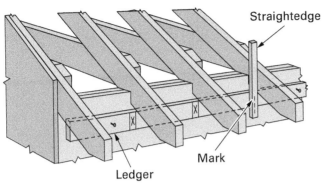

Figure 78 Marking lookout locations on a ledger.

Step 4   Trim the corner lookouts to match the slope of the rafters (*Figure 79*).

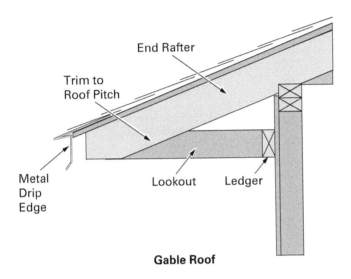

Gable Roof

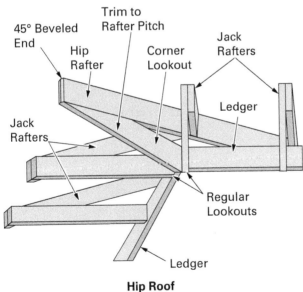

Hip Roof

Figure 79 Determining the length of corner lookouts.

Step 5   Attach the lookouts to the ledger, then install the assembly, nailing through the sheathing into studs where possible.

Step 6   Once the lookouts are in place, the soffit and fascia boards can be attached (*Figure 80* and *Figure 81*).

Step 7   The cornice enclosure must be cut to fit, and the cornice end piece is grooved to fit inside the cornice enclosure (*Figure 82*).

Figure 80 Soffit installation.
Source: Lex20/Getty Images

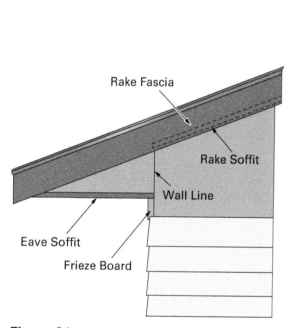

Figure 81 Rake soffit cut to wall line.

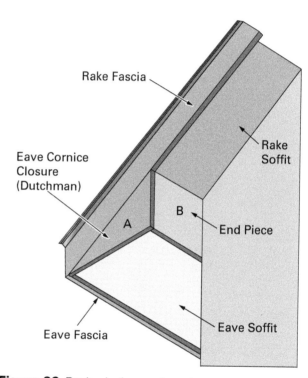

Figure 82 Boxing in the cornice return.

Sealants and Cleanup

In general, sealant (*Figure 83*) is applied around doors, windows, and gables where the siding meets wood or metal, except where accessories are used to make sealant unnecessary. Sealant may also be needed where siding or siding accessories meet brick or stone around chimneys and walls. Neatly apply sealant around faucets, meter boxes, and other panel cutouts.

It is important to get a deep sealant bead that is $1/4$" minimum in depth, not just a wide bead. To achieve this, cut the plastic sealant cartridge tip straight across rather than at an angle. Move the gun smoothly as you apply even pressure on the trigger. A butyl type of sealant is preferred, as it has greater flexibility, and most producers supply color-matching options for siding and accessories. Do not depend on sealant to fill gaps more than $1/8$" wide, as the expansion or contraction of the siding may cause the sealant to crack.

Figure 83 Applying sealant.
Source: valentyn semenov/Alamy Images

On remodeling projects, reinstall all fixtures, brackets, downspouts, and similar items that were removed. Accessories that were not replaced, such as vents or service cables, may be painted to match the new siding color. Most manufacturers have matching touch-up paint formulas, which can be purchased at a local paint store. Maintain your work area. Remove all scrap pieces, cartons, nails, and other materials, and leave the jobsite neat and tidy each day.

3.0.0   Section Review

1. A key element of successful siding installation layout is establishing a straight _____.
   a. nailing pattern
   b. reference line
   c. siding joint stagger
   d. downspout

2. One of the most versatile interlocking board siding profiles that can be installed horizontally, vertically, and diagonally is called _____.
   a. lap
   b. Dolly Vardan
   c. tongue-and-groove (T&G)
   d. clapboard

3. The two most common types of cornices are the closed cornice and the _____ cornice.
   a. custom
   b. Bailey
   c. round
   d. box

Module 27204 Review Questions

1. When using a step ladder, you should never stand above _____.
   a. the safety labels
   b. the second uppermost step
   c. 4'
   d. the reference line

2. Per OSHA, a scaffold must be equipped with top rails, midrails, and toeboards, or workers must use a fall arrest system if it is above _____.
   a. the walking surface
   b. 10'
   c. eye level
   d. 6'

3. The most dangerous hazard from working with fiber-cement materials is _____.
   a. sharp edges
   b. their lack of durability
   c. silica dust
   d. heat retention

4. A nonloadbearing material beneath eaves that encloses and protects the space between a roofline and exterior wall is known as a _____.
   a. soffit
   b. starter strip
   c. rake wall
   d. corner post

5. The main purpose of building wrap or weather-resistant barrier (WRB) is to _____.
   a. provide the exterior with a rustic appearance
   b. improve structural fire-resistance
   c. provide padding for vinyl and metal siding materials
   d. protect the exterior sheathing and allow water vapor to pass from inside a structure to the exterior

6. The triangular section of a wall that is enclosed by the edges of a pitched roof is called a _____.
   a. cornice
   b. gable
   c. soffit
   d. drip cap

7. The most common flashing materials reserved for structures near coast regions that require extra corrosion-resistance due to the need to protect against the salty air, or on high-end projects and historic restorations, are stainless steel and _____.
   a. rubberized membrane
   b. aluminum
   c. copper
   d. fiber cement

8. The two most common tools for establishing a straight reference line are the water level and _____.
   a. framing square
   b. 6' level
   c. siding gauge
   d. laser level

9. You can provide a nailing surface for vertical siding by installing _____.

    a. flashing
    b. horizontal blocking
    c. WRB
    d. vertical furring

10. The two basic methods of shingle exterior wall application are single course and _____ course.

    a. side
    b. split
    c. double
    d. fancy shake

11. The most common siding fasteners are siding nails ranging between 6d and 10d sizes and _____.

    a. medium-crown staples
    b. manufacturer adhesive tape
    c. sealant
    d. brad nails

12. A common downside of vinyl, which has led many designers and craft professionals to move toward fiber-cement alternatives, is its _____.

    a. limited availability
    b. complex installation process
    c. lack of profile options
    d. low insulation value

13. The major advantages of fiber-cement siding over wood siding are that it is decay-resistant, noncombustible, and _____.

    a. comes in various profiles
    b. termite-proof
    c. does not create dust during cutting operations
    d. more environmentally friendly

14. The two most common types of cornices are the closed cornice and the _____ cornice.

    a. custom
    b. Bailey
    c. round
    d. box

15. Joists that extend past the exterior wall of a structure to support roof loads and provide nailing surfaces for fascia are called _____.

    a. lookouts
    b. soffits
    c. corner caps
    d. sills

Answers to Odd-Numbered Module Review Questions are found in *Appendix A*.

Answers to Section Review Questions

Answer	Section	Objective
Section One		
1. c	1.1.0	1a
2. a	1.2.3	1b
Section Two		
1. b	2.1.0	2a
2. d	2.2.0	2b
3. a	2.3.3	2c
Section Three		
1. b	3.1.0	3a
2. c	3.2.4	3b
3. d	3.5.0	3c

MODULE 27214

Mass Timber Construction

Source: sockagphoto/Shutterstock

Objectives

Successful completion of this module prepares you to do the following:

1. Describe the use of mass timber in the construction industry.
   a. Identify the building codes related to the use of mass timber.
   b. Examine the benefits of mass timber in different types of construction.
   c. Categorize the types of mass timber.
2. Summarize building techniques and tools used with mass timber.
   a. Explain specific handling, tool, and hardware requirements for mass timber.
   b. Recognize mass timber framing systems.
3. Describe common construction phases in the mass timber building process.
   a. Provide a brief explanation of the importance of early design phases in mass timber construction.
   b. Describe common safety considerations during the construction phase.
   c. Explain mass timber finishing techniques used during the post-construction phase.

Performance Tasks

This is a knowledge-based module. There are no Performance Tasks.

Overview

Mass timber construction has become increasingly popular in the US. More companies are manufacturing mass timber products and constructing mass timber buildings than ever before. To go along with this increase, there have been many code changes specific to mass timber. Mass timber products can be used in tall buildings, schools, hotels, multifamily residential buildings, public/institution buildings, and healthcare facilities. Because most mass timber products are prefabricated, there are specific guidelines for transportation, handling, and installation. This module will give you an overview of mass timber codes, types of buildings, benefits of mass timber, and framing systems used with mass timber construction.

Digital Resources for Carpentry

Scan this code using the camera on your phone or mobile device to view the digital resources related to this craft.

1.0.0 Mass Timber Construction

Performance Tasks

There are no Performance Tasks in this section.

Objective

Describe the use of mass timber in the construction industry.

a. Identify the building codes related to the use of mass timber.

b. Examine the benefits of mass timber in different types of construction.

c. Categorize the types of mass timber.

Mass timber is a relatively new term in the construction industry. It represents traditional large wood building elements called *heavy timber* along with engineered wood products created by gluing or fastening wood layers together to create panels, beams, and columns. Mass timber includes cross-laminated timber (CLT), glued laminated timber (glulam or GLT), laminated veneer lumber (LVL), parallel strand lumber (PSL), laminated strand lumber (LSL), nail-laminated timber (NLT), dowel-laminated timber (DLT), mass plywood panels (MPP), and certain heavy timber (HT) (*Figure 1*).

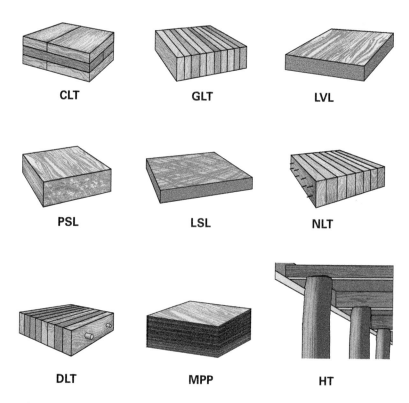

CLT **GLT** **LVL**

PSL **LSL** **NLT**

DLT **MPP** **HT**

Figure 1 Types of mass timber.

Mass timber products can be used to construct walls, floors, and roofs, or they can be used as accents. If mass timber is only used as an accent, the structure is not considered a mass timber building. A construction project is only considered a mass timber building when the primary load-bearing structure is made of either engineered or solid wood products with specific minimum dimensions as prescribed by the building code.

Mass timber was first used in Europe for construction projects in the mid-1990s. Since the early 2000s, its use has increased in Europe and spread to Canada and the US. In 2009, a signature mass timber project was completed in Canada for the 2010 Olympics. In 2017, the University of British Columbia built an 18-story mass timber student residence called Brock Commons. In 2016, Albina Yard, the first mass timber office building in the US, was completed in Portland, Oregon. Since then, mass timber construction has continued to increase throughout the US.

1.1.0 Building Codes

While traditional heavy timber (HT) construction has been part of the earliest building codes, a new definition for mass timber first appeared in the 2015 *International Building Codes*® (*IBC*®). With the increase in the use of mass timber tall building construction, the International Code Council® (ICC®) recognized the need to develop more in-depth codes specifically for mass timber. In December 2015, the ICC® developed the Tall Wood Building (TWB) Ad Hoc Committee with the specific purpose of exploring the science of tall wood buildings and developing potential code changes for tall wood buildings. The committee was made up of subject matter experts such as architects, structural engineers, fire protection engineers, fire officials, building officials, and building material representatives. In March 2016, the committee began researching, investigating, and testing mass timber products and construction. As a result, the ICC® TWB Ad Hoc Committee submitted 17 proposed code changes. The ICC® membership reviewed the proposed changes and accepted all of them as part of the 2021 *IBC*®. In addition to these code changes, the ICC® also developed a Tall Mass Timber Special Inspector curriculum to train and certify individuals specifically for mass timber building special inspections.

The new code provisions are based on the following changes addressing requirements for the three new types of mass timber construction, allowable building size limits, and fire protection provisions:

- *IBC*® Section 602.4 — Type of Construction (G108-18)
- *IBC*® Section 703.6 — Performance Methods for Fire Resistance from Noncombustible Protection (FS5-18)
- *IBC*® Section 722.7 — Prescriptive Fire Resistance from Noncombustible Protection (FS81-18)
- *IBC*® Section 703.7 — Sealants at Edges (FS6-18)
- *IBC*® Section 718.21 — Fire and Smoke Protection (FS73-18)
- *IBC*® Section 403.3.2 — High-Rise Sprinkler Water Supply (G28-18)
- *IBC*® Section 701.6 — Owners' Responsibility (F88-18)
- *IFC*® Section 3308.4 — Fire Safety During Construction (F266-18)
- *IBC*® Table 504.3 — Building Heights (G75-18)
- *IBC*® Table 504.4 — Number of Stories (G80-18)
- *IBC*® Table 506.2 — Allowable Area (G84-18)
- *IBC*® Section 3102 — Special Construction (G146-18)
- *IBC*® Appendix D — Fire Districts (G152-18)
- *IBC*® Sections 508.4 and 509.4 — Fire Barriers (G89-18)
- *IBC*® Table 1705.5.3 — Special Inspections (S100-19)
- *IBC*® Section 110.3.5 — Connection Protection Inspection (ADM35-19)
- *IBC*® Section 2304.10.1 — Connection Fire Resistance Rating (S170-19)

NOTE

Building codes vary from jurisdiction to jurisdiction. Because of the variations in code, check the approved construction documents and/or the local jurisdiction code whenever regulations are in question.

NOTE

Prior to the 2021 *IBC*®, heavy timber was classified as Type IV construction. With the addition of the mass timber construction types, heavy timber became Type IV-HT, but the code requirements stayed essentially the same.

More information about these code changes is available at *https://www.iccsafe.org/.* The three new Type IV construction types are shown in *Table 1* and *Figure 2.*

TABLE 1 Mass Timber Building Code Requirements

Type	Stories (Max)	Building Height (Max)	Fire Protection of Mass Timber Elements	Fire-Resistance Rating
Type IV-A	18	270'	• Fully protected mass timber elements • No exposed mass timber permitted	• 3 hours for bearing walls and structural frame construction • 2 hours for floor construction • 1.5 hours for roof construction
Type IV-B	12	180'	• Protected exterior and limited exposed interior mass timber • Limited exposed mass timber permitted	• 2 hours for bearing walls, structural frame, and floor construction • 1 hour for roof construction
Type IV-C	9	85'	• Protected exterior and exposed mass timber interior • Exposed mass timber permitted except at shaft walls, within concealed spaces, or on the exterior side of exterior walls	• 2 hours for bearing walls, structural frame, and floor construction • 1 hour for roof construction

Adapted from 2021 *IBC*®

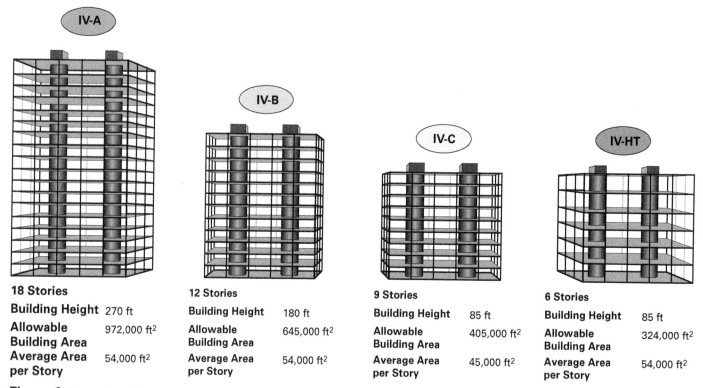

IV-A

18 Stories

Building Height 270 ft

Allowable Building Area 972,000 ft²

Average Area per Story 54,000 ft²

IV-B

12 Stories

Building Height 180 ft

Allowable Building Area 645,000 ft²

Average Area per Story 54,000 ft²

IV-C

9 Stories

Building Height 85 ft

Allowable Building Area 405,000 ft²

Average Area per Story 45,000 ft²

IV-HT

6 Stories

Building Height 85 ft

Allowable Building Area 324,000 ft²

Average Area per Story 54,000 ft²

Figure 2 Type IV construction.
Source: Adapted from 2021 *IBC*®

Combustible light-frame materials (including fire-retardant treated wood—FRTW) are specifically and completely excluded from these new types of construction. *IBC*® defines light-frame construction as "construction whose vertical and horizontal structural elements are primarily formed by a system of repetitive wood or cold-formed steel framing members." Light frame wood stud or joist assemblies are prohibited, including using light frame wood to fur out concealed spaces. Interior and exterior load-bearing and nonbearing walls must be mass timber or noncombustible construction.

1.2.0   Types of Construction and Benefits

Mass timber is being used in many building types that would have traditionally used steel, concrete, or masonry. Many building projects use only mass timber whereas others combine mass timber with concrete or steel in hybrid projects.

In the US, the use of mass timber construction for business offices, public/institution buildings, schools, multifamily residential buildings, hotels, healthcare facilities, and other buildings continues to increase. While code changes certainly made it easier to utilize mass timber for construction projects, they are not the only reasons for the increase. Expanding the use of mass timber will have environmental benefits and provide economic opportunities to disadvantaged rural communities with timber resources. It can make possible significant energy efficiency benefits, address construction labor shortfalls, and shorten construction schedules. The use of mass timber also has the potential to provide needed fire safety benefits, including wildland fire mitigation and more fire-safe construction sites.

The following advantages of mass timber continue to promote its use in many different types of construction projects:

- *Strength and durability* — Engineered wood products have a high strength-to-weight ratio. Fewer load-bearing walls means buildings can have more spacious interiors (*Figure 3*). Lighter materials mean reduced foundation sizes.

Figure 3 Open-plan construction.
Source: Simon Turner/Alamy

- *Environmental impact* — Wood is a renewable resource. Sustainable forestry practices promote selective cutting of trees to allow others to continue to grow and planting new trees to replace those that have been harvested (*Figure 4*). Mass timber walls may fulfill requirements that wall systems have a heat capacity exceeding limits that qualify as mass walls for energy consumption, thus reducing energy use. For energy code compliance, the definition of *mass* is different than the definition used in fire-resistive requirements. Wood also captures and stores carbon in a process known as carbon sequestration, so it helps reduce the amount of carbon dioxide in the atmosphere.

- *Waste and reclamation* — Mass timber products produce less waste because of how they are manufactured. Prefabricated mass timber products can be disassembled and reused or refabricated at the end of a building's life cycle. Wood is also biodegradable, so mass timber can be returned to the environment to decompose at the end of its useful life.

Source: Aardvark/Alamy

Figure 4 Sustainable forestry.
Source: Dietrich Leppert/Shutterstock

Figure 5 Biophilic design.
Source: Cannon Photography LLC/Alamy

- *Construction efficiency and safety* — Prefabricated mass timber products mean shorter project timelines. More work performed off-site during prefabrication means there is less cutting, fewer workers, and fewer temporary structures needed on-site. This helps to improve accuracy and safety on the jobsite.

- *Fire resistance* — Mass timber has lower thermal conductivity than steel, which means it transfers heat more slowly and is a better insulator. When exposed to fire, mass timber chars on the outside, forming a protective layer for the interior wood. Many tests have been conducted on mass timber to prove its fire-resistive properties and establish standards for construction.

- *Market value* — Some research studies have shown that buildings constructed of mass timber have increased marketability and better leasing rates.

- *Health and aesthetics* — People have a natural motivation to connect with nature, called *biophilia*. Mass timber construction utilizes biophilic design to incorporate wood and natural light to create a warm, inviting atmosphere (*Figure 5*). Some of the many biophilic benefits of mass timber buildings include greater satisfaction and productivity, lower heart rates and blood pressure, reduced stress and anxiety, reduced healing times in healthcare settings, positive morale, a feeling of well-being, and increased social interaction.

1.3.0   Types of Mass Timber

Mass timber as defined in the *IBC*® includes: "Structural elements of Type IV construction primarily of solid, built-up, panelized or engineered wood products that meet minimum cross section dimensions of Type IV construction." This identification provides a single term to represent various sawn and engineered wood products including structural composite lumber (SCL), glulam, and CLT.

Minimum dimensions for mass timber elements are included in the *IBC*® and are summarized in *Table 2*. Larger dimensions than the minimums may be required due to structural or fire design considerations.

TABLE 2 Minimum Dimensions of Heavy Timber Structural Members

Supporting	Min. Nom. Solid Sawn Size		Min. Glulam Net Size		Min. SCL Net Size	
	Width	Depth	Width	Depth	Width	Depth
Floor or Floor/Roof	8"	8"	6¾"	8¼"	7"	7½"
	6"	10"	5"	10½"	5¼"	9½"
Roof Only	6"	8"	5"	8¼"	5¼"	7½"
	6"	6"	5"	6"	5¼"	5½"
	4"	6"	3"	6⅞"	3½"	5½"

CLT Minimum Thicknesses: Floors = 4"; Roofs = 3"; Exterior Walls in Type IV = 4" with limitations

1.3.1 Cross-Laminated Timber (CLT)

Cross-laminated timber (CLT) is a prefabricated engineered wood product consisting of at least three layers of solid-sawn lumber or structural composite lumber. The adjacent layers are cross-oriented and bonded with structural adhesive to form a solid wood element like the one shown in the floor and walls in *Figure 6*. Panels are prefabricated based on the project design and arrive at the jobsite with windows and doors pre-cut. While sizes vary by manufacturer, CLT panels can range from 2' to 10' wide and be up to 70' long and 20" thick. These panels are lightweight, yet very strong, and are often used for long spans in walls, floors, and roofs.

Figure 6 CLT construction.
Source: holivideo/Getty Images

1.3.2 Glued Laminated Timber (Glulam or GLT)

Glulam is manufactured from lengths of solid, kiln-dried lumber that are glued together with moisture-resistant adhesive. The adjacent layers run parallel to each other. Glulam is popular in architectural applications where exposed beams are used. Glulam is available in three appearance grades: industrial, architectural, and premium. Industrial grade is used in open buildings such as warehouses and garages where appearance is not a priority or where beams are not exposed. Architectural grade is used where beams are exposed, and appearance is important. Premium, the highest grade, is used where the highest-quality appearance is needed (*Figure 7*).

Glulam is already a familiar product in commercial structures and integrates well in mass timber construction. It can be used as beams or columns or laid on edge to form horizontal panels similar to CLT, NLT, or DLT. In that application, the term GLT is often used.

Glulam members are used for many applications, including ridge beams; basement beams; window and door headers; stair tread, supports, and stringers;

Figure 7 Glulam construction.
Source: Roberto/Getty Images

a cantilever; and a vaulted ceiling. Because glulam members are laminated, they can be formed into arches and a variety of curved configurations spanning widths of 300' or more without using support columns.

1.3.3 Laminated Veneer Lumber (LVL)

Like plywood, LVL (*Figure 8*) is made from laminated wood veneers. Douglas fir and southern pine are the primary sources. Thin ($^1/_{10}$" to $^3/_{16}$") sheets are peeled from the trees in widths of 27" or 54". The veneers are laid up in a staggered pattern with the veneers overlapping to increase strength. Unlike plywood, the grain of each layer runs in the same direction as the other layers. The veneers are bonded with an exterior-grade adhesive, then pressed together and heated under pressure.

LVL is used for floor and roof beams and for the support members called headers that are used over window and door openings.

LVL Cutout

LVL Roof Construction

Figure 8 LVL construction.
Sources: Tim Cuff/Alamy (left); Vasily Ulyanov/Alamy (right)

1.3.4 Parallel Strand Lumber (PSL)

PSL is made from long strands of Douglas fir and southern pine. The strands are about $^1/_8$" or $^1/_{10}$" thick and are bonded together with waterproof adhesive in a special heating process. PSL has a high length-to-thickness ratio, or bending strength, which makes it very good for long-span beams and heavily loaded structural columns.

1.3.5 Laminated Strand Lumber (LSL)

LSL is manufactured from small logs of almost any species of wood. Aspen, red maple, and poplar that cannot be used for standard lumber are commonly used for LSL. In the manufacturing process, the logs are cut into short strands, which are bonded together and pressed into blocks (billets) up to 8' wide and 40' long. LSL will not support as large a load as a comparable size of PSL because PSL is made from stronger wood.

1.3.6 Nail-Laminated Timber (NLT)

NLT has been around for a long time and is traditionally defined in the code as mechanically laminated decking. It is made by stacking layers of dimensional lumber on edge and fastening them together with nails or screws. NLT can be used to create floors, roofs (including curved roofs), walls, and elevator shafts. Because of the nails or screws that hold it together, it is not a favored option for any span that needs openings cut in it.

1.3.7 Dowel-Laminated Timber (DLT)

DLT is similar to NLT in that it does not use adhesives. For DLT, pieces of dimensional softwood lumber are laid on edge next to each other with holes through all pieces so that hardwood dowels can be friction-fitted through all of them. The differences in the moisture content between the softwood and hardwood cause shrinking and swelling, which locks the panels together. DLT can be used for walls, floor and roof construction, stairs, elevator shafts, and to create curved roof structures similar to NLT.

1.3.8 Mass Plywood Panels (MPP)

MPP is one of the newest mass timber products on the market. These panels are made from thin layers of veneer that are glued together as a combination of face and core panels. The face and core panels have different ply configurations to give the MPP greater stiffness and stability. Because of the way they are made, MPP can use smaller trees and 20% less wood fiber than CLT yet have the same structural specifications. The panels vary in size, but can be up to 12' wide, 48' long, and 1' thick.

1.3.9 Heavy Timber as a Type of Mass Timber

In certain instances, structural round timber (SRT) and sawn heavy timber (SHT) can be considered mass timber. SRT is large solid logs of wood used for beams, floors, columns, and roof assemblies (*Figure 9*). SHT is solid sawn lumber often used in post and beam construction. Heavy timber is classified as Type IV-HT and must meet the dimensional requirements of *IBC*® Section 2304.11.

Figure 9 Craig Thomas Discovery Visitor Center, Grand Teton National Park.
Source: Lee Rentz/Alamy

1.0.0   Section Review

1. What is the maximum height permitted for mass timber construction in the *IBC*?
   - a. 85'
   - b. 120'
   - c. 180'
   - d. 270'

2. What is biophilic design in mass timber construction?
   - a. Prefabricating materials off site
   - b. Increasing seismic performance
   - c. Incorporating wood and natural light
   - d. Decreasing foundation size

3. Which type of mass timber is fabricated with cross-oriented layers bonded with structural adhesive?
   - a. GLT
   - b. NLT
   - c. CLT
   - d. PSL

2.0.0 Building Techniques and Tools

Performance Tasks	Objective
There are no Performance Tasks in this section.	Summarize building techniques and tools used with mass timber. a. Explain specific handling, tool, and hardware requirements for mass timber. b. Recognize mass timber framing systems.

Mass timber is a specialized building process with considerable potential for growth as society moves toward a greener and more sustainable future. Many of the core principles of carpentry will still be applicable with mass timber construction. However, most of a mass timber project is coordinated and finalized during the design phase, allowing smaller teams of specialized craft professionals to complete their scope in a fraction of the time it would take to construct a similar project using traditional building methods (*Figure 10*).

Figure 10 Mass timber construction.
Source: imageBROKER.com GmbH & Co. KG/Alamy

Depending on the type of mass timber project you are working on, you will likely use many of the same tools discussed in the rest of this Advanced Carpentry curriculum. Because most of the panels and beams you'll be working with are prefabricated, few modifications will be needed if the project team has ensured that all structural members will align within tolerances (*Figure 11*). Although there is often more than one way to complete a task, the following sections will include an introduction to the most common material handling and installation techniques in the field.

Figure 11 Mass timber beams.
Source: Opreanu Roberto Sorin/Alamy

2.1.0 Material Handling, Tools, and Hardware

As with any other form of construction, mass timber and engineered wood construction projects can be completed using various methods. However, the nature of prefabrication results in minuscule tolerances and limited field modification. Therefore, the preconstruction planning phase is vital to ensure that all the pieces fit together to construct a structurally strong and aesthetically pleasing building (*Figure 12*).

Figure 12 Mass timber aesthetics.
Source: Peter Cook-VIEW/Alamy

Regardless of your individual responsibilities on a given project, as a craft professional you need a basic understanding of mass timber construction processes from start to finish. This knowledge will help you expand your skill set and continue to rise in the ranks of mass timber specialists.

2.1.1 Material Handling and Transportation

Material handling is one of the most important aspects of mass timber construction because damaging large, prefabricated structural members could delay a project for months, adding potentially millions of dollars to the budget while a replacement piece is ordered, fabricated, and shipped. Many engineered posts, beams, and panels are exposed, so even small damage will ruin the appearance of the final product. Craft professionals must take great care when receiving, organizing, handling, and installing these high-cost, essential building materials (*Figure 13*).

Figure 13 Installing mass timber panels.
Source: Michael Doolittle/Alamy

Transportation is another complex piece of the mass timber puzzle. Each project poses unique challenges, and project teams are responsible for selecting the most effective strategy to optimize productivity, reduce trade stacking, and ensure the installation team receives each unique delivery package on time and in the correct sequence. The following three material transportation techniques are commonly used to get engineered materials from manufacturers to the jobsite:

- *Just-in-Time (JIT) Delivery* — This delivery strategy is the most common for mass timber projects, especially those with limited laydown areas for storing materials. JIT deliveries require strict adherence to the project schedule and constant cross-team collaboration to ensure that the site is prepared and ready for shipments before they arrive. This also means that other trades must complete their scope and clean their respective work areas to avoid the risk of hindering lifting and installation operations. JIT deliveries only allow for a short window of vehicle access, rigging, and installation all in one shot (*Figure 14*).

- *One-Time Shipment* — One-time shipments are common for large-scale timber projects or hybrid construction projects that require only a small amount of engineered wood materials, as is the case with artistic architectural projects with mass timber facades (*Figure 15*). This delivery technique requires a detail-oriented focus on material protection because building materials may

Figure 14 Mass timber transportation.
Source: Vitpho/Shutterstock

Figure 15 On-site storage.
Source: nieriss/Shutterstock

be stored for weeks or months. Inadequate material protection will likely result in common damages like cracking, cupping, and delamination from shrinkage caused by rapid moisture content fluctuations.

- *Phased Delivery* — This transportation technique is basically a hybrid between one-time and JIT deliveries. Scheduled packages of engineered wood panels, beams, and posts are shipped by ship, freight rail, and flatbed trucks to the jobsite to be installed in specific building sectors. Phased deliveries may contain necessary materials for an entire floor and will therefore require more storage area than a typical JIT delivery.

Around the World

Mass Timber Logistics

Although North American manufacturers are quickly adapting to this new and exciting industry, and the North American continent is rich in sustainable lumber, our society is somewhat late in adopting mass timber as a viable alternative to traditional steel and concrete structures. For this reason, many projects still rely on foreign manufacturers to fulfill orders for engineered wood building materials.

Some of the largest producers of CLT and other common mass timber materials include Germany, Sweden, Austria, and Japan. This distance between the manufacturers and jobsites must be a major logistical focus for any wood structure project because the initial lead times will often take three to four weeks. Because every product is custom-built to the specifications of the project, building components that are missing in transit, damaged in storage or handling, or fabricated to the wrong dimensions will cause a costly hiccup in the project schedule. The age-old carpenter's adage "measure twice, cut once" has never been more true than when working on these complex projects.

Sources: Agencja Fotograficzna Caro/Alamy (left); Andrey Guryanov/Alamy (right)

NOTE

Many projects may require extra materials to be ordered for field modifications. Although most aspects of mass timber projects should meet exact dimensions and tolerances, building is never an exact science. These engineered wood components must lay on concrete slabs, connect to steel and concrete structural pieces, and attach to hardware locations—all of which may not meet the necessary tolerances of the original design.

Did You Know

Building Information Modeling (BIM) + Augmented Reality (AR) = The Future

Mass timber is a pioneering building model taking its first steps toward a more streamlined, sustainable future of construction. Many mass timber building components are prefabricated to tolerances ranging between mere millimeters and $\frac{1}{16}$". The project team and installers must ensure that any hybrid systems meet the necessary dimensions for each panel and beam to come together as perfectly as possible.

Most of this quality control can be accomplished using technology throughout the construction process. Building information modeling (BIM) creates a workable 3D model of the structure that allows teams to quickly filter every level down to the nut and bolt in question. Teams using BIM can enjoy faster project coordination between subcontractors and manufacturers, as well as improved clarity on minute details. Emerging augmented-reality technologies are taking this concept a step further with wearable goggles that can create heads-up displays (HUD) for workers in the field, allowing craft professionals to visualize every facet of construction in the field. This convenient technology will be especially helpful for mass timber specialists as they track necessary changes to structural panels for mechanical, electrical, and plumbing (MEP) chases and other alterations.

Source: ME Image/Shutterstock

2.1.2 Moisture Protection

One of the greatest weaknesses of mass timber construction is that wood products are not as resilient against moisture as traditional concrete and steel materials. Therefore, developing and implementing a material storage and protection plan is essential to reduce the risk of damaging vital building components.

First, identify the weak points where moisture has a greater chance of penetrating into the wood. These weak points include end grains, floor-to-vertical connections, exterior wall openings, and prefabricated chases (*Figure 16*). Once moisture penetrates CLT and other engineered products, it can interact with heat from the sun, causing a hygroscopic effect. This can lead to wood panels and beams cupping, cracking, or warping as materials dry and shrink, which can cause serious setbacks with any building style including mass timber framing systems.

Figure 16 Identifying weak points.
Source: Flystock/Shutterstock

Because mass timber projects require precise structural connections, even small changes to the intended form and dimensions of a wood component could render it unusable, forcing the project team to request information on adequate field modifications from the engineer of record (EOR), manufacturer, and architects. They may even be required to replace the damaged component entirely. This is another reason why it's vital to coordinate between all trades and manufacturers during the preconstruction phase to ensure that each vertical and horizontal member is prefabricated to accommodate the necessary space and openings for mechanical, electrical, and plumbing (MEP) and other essential systems (*Figure 17*).

Moisture can also cause watermarks on exposed surfaces, and extreme water damage to wood materials could increase the risk of mold and mildew. At the very least, these watermarks must be fixed to produce a clean aesthetic look, and like any other form of remediation or touch-up work, these tasks can cause schedule delays and unintended labor costs.

Most manufacturers will ship wood products wrapped in a fleece-based breathable membrane or similar packaging to combat these issues. Leaving this protective wrap in place will help keep your building components safe from atmospheric conditions during transport, site delivery, and site storage (if applicable). However, these wraps do little to combat moisture damage once installed. Other water-damage reduction strategies must then be developed once materials are in place. Protective materials can be applied around columns and high-traffic areas to reduce the risk of impact damage (*Figure 18*).

Figure 17 Prefabricated MEP openings.
Source: D Watts/Alamy

Figure 18 Protective materials.
Source: Jarama/Shutterstock

Vertical structural members are often less affected by atmospheric conditions than horizontal floors because the runoff effect will reduce the amount of pooling. Your team should devise a strategy for mitigating standing water on any exposed floors. Supervisors and project leads should be aware of changing weather reports and take necessary precautions for operations when heavy rains or snow are expected in the daily forecast (*Figure 19*).

Limiting moisture exposure is the best way to ensure your mass timber materials function as intended. Specialized tape, sealant, and weather barrier can help protect weak points, such as the end grains of CLT, from saturation. This strategy may be more difficult to achieve in regions that experience regular rainfall for most of the year.

If your mass timber components become wet, avoid using strong heaters, fans, and other drying methods you would typically use for standard dimensional lumber. These strategies could cause distortions from rapid temperature

Figure 19 Weather changes.
Source: Bannafarsai_Stock/Shutterstock

and moisture changes. Your best bet is a slow, steady drying process with adequate air circulation and regular moisture meter measurements.

2.1.3 Rigging Mass Timber

As a craft professional specializing in mass timber construction, you must learn the safety planning and rigging techniques for common engineered wood materials of various dimensions and weights. Luckily for certified riggers trained in structural steel and concrete tilt-up operations, the rigging and lifting techniques, tools, and mathematical calculations remain the same (*Figure 20*). For a deeper dive into crane safety and operations, you should familiarize yourself with OSHA 1926 standards—most of which are covered in NCCER's Mobile Crane Operations Curriculum.

Figure 20 Rigging.
Source: Michael Doolittle/Alamy

One of the most common mass timber lifting maneuvers is rigging and moving flat panels (*Figure 21*). These panels can be delivered with housed or unhoused transport anchors. These transport anchors are sometimes known as "screw-hoist systems" because they are fastened into place with self-tapping screws by the installers on-site or preinstalled by the manufacturer. Most anchors require between four and twelve long, self-tapping screws, and the number depends on the load capacity.

Another common hoisting hardware option for flat panel operations is the yoke anchor (*Figure 22*). A yoke anchor is installed at every corner of the panel, spaced equally from the edge to support the load's center of gravity evenly. Because many of these anchors surpass the necessary 5000 lb strength capacity, they can also double as OSHA-compliant fall-arrest anchors for installers working near exposed edges of elevated work platforms. However, it is vital that you discuss this fall-protection method with your project's safety team while developing your team's task-specific safety plan.

Figure 21 Moving flat panels.
Source: sockagphoto/Shutterstock

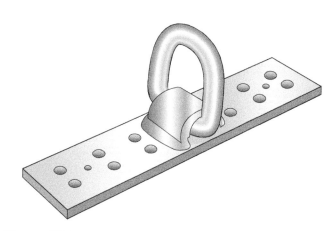

Figure 22 Yoke anchor.

Each yoke system faces diagonally toward the center of the load, so the loop aligns with the sling axis at a minimum 60° sling angle. You can also use heavy-duty yoke anchors on vertical wall panel lifts by installing the anchors on the top two corners with baseplates installed in vertical orientations (*Figure 23*).

Figure 23 Vertical panel installation.
Source: Michael Doolittle/Alamy

Because mass timber construction is a streamlined building process with minuscule tolerances on prefabricated materials, your team can save considerable time by grouping similar lifts in phases. You can expedite the process further by laying out a single panel for anchor installation, and then crafting job-built jigs to simplify repetitive lifts (*Figure 24*).

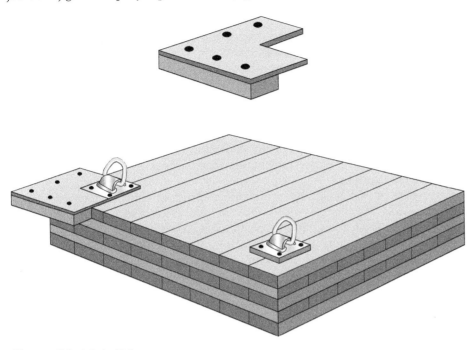

Figure 24 Job-built jigs.

Some project teams and manufacturers choose to skip the step of installers fastening anchors on site and will instead choose an integrated lifting system that is built into the panel itself. The speed and convenience of these systems significantly expedite rigging and installation operations, which makes them the perfect choice for JIT deliveries with short delivery windows.

These prefabricated anchor points typically include a threaded bolt, topped with a loop, sling, hook, or eyelet that is tightened into tube steel that runs through the depth of the panel. The bottom end is capped with a steel base plate for secure transportation. This base plate is either welded to the tube steel or tightened securely in place with a heavy-duty nut for threaded rods (*Figure 25*). For floor panels, once the material is in place, installers can quickly disassemble the system and repair any holes before pouring the protective slab. Otherwise, some housed systems can be left in place.

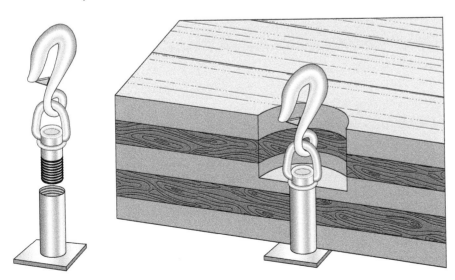

Figure 25 Anchor points.

Figure 26 Out-of-service sling.
Source: King Rope Access/Shutterstock

Lifting slings are commonly used as quick and efficient rigging devices for large panels with openings, like what you might find on exterior walls with windows or floor panels with prefabricated chases. They are also one of the best tools for lifting long glulam beams. Nylon slings are the best choice for rigging members with exposed surfaces because they will not scratch the finished product as much as chains and steel ropes. Just ensure they are regularly inspected by a competent person on your team and tagged/taken out of service if any fraying or other damages are found that may lead to a system failure and potentially life-threatening struck-by accidents (*Figure 26*).

A load spreader or spreader bar may be another useful device if your project requires lifting and moving massive panels. Spreaders consist of a long bar holding spaced wire rope slings that convert heavy loads into compressive forces through the bar and tensile forces through the slings, effectively spreading the load and center of gravity (*Figure 27*). This spreading of weight allows the crane operator more control over the loads and allows the rigging and installation team more maneuverability as they manipulate the beams and panels into place with attached tag lines.

Regardless of which rigging and lifting techniques your team chooses, it is everyone's duty to never sacrifice safety for production. Cutting corners at any stage, whether it be equipment inspections, safety planning, or load calculations, can lead to serious accidents resulting in the loss of property and life. Take the time to ensure that everyone understands their role in this high-risk operation. If you see something that appears to be unsafe, say something immediately. It could mean the difference between life and death.

Figure 27 Spreader bar.
Source: Red_Shadow/Shutterstock

2.1.4 Common Tools and Tool Safety

The most effective tool in your toolbox as a mass timber installer is a low rotation-per-minute (RPM) drill with exceptionally high torque. Low-RPM/high-torque drills are essential for driving long fasteners into engineered wood components. A high-RPM drill or impact driver will likely break longer screws, leading to several complications and incorrect fastening of vital hardware and timber-to-timber connections (*Figure 28*).

Figure 28 Cordless compact drill/driver.
Source: Rigid Tool Company

Several battery-powered options for these tools are available, but for extremely long screws, you should select a model with a hand-grip attachment and drive guide (*Figure 29*) that offers more control, allowing you to drive fasteners straight and true.

Figure 29 Cordless drill with handle attachment and drive guide.
Sources: Rigid Tool Company (left), Mitch Ryan (right)

To avoid many of the issues of limited battery life and eliminate the need for battery charging stations on every floor, your team can choose a corded option if your work area can accommodate spider boxes or other (semi-mobile) power connections (*Figure 30*). This choice may expedite your work and circumvent many logistical problems faced by installers on the Ascent building, currently the tallest completed mass timber tower project in the world.

Workers on the Ascent project have said that traditional battery-powered drills and impact drivers were one of the greatest drawbacks to their workflow. Each floor required a team of installers to drive nearly 7,000 long fasteners, and batteries would drain after completing less than ten before needing a recharge. Many industry leaders recommend avoiding the use of impact drills for long

NOTE

All other safety considerations for tools used in mass timber construction should align with principles found in NCCER's *Core*. As a rule of thumb, craft professionals should study the manufacturer's instructions and receive proper training on operations, maintenance, and all other vital information for each tool they use to complete their scope in the field.

self-tapping screws because the quick starting-and-stopping action can cause damage to the fastener and reduce overall structural integrity.

An important safety aspect of working with hand tools at great heights is ensuring that you take all the necessary precautions to avoid dropping your equipment on people below. Every jobsite must have the proper guardrails, toe-kicks, and safety netting in place, but it doesn't hurt to take an extra step of caution when working near the edges of the structure. All craft professionals can use tool lanyards when working with hand tools on multi-story projects. These accessories attach seamlessly to an installer's fall-protection harness and add an extra level of protection for working at extreme heights (*Figure 31*).

Figure 30 Corded hammer drill.
Source: Rigid Tool Company

Figure 31 Installer with tool lanyard attached to safety harness.
Source: Photobac/Shutterstock

2.1.5 Class 1 Connections

Mass timber panels are manufactured to either lap, join, or butt together to form a tight connection, made structurally sound by approved fasteners. Class 1 connections include any wood-to-wood joining without the need for hardware. This can include various types of mortise-and-tenon-style joints or dowelled post-to-beam connections on small structures, but for commercial structures, the most common connections are rabbeted, cleated, and splined.

Rabbeted joints are the most common for corner wall connections. Corner panels are manufactured with half-laps that marry together to form a tight 90° intersection. This method is superior to a traditional butt joint because installers can fasten long self-tapping screws through the overlap and securely fasten the exterior of one panel into the end grain of the other (*Figure 32*).

The Structural Engineer of Record (SER) may also decide on several situations where a cleated joint is the most effective connection. These typically occur at T-shaped wall intersections but may also be effective for floor-to-wall connections where the panel slides on top of a sill plate or when continuous vertical wall panels must fit together in a tongue-and-groove pattern (*Figure 33*).

One of the most commonly used Class 1 connections in mass timber floor and roof panel intersections is the spline joint (*Figure 34*). Splines are narrow rips of wood structural panel (WSP) or LVL that fit within grooves between adjoining panels, effectively taking the place of a half-lap joint. A spline is set in place and fastened with the specified gap tolerances and nailing pattern to allow for shear movement and seismic shifting of the structure. In common

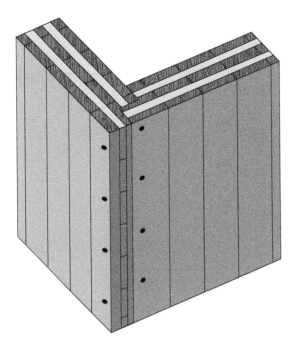

Figure 32 Rabbet joint.

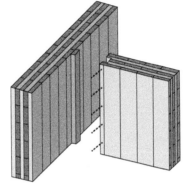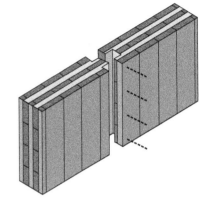

Figure 33 Cleated joint.

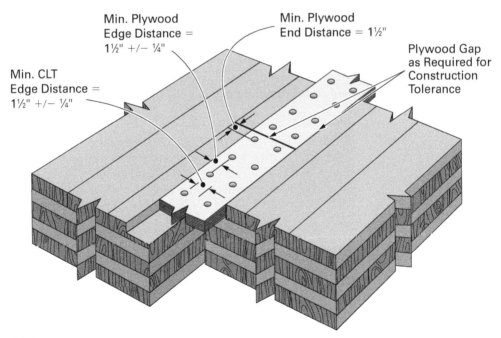

Min. CLT
Edge Distance =
1½" +/− ¼"

Min. Plywood
Edge Distance =
1½" +/− ¼"

Min. Plywood
End Distance = 1½"

Plywood Gap
as Required for
Construction
Tolerance

Figure 34 Spline joint.

diaphragm assemblies, these gaps may need to be applied with a fire-stopping sealant unless the floor is to be covered with a concrete slab.

These splines can be applied on the surface or fit internally as the panels fit together (*Figure 35*). Surface splines are far easier to fasten and have more flexibility with tolerances; however, an internal spline may be the better option when the design requires hidden connections between panels or when one side is inaccessible. Regardless of what spline is used, many panel-to-panel seams are finished with a manufacturer-approved, peel-and-stick weather-resistant membrane to reduce moisture penetration and fill gaps for future floor coverings.

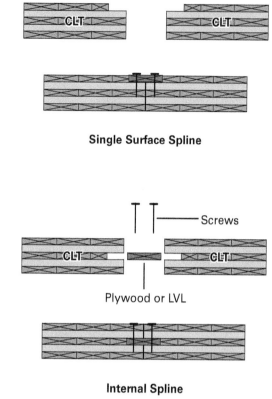

Figure 35 Splines.

2.1.6 Class 2 Connections

Class 2 connections require strong metal hardware to complete the wood-to-wood connection and evenly hold and distribute load and shear forces, especially in lower levels that experience the most force. Due to the necessary gaps between metal and wood components to accommodate natural movement, additional fire protection will likely be required to meet code specifications.

Metal Brackets and Plates

Metal brackets and plates make up the majority of Class 2 connections and they are manufactured to serve a wide range of purposes on a mass timber project. Angle brackets can act as supports and connection points for beams and stairs, as well as floor-to-wall connections for diaphragms and shear walls (*Figure 36*).

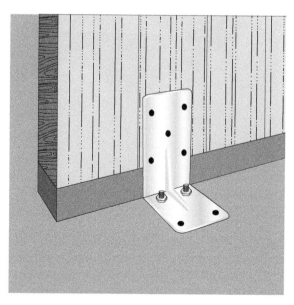

Figure 36 Angle bracket.

Depending on the load or fire resistance needed and the aesthetic intent of the project, some connections may require the use of hidden brackets and knife plates that fit in a prefabricated or field-cut slot inside a mass timber component, before being secured with fasteners from the outside (*Figure 37*).

Figure 37 Prefabricated fastener slots.
Source: Arch Steve/Shutterstock

Anchors and Tension Straps

Anchors or "hold-downs" are common hardware options for light-frame construction, but they also play a significant role in mass timber designs. They provide resistance against overturning moments caused by lateral loads such as wind or seismic action and are applied to handle vertical uplift forces to the foundation. Continuous anchor tie-down systems are common for interior shear wall framing on multi-level hybrid structures that implement light-frame construction, while floor-to-wall/panel-to-panel connections are typically completed with individual hold-downs (*Figure 38*).

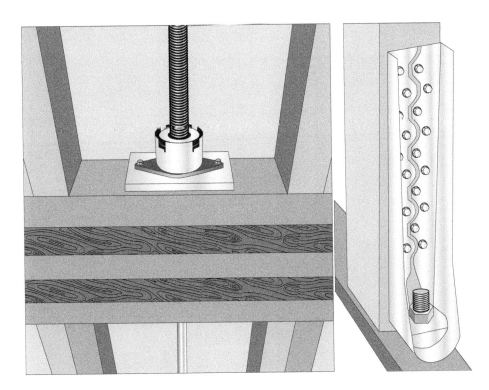

Figure 38 Anchor tie-downs.

Tension straps, commonly known as "drag straps," resist tension forces between two structural components (*Figure 39*). In mass timber design, these metal straps are installed between adjoining exterior wall panels or between mass timber and another building material. They may also act as bridges between one exterior wall and the wall of the floor below. Because CLT panels are so large, most designs will require at least 7-gauge drag strap thicknesses.

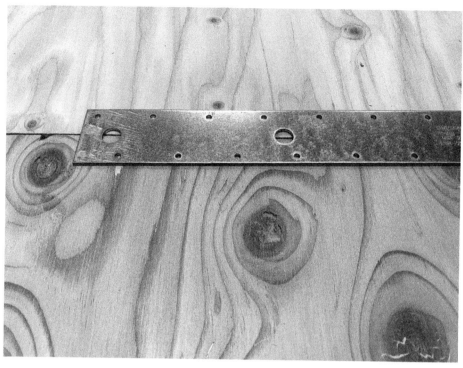

Figure 39 Tension straps.
Source: Mitch Ryan

Beam Hangers and Beam Seats

Beam hangers (*Figure 40*) and beam seats are an essential part of post-and-beam construction because they provide a secure connection to posts and girders while still providing the proper standoff to allow mass timber structural members to

Figure 40 Beam hangars.
Source: Charles Stirling/Alamy

swell, shrink, and move with seismic and shear forces within tolerances. Beam seats also provide the benefit of creating the necessary material separation between wood and steel or concrete components, reducing the risk of water damage to the wood and improving the strength and longevity of the building's structural design.

2.1.7 Class 3 Connections

Class 3 connections require the use of custom, proprietary hardware that is crafted to meet specific requirements and perform individual functions. Many of the proprietary hardware options in the mass timber field, such as concealed beam hangers (CBH), accommodate quick and simple installation between beams and panels with the hardware prefabricated or field-fitted within the structural members. This hidden design creates a seamless appearance without the need for face-mounted hardware that detracts from the natural aesthetic (*Figure 41*). These hidden connection systems are also required to have fire protection design, and wood cover is one way of achieving this.

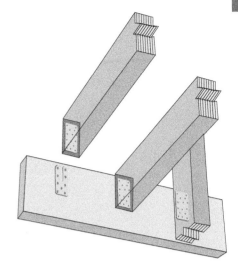

Figure 41 Concealed beam hanger.

Fasteners

Nails are not used in mass timber construction as often as in traditional light-frame wood construction due to the way that engineered wood panels are formed. Because mass timber members are manufactured by several pieces of wood being glued and pressed together, nails are typically used as specified for hardware connections, such as metal plates and brackets, applied to the panel face. Any nails applied to the end grain will not be able to resist withdrawal forces caused by the natural movement of shear and uplift forces. Therefore, self-tapping screws of varying sizes are the preferred fastener choice for connecting mass timber components (*Figure 42* and *Figure 43*).

Figure 42 Fasteners.
Source: Timelynx/Shutterstock

Figure 43 Mass timber fasteners.
Source: Mitch Ryan

2.2.0   Framing Systems

Mass timber construction allows for various building designs, from small, modular structures to massive, multipurpose high-rises. Because mass timber is a budding industry in North America, project teams are still learning the strengths, weaknesses, and logistical challenges of this type of construction. The following section will discuss common framing systems used with mass timber, as well as a brief introduction to how these products can be used in hybrid systems with steel and concrete.

2.2.1 Platform Construction

Platform construction, in which floor panels rest on loadbearing wall panels, is one of the easiest and most straightforward framing systems to learn because it aligns with traditional building principles that are familiar to most craft professionals. The main difference is the designs select large, engineered wood panels, columns, and beams to carry the load typically reserved for masonry, concrete, and stick-framed walls.

2.2.2 Post-and-Beam Construction

The three main components of a typical post-and-beam mass timber framing system are vertical columns, horizontal beams in one or multiple directions, and a mass timber floor or roof panel that sits on top. Differing from the post-and-plate style, this floor panel only spans along its primary strength axis between beams. However, due to its simple, sturdy design, post-and-beam remains the most common framing system in mass timber construction (*Figure 44*).

2.2.3 Post-and-Plate Construction

One of the major differences between post-and-plate (*Figure 45*) and post-and-beam mass timber construction is the floor or roof panels span in both major and minor strength axes, allowing the structure to distribute loads evenly on large columns and avoid the need for supporting beams. This framing technique improves finished ceiling heights and accommodates spacious floor plans for multifamily residential common spaces and open-concept commercial offices.

Figure 44 Post-and-beam construction.
Source: Jarama/Shutterstock

Figure 45 Post-and-plate construction.
Source: sockagphoto/Shutterstock

However, one of the main drawbacks of doing away with beams is that designs must reduce column spacing and potentially add proprietary connection devices to provide adequate structural load bearing (*Figure 46*).

2.2.4 Balloon Construction

Balloon framing is an effective strategy for quickly constructing a low-rise mass timber structure. A balloon-framed timber building consists of continuous wall panels (up to 3 stories tall) with drop-in beams and engineered floor panels. The speed of construction is a game changer when building in high-moisture regions where the installation windows are shorter between substantial rainfall. Balloon framing increases floor-to-floor height capabilities, accommodating tall ceilings in open floor plans, but also adds stress to connection points. Although more research is needed to test the seismic, thermal, and sound-dampening efficiency of this framing method,

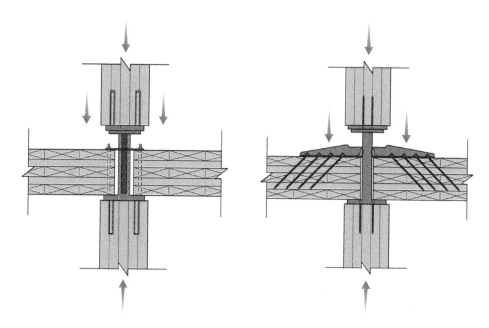

Figure 46 Proprietary column-to-column connections.

the combination of swift applications with prefabricated structural members opens the door to a new realm of possibilities.

Mass timber balloon construction is also making headway as a fast and effective alternative to low-rise concrete tilt-up designs and shaft construction for buildings up to 12 stories or 180'. Because mass timber panels can create an easy-to-install fire-resistant assembly for shaft enclosures, more design teams are realizing the potential for mass timber solutions that meet continuity requirements detailed in *IBC*® Section 707.5. However, these teams must also contend with necessary shaft penetrations and designs where shaft walls are also exterior walls. Ensuring continuity through these vital fire-barrier structures requires creative solutions.

As a craft professional working on a balloon-framed mass timber project, you are not expected to memorize codes word for word, but you need a basic understanding of shaft wall construction and common code requirements. This knowledge will give you the necessary resources to ensure your work will meet industry standards and protect future occupants in case of an emergency.

2.2.5 Hybrid Mass Timber Systems

Both platform and balloon framing techniques can benefit from the use of hybrid systems using concrete, steel, or a combination of both. Mass timber can improve the sustainability of a project because it is a carbon-sequestering building material that requires less water consumption than concrete and steel. Its high strength-to-weight ratio can also add vital strength for a fraction of the load. This allows project teams to add the utility of mass timber to their design toolbox to solve a wide range of construction challenges.

Mass Timber with Concrete

One of the most common combinations for hybrid mass timber structures is engineered wood and concrete (*Figure 47*). Reinforced concrete is always the primary foundation slab, and often the core as well. Because mass timber floors do not perform well against wear and tear and sound attenuation, each floor is often covered with an acoustical sound mat and concrete- or gypsum-based top slab.

One major challenge of mass timber/concrete hybrids is scheduling. Project teams need precise coordination between manufacturers and subcontractors. The unpredictability of concrete can often cause trade stacking and schedule delays as each delivery/installation phase is dependent on the cure time of each subsequent slab pour. Another consideration is the need to avoid direct contact between the materials, because concrete provides a conduit for moisture, leading to water damage and various structural issues.

Figure 47 Mass timber and concrete.
Source: sockagphoto/Shutterstock

Did You Know?

Project Highlight: Mississippi Building by Waechter Architecture

Portland, Oregon, a small metropolis in the Pacific Northwest, is quickly becoming an epicenter for the mass timber movement in North America. Surrounded by acres of sustainable forests and sprawling urban areas, teeming with an eco-conscious population in need of housing and retail space, the City of Portland is perfectly positioned to become the industry testing ground for innovative designs.

Portland's award-winning Mississippi Building was designed and built by Waechter Architecture. The 9,550 ft^2, balloon-framed, mass timber construction was completed in the spring of 2022. Aside from the radiant concrete floor slabs, MEP systems, structural hardware,

and fasteners, most of the structure is made entirely of CLT panels and exposed glue-lam beams (sourced and fabricated by KLH) without the need for additional fire protection. Even the elevator shaft and drop-in stairs were constructed using prefabricated engineered wood panels. The interior walls are also exposed, and the innovative design emanates a rich, natural aesthetic.

Trailblazing designs like Waechter Architecture's Mississippi showcase what's possible with mass timber construction. Further study and increased investment in the field point the way to a brighter, more sustainable future for the construction industry.

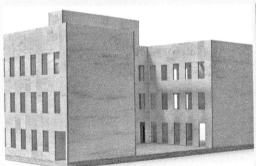

Source: Waechter Architecture (left)

Mass Timber with Steel

Steel is one of the most complementary materials for mass timber—both structurally and aesthetically. Steel can help carry areas of the building that experience concentrated forces, such as braces, columns, and trusses. The combination of steel with mass timber framing systems allows for longer spans, increased column spacing, and improved seismic resilience (*Figure 48*).

Figure 48 Mass timber and steel.
Sources: Paul Christian Gordon/Alamy (left); Opreanu Roberto Sorin/Alamy (right)

The visual juxtaposition of wood and steel structural members can also create elaborate aesthetics for designers hoping to achieve an elegant, industrial look. Some of the most common mass timber design elements you'll find in new structures are long, curved CLT beams that serve both structural and interior design functions, as well as engineered wood components that can be implemented as nonstructural focal points of modern facades and decorative curtain wall systems (*Figure 49*).

Figure 49 Curved CLT beams.
Sources: Hufton+Crow-VIEW/Alamy (left); Alan Paterson/Alamy (right)

2.0.0 Section Review

1. Which class of connections for mass timber panels includes any wood-to-wood joining without the need for hardware?
   a. 1
   b. 2
   c. 3
   d. 4

2. In which type of mass timber framing do the floor or roof panels span in both major and minor axis strength, allowing the structure to distribute loads evenly on large columns and avoid the need for supporting beams?
   a. Post-and-beam
   b. Balloon
   c. Post-and-plate
   d. Hybrid

3.0.0 Mass Timber Building Process

Objective

Describe common construction phases in the mass timber building process.

a. Provide a brief explanation of the importance of the early design phases in mass timber construction.

b. Describe common safety considerations during the construction phase.

c. Explain mass timber finishing techniques used during the post-construction phase.

Performance Tasks

There are no Performance Tasks in this section.

Mass timber construction is similar to other structural projects in several ways, but one of the most vital differences is the strong emphasis on early coordination for every subsequent phase to go as smoothly as possible. This section will provide a brief overview of some of the most important steps in the mass timber process, from systems and detail designs to material handling, installing, and finishing.

3.1.0 Phases of Construction

Every construction project begins in the conceptual stage where architects, engineers, and general contractors come together to solve a problem: What do we build and how do we build it to fulfill the potential needs of future occupants? To do this, many different teams with specialized skill sets must cooperate to turn an idea into a functional and aesthetically pleasing final product.

It is simple to think of a building like a human body. Each project begins with the structural design that forms the bones of the structure. Connection details provide the muscles and ligaments that keep the structural bones in the correct configuration to allow the building to stand upright without falling due to its own weight or outside forces (*Figure 50*).

Next, architects, engineers, and the project team work together to design MEP systems that act as the structure's vital organs and circulatory system, transporting power, air, and water wherever it is needed throughout the building. This crucial step is where mass timber designs differ from more traditional construction processes. Because most mass timber components are prefabricated, project teams must develop creative and innovative solutions to ensure that structural designs accommodate MEP cutouts and chases before shop drawings are sent to the manufacturer for production.

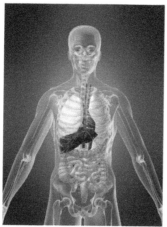

Figure 50 Anatomical comparison between human body and construction process.
Sources: SciePro/Shutterstock (left); Lifestyle Graphic/Shutterstock (right)

3.1.1 Design Phase

Mass timber projects save costs with a "more work upfront, less later" approach. These savings hinge on the project team's accuracy and thoroughness and ability to coordinate between designers, subcontractors, and manufacturers to ensure that nearly every phase of construction is planned down to the individual bracket and screw before prefabrication (*Figure 51*).

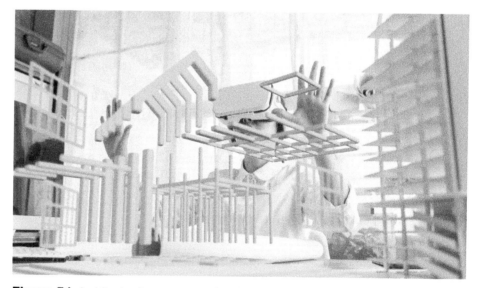

Figure 51 Architect using augmented reality during the design phase.
Source: ME Image/Shutterstock

Architectural and structural drawings are typically up to 90% completed before the project team hands them off to MEP subcontractors. Electricians, plumbers, HVAC professionals, and other essential trades then overlay those drawings with their submittals to deduce whether they have the necessary clearance for conduit, pipes, and ductwork to meet code.

If the clearance is insufficient, additional modifications to panels, beams, or chases must be made at this time. These modifications may be difficult (or impossible) to perform in the field, leading to costly schedule delays and projects potentially going over budget while replacement mass timber components are designed, ordered, manufactured, and shipped to the jobsite (*Figure 52*).

Low-rise buildings, residential homes, and bridges (*Figure 53*) will not require as much project coordination as mass timber tower projects. Due to the pre-planned nature of the construction process, however, every mass timber project will require early input from all contractors involved during the design phase.

Figure 52 Mass timber building during construction.
Source: Flystock/Shutterstock

Figure 53 Mass timber bridge.
Source: Patti McConville/Alamy

3.1.2 Site Planning and Logistics

Site planning and logistics during construction will be some of the most important tasks to keep the day-to-day workflow going. Project management software like Bluebeam® and Procore® will help teams coordinate requests for information (RFIs), change orders, and any other project and schedule changes, but it is the responsibility of each project team to plan ahead and mitigate potential setbacks (*Figure 54*).

Two of the most difficult logistical challenges of mass timber construction are crane operations and constant coordination between subcontractors to avoid trade stacking when possible.

Tower Cranes Versus Mobile Cranes

Depending on the project and site layout, project teams must decide whether to use a tower crane or mobile crane (*Figure 55*) to lift heavy mass timber panels, beams, and posts into position. Each choice has pros and cons that must be considered before finalizing the decision.

Figure 54 Project planning meeting.
Source: Gorodenkoff/Shutterstock

Figure 55 Mobile crane and tower crane on construction site.
Source: Bannafarsai_Stock/Shutterstock

A tower crane is typical for high-rise structures, and although it requires considerably more time and labor to erect in the early stages of the project, the investment pays off for designs that require daily lifts. Tower cranes also offer a much higher lifting capacity for massive loads, and because the crane is erected above the roof level, you will never need to worry about reaching high-level lifts.

Mobile cranes are excellent alternatives to tower cranes for low-rise structures and projects that do not have the space for a tower crane. Because mobile cranes provide quick and simple set-up, these equipment options are common for JIT deliveries and flexible lift schedules.

Avoiding Trade Stacking

Every project will have some level of trade stacking where productivity is reduced by overlapping scopes, but hybrid mass timber structures require more rigid installation sequences compared to more traditional construction methods. Mass timber products are not as moisture-resistant as concrete and steel structural members when exposed to the elements, so deliveries, crane operations, and material-protection strategies (before and after installation) must align with project schedules exactly to protect exposed components and guarantee structural integrity (*Figure 56*).

Figure 56 Unwrapped delivery of CLT panels.
Source: Flystock/Shutterstock

Installing mass timber materials is straightforward because nearly every piece is prefabricated to fit into an exact connection. However, this precision puts more stress on concrete and steel subcontractors to match the minimal tolerances needed. Minor mistakes or delays at even the foundational slab level could have drastic consequences and quickly put the entire project behind schedule, forcing crane operators and mass timber installers to wait for issues to be corrected before completing their scopes (*Figure 57*).

Figure 57 Workers pouring concrete for a foundation slab.
Source: bubutu/Shutterstock

3.2.0 Safety Considerations for Mass Timber

Safety is the most important aspect of any construction project, but the rapid installation of heavy mass timber materials can produce additional risk during crane operations and overhead work. As discussed in this module, mass timber products are more fire-resistant than light-frame lumber. However, even with the charring effect adding an extra level of protection, it's always important to remember fire safety guidelines, especially in hybrid structures that require welding and other spark-emitting work.

Although each general contractor will provide a safety manager or safety team, it is everyone's responsibility to act as a safety manager on site, regardless of seniority or job title. If something seems unsafe, it likely is. Never hesitate to ask a supervisor or safety professional for clarification and ensure that all team members comprehend and follow site- and task-specific safety plans.

3.2.1 Crane Operations and Overhead Work

Crane operations require extreme safety measures to ensure proper precautions are taken at every stage of the lift. Qualified crane operators must understand the limitations of their equipment and immediately identify safety hazards with daily inspections and constant investigation of the jobsite environment. Safety managers and certified riggers must complete an additional inspection of all connections, hooks, and rigging before each lifting operation begins (*Figure 58*). For a deeper dive into crane safety and operations, you should familiarize yourself with OSHA 1926 standards, most of which are covered in NCCER's Mobile Crane Operations Curriculum.

Figure 58 Crane hook ready for inspection.
Source: Toomko/Shutterstock

As a mass timber installer, you will go through intensive rigging training to receive your certification and become a safety-focused asset to your team. However, if you have just begun your journey in the industry, here are a few basic guidelines to keep you and others out of harm's way:

- Keep both the rigging and installation locations organized and free of unnecessary tools or materials. These are fast-paced operations and cluttered work areas can lead to tripping and fall hazards.

- Always watch the load and stabilize it with a guideline when it is within reach. Although crane operators are highly trained professionals, loads can swing or roll with little warning. Never place yourself in pinch-points between the load and a stationary object.

- Never stand directly beneath a load. Cordon off the path of travel between the initial lift and the installation area to minimize the risk of someone unknowingly walking beneath the load. Use appropriate safety rope, danger tape, and clear signage to mark the areas (*Figure 59*). Further considerations may be required if mobile cranes are forced to be set up near or on public roadways.

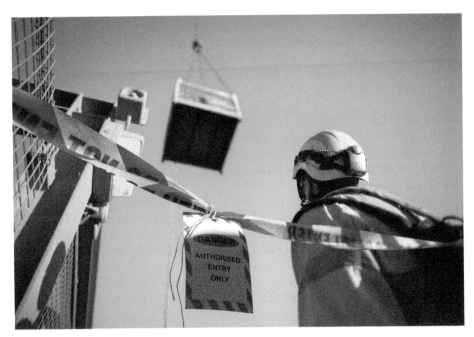

Figure 59 Worker maintaining visual on overhead load.
Source: King Ropes Access/Shutterstock

Guidelines for Overhead Work

Mass timber projects must follow the same OSHA safety regulations as other building types. The emphasis for any high-rise mass timber project safety plan must be to minimize two of the highest risks that lead to injuries and death in elevated work: falls and struck-by incidents.

OSHA regulations require some form of fall protection for craft professionals working above 6' near an exposed edge. Exposed edges are common as each mass timber panel floor system is completed, so installers must wear an OSHA-approved personal fall arrest system (safety harness) while they complete their tasks.

These fall arrest systems are no longer required once a guard and/or safety net system is installed at every exposed edge. Systems with top rails, mid-rails, and toeboards protect workers from falling over the edge, while also protecting people below from falling tools and materials (*Figure 60*).

Figure 60 Mass timber building with a guard system installed.
Source: sockagphoto/Shutterstock

Lastly, your team must ensure that each vertical panel and column is braced with adjustable braces for two reasons. First, it allows you to adjust the component into plumb before the adjacent panel or floor above is placed. Second and more importantly, it stabilizes the piece and reduces the risk of it being pushed off the edge by outside forces.

3.2.2 Fire Safety During Construction

Although current research has proven the fire-resistant capabilities of mass timber products due to the charring effect, Types IV-A, IV-B, and IV-C mass timber construction projects must follow specific safety protocols once the structure is built up to the seventh floor.

For example, the *International Fire Code®* (*IFC®*) states that a layer of non-combustible protection must be installed on mass timber interior walls and ceilings beginning on four levels below active construction. This is only required where *IBC®* Section 602.4 requires noncombustible protection over mass timber.

Note that Type IV-C and portions of Type IV-B might not require interior noncombustible fire protection, and therefore, a single layer of noncombustible protection is not required to be installed during construction. The structure's exterior siding or cladding must also be installed beginning on four levels below active construction regardless of the interior protection requirements.

Additional requirements for fire safety during construction are required by *IBC®* Chapter 33. The *IFC®* also includes the following:

- Heaters must be labeled, installed, and maintained in accordance with *IFC®* Section 3304.

- Spark-emitting tasks such as welding or electrical work may require a qualified fire watch to reduce the risk of structural fires during construction (*IFC®* Section 3305) (*Figure 61*).

Figure 61 Welder working with a fire extinguisher within reach.
Source: Yuthtana artkla/Shutterstock

- Smoking is permitted only in approved areas with posted signage (*IFC®* Section 3305.1).

- An approved fire-protection water supply and fire-fighting vehicle access must be provided within 100 feet of temporary or permanent fire department connections (*IFC®* Section 3311) (*Figure 62*).

Figure 62 Firefighters responding to a structure fire.
Source: Ted Pendergast/Shutterstock

3.3.0 | As-Builts, Finishes, and Repairs

Mass timber prefabrication takes a lot of the guesswork out of the installation process. However, no plan is perfect, and even trusted manufacturers can make mistakes that must be addressed in the field. Necessary changes to the project design or MEP layout, as well as complex connections or mismatched tolerances between structural members and adjacent materials, can create challenges that require innovative solutions. This section briefly describes guidelines for as-builts, finishes, and material repairs for mass timber structures.

3.3.1 As-Builts

As-builts and field modifications are much less common in mass timber, but they may be necessary to fix fabrication issues, such as incorrect dimensions and pre-installed hardware placed in the wrong locations. Consistent quality control (QC) methods like surveys and spot checks ensuring all horizontal and vertical components are square, level, and plumb will reduce the timely and costly risk of material replacements (*Figure 63*). But even then, your team may be forced to perform field adjustments approved by the engineer of record and supervising general contractor.

You may be able to complete some tasks with common carpentry tools, but for accurate and efficient cuts, you may need to rely on specialized tools capable of sawing through massive beams and panels. The beam saw and 24" circular saw are two of the most popular options, but these large-bladed tools require an additional level of skill and training to safely operate (*Figure 64*).

NOTE

Depending on the location of the modified mass timber component, you may be required to apply sealant once the cut is cleaned of any dust or debris. This is especially important at weak points like the end grains where moisture can seep into the core, causing water damage and potential delamination.

Figure 63 Workers surveying construction site.
Source: Sorn340 Studio Images/Shutterstock

Beam Saw (Chainsaw)   **Large Circular Saw**

Figure 64 Saws for cutting mass timber.
Sources: Kekyalyaynen/Shutterstock (left); samfern/Shutterstock (right)

3.3.2 Mass Timber Finishes

Mass timber is typically prefinished to protect the wood from moisture and maintain a rich, natural appearance long-term. This is a vital aspect for any beams, posts, or panels that are exposed to the elements (*Figure 65*).

Finishes also maintain the structure of the engineered wood product itself. Because most mass timber is produced from softwood lumber species, unfinished products may experience seepage of tannins, pitches, and resins, which tend to bleed through the surface, causing discolorations on lighter wood. The following procedures can help with other aesthetic considerations:

- Maintain separation between metal and wood components and remove any metal shavings from drilling plates and hardware.

- Wipe off excess oils from high-strength connections immediately after installation.

- Paint utilities connected to mass timber products before they are attached to reduce the potential for overspray on exposed wood finishes.

Figure 65 Stadium with exposed mass timber beams.
Source: Hufton+Crow-VIEW/Alamy

If damage or as-built modifications require resealing, do so in a well-ventilated area while wearing the proper PPE. Depending on the amount of touch-up work needed, you could apply manufacturer-approved sealant or finish with a brush, roller, or sprayer (*Figure 66*).

Figure 66 Finish sprayer.
Source: MDV Edwards/Shutterstock

NOTE

Touch-ups are time-consuming and labor-intensive. Excessive damage can result in considerable schedule and project cost setbacks. The best way to avoid these tasks is to focus on a material protection plan. This plan can include wrapping columns with a breathable weather-resistive barrier or surrounding exposed surfaces with cones and caution tape to deter workers from denting or marring finished surfaces.

Repairing Mass Timber Products

Some failing glulam beams and other loadbearing members may require total replacement or engineered solutions, such as shear dowel reinforcement or the addition of post-tension mechanisms. However, minor repairs like dents, discoloration, scratches, and screw holes can be filled with a manufacturer-approved sealant, sanded, and refinished with a color-matching finish.

Sources: Miriam Doerr Martin Frommherz/Shutterstock (left); Flystock/Shutterstock (right)

3.0.0   Section Review

1. What is the most crucial construction phase for cross-team coordination to ensure accurate fabrication of mass timber components?
    a. Design phase
    b. Close-out phase
    c. Post-construction phase
    d. Installation phase

2. What should a cordoned area include to protect workers from walking beneath a load during hoisting operations?
    a. Fire extinguishers
    b. Safety rope, danger tape, and clear signage
    c. A certified laborer
    d. Storage for extra materials

3. What should a craft professional do after modifying a mass timber component to reduce the risk of moisture damage?
    a. Leave it in the elements to climatize naturally
    b. Install it immediately
    c. Seal the end grains with a manufacturer-approved sealant
    d. Pre-install the component with the connecting hardware

Module 27214   Review Questions

1. True or False: A construction project that uses mass timber as an accent is considered a mass timber building.
    a. True
    b. False

2. Which mass timber type requires a 3-hour fire-resistance rating for bearing walls and structural frame construction?
    a. Type IV-HT
    b. Type IV-C
    c. Type IV-B
    d. Type IV-A

3. Which mass timber type has a fire-protection requirement that allows exposed mass timber except at shaft walls, within concealed spaces, or on the exterior side of exterior walls?

    a. Type IV-HT
    b. Type IV-A
    c. Type IV-C
    d. Type IV-B

4. What is the maximum height for Type IV-B construction?

    a. 75'
    b. 85'
    c. 180'
    d. 270'

5. How does mass timber compare to steel in terms of thermal conductivity?

    a. Mass timber transfers heat more slowly and is a better insulator.
    b. Steel has lower thermal conductivity than mass timber.
    c. Steel transfers heat more slowly, making it a better insulator.
    d. Mass timber transfers heat faster, making it a better insulator.

6. Which of the following is an environmental advantage of mass timber?

    a. Biophilic design
    b. Carbon sequestration
    c. Fire resistance
    d. Increased market value

7. Which of the following is a benefit of prefabricating mass timber products for a construction project?

    a. Shorter project timelines
    b. More cutting on site
    c. More workers on site
    d. Decreased seismic performance

8. Which mass timber product can be formed into arches and curved configurations spanning 300' or more without support columns?

    a. CLT
    b. LVL
    c. DLT
    d. GLT

9. Which mass timber product is well suited for beams and columns?

    a. CLT
    b. PSL
    c. NLT
    d. MPP

10. Which mass timber product is bonded together and pressed into blocks (billets)?

    a. LSL
    b. NLT
    c. DLT
    d. HT

11. Which two mass timber products are *NOT* made with adhesive?
    a. CLT and MPP
    b. GLT and LSL
    c. PSL and LVL
    d. NLT and DLT

12. Which delivery method is the most efficient for a project site without adequate material storage or laydown areas?
    a. One-time shipment
    b. Just-in-time delivery
    c. Overnight delivery
    d. Trade stacking delivery

13. Which of the following is a weak point in mass timber construction where moisture has a greater chance of penetrating into the wood?
    a. Interior wall openings
    b. Columns more than 8" wide
    c. End grains
    d. Beams more than 30' long

14. What determines how many self-tapping screws will be required for the transport anchors when rigging and moving mass timber panels?
    a. Load capacity of the anchors
    b. How far it will be moved
    c. Type of mass timber used for the panels
    d. Framing system being used

15. Which type of mass timber framing consists of continuous wall panels (up to 3 stories) with drop-in beams and engineered floor panels?
    a. Post-and-plate
    b. Hybrid
    c. Post-and-beam
    d. Balloon

16. What is one of the most vital differences between mass timber construction and other structural projects?
    a. Mass timber construction requires less work upfront.
    b. Mass timber construction can accommodate changes more easily.
    c. Mass timber construction requires a strong emphasis on early coordination.
    d. Mass timber construction incorporates MEP systems at the end of the construction process.

17. Which type of crane is best for low-rise structures and projects that have limited space for crane setup?
    a. Tower
    b. Mobile
    c. Aerial
    d. Overhead

18. How frequently must connections, hooks, and rigging be inspected?
    a. At the beginning of each shift
    b. At the beginning of each work week
    c. As often as the safety manager thinks is necessary
    d. Before each lifting operation begins

19. At what point must a layer of noncombustible protection be installed on mass timber walls and ceilings in Type IV-A, IV-B, and IV-C construction?

    a. On all levels above the seventh floor as they are built
    b. Beginning four levels below active construction once the seventh floor is built
    c. Beginning two levels below active construction once the seventh floor is built
    d. After the entire structure is built

20. Which type of saw would be used to cut mass timber products to make field modifications?

    a. Jig
    b. Miter
    c. Beam
    d. Band

Answers to Odd-Numbered Module Review Questions are found in *Appendix A*.

Answers to Section Review Questions

Answer	Section	Objective
Section One		
1. d	1.1.0; Table 1	1a
2. c	1.2.0	1b
3. c	1.3.1	1c
Section Two		
1. a	2.1.5	2a
2. c	2.2.3	2b
Section Three		
1. a	3.1.1	3a
2. b	3.2.1	3b
3. c	3.3.1	3c

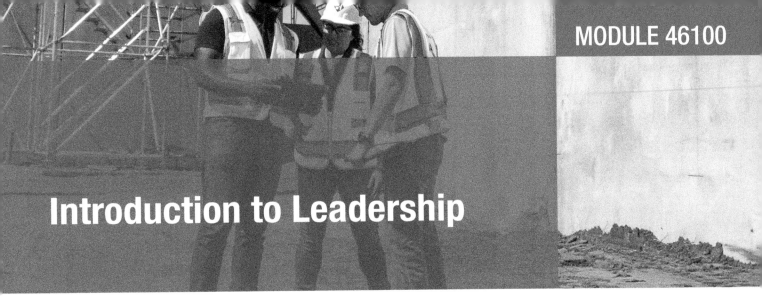

Introduction to Leadership

Objectives

Successful completion of this module prepares you to do the following:

1. Explain business and organizational structures in the construction industry.
   a. Understand organizational charts.
   b. Describe organizational roles and relationships, including job descriptions and responsibility, authority, and accountability.

2. Describe frontline leadership and communication skills.
   a. Describe the qualities and role of a leader, including leadership styles and ethical leadership.
   b. Explain the communication process and distinguish between verbal, nonverbal, written, and visual communication types.
   c. Compare problem-solving and decision-making.

3. Identify key skills for team leadership and motivation.
   a. Choose strategies to motivate individuals and a team.
   b. Describe key approaches to building and leading teams.
   c. Recognize how to approach leadership challenges, such as transitioning from peer to leader, addressing performance issues, and resolving conflict.

4. Identify a frontline leader's safety leadership responsibilities.
   a. Assess the impact of incidents.
   b. Explain the purpose of OSHA and describe the role of OSHA in administering worker safety.
   c. Describe the importance of safety and health programs.
   d. Explain the frontline leader's role in incident preparedness and response.
   e. Recognize the importance of supporting your crew's total health.

5. Demonstrate an understanding of construction project phases, project planning and scheduling, field reporting, and cost and resource management.
   a. Identify construction project phases.
   b. Summarize crew planning and scheduling responsibilities, including creating lookahead schedules.
   c. Describe the frontline leader's role in field reporting.
   d. Explain approaches to managing costs and controlling project resources, such as materials, tools, and equipment.

Performance Tasks

Under supervision, you should be able to do the following:

1. Organize and deliver a safety-related toolbox talk, applying communication skills.
2. Complete a job safety analysis.
3. Develop and present a lookahead schedule.

Overview

As a frontline leader, you must be able to communicate effectively, provide direction to your crew, plan and schedule work, manage costs, and ensure the jobsite is safe for everyone. Whether you are already a frontline leader or want to become one, this module will help you learn more about the requirements and skills needed to succeed.

Digital Resources for Carpentry

Scan this code using the camera on your phone or mobile device to view the digital resources related to this craft.

NCCER Industry–Recognized Credentials

If you are training through an NCCER-accredited sponsor, you may be eligible for credentials from NCCER. The ID number for this module is 46100. Note that this module may have been used in other NCCER curricula and may apply to other level completions. Contact NCCER at 1.888.622.3720 or go to **www.nccer.org** for more information.

You can also show off your industry-recognized credentials online with NCCER's digital credentials. Transform your knowledge, skills, and achievements into badges that you can share across social media platforms, send to your network, and add to your resume. For more information, visit **www.nccer.org**.

1.0.0 Business and Organizational Structures

Performance Tasks

There are no Performance Tasks in this section.

Objective

Explain business and organizational structures in the construction industry.

a. Understand organizational charts.

b. Describe organizational roles and relationships, including job descriptions and responsibility, authority, and accountability.

Crew leader: A person appointed leader of a team or crew, usually an experienced craft professional who has demonstrated leadership qualities. Also known as *frontline leader, frontline supervisor, lead person,* or *foreman.*

Superintendent: An on-site supervisor who is responsible for one or more crew leaders or frontline supervisors.

Project manager: Person responsible for managing one or more projects.

When a crew is assembled to complete a job, one person is appointed the leader. Different companies give this role different titles, such as **crew leader**, *frontline supervisor, lead person,* or *foreman.* In this module, this role is called the *crew leader* or *frontline leader.* A crew leader or frontline leader is often an experienced craft professional who has demonstrated leadership qualities.

A **superintendent** is essentially an on-site field supervisor who is responsible for one or more crew leaders or frontline supervisors. A **project manager** may be responsible for managing one or more projects. This training will concentrate primarily on the supervisory role of the frontline leader. These individuals work with others on projects as part of an organization.

1.1.0 Organizations and Organizational Structures

Organization: A group of people working together.

Organizational chart: A diagram that shows how the various management and operational responsibilities relate to each other within an organization. Named positions are ranked top to bottom, from those functional units with the most and broadest authority to those with the least. Lines connect positions and indicate chains of authority and other relationships.

An **organization** is a group of people working together. Businesses, clubs, and associations are examples of types of organizations. On a jobsite, the relationships among the people within the company or project create a type of organization.

1.1.1 Organizational Charts

A formal organization will have an **organizational chart** that shows all of the positions in the organization and how those positions are related. A detailed organizational chart, or org chart, may also show the name of the person in each position. *Figure 1* shows an example of an organizational chart for a construction company. Each position on that organizational chart represents an opportunity for career advancement for a crew leader.

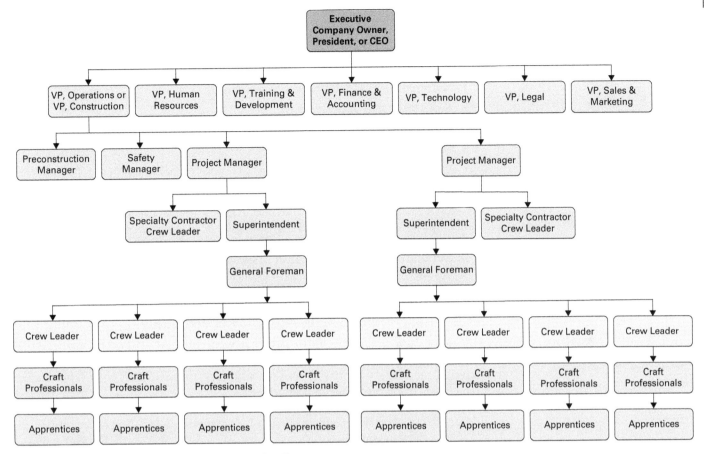

Figure 1 Sample organization chart for a construction company.

As shown in *Figure 1*, most companies have separate groups or departments that focus on running specific functions. Smaller companies may have one or two individuals who perform multiple functions. Larger companies will often create specific roles or departments to handle more complex projects and larger teams.

Companies may have additional roles or teams, such as architectural and engineering design functions. Depending on the size of these groups, they may become separate departments.

NOTE

Organizational charts vary widely based on the size of the company and the specific construction sector in which they operate. For example, an industrial company might have crew leaders who report to a plant manager. Some larger companies will have a superintendent per shift, who will report to a project superintendent, who reports to the project manager. If you are interested in learning more about your company's organizational chart, ask your supervisor or manager.

1.2.0 Organizational Roles and Relationships

As a crew leader, you will assign others the responsibility to perform specific jobs or tasks. Your crew and team members will perform best if the following is true:

- They have a clear job description that explains their job duties.
- They have the required skills and abilities to perform the task.
- They know who has **authority**, **responsibility**, and **accountability**.
- They have a clear understanding of expectations.
- They are trained on policies and procedures.
- They have all needed tools, equipment, and material.

1.2.1 Job Descriptions

As part of the hiring process, most companies give each employee a **job description**. A job description is a statement that describes the duties and responsibilities of a position with a company. A **duty** is a work task an employee is expected to perform. Responsibility involves accepting a task and completing it to the best of one's ability.

Authority: The right to act or make decisions in carrying out an assignment.

Responsibility: The act of accepting a task and completing it to the best of one's ability.

Accountability: In relation to job completion, accepting responsibility and being personally answerable for the results.

Job description: A description of the scope and responsibilities of a worker's job so that the individual and others understand what the job entails.

Duty: A work task an employee is expected to perform.

A job description should include enough detail so that an employee understands their job. It also should have any information needed to evaluate the employee's job performance. A job description should contain at least the following:

- Job title
- The supervisor to whom the position reports
- A general position description
- Specific duties and responsibilities
- Required and desired qualifications, including certifications and licenses
- Other information, such as working conditions, job type, hours, pay ranges, and more

Job descriptions set a standard for an employee. They outline the tasks that an employee should handle. After reviewing a job description, a new employee should understand all of the duties and responsibilities of the job. Having a clear understanding of the job makes it easier for new employees to work well with the rest of the crew while they gain experience. Clear job descriptions help crew leaders to judge performance and identify training needs for new employees.

1.2.2 Responsibility, Authority, and Accountability

A good crew leader needs to delegate work to their team, but always stay accountable for the outcome. When you assign a crew member work, you are giving them the authority to do that task. *Authority* is the right to act or make decisions in carrying out an assignment.

When the employee takes on a job or assignment, they take on responsibility. As noted previously, *responsibility* involves accepting a task and completing it to the best of one's ability.

In relation to job completion, *accountability* is accepting responsibility and being personally answerable for the results.

Each project or company has differences in the type and amount of authority, responsibility, and accountability a supervisor or worker has. As a crew leader, you must give sufficient responsibility and authority to allow employees to be successful at their job activities. You then must hold employees responsible for completing those assigned activities.

As shown in *Figure 2*, even though you may delegate authority and responsibility to crew members, the overall accountability for tasks assigned to a crew always rests with the frontline leader or crew leader.

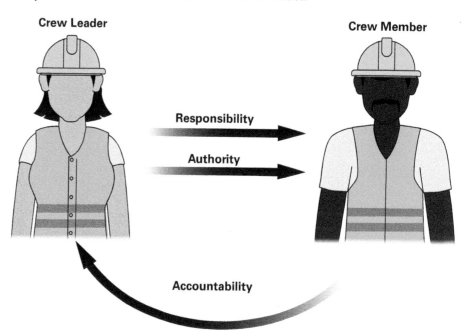

Figure 2 Accountability always rests with the frontline leader.

1.2.3 Policies and Procedures

Most companies have formal policies and procedures established to help crew leaders carry out their duties.

- A **policy** is a general statement establishing guidelines for a specific activity. Examples include policies on vacations, breaks, workplace safety, and checking out tools.

- **Procedures** are formal instructions to carry out and meet policies. For example, a procedure written to implement a policy on workplace safety would include guidelines for reporting incidents and following general safety practices.

A crew leader must understand any company policies and procedures that relate to their work, especially safety practices. The **Occupational Safety and Health Administration (OSHA)** is a national public health agency that is part of the US Department of Labor. To promote a safe and healthy work environment, OSHA issues standards and rules for working conditions, facilities, equipment, tools, and work processes. When OSHA inspectors visit a jobsite, they often question crew leaders and employees about the company safety policies. During incident investigations, the inspector will verify if the responsible crew leader knew and followed the applicable company policy.

Policy: A general statement establishing guidelines for a specific activity.

Procedures: Formal instructions to carry out and meet policies.

Occupational Safety and Health Administration (OSHA): A national public health agency that is part of the US Department of Labor. To promote a safe and healthy work environment, OSHA issues standards and rules for working conditions, facilities, equipment, tools, and work processes.

1.0.0 Section Review

1. Which of the following is shown on an organizational chart?
   a. How to organize materials on a jobsite
   b. How positions in an organization are related
   c. How prevailing wages are determined
   d. How to coordinate work with other crafts

2. Although you may delegate authority and responsibility for tasks to crew members, as a frontline leader, you have the overall _____.
   a. policy
   b. accountability
   c. job description
   d. authority

2.0.0 Leadership and Communication

Objective

Describe frontline leadership and communication skills.

a. Describe the qualities and role of a leader, including leadership styles and ethical leadership.
b. Explain the communication process and distinguish between verbal, nonverbal, written, and visual communication types.
c. Compare problem-solving and decision-making.

Performance Tasks

There are no Performance Tasks in this section.

In addition to being an experienced craft professional, a frontline leader also must demonstrate leadership qualities and strong morals. They must be able to communicate effectively, solve problems, and make good decisions. Frontline leaders are generally promoted up from a work crew. A worker's ability to accomplish tasks, get along with others, meet schedules, and stay within the budget have a significant influence on the selection process. The frontline leader must lead the team to work safely and provide a quality product.

This section explains the importance of developing and improving leadership skills as a frontline leader. It will cover the qualities and roles of a leader, as well as effective ways to communicate with coworkers and employees at all levels, solve problems, and make decisions.

2.1.0   The Qualities and Role of a Leader

There are many ways to define leadership. One simple definition of leadership is the ability of a person to influence others to achieve a common goal. A leader is someone who sees how to accomplish or improve things. A leader then rallies people to move toward that vision. Effective leadership requires communicating ideas effectively to engage others to act accordingly.

Leaders often have power based on a role or title. However, leadership is much more than power. An effective leader inspires others to act, while also directing the way that they act. A leader must have respect, so others follow their directions. A leader also must have the critical thinking skills to know the best way to use available resources.

Motivate: To make someone feel determined or enthusiastic or to make someone behave in a particular way.

Some people develop leadership qualities as part of their upbringing. Others may work hard to develop the qualities that **motivate** others to follow and perform. Leadership is a skill you can learn through experience and training, as well as hard work, focus, and practice.

2.1.1 Functions of a Leader

What is the purpose or function of a leader? The answer will vary based on the specific project, the work environment, the team they lead, and the tasks they must perform.

As shown in *Table 1*, a frontline leader has certain functions that apply to almost every work situation.

TABLE 1   Frontline Leadership Functions

Leadership Duties and Responsibilities	Leadership Functions
Planning the Work	Organize, plan, staff, direct, and control work
Safety and Compliance	Ensure that the team understands and follows company policies and procedures
Empathy	Be sensitive to the needs and qualities of a diverse workforce
Motivating and Coaching	Give team members the confidence to make decisions and take responsibility for their work
Conflict Resolution	Maintain a cohesive team by resolving tensions and differences
Professionalism	Represent the team or group and represent the company or organization
Accountability	Accept responsibility for the successes and failures of the team's performance

2.1.2 Leadership Traits and Behavior

Trait: A distinguishing quality of a person.

Behavior: The way a person responds to a specific situation.

As previously noted, leadership is a skill you can learn and develop through experience and training. As you continue to grow your leadership skills, remember that while each leader has some unique approaches, the most effective leaders share some common traits and behaviors. A **trait** is a distinguishing quality of a person. A **behavior** is the way a person responds to a specific situation.

Effective leaders share the following traits and behaviors:

- Follows company policies and procedures
- Has necessary technical knowledge

- Can plan, prioritize, and organize work
- Leads by example
- Is fair and ethical
- Solves problems, makes decisions, and assumes responsibility
- Communicates effectively
- Can promote an idea
- Is enthusiastic and shows strong initiative
- Is able to motivate individuals and teams
- Is loyal to their company and crew
- Extends respect and trust to all team members
- Can teach others and learn from others

First and above all, excellent leaders lead by example. They work and live by the standards that they set for their crew members. A leader will not ask someone to complete a task that they are not willing to do themselves.

Leaders tend to have a high level of drive and determination, as well as a persistent attitude. When faced with obstacles, effective leaders don't get discouraged. Leaders identify the potential problems and make plans to overcome them. They then work toward achieving the intended goal. In the event of failure, effective leaders learn from their mistakes and apply that knowledge to future situations. They also learn from their successes.

Respect is an important aspect of leadership. Leaders earn respect by being fair to employees and by listening to their complaints and suggestions. Using appropriate incentives and **rewards** to motivate a crew helps leaders earn respect. In addition, frontline leaders who have a positive attitude tend to gain the respect of their team and their peers. A **peer** is someone who is in a position or level in an organization equal or similar to yours.

Trust is an equally important part of being a great frontline leader. Trust is a group's belief that their leader is committed, capable, and will do what's best for all involved. To build trust, effective leaders create interactions and conversations with the team. They treat everyone on the team as an individual.

Leaders are good communicators who lay out the goals of a project to their crew members clearly. To do this, you may need to overcome language barriers, bias, or differences in personalities to ensure that each member of the crew understands the established goals of the project.

Effective leaders can motivate their team to work to their full potential. They will work to develop crew members' skills. They encourage them to improve and learn so they can contribute more to the team effort. Effective leaders strive for excellence from themselves and their team, so they work hard to provide the necessary skills and leadership.

It is important for leaders to have an expert knowledge of the activities they supervise. This is important because the crew members need to know that they have someone to turn to when they have a question or a problem, need some guidance, or must make modifications or changes on the job.

In addition, leaders must possess organizational skills. They know what needs to be accomplished and they use their resources to make it happen. Because they can't do it alone, leaders require the help of their team members to share in the workload. Effective leaders delegate work to their crew members. They also implement company policies and procedures to complete the work safely, effectively, and efficiently.

Finally, outstanding leaders have the experience, authority, and self-confidence that allow them to make decisions and solve problems. To accomplish their goals, leaders must be able to calculate risks, take in and understand information, set a course of action, make decisions, and assume the responsibility for those decisions.

Rewards: Tangible items given to an employee as recognition for good work.

Peer: Someone who is in a position or level in an organization equal or similar to yours.

2.1.3 Leadership Styles

Many terms are used to describe different leadership styles. As a frontline leader, you will never lead the exact same project twice. Having a leadership style that is also adaptable is important for you, the project, and your company's success.

Situational leadership is a leadership approach that involves adjusting the way you lead to reflect the specific situation or task and the needs of the group. The main idea behind situational leadership is that there is no "one size fits all" or best style of leadership. Instead, you must adapt or adjust your leadership style to the situation.

As shown in *Figure 3*, situational leadership provides a model with four primary leadership styles:

- *Directing:* The leader tells a team what to do and how to do it.
- *Coaching:* The leader coaches individuals to improve skills. The leader encourages more back-and-forth between leaders and the team.
- *Supporting:* A leader supports the group, participating to help increase motivation. The leader offers less direction. Instead, the leader has team members take a more active role in coming up with ideas and making decisions.
- *Delegating:* A leader allows other people to own the task, and they facilitate it happening. Team members make most of the decisions and take most of the responsibility for what happens.

Situational leadership: A leadership style that involves adjusting to the specific situation or task and the leadership needs of the group.

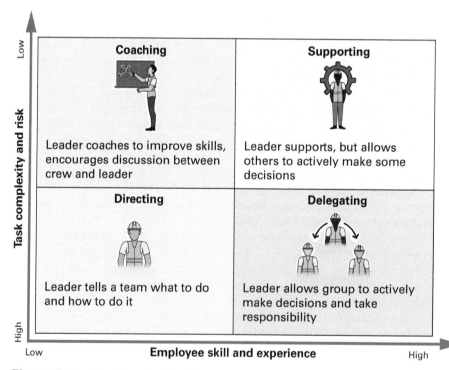

Figure 3 Situational leadership styles.

Knowing which leadership style to apply depends on several key factors:

- *Team relationships:* Consider your team, how well they work together, and how productive they are. A group that is not efficient or productive might benefit from a Directing style. That style emphasizes direction, organization, and clearly defined roles. On the other hand, a highly cohesive group might benefit from a Supporting style, which gives the team more inputs on decisions.
- *Team skills:* Consider the skills and abilities of the individuals completing the task. Giving a task to a team or team member who lacks the skills to complete the task is risky, is dangerous, and may end in failure.

- *Work tasks:* Consider the actual work or tasks at hand. You must have an expert knowledge of the tasks and activities to ensure they are done well, completely, and in the safest way possible. You may need to coach or train workers who do not have experience to help them complete the task.

- *Work situation:* Consider the situation you're facing. In some cases, the situation itself will override all other factors.

For example, if the crew does not have enough experience for the job ahead, you may want to use a *Directing* style. A Directing style works with people and teams who have limited experience or skill performing a task. It also works when team members are insecure or unmotivated to try a task. The Directing style requires your close supervision to spot signs of progress, which you can use to motivate ongoing development. If a job task or situation is very risky, a Directing style can help reduce the risk by providing clear and very specific direction.

If your team is working well and showing strong progress on work tasks, you might use a *Coaching* style. You still direct how and when specific tasks are completed, but you also discuss why the task matters and where it fits into the overall project or job. With this style, you recognize the enthusiasm, interest, and commitment of the team. You also reinforce the importance of learning and gaining task-related experience.

You may want to use a *Supporting* style if the team is reasonably skilled and work tasks are not complex, but commitment and attitude are an issue. The goal of the Supporting style is to create alignment. As a leader, you can discuss a person's commitment by asking open-ended questions. These questions should help you understand what is behind the performance challenges and create possible solutions. The Supporting style of leadership also works well when work is creative, as it is helpful to brainstorm and exchange ideas with crew members.

The *Delegating* style works best with an experienced, strong, committed crew on a well-defined project. The Delegating style of leadership is also effective when jobs have repetitive tasks that require few decisions. To use the Delegating style, however, your company must give you enough authority as the frontline leader to do the job.

Effective leadership takes many forms. The correct style for a particular situation depends on the team and the work you need to accomplish. Knowing the factors to consider can help you decide which type of leadership to apply in which situation. This flexibility allows you to meet each situation with the leadership style that brings out the best in your team and gives them the highest chance of success.

2.1.4 Ethics in Leadership

Ethics are the moral principles that guide a person's actions when dealing with others. *Figure 4* outlines three basic types of ethics. Frontline leaders must maintain strong ethics and the highest standards of honesty and legality. Every day, a leader makes decisions that have fairness and moral implications. When you make an unethical decision, it reflects poorly on you. It also impacts your team, other project team members, and the company as a whole.

As a leader, you will find yourself in situations where you will need to weigh the ethical results of a choice or decision. For example, imagine a crew member shows symptoms of heat exhaustion. Do you keep that employee working just because the superintendent says the project is behind schedule? If you are aware that your crew did not properly erect the reinforcing steel, do you stop the pour and correct the situation? In both cases, you must act ethically as a leader and make the right choice. If supervision ever asks a frontline leader to carry out an unethical decision, that individual must inform the next level of authority of the unethical nature of the issue.

Ethics: The moral principles that guide an individual's or organization's actions when dealing with others.

NOTE

Anyone who is aware of illegal activity can face legal concerns simply by not acting. This is true even if the activities were requested by one's supervisor.

 Business or legal ethics	 Professional or balanced ethics	 Situational ethics
Adheres to all laws and published regulations related to business relationships or activities.	Carries out all activities in such a manner as to be honest and fair to everyone under one's authority.	Relates to specific activities or events that may initially appear to be a gray area. For example, you may ask yourself, "How will I feel about myself if my actions were going to be published in the news or if I need to justify my actions to my family, friends, and colleagues? Would I still do the same thing?"

Figure 4 Types of ethics.

2.2.0   Communication

> **Communication**: The process of exchanging facts, information, and ideas. Communication comes in many forms, including verbal, written, visual, and nonverbal.

Successful crew leaders learn to communicate well with people at all levels of an organization. **Communication** is the act of accurately and effectively sharing facts, feelings, and opinions with another. Simply stated, communication is the process of exchanging information and ideas. Communication comes in many forms, including verbal, written, visual, and nonverbal.

2.2.1 Why Communication Matters

Effective communication on a jobsite or project is critical for many reasons (*Figure 5*). First and foremost, clear communication is a key part of safety. Everyone on a jobsite must receive and understand communications about safety policies and procedures. If this does not happen, it creates safety issues for everyone in the group.

Miscommunication is also expensive (*Figure 5*). One study indicated that over half of all rework is due to poor communication and bad information. If communication on a construction site fails, it can cause delays and quality issues for the

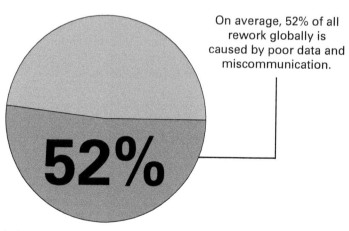

On average, 52% of all rework globally is caused by poor data and miscommunication.

52%

Figure 5 Industry impacts of miscommunication in the construction industry.

project. It can also result in damage to equipment or property. The good news is that communication is a skill that can be taught, practiced, and improved.

2.2.2 The Communication Process

Clear communication has three basic steps, as shown in *Figure 6*.

1. *Sender sends message to receiver.* A sender sends a message to a receiver, as a verbal, written, visual, or nonverbal communication. A message can be a set of directions, an opinion, or a personal message such as praise or a warning. Whatever it is, a message is an idea or fact that the sender wants the receiver to know.
2. *Receiver interprets message and repeats the message to provide feedback.* The receiver is the person who takes in the message. When the receiver gets the message, they figure out what it means by listening, viewing, or reading, and then repeat the message to provide feedback. *Feedback* is the receiver's communication back to the sender in response to the message. Feedback is a very important part of the communication process. It shows the sender how the receiver interpreted the message and if the receiver understood it as intended. In other words, feedback is a checkpoint to make sure the receiver and sender are on the same page.
3. *Sender validates that receiver correctly understood the message.* The sender confirms the receiver heard and understood the message.

This process is called **three-way communication**. Three-way communication is also sometimes called *repeat-back communication*.

Three-way communication is a simple way to make sure that communication is clear. If this process is easy, why is good communication so hard to achieve? When we try to communicate, many things can get in the way. These obstacles or **distractions** are anything that disrupts the communication flow between sender and receiver. Also called *noise*, distractions can include poorly worded messages, emotions, interpretations of a word, or the actual noise from a construction site. Distractions can happen at any point in the communication process.

Three-way communication: A form of communication with an initial message from a sender, which is then repeated by the receiver to confirm accuracy, and then validated by the sender. Also called *repeat-back communication*.

Distractions: Anything that disrupts the communication flow between sender and receiver. Also called *noise*.

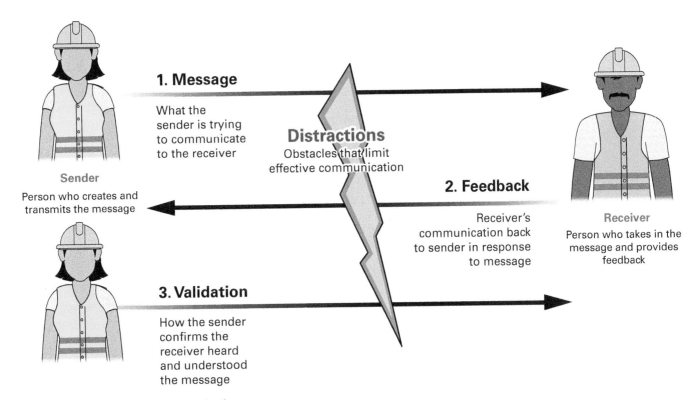

Figure 6 The three-way communication process.

Verbal	Written
• Spoken or oral communication • Examples: face-to-face conversations, telephone call, videoconference, voice chat	• Messages or information sent using words • Can be digital or physical • Examples: text message, email, training guide, weekly report, change request, purchase order

Communication Types

Visual	Nonverbal
• Using visual elements, such as photos, graphics, charts, colors, and fonts to convey messages • Examples: sketches, diagrams, graphs, photos, construction drawings, signs	• Things the speaker and the receiver see or hear while communicating • Examples: appearance, personal environment, use of time, facial expressions, body movements, hand gestures, tone of voice, and eye contact

Figure 7 Types of communication.

As noted in *Figure 7*, there are four different types of three-way communication. The following sections review these different types of communication and discuss ways to increase the effectiveness of your communication, regardless of type.

2.2.3 Verbal Communication

Verbal communication: Messages that are sent using voice or other audible sounds. Also called *spoken communication* or *oral communication*.

Verbal communication includes messages that are sent face-to-face or by telephone, videoconferencing, voice chat, or other methods. Verbal communication is also called also *spoken* or *oral* communication.

With verbal communication, the way you speak is as important as the message itself. As a leader, you should assess your audience before you speak. Even in a meeting as standard as a daily huddle, using clear verbal communication sets your crew up for a successful day. Consider each person's knowledge of the topic and any language, gender, or cultural differences. Do not talk down to anyone in a group. Everyone, whether in a senior or junior position, deserves respect and courtesy.

You should also be sure to use words and phrases that the receiver can understand. For example, avoid technical language, acronyms, or slang if the receiver does not know these terms. In addition, speak at a reasonable pace, which gives the audience time to understand one point at a time.

If you are receiving a message, be sure to listen actively and avoid distractions that interfere with the delivery of the message. There are many barriers to effective listening, particularly on a busy construction jobsite.

As a receiver, you can show that you are actively listening. You can accomplish this without saying a word. Examples include maintaining eye contact, nodding your head, and taking notes. If you are not clear on the message, ask questions to get additional explanation or information. You can also provide feedback by using your own words, or *paraphrasing*, to repeat a message back to the sender. This lets the sender know that you understood the message.

2.2.4 Written or Visual Communication

Written communication: The process of sending messages or information using words, whether digitally or physically.

Much of the information on a job or project is in written or visual form. **Written communication** is the process of sending messages or information using words, whether digitally or physically. Examples include everything from a text message to an email to a printed training guide. A lot of project documentation is written, including weekly reports, requests for changes, purchase orders, and contracts. Many of these items are written so they can form a permanent record for business and historical purposes.

Visual communication: The process of using visual elements, such as photos, graphics, colors, and fonts, to convey messages.

Visual communication is the process of using visual elements to convey messages. Photos, diagrams, graphics, charts, and colors are all types of visual communication. Visual communication is widely used because it is an effective way to convey information quickly. Traffic signs, for example, use standard visual symbols because they are clear and well-understood. Many messages on a job also use visual communication. Examples include the project plans, drawings, or a video of a worksite incident. Visual communication can help workers who speak different languages or have limited language skills.

Written and visual communication used together are even more clear. For example, tags and signs on a project (*Figure 8*) use both written words and visual elements to help make them clear to all workers on a jobsite.

Information Sign

Caution Sign

Warning Sign

Safety Sign

Danger Sign

Figure 8 Communication tags and signs combine written and visual communication.
Source: Courtesy of AccuformNMC.com LLC

When writing or creating a written or visual message, you first should assess the receiver (the reader). The receiver must be able to read the message and understand the content. If they cannot, the communication process will be unsuccessful. Always consider the actual meaning of words or diagrams and how others might interpret them. *Table 2* and *Table 3* provide checklists to help make your written and visual communications effective.

TABLE 2 Checklist for Effective Written Communication

Did You . . .?	Yes/No
Assess your audience. Consider existing knowledge. Consider any language, gender, or cultural differences.	
State the purpose of the message clearly.	
Present the information in a logical manner.	
Provide an adequate level of detail.	
Avoid emotion-packed words or phrases.	
Avoid making judgments, unless asked to do so.	
Avoid using unfamiliar acronyms or abbreviations.	
Avoid using technical language unless you know your audience knows those terms.	
Make sure that the communication is clear and your handwriting is readable.	
Proofread your work. Check for typos, spelling errors, and grammatical errors. Do not assume spell-check or autocorrect got it right.	
Be prepared to provide a verbal or visual explanation if needed.	

TABLE 3 Checklist for Effective Visual Communication

Did You ...?	Yes/No
Avoid complex graphics. (Simple is better.)	
Use colors and graphical elements to highlight key information.	
Ensure that the visual is large enough to be useful.	
Present the information in a logical order.	
Provide an adequate level of detail and accuracy.	
Provide a legend or written description if you include symbols.	
Be prepared to provide a written or verbal explanation of the diagram if needed.	

Nonverbal communication: Things the speaker and the receiver see or hear while communicating. Examples include facial expressions, body movements, hand gestures, tone of voice, and eye contact.

Did You Know?

The Major Impact of Nonverbal Communication

Over the years, many studies have tried to measure how much of all communication is nonverbal. The specific numbers vary by study but indicate that between 55 to 93 percent of all communication is nonverbal. This means that your nonverbal signals send an even more powerful message than your verbal communication. Always remember that people might not remember what you say, but they do remember how you make them feel. Nonverbal cues have a major impact on how you make people feel when you communicate.

Did You Know?

Communication Is Number One

Research shows that the most common thing a supervisor does daily is communicate. They may do this as either a sender or receiver of messages. Eighty percent of a typical supervisor's day is spent communicating through writing, speaking, listening, or using body language. Of that time, studies suggest that around 20 percent is written communication, while 80 percent involves verbal communication.

2.2.5 Nonverbal Communication

Unlike verbal or written communication, **nonverbal communication** does not involve spoken or written words or images. Nonverbal communication refers to things the speaker and the receiver see or hear while communicating. This includes appearance, personal environment, use of time, and body language. Examples include facial expressions, body movements, hand gestures, tone of voice, and eye contact.

Nonverbal communication can provide an external sign of someone's inner emotions. It often occurs at the same time as verbal communication. You can express feelings in many ways without realizing it. You may show you are happy by smiling and giving someone a warm handshake. You may show you are angry by raising your voice.

People observe your physical, nonverbal cues when you are communicating. In fact, nonverbal signals can have more impact than your verbal communication. Be aware of nonverbal cues to avoid miscommunication based on your posture, facial expression, or other signals. For example, if you have your arms crossed in a defensive posture while speaking, the person might assume you are angry, even if you are not.

On a jobsite, nonverbal communication is also a matter of safety. For example, a major aspect of rigging safety is having clear communication between a crane operator and the designated signal person on the ground. This communication relies on common signals, including nonverbal hand signals. The Emergency Stop signal used in rigging operations is a nonverbal communication. Anyone on the ground within sight of the crane operator can use this signal in the event of an emergency (*Figure 9*).

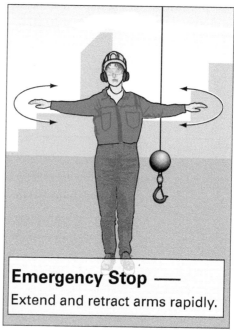

Emergency Stop —
Extend and retract arms rapidly.

Figure 9 The Emergency Stop signal is an example of nonverbal communication.

2.2.6 Additional Communication Considerations

As a leader in the field, the way that you communicate has significant impact on the safety, quality, and productivity of your group. Regardless of your communication method, remember that everyone communicates differently. In some cases, you might need to communicate in multiple ways. You might also need to adjust your level of detail to ensure that everyone understands the intended meaning. For instance, a visual learner may need a map to comprehend directions. Another team member might just need a verbal explanation of which turns to take to a destination. *Figure 10* shows an example of ways to tailor the message to your audience.

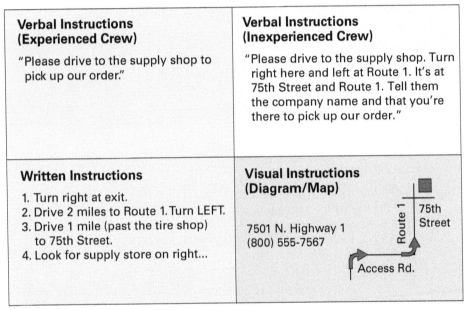

Figure 10 Assess your audience and then tailor your message.

The construction industry is a diverse and multicultural one. It is quite common to have a crew with people from different cultural backgrounds working together. Language and cultural differences can make communication even more challenging. Even if employees speak the same language, different understandings of words or symbols can create confusion or cause offense.

An important part of effective communication is to remember that people are different and to adjust your communication style to meet the needs of the person or people receiving your message. This means field leaders need to invest the time to learn how to communicate well with each individual. This requires you to know each crew member, recognize how each person learns, how each person delivers feedback, how each person works with others, and who they are as individuals, not just coworkers.

2.3.0 Problem-Solving and Decision-Making

Problem-solving and **decision-making** are a large part of every field leader's daily work. They are a part of life for all supervisors, especially in fast-paced, deadline-oriented industries like construction.

What is the difference between problem-solving and decision-making? Problem-solving is a process used to identify possible solutions to a situation. Decision-making refers to the action of making a choice. Decision-making involves starting or stopping a plan or choosing a different plan of action as

Problem-solving: A process used to identify possible solutions to a situation.

Decision-making: The action or process of making a choice. Decision-making involves starting an action, stopping one, or choosing a different plan or action.

appropriate for the situation. A decision may be complicated and involve many steps, but it is simply a choice.

The key difference between problem-solving and decision-making is that solving problems is a process, while making decisions is an action based on experience or information learned during the problem-solving process.

2.3.1 Problem-Solving

The ability to solve problems is an important skill in any workplace. It is especially important for craft professionals, whose workday is often not predictable or routine. Your group will look to you as a leader to solve problems regularly.

Figure 11 shows a five-step process for solving problems, which you can apply to any situation.

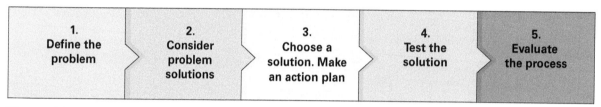

Figure 11 The problem-solving process.

Step 1 *Define the problem.* Defining a problem is not as easy as it sounds. Thinking through the problem often uncovers additional problems. Also, drilling down to the facts of the problem may mean setting aside your own biases or presumptions toward the situation or the individuals involved.

Step 2 *Consider possible solutions to the problem.* A problem may have more than one solution. Think through a number of solution options. As you do this, get inputs from people most affected by the problem, those who must correct the problem, and those who will be most affected by any potential solution.

Step 3 *Choose the solution that seems best and make an action plan.* After you consider several options to solve the problem, you must think through each possible solution and pick the best one. Again, you do not have to do this alone. Ask for feedback from others who have a stake in solving the problem well. Keep an open mind. The best solution might be taking pieces of two different solutions and combining them to create a new solution.

Step 4 *Test the solution to determine whether it actually works.* A solution might sound perfect before you try it but turn out not to be effective when you do. Alternatively, you might discover from trying to apply a solution that it is acceptable with a little change. If a solution doesn't work, think about how you could improve it, and then test your new plan.

Step 5 *Evaluate the process.* After your solution is in place, review the steps you took to discover and implement the solution. Could you have done anything better? The next time you are faced with a similar situation, you can use this experience to help solve the problem more effectively.

This problem-solving process can apply to almost any situation. *Figure 12* walks through a scenario to show how problem-solving works in practice. A

Problem Scenario

You are leading a team assigned to a new shopping mall project. The project will take about 18 months to complete. The only available parking is half a mile from the jobsite. The crew must carry heavy toolboxes and safety equipment from their cars and trucks to the work area at the start of the day, and then carry them back at the end of their shifts.

1. Define the problem	2. Consider problem solutions	3. Choose a solution Make an action plan	4. Test the solution	5. Evaluate the process
1. **Problem:** Workers are wasting time and energy hauling their tools and equipment to and from the worksite.	2. **Problem solution options:** Think about different ways to solve the problem. Several workers have proposed solutions: • Install lockers for tools and equipment closer to the worksite. • Have workers drive up to the worksite to drop off their tools and equipment before parking. • Bring in another construction trailer.	3. **Solution choice and action plan:** The worksite superintendent doesn't want an additional trailer in your crew's area. You and the team agree the shuttle service is the best option. It solves the time and energy issue *and* workers can keep their tools with them. The project supervisor plans to arrange a large van and driver to provide the shuttle service. **Note that this is a decision point in the problem-solving process.**	4. **Test the solution:** At first, the solution works, but a new problem arises. Everyone starts and leaves at the same time. The van can only hold half of the team and their equipment. To solve this problem, the supervisor schedules shuttle pickups every 15 minutes and adjusts work schedules so workers board the van at two different times. **Note that this is also a decision point in the problem-solving process.**	5. **Evaluate the process:** This process gave management and workers a chance to express an opinion and discuss the various solutions. Everyone feels pleased with the process and the solution.

Figure 12 Applying the problem-solving process, with decision points.

key point to note is that it's important to receive input from anyone in your group affected by the problem or the solution. Each solution will have pros and cons, so it's important to receive input from the workers affected by the problem. For example, workers will probably object to any plan (like the drop-off plan) that leaves their tools vulnerable to theft. Gaining the inputs of others will help you gain the trust and respect of your team.

2.3.2 Decision-Making

Figure 12 also shows where decisions are made as part of the problem-solving process. Decision-making is a choice made by using one's judgment. This judgment might be based on experience or on information learned during a problem-solving process. In general, you will have to make three types of decisions as a field leader, as shown in *Figure 13*.

No matter what the type, some decisions are simple or can be made based on past experiences. Other decisions are more difficult to make. These decisions require more careful thought about how to carry out an activity by using problem-solving techniques.

Note that decisions do have impacts, which you will need to consider. If you choose to use a shuttle to transport workers to a jobsite in shifts, as shown in *Figure 12*, using the shuttle may create a safety liability while workers are transported.

Decision Type	Description	Examples
Yes or no decisions  ? ✓ ✗	Decisions in which the answer is yes or no. Also called *whether* decisions, because you have to decide whether to take an action or not.	• Whether to hire a van to use as a shuttle service • Whether or not to promote a team member • Whether to alert a supervisor to an unsafe condition • Whether to proceed with a concrete pour
Options decisions 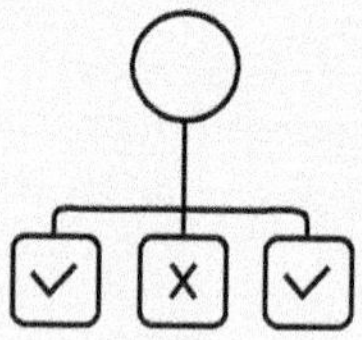 ○ ✓ ✗ ✓	Decisions that require you to choose from a group of options. Also called *which* decisions, because you have to choose which option to take.	• Which of the three options will best address the problem of workers having to haul tools and equipment to and from the worksite? • Which toolbox talk would be most important to deliver to the team today? • Which team member should I have drive to the supply shop to pick up the order?
Contingent decisions 	Decisions already made but on hold, until other specific events happen. These decisions can be very powerful because you can "pre-decide" what you will do in the event those events occur.	• If Carlo is late again two more times, I will set up a performance conversation. • If the materials arrive early, we will store them in the back of the jobsite. • If the team completes this task on time, I will find a clear way to recognize and reward the success.

Figure 13 Types of decisions.

2.0.0 Section Review

1. Which of the following is *not* a trait or behavior of effective frontline leaders?
   a. They lead by example.
   b. They are good communicators who lay out project goals to their crew.
   c. They motivate the team to work to their full potential.
   d. They have limited knowledge of the activities they supervise.

2. Which of the following is *not* a good practice to ensure effective communication?
   a. Speaking clearly at a reasonable pace
   b. Using highly technical language, acronyms, and slang
   c. Using visual cues, such as sketches or diagrams, to clarify instructions
   d. Using three-way communication to validate that a message was clear

3. While problem-solving is a process used to identify one or more possible solutions to a problem, the action of making a choice is called _____.
   a. decision-making
   b. a problem statement
   c. a decision statement
   d. action planning

3.0.0 Team Leadership and Motivation

Performance Tasks	Objective
There are no Performance Tasks in this section.	Identify key skills for team leadership and motivation. a. Choose strategies to motivate individuals and a team. b. Describe key approaches to building and leading teams. c. Recognize how to approach leadership challenges, such as transitioning from peer to leader, addressing performance issues, and resolving conflict.

In your role as a frontline leader, you will have incredible impact on the individuals on your team. Motivating individuals and building successful teams are important and challenging tasks for any leader. You will also face leadership challenges as a new or developing leader, including managing performance issues, handling conflict, and leading team members who were once your peers. In this section, you will learn strategies and approaches to succeed as a team leader.

3.1.0 Motivating Individuals

The ability to motivate others is a key skill that leaders must develop. To motivate is to make someone feel determined or enthusiastic or to make someone behave in a particular way. On a jobsite, a leader uses authority and influence to encourage individuals to put in their best effort to accomplish the project goals. Lack of motivation affects job quality and productivity, reducing profits and growth for companies, businesses, contractors, or organizations that rely on people to complete the work.

As a supervisor or crew leader, it is important to understand what motivates each individual on your team. Different things motivate different people in different ways. This means there is no single best approach to motivating all of your crew members, because what motivates one crew member may not motivate another. In addition, what works to motivate a crew member once may not motivate that same person again in the future.

How can you know what motivates the members of your project team? Effective communication skills, specifically active listening, are a key part of getting to know each person on your team. You can also learn more about what motivates an employee by looking at factors in the employee's performance. Examples of factors that may indicate an individual's motivation include the level or rate of unexcused absenteeism, the number of complaints, and the quality and quantity of work produced.

Generally, the factors that motivate individuals are the same as those that create job satisfaction. These include the following:

- Clear expectations and goals from leaders
- Training to help individuals be successful
- Consistent sense of accomplishment and growth
- Recognition and rewards
- A good work environment and culture

Accomplishment: An individual's need to set challenging goals and achieve them.

A crew leader's ability to provide strong support in these areas increases the likelihood of high morale within a crew. Morale refers to an individual's emotional outlook toward work and the level of satisfaction gained while performing the jobs assigned. High morale means that employees will be motivated to work hard, and they will have a positive attitude about coming to work and doing their jobs. A motivated team is also generally a more productive and safer team, which can have a positive effect on project success.

Morale: An individual's emotional outlook toward work and the level of satisfaction gained while performing the jobs assigned.

3.1.1 Clear Expectations and Goals

As a leader, it is important for you to clearly communicate your expectations. A lack of clear expectations can leave your team feeling lost and demotivated. To get the best performance from your team and to motivate them, you need

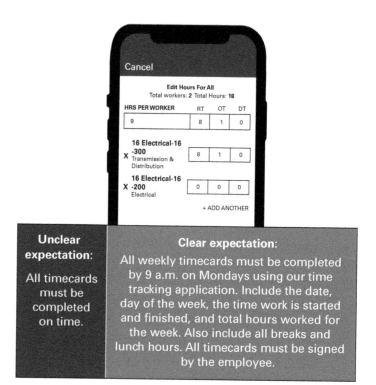

Figure 14 Examples of unclear and clear expectations.

to set clear goals for completing tasks. You also need to ensure that everyone on the team understands the deadlines for those tasks. *Figure 14* includes an example of clear expectations.

Set achievable short-term and long-term goals, including tasks to be done and expected timeframes. Having short-term goals gives you a chance to provide regular feedback, positive praise, and rewards. However, because short-term goals can change, also setting long-term goals gives the team additional opportunities for success. Do not set workers up for failure, as this leads to frustration, and frustration can lead to reduced productivity.

In addition to setting clear goals and expectations, you must also deliver timely feedback. People want and need feedback and positive reinforcement for their contributions and efforts. However, telling someone they did a good job last year or criticizing them for work from a month ago is meaningless. When sharing timely feedback, always include specifics to help the individual know what they did well or where they could improve.

You always should share negative feedback in private. When delivering negative feedback, approach the problem as a coaching moment, working with the individual to understand their challenges and working with them to correct the problem. Every failure or miss can be an opportunity for you to help that individual grow, if the conversation and feedback are constructive.

Positive feedback can be shared privately or publicly. When you praise a person in front of their team, it may also encourage other members of the team. Praise can be a valuable tool to reinforce behaviors that you want. (Not all people like to be publicly praised, so consider that before you offer public praise.)

3.1.2 Accomplishment and Growth

Accomplishment refers to an individual's need to set challenging goals and achieve them. A sense of accomplishment motivates most people. That is why setting and working toward recognizable goals tends to make employees more productive. A person with a sense of achievement is likely to be better motivated and help to motivate others.

A person who can align a task to their sense of meaning is much more likely to complete it. Frontline leaders can help their team feel a sense of accomplishment by keeping the work challenging and interesting. Workers who are completing the same task every day or performing work above or below their skill level can quickly become demotivated. For many individuals, overcoming the challenge of new or difficult assignments is much more satisfying than performing repetitive, boring tasks.

Remember, however, that too many or significant changes in work may have a negative impact on morale, because people often prefer some consistency and predictability in their work. Crew members react to change in many different ways and reasons. For some individuals, change creates anxiety and uncertainty, which can result in a negative attitude toward the work. To help manage change with your crew, remember to share why the change is occurring and explain what the employees and company will gain.

As a leader, you must ensure the individuals on your team feel that their skills and abilities are valued and make a difference. Employees who do not feel valued tend to have performance and attendance issues. Leaders should work to ensure the following is true for every crew member:

- They feel like an important part of the team with a skill set that contributes to the project success.
- They know that every position helps them build skills and experience that will last throughout a career.
- They have a chance to learn new things and develop their skill sets.
- They see opportunities to grow and progress within the company.

Professional and personal growth are other critical factors in motivation. **Professional growth** refers to gaining new skills and work experience that can help you reach a goal in your career. It is important that employees know that they aren't limited to their current jobs. Let them know that they have a chance to grow with the company and to be promoted as acknowledgment for excelling in their work when such opportunities occur.

Professional growth: Gaining new skills and work experience that can help one to reach a goal in one's career.

Whenever possible, show your team that you value their professional growth by giving them more advanced responsibilities when they are ready to perform them safely. People who are presented with challenging tasks often are motivated to work harder and improve. This approach is a win–win since the construction project benefits from a higher level of productivity, while the worker experiences professional growth.

Effective leaders encourage each of their crew members to work to their full potential. In addition, they share information and skills with their employees to help them advance within the organization.

Personal growth is about improving your talents and potential, both in and out of the workplace. For example, a team member may find it very rewarding to master a new competency on the job. Experiencing this type of personal growth at work helps to motivate, reduce boredom, and spark enthusiasm for additional growth.

Personal growth: Improving one's talents and potential, both in and out of the workplace.

Personal growth also takes place outside the workplace. Crew leaders should encourage their employees to identify goals for themselves both professionally and personally and then support their efforts to accomplish those goals. As an example, training can be a way to improve both professional and personal skills, such as time management, communication, financial management, and more.

3.1.3 Training and Upskilling

Another key factor in motivating individuals—and ensuring productive and safe projects—is providing sufficient training and opportunities for **upskilling.** Upskilling is a type of training that provides someone with more advanced skills through additional education. Upskilling provides employees with the tools and knowledge they need to perform their current job more effectively, more efficiently, or in a more modern way.

Upskilling: A type of training that provides someone with more advanced skills through additional education.

As people enter the construction workforce, many will have completed training programs but may not have the specific practical experience needed to complete a task on a project site. When a new employee joins your crew, you will need to take an active role in confirming their skills and helping with on-the-job training if there are gaps. A frontline supervisor on a construction jobsite only has a short time to evaluate their crew members on what tasks they can do or how much experience the crew member actually has.

In the first week, with any new employee, you should ensure that you train the new employee on any jobsite processes and allow time to do a task verification (*Figure 15*). After you have established the skills the individual has, you can work on a plan to leverage those skills while establishing new ones. If you identify gaps, you can create a training plan for that individual to address those areas.

3.1.4 Recognition and Rewards

Recognition: Acknowledgement of an employee's accomplishments through verbal and nonverbal communications.

Positive feedback is a form of **recognition**. Recognition acknowledges an employee's accomplishments through verbal and nonverbal communications. Individuals want to have good work appreciated, applauded, and acknowledged by others. You can accomplish this by thanking employees, offering an award for Employee of the Month, or giving tangible or monetary awards.

The following are tips to remember for giving recognition and praise:

- Be available on the jobsite so that you see good work as it is happening.
- Provide positive feedback publicly or in private.
- Give recognition and praise only when truly deserved. People quickly recognize insincere praise and will lose respect for the person delivering it.
- Encourage improvement by showing confidence in the ability of the crew members to do above-average work.
- Use goal setting as a tool to build confidence and create opportunities for praise.

Rewards are tangible recognition for good work. Rewards can be monetary, such as opportunities for raises, bonuses, or other incentive pay. Other types of rewards might include company gear, vouchers for new boots, gift cards, a team outing at the end of the week, or the best spot in the parking area for a month.

3.1.5 Good Work Environment and Culture

Having a good work environment and culture is critical for creating strong employee motivation and high morale. As a leader, you can create a good work environment and culture for your team in several ways.

Treat employees well. Be considerate, kind, and respectful. Treat people fairly. Mentor your employees. Coaching and supporting employees boosts their self-esteem, their self-confidence, and their motivation. Remember that small things matter. Having clean lunch areas, good housekeeping, and a comfortable workplace communicates that you respect the group.

Figure 15 When a new employee starts, be sure to evaluate the crew member's skills.
Source: kali9/E+/Getty Images

Communicate actively. As a leader, you should make yourself available for questions, opinions, and ideas. Actively engaging and communicating with workers plays a crucial role in motivating your team, as it shows that you appreciate and value their feedback. Involve employees in discussions so that they feel their opinions are valued. This, in turn, leads to a sense of pride and active participation. Making yourself available also prevents team members from feeling isolated from leadership.

When you communicate, be positive. Approaching your leadership responsibilities by threatening employees with negative consequences can result in higher employee turnover instead of motivation.

As much as possible, maintain a sense of humor, especially toward your own failings as a human being. No one is perfect. Your employees will appreciate that and be more motivated to encourage you.

Help individuals manage stress. In a job that can be demanding, fast-paced, and tiring at times, managers in the construction industry must take measures to ensure that their team is not exposed to undue stress. Work with your supervisors to ensure you have a safe and reasonable work environment. This means not only physical spaces, but also policies and procedures. For example, allowing workers to take time off to deal with personal matters or emergencies shows that the company prioritizes their well-being. Engage with your supervisors to help address resources and to organize shifts in such situations to ensure that project progress and construction costs are not negatively impacted.

Lead by example. Employees will be far more motivated to follow someone who is willing to do the same things they are required to do. Being a good leader and motivator means leading by example. Employees will not be motivated to work if upper management does not display the same level of commitment to the job that they are asked to have. For example, if you ask everyone to work late, but then leave early to go see a movie, this will be noticed and may create feelings of resentment and decrease team morale. As a leader, you can inspire your team by displaying the behaviors you would like to see from them, including working hard, showing personal and professional integrity, and having a can-do attitude.

3.2.0 Leading Teams

As a crew leader, you will need to lead a team of individuals from a range of backgrounds and experiences and create a high-performing team. To do so, you will need to delegate work to that team and implement policies and procedures that support the team. You will also need to address challenges like language barriers and **harassment**.

Harassment: Unwelcome conduct that is based on race, color, religion, sex, national origin, age, disability, or genetic information. Harassment includes verbal or physical conduct that impacts a person's work performance or creates an intimidating, hostile, or offensive work environment.

3.2.1 Delegating to Your Team

Once the various activities that make up the job have been determined, the crew leader must identify the person or persons who will be responsible for completing each activity. This requires that the crew leader be aware of the skills and abilities of the people on the crew. Then, you must put this knowledge to work by matching the crew's skills and abilities to specific tasks needed to complete the job.

After matching crew members to specific activities, you must then delegate the assignments to the responsible person(s). Generally, when delegating responsibilities, the crew leader verbally communicates directly with the person who will perform or complete the activity.

When delegating work, remember to do the following:

- *Delegate work to a crew member who can do the job properly.* If it becomes evident that the worker doesn't perform to the desired standard, either teach the crew member to do the work correctly or turn it over to someone else who can (without making a public spectacle of the transfer).

- *Make sure crew members understand what to do and the level of responsibility.* Be clear about the desired results, specify the boundaries and deadlines for accomplishing the results, and note the available resources.

- *Identify the standards and methods of measurement for progress and accomplishment.* Also identify the consequences of not achieving the desired results. Discuss the task with the crew member and check for understanding by asking questions. Allow the crew member to contribute feedback or make suggestions about how to perform the task in a safe and quality manner.

- *Give the crew member the time and freedom to get started.* Don't make them feel pressured by too much supervision. When making the work assignment, be sure to tell the employee how much time there is to complete it and confirm that this time is consistent with the job schedule.

- *Check in on progress.* Be sure to periodically check in on the crew member's progress. This will allow you to see how the task is going and allow the crew member to ask questions. Checking in along the way will also help you clear up any misunderstandings early and avoid rework.

- *Examine and evaluate the result.* After the task is complete, ask the crew member for their comments on how the work went. Provide your own feedback. Having this discussion will help you know what kind of work to assign that crew member in the future. It will also provide a means of measuring your effectiveness in delegating work.

There may be a time when someone else in the company issues written or verbal instructions to a crew member or your crew without going through you as the crew leader. If this happens, work with that individual to understand why they took that approach. Then, help that person understand why it is important for them to work through you for this type of instruction, whenever possible, so that you can manage the work of the entire crew safely and productively.

3.2.2 Implementing Policies and Procedures

Every company establishes policies and procedures that crew leaders are expected to implement and employees are expected to follow. Company policies and procedures are essentially guidelines for how the organization does business. They can also reflect organizational philosophies, such as putting safety first or making the customer the top priority. Examples of policies and procedures include safety guidelines, credit standards, and billing processes.

The following tips can help you effectively implement policies and procedures:

- Learn the purpose of each policy. This will help you follow the policy and apply it appropriately and fairly.

- If you're not sure how to apply a company policy or procedure, check the company manual or ask your supervisor.

NOTE

Try to obtain a supervisor's policy interpretation in writing or print out their email response so that you can append the decision to your copy of the company manual for future reference.

- Always follow company policies and procedures. Remember that they combine what's best for the customer and the company. In addition, they provide direction on how to handle specific situations and answer questions.

Crew leaders may need to issue directions to their crew members. A *direction* is a form of communication that initiates, changes, or stops an activity. Directions may be general or specific, written or verbal, formal or informal. It is up to you how you issue directions, but the policies and procedures of the company may govern your choice.

When issuing directions, do the following:

- Make them as specific as possible. Avoid being general or vague unless it is impossible to foresee all the circumstances that could occur in carrying out the directions.

- Recognize that you do not have to write directions for simple tasks unless your company requires this.

- Write directions for more complex tasks, tasks that will take considerable time to complete, or tasks that are permanent (standing) orders.

- Consider what is being said, the audience to whom it applies, and the situation under which it will be implemented to determine the appropriate level of formality for the direction.

3.2.3 Addressing Language Barriers

Leading a diverse team also means addressing language barriers. Millions of US workers speak languages other than English. Jobsites where two or more languages are used are common. In addition, even among English speakers, accents and dialects can present communication barriers.

As you have learned, any barrier in communication keeps people from being able to understand each other clearly. If a person does not fully understand the task a frontline supervisor gives them, the task may be completed incorrectly, which can cause delays, cause additional expenses, or create an unsafe situation. As a supervisor of a crew, you need to know that each crew member understands the work to be done. The following tips will help when communicating across language barriers:

- Be patient. Give workers time to process the information in a way they can comprehend.
- Avoid humor. Humor is easily misunderstood. The worker may misinterpret what you say as a joke at the worker's expense.
- Do not assume an individual is not intelligent simply because they don't understand what you are saying.
- Speak slowly and clearly. Avoid the tendency to raise your voice.
- Use face-to-face communication whenever possible. Other forms of communication can be more difficult when a language barrier is involved.
- Use the plans and drawings or make sketches to ensure people understand clearly.
- For critical written instructions, ensure the individual's literacy level matches the level of difficulty and language of the document. A person may speak clearly in a second language, but struggle with written understanding.
- Communicate in multiple ways or adjust your level of detail or terminology to ensure that everyone understands the meaning as intended.
- Confirm understanding. Ask for feedback with open-ended questions that require more than yes-or-no answers. Use three-way communication and have an individual repeat back what they heard to ensure mutual understanding and prevent miscommunication. If a worker is not fluent in English, ask the person to demonstrate their understanding through other means.

Many organizations offer language classes for supervisors and employees to help create more effective communication on jobsites. Whenever possible, take advantage of these training opportunities and encourage your crew to do so.

3.2.4 Preventing Harassment

As a frontline supervisor of a team, you must be aware of harassment as a potential challenge. As defined by the US Equal Employment Opportunity Commission (EEOC), harassment is unwelcome conduct that is based on race, color, religion, sex, national origin, age, disability, or genetic information. Harassment includes verbal or physical conduct that impacts a person's work performance or creates an intimidating, hostile, or offensive work environment.

The following are examples of activities that might qualify as harassment:

- Telling offensive jokes
- Calling someone offensive names
- Making verbal or physical advances or assaults

- Making threats or intimidating others
- Displaying offensive objects or pictures

Harassment becomes unlawful when a person must endure the offensive conduct to remain employed or the conduct is severe or common enough to create an intimidating, hostile, or abusive work environment. While not all acts of harassing behavior may be serious enough to be considered a violation of the law, all such behavior destroys teamwork and negatively affects jobsite safety and productivity.

Anyone can be a victim of harassment. Historically, many have thought of harassment as an act perpetrated by someone in power over another. However, a harasser can be a supervisor, a supervisor in another area, a coworker, or any other person in the work environment.

The EEOC enforces sexual harassment laws within industries. When investigating allegations of harassment, the EEOC looks at the whole record, including the circumstances and the context in which the alleged incidents occurred. Supervision may hold the crew leader responsible if they were aware of sexual harassment and did nothing to stop it. You should not only take action to stop harassment, but also should serve as a good example for the rest of the crew.

Prevention is the best way to eliminate harassment in the workplace. Your employer will have a harassment policy that clearly communicates zero tolerance for harassment and outlines a complaint or grievance process. This process should include taking immediate and appropriate action when an employee complains. This anti-harassment policy should apply to all employees (including managers, supervisors, and coworkers) and nonemployees (such as suppliers, owners, vendors, and others). Creating an open and inclusive work environment will discourage harassment.

As a leader, you must understand and uphold your employer's policies and procedures on harassment. If you believe that someone has violated the harassment policy, immediately notify your supervisor. If you are not comfortable reporting the incident to your immediate supervisor, you may report the complaint to any employee in a supervisory position or Human Resources. If you are not sure how to proceed in a situation, reach out to your Human Resources partners for guidance.

3.3.0   Leadership Challenges

As the leader of a team, you will need to manage leadership challenges and opportunities every day on the job. While the specific challenges will vary based on your team, the project, and other conditions, you must learn to address these when they arise. Such challenges can include navigating your initial transition from peer to leader, managing performance issues with your crew, or resolving conflicts that arise on the jobsite.

3.3.1 Transitioning from Peer to Leader

Making the transition from crew member to a crew leader can be difficult, especially when the new position involves overseeing a group of former peers. When you move from being a peer to a leader, your relationship with your former team members changes in important ways.

When you become a crew leader, you are no longer responsible for your work alone. You are now accountable for the work of a diverse team of individuals with varying skill levels, personalities, work styles, and cultural and educational backgrounds.

Crew leaders must learn to put personal relationships aside and work for the common goals of the entire crew. The crew leader can overcome these problems by working with the crew to set mutual performance goals and by freely communicating with them within permitted limits. Use their knowledge and strengths along with your own so that they feel like they are key players on the team. *Table 4* lists some additional suggestions on how to make the transition from peer to leader easier.

TABLE 4 Making the Transition from Peer to Leader

Strategy	Tactic
Reset boundaries with existing friends	Expectations and boundaries must be set with existing friends. Having a strong, professional working relationship is important, but you may need to reduce time spent with old peers outside of the project. Be clear and explain that, on the job, business comes before personal relationships.
Demonstrate that you have earned this leadership role	Quickly demonstrate good leadership for your team to earn employees' trust and respect. Be fair and consistent in everything you do. The perception of favoritism will tear a crew apart and negatively impact the outcome of any project.
Make the best decisions for the project and your company	Always make decisions that are best for the business. While it can be hard to make a decision that impacts someone who was a peer and is a friend, don't compromise to make a popular decision.

3.3.2 Managing Performance Issues

At the end of the day, if someone is simply not being accountable and completing their work, it might be time for a difficult conversation. In the workplace, a difficult conversation is one in which you have to manage emotions and information in a sensitive way to deal with a workplace issue. A difficult conversation may involve the following:

- Communicating business decisions
- Discussing poor performance or behavior
- Addressing conflict
- Handling complaints
- Ending employment or other changes in job status

As a leader and manager, you will need to have a difficult conversation at some point in your career. These conversations are never enjoyable, but you can handle them effectively by following some basic steps.

Identify the problem to resolve. Ensure that you have a clear understanding of the problem that you want to resolve. Before you start preparing for the conversation, ask: Why do you want to have the conversation? What do you want to achieve? Is the outcome you're looking for realistic?

Gather the information you need. Make sure you know the facts of the situation before approaching the employee. Bring any documents you will need to the meeting. For example, if you are discussing how an employee is late to work, you will want to have relevant policies and tardiness data at hand. If you are not sure about any internal workplace policies and procedures, you can always seek advice from a supervisor or Human Resources manager.

Think about the employee's point of view. Before you speak, you should think about the matter from your employee's point of view. Engage in the conversation with an open mind and truly seek to understand the situation from the employee's point of view.

Get the timing right. Try to schedule the conversation as soon as possible, but not if emotions are running high. Choose a time and place appropriate for the discussion, where you will not be rushed or interrupted. Generally, avoid having negative discussions at the end of the week and, if an employee needs time to process the conversation, offer to take a break and meet again later that day or the next.

Prepare what you are going to say. Write down the key points you need to cover, so that you can keep your conversation on track and stay in control. While it is important to prepare notes, avoid preparing a rehearsed script, as it may hamper your ability to listen effectively, react accordingly, or find alternative solutions.

Give the employee time to prepare. Make an appointment with the employee and provide some context to give them some time to prepare. Remember that knowing this conversation is coming may be stressful for the employee, so phrase the approach carefully and don't leave it too long. Examples of ways to phrase the appointment might be:

> "I'd like to talk about a series of quality issues I'm noting and get your point of view. Can you come see me tomorrow?"

"I think we have different perceptions about what it means to have a clean work area. I'd like to hear your thoughts on this. Are you free this afternoon?"

Seek support if you need it. If you're not comfortable having the conversation one-on-one with the employee, enlist help from your manager or Human Resources person.

After the conversation, be sure to take the following steps:

- *Document the discussion.* Write down any agreements that were reached, and indicate the date, time, and who was present. This will be helpful if you need to revisit the discussion at a future date. Your *job diary*, which you will learn about later, is a useful tool here.

- *Follow up with the employee.* Confirm what you had agreed in writing and send these to the employee. Be sure to give the employee time to actually act upon the steps they agreed to.

- *Take some time to reflect on the matter.* Learn from the experience and think about what you could do differently next time. Debrief with your supervisor or Human Resources if you feel you need to discuss the matter.

Also, remember that you should never discuss this matter with other staff members. You must treat the employee the same as you had prior to the conversation and remain completely professional.

If you do need to have this type of difficult conversation, do it as soon as you start to have concerns or see early signals of conflict. The worse a situation becomes, the harder it can be to manage and resolve. In the meantime, your project and the rest of the crew are being affected.

As previously noted, the best communication strategy is to have an open-door policy or informal one-on-one conversations so that issues can be dealt with as naturally as possible. Be approachable and keep in touch with your team, so they are more likely to come to you with problems before they escalate.

3.3.3 Conflict Resolution

Conflicts will happen during a project. When you have many groups, including owners, general contractors, and specialty contractors all working together to complete a project, disagreements are likely to happen. Conflicts can arise within your crew as well. Each person or group might have a different view or opinion on how something should be done. Those differing opinions often lead to conflicts.

As a leader of the team, you will have many opportunities to look for positive and proactive means to resolve conflict. Resolving conflicts can lead to a better understanding or a new way of doing things that improve the overall project. If you handle conflict well, it will further earn the respect and trust of your crew. The following are some strategies to use when addressing conflict:

Recognize that conflict is normal. Not all conflict is bad. Conflicts on the construction site can be good. If the individuals engaged in the conflict are mature, conflict can improve processes and approaches by creating conversation about new ways to do work.

Do not ignore the conflict. Take action early to avoid further complications down the line. Even a small dispute that may appear inconsequential can balloon into a larger problem if it goes unresolved. If you allow the parties involved to work it out on their own, you still need to follow up and make sure that there aren't any lingering issues.

Understand the real issue behind the conflict. You must understand the real issues causing the conflict to determine the best solution. Conflicts are often caused by different interpretations of plans and specifications, change orders, scheduling, and poor communication. Work stress can also be an underlying issue. Conflicts, for example, may happen more often later in the day or in a work shift, due to fatigue. Use your active-listening skills to listen to everyone involved. This shows everyone that their opinions matter and that you are committed to working with them to resolve the issue fairly.

Remain neutral and focus on business, not people. Conflicts can escalate quickly from a simple disagreement to a shouting match. If appropriate, call for a short break before starting the discussion again. Remind everyone that resolving conflicts is about making the best business decision for the project, not personal feelings. Both parties should sit down and stay focused on developing a solution that is in the best interest of the project.

Compromise and work with the team. Conflicts rarely have a clear-cut right and wrong. As you work to resolve the conflict, you will need to encourage others to be willing to compromise and work together to solve the problem. The effort here must stay focused again on the desired business result: "What is the best decision for this situation?" Construction is all about cooperation and collaboration. However, if the conflict involves safety rules and regulations or adhering to building codes, any solution must follow the proper policy, procedure, or regulation.

Avoid conflict through clear communications. Poor communication on the jobsite often leads to conflicts. Make sure that you clearly communicate expectations with everyone on the project. Keeping everyone up to date with any changes to the project or schedule can help avoid conflict.

Communicate the resolution and follow up. Resolving conflicts may not result in a win for everyone. It is important that once resolution is reached, everyone gets an explanation and is clear on what is expected moving forward. They might not always agree with the decision. Having a clear expectation will help others accept the decision. Whenever an agreement is reached, follow up to determine if both sides are working as discussed or if issues still exist. You also will remind everyone that, although the conflict might have been difficult, the resolution produced a good outcome for the project.

If you cannot resolve a conflict on your own, always reach out to your supervisor for guidance. Your supervisor or another senior leader can make a decision and communicate it to the appropriate individuals.

Remember, the goal of any construction project is to see the successful completion of the job on time and within budget. When conflicts arise, it can threaten the goal of on-time and on-budget job completion. As the leader of your crew, you must learn to address and resolve conflicts, so that the project work can proceed. Resolving conflicts can be challenging. It all comes down to making the best business decision to keep the project moving forward.

3.0.0 Section Review

1. Which of the following is *not* a way to motivate a team?
   a. Providing training and helping individuals feel a sense of accomplishment and growth
   b. Setting expectations without clear short-term and long-term goals
   c. Offering recognition and rewards when appropriate
   d. Having a good work environment and culture

2. One sign of a high-performing team is _____.
   a. the leader cannot delegate work to the team
   b. harassment is common on the worksite
   c. everyone on the team follows policies and procedures
   d. language barriers prevent effective communication

3. As a frontline leader, steps to successfully resolving conflict may involve any of the following *except* _____.
   a. remaining neutral and focusing on the business, not on people
   b. compromising and working with the team
   c. ignoring the conflict
   d. understanding the real issues causing the conflict

4.0.0 Safety and Safety Leadership

Performance Tasks

1. Organize and deliver a safety-related toolbox talk, applying communication skills.
2. Complete a job safety analysis.

Objective

Identify a frontline leader's safety leadership responsibilities.

a. Assess the impact of incidents.
b. Explain the purpose of OSHA and describe the role of OSHA in administering worker safety.
c. Describe the importance of safety and health programs.
d. Explain the frontline leader's role in incident preparedness and response.
e. Recognize the importance of supporting your crew's total health.

Hazards: Conditions or activities that, if left uncontrolled, have the potential for harm.

Incidents: Unplanned events that result in property damage, fatalities, injuries, illnesses, or near misses.

Craft professionals routinely face **hazards**. Examples of these hazards include falling from heights, working on scaffolds, using cranes in the presence of power lines, operating heavy machinery, and working on electrically powered or pressurized equipment. Despite these hazards, applying preventive safety measures can significantly reduce the number of **incidents**.

Safety matters because people matter. As a crew leader, some of your most important responsibilities are to enforce the company's safety program, make sure that all workers are performing their tasks safely, and ensure that everyone on the jobsite remains safe and goes home healthy (*Figure 16*). You must recognize safety risks and work to eliminate or reduce them.

To be successful, the crew leader should do the following:

- Be aware of the human and business impacts of incidents.
- Understand all federal, state, and local governmental safety regulations applicable to your work.
- Be the most visible example of the best safe work practices.
- Be involved in training workers in safe work methods.
- Conduct training sessions.
- Get involved in safety inspections and incident investigations.

Crew leaders are in an important position to ensure that their crew members perform all jobs safely. Providing employees with a safe working environment by preventing incidents and enforcing safety standards will go a long way toward maintaining the job schedule, ensuring a timely job completion within budget, and keeping your team healthy.

Figure 16 Committing to safety is a key part of a crew leader's role.
Source: RainStar/E+/Getty Images

4.1.0 The Impact of Incidents

An incident is an unplanned event that results in property damage, a fatality, injury, illness, or a near miss. A near miss is an unplanned event that does not result in personal injury but may result in property damage or is worthy of recording.

Incidents including work-related injuries, sickness, and fatalities cause immense suffering for workers and their families. Resulting project delays and budget overruns can cause huge losses for employers, and worksite incidents damage the overall morale of everyone on a project or jobsite.

Businesses lose billions of dollars every year because of on-the-job incidents. As just one measure of the business impact of incidents, construction companies in the United States pay over $1 billion per week in **workers' compensation** for nonfatal injuries. Workers' compensation, often called *workers' comp*, is insurance that companies have to cover medical expenses, lost wages, and other costs for employees who are injured or become ill while completing work for their job. It also pays death benefits to families of employees who are killed on the job.

Preventing injuries means protecting the lives of workers, while reducing significant costs for their employers. It is estimated that the full impact of workplace injuries and fatalities cost companies approximately $151 billion per year. These costs affect the individual employee, the company, and the construction industry as a whole.

Organizations encounter both direct and indirect costs associated with workplace incidents.

- Direct costs are the money companies must pay out to workers' compensation claims and sick pay.
- Indirect costs are all of the costs a company must account for as the result of a worker's injury or death, such as time off from work.

The direct and indirect costs of incidents are comparable to the visible and hidden portions of an iceberg, as shown in *Figure 17*. The visible part of the iceberg represents direct costs, some of which are covered by insurance. The hidden part of the iceberg represents the indirect costs, which are a greater financial burden than the direct costs.

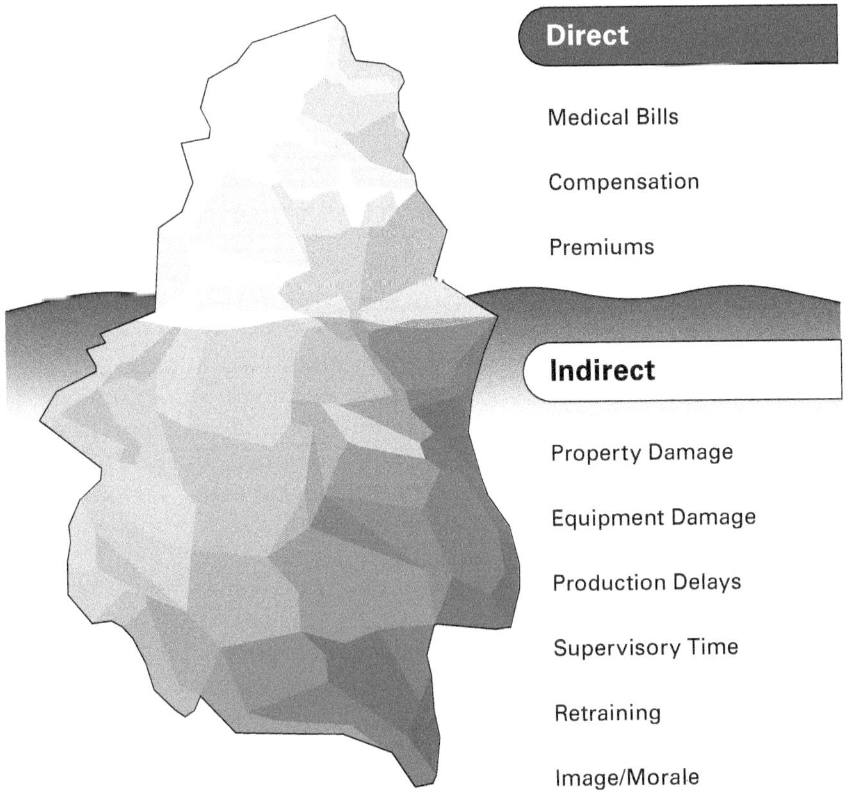

Workers' compensation: Insurance that companies have to cover medical expenses, lost wages, and other costs for employees who are injured or become ill while completing work for their job.

Direct

Medical Bills

Compensation

Premiums

Indirect

Property Damage

Equipment Damage

Production Delays

Supervisory Time

Retraining

Image/Morale

Figure 17 Incidents create both direct and indirect costs.

4.2.0 OSHA

To reduce safety and health risks and the number of injuries and fatalities on the job, the federal government has enacted laws and regulations, including the Occupational Safety and Health Act of 1970 (OSH Act of 1970). This law created the Occupational Safety and Health Administration (OSHA), which is part of the US Department of Labor. OSHA also provides education and training for employers and workers. Through the administration of OSHA, the US Congress seeks to ensure safe and healthful working conditions for workers by setting and enforcing standards and by providing training, outreach, education, and assistance.

OSHA has many functions. OSHA issues standards and rules for working conditions, facilities, equipment, tools, and work processes. It does extensive research into occupational incidents, illnesses, injuries, and deaths to reduce the number of occurrences and adverse effects. In addition, OSHA regulatory agencies conduct workplace inspections to ensure that companies follow the standards and rules. The agencies also have the right to take legal action if companies are not in compliance with OSHA standards and rules.

OSHA has established significant monetary fines for the violation of its regulations. *Table 5* lists the maximum OSHA penalty amounts as of 2022. In addition to the fines, there are possible criminal charges for willful violations resulting in death or serious injury. The attitude of the employer and their safety history can have a significant effect on the outcome of a case.

TABLE 5 OSHA Maximum Penalty Amounts

Type of Violation	Maximum Penalty
Serious	$14,502 per violation
Other-Than-Serious	
Posting Requirements	
Failure to Abate	$14,502 per day
Willful or Repeated Violation	$145,027 per violation

4.3.0 Safety and Health Programs

Under the OSH Act of 1970, employers have a responsibility to provide a safe workplace. Your employer must set up a program to manage workplace safety and health and to reduce work-related injuries, illnesses, and fatalities. The program must be appropriate for the conditions of the workplace. It should consider the number of workers employed and the hazards they face while at work. On a project site, a hazard usually is a condition or activity that, if left uncontrolled, has the potential for harm.

One way employers help create a safe working environment is to create a **construction safety plan**. A construction safety plan is a document that outlines the procedures, rules, and regulations that are or will be put in place to protect workers during a construction project. The safety plan should include steps to prevent incidents and respond if they occur, including emergency medical services and processes to review and report incidents. For some projects, a **site-specific safety plan (SSSP)** will include a focused set of information that accounts for a jobsite's unique hazards. Safety plans should be created before a construction project is underway, but also should be updated on a regular basis as conditions on the site change.

Construction safety plan: A document that outlines the procedures, rules, and regulations that are or will be put in place to protect workers during a construction project.

Site-specific safety plan (SSSP): A safety plan that includes a focused set of information that accounts for a jobsite's unique hazards.

4.3.1 Education and Training

All employees should receive a safety policies and procedures manual and initial training when they are hired. A company's safety manual should cover the company's safety expectations and the procedures that must be followed to create a safe work environment and meet OSHA requirements. During orientation, appropriate company staff should guide new employees through the safety policies and procedures manual, focusing on the sections most important for their job. An employer must ensure that any barriers to worker understanding, such as language barriers, are removed or addressed.

In addition to formal training, safety is a topic you need to discuss with your crew every day on the jobsite. You need to help employees be aware of and avoid the hazards to which craft professionals are exposed. As a leader, you should plan to conduct frequent safety trainings and meetings with your crew, including holding toolbox talks (*Figure 18*). These meetings are important to building a strong safety culture and reinforcing your company's commitment to protecting your workers.

Toolbox talks are one way to effectively keep all workers aware and informed of safety issues and guidelines. Toolbox talks, sometimes referred to as *tailgate meetings* or *safety briefings*, are short, informal safety meetings held at the start of each day or shift. Toolbox talks are 5- to 10-minute meetings that review specific health and safety topics. *Table 6* provides a checklist for delivering effective safety briefings or toolbox talks.

Always document your toolbox talks or other safety training. Even if certain OSHA standards do not require documentation of safety training, keep a record of each safety session signed by all attendees. In the record, include the date, a summary of the training session, the trainer, and attendees.

Toolbox talks: Short, informal safety meetings held at the start of each day or shift. Sometimes referred to as *tailgate meetings* or *safety briefings*.

Figure 18 Safety management software can help companies deliver and track worker training. Some of the solutions help you schedule, track, and deliver toolbox talks.

TABLE 6   Delivering Effective Safety Briefings or Toolbox Talks

Did You . . .?	Yes/No
Schedule time in advance and create time to allow workers to participate.	
Speak clearly and directly at a natural pace.	
Try to find a spot free of noise and other distractions. If the workers cannot hear you talking, or are distracted by other activities in the area, they won't be focusing on your talk.	
Use props to help make your point and provide visual communication cues. For example, have a safety harness with you for your talk on fall protection.	
Reinforce safety basics, focus on high-risk scenarios, and inform workers about changes to the jobsite and working conditions that may have occurred since their last shift.	
Discuss any incidents or injuries that have occurred and how they could have been prevented.	
Focus the discussion on active work. For example, do not hold a toolbox talk on sitework when electrical work is the only thing being done that day.	
If appropriate, break out a toolbox talk into smaller groups to focus on individual crafts or work activities for certain individuals.	
Focus on the positive, too. Highlight safe working practices. Provide positive reinforcement and applaud workers following safety rules and regulations.	
Always give workers an opportunity to ask questions at the end of the toolbox talk.	

One-on-one or small group training is also a critical part of keeping everyone on the project safe. When you assign an inexperienced employee a new task, you must ensure that the employee can do the work in a safe manner. You can accomplish this by providing safety information or training for groups or individuals.

When assigning an inexperienced employee a new task, do the following:

- Define the task.
- Explain how to do the task safely.
- Explain what tools and equipment to use and how to use them safely.
- Identify the necessary personal protective equipment and train the employee in its use.
- Explain the nature of the hazards in the work and how to recognize them.
- Stress the importance of personal safety and the safety of others.
- Review Safety Data Sheets that may be applicable.

4.3.2 Hazard Identification, Assessment, Prevention, and Control

Crew leaders and workers play an important role in identifying and reporting hazards. Hazards are often identified in job safety or job hazard analyses, which also document how the hazards will be eliminated or mitigated. A **job safety analysis (JSA)**, or *job hazard analysis (JHA)*, is a technique that focuses on job tasks as a way to identify hazards before they occur. It focuses on the relationship between the worker, the task, the tools, and the work environment. After you identify uncontrolled hazards, you will take steps to control them by eliminating or reducing them to an acceptable risk level.

Figure 19 shows an example of a form used to conduct a JSA. In a JSA, the task at hand is broken down into its individual parts or steps and then each step is analyzed for its potential hazards. Once a hazard is identified, certain actions or procedures are recommended that will correct that hazard. For example, during a JSA, it is determined that using a chain hoist to install a pump motor in a tight space would be safer than having a worker do it manually. By using the chain hoist, the chance that the worker's hand would get crushed during installation is reduced. Using the JSA process would help protect the worker from injury.

Job safety analysis (JSA): A technique that focuses on job tasks as a way to identify hazards before they occur. It focuses on the relationship between the worker, the task, the tools, and the work environment. Also known as *job hazard analysis (JHA)* or *activity hazard analysis (AHA)*.

JOB SAFETY ANALYSIS

TITLE OF JOB OR TASK

TASK	START	END	HAZARDS	CONTROLS
1.				
2.				
3.				
4.				
5.				
6.				
7.				
8.				

Required Training: Required Personal Protective Equipment (PPE):

Job Name: _____   Weekly Vehicle Checklist: ____ Tire Pressure ____ Transmission Fluid
Job Number: _____   ____ Oil   ____ Lights
Supervisor: _____   ____ Air Filter   ____ Wkly Mileage
Date: _____

Name of Employees:

PRINT NAME	SIGN NAME	TOTAL HOURS

Figure 19 Job safety analysis form.

A job safety analysis can also be used during planning to ensure that safety is planned into the job. Taking this approach helps ensure that safety does not get compromised for the sake of schedule and productivity.

Crew leaders must protect workers from existing or potential hazards in their work areas. It is the crew leader's responsibility to determine what working conditions are unsafe and to inform employees of hazards and their locations. In addition, they should encourage their crew members to tell them about hazardous conditions. To accomplish this, crew leaders must be present and available on the jobsite.

4.3.3 Management Leadership and Employee Participation

A culture of commitment to safety and a strong safety and health program require management leadership and employee participation. As a frontline leader, the best way for you to encourage employee participation in safety and health is to lead by example. Your behavior sets standards for your crew members. If you cut corners on safety, you are telling your crew they may do so as well.

You should always be sure to do the following:

- Stay actively involved in your company's safety and health program.
- Exceed expectations when it comes to safety. Help others follow the rules and ask your crew if there are safer ways to do a task.
- Host safety training and meetings. Discuss safety at the start of each shift. Have regular toolbox talks. Include safety in all meetings, even if the primary purpose is schedule, cost, or production related.
- Be a visible participant on safety walks.
- Review and provide feedback on hazard analyses.
- Always follow up on safety concerns raised by the workers.

CAUTION

Team members often "follow the leader" when it comes to unsafe work practices. Sometimes, supervisory personnel engage in unsafe practices and take more risks because they are more experienced. However, inexperienced or careless workers who take the same risks won't be as successful in avoiding injury. As a leader, you must follow all safety practices to encourage your crew to do the same.

Your employer will use their policies and procedures for incident investigation and documentation. Be sure to follow any of these guidelines, when the situation arises.

4.4.0 | Incident Preparedness and Response

While the goal is to have zero incidents on a worksite, you and your team must understand how to respond quickly to a safety incident. If an incident occurs, you should follow the safety policies and procedures in the construction safety plan. As a frontline leader, you are responsible for making sure everyone on your team understands the safety plan. You also may have reporting and investigation responsibilities after an incident.

4.4.1 Administering First Aid

When someone is hurt or has a medical event, your first goal is ensuring the best outcome: minimizing the health impact to the person. **First aid** is help given to a sick or injured worker on the worksite immediately after an injury occurs. The primary purpose of first aid is to provide immediate and temporary medical care to employees involved in incidents, as well as employees experiencing non-work-related health emergencies, such as chest pain or breathing difficulty.

Many companies have very specific policies on administering first aid. If you are unsure of these policies, ask your supervisor and your safety team for guidance. You need to know how to handle these situations before an incident occurs. A quick response can be the difference between life and death. The victim of an injury or sudden illness at a jobsite may be at a remote location. The site may be far from a rescue squad, fire department, or hospital, presenting a problem in the rescue and transportation of the victim to a hospital.

Your employer must ensure prompt first aid treatment for any injured employees. To do this, they will provide access to readily available first aid kits. You and everyone on your team should be aware of the location and contents of first aid kits available on the jobsite. Many jobsites also have an **AED (automated external defibrillator)** that is used to help those experiencing sudden cardiac arrest (*Figure 20*). Many lives have been saved due to a prepared team who knew where the AED device was stored and how to use it.

First aid: Help given to a sick or injured worker on the worksite immediately after an injury occurs.

AED (automated external defibrillator): A device used to help those experiencing sudden cardiac arrest.

Figure 20 Many jobsites have an AED to help those experiencing sudden cardiac arrest. You should always know where AED devices are stored.
Source: PardiMer/Shutterstock

Your company will have an on-site medical professional or at least one member of the workforce who is trained and certified in first aid, including the use of an AED and CPR. **CPR (cardiopulmonary resuscitation)** is a lifesaving technique that is used when someone's heartbeat or breathing has stopped. The benefits of having personnel trained in first aid at jobsites include the following:

- Immediate and proper treatment of minor injuries. First aid may prevent minor injuries from developing into more serious conditions. OSHA does not require companies to record when simple first aid is used to address the injury or illness.

- Ability to determine if the injured person requires professional medical attention.

- Ability to help stabilize an injured person before professional medical care arrives.

- Help to eliminate or reduce medical expenses, lost work time, and sick pay.

If further medical treatment is needed, preparedness and a rapid communication process will mobilize the right people as soon as possible. Emergency numbers should be posted on the jobsite. As a leader, you should be very familiar with the construction safety plan and work with your Human Resources and safety leaders to ensure it stays updated. You need to locate your local off-site medical provider for emergency or nonemergency care prior to an incident occurring.

CPR (cardiopulmonary resuscitation): A lifesaving technique that is used when someone's heartbeat or breathing has stopped.

4.4.2 Reporting and Investigating Incidents

As a frontline leader, you must know and follow your company's incident reporting processes. You also must ensure your team is following those policies. Your team should know from their training to immediately report any unusual event, incident, or injury. The policies and definitions around reporting should be included in safety manuals and construction safety plans.

If your employer is subject to the recordkeeping requirements of the OSH Act of 1970, it must maintain records of all recordable occupational injuries and illnesses using standard OSHA forms. In the event of an incident, the employer also is required to investigate the cause of the incident and determine how to avoid it in the future. According to OSHA regulations, the employer must investigate each work-related death, serious injury or illness, or incident having the potential to cause death or serious physical harm. The employer should document any findings from the investigation, as well as the action plan to prevent future occurrences. The company should complete these actions immediately, with photos or video if possible.

As a crew leader, you may be involved with an incident investigation. If required by company policy, you also will need to make a formal investigation and submit a report after an incident. An investigation looks for the causes of the incident by examining the circumstances under which it occurred and talking to the people involved. Investigations are perhaps the most useful tool in the prevention of future incidents. As shown in *Figure 21*, an incident investigation has four major parts.

NOTE

The American Red Cross, Medic First Aid, and many other local organizations provide basic and advanced first aid courses at nominal costs. These courses include both first aid and CPR. The local area offices of these organizations can provide further details regarding the training available. CPR certifications must be renewed every two years.

NOTE

Three OSHA forms are used for injury and illness recordkeeping:

- OSHA Form 300, Log of Work-Related Injuries and Illnesses
- OSHA Form 300A, Summary of Work-Related Injuries and Illnesses
- OSHA Form 301, Injury and Illness Incident Report

All of these forms are available online at the OSHA website. OSHA has also launched their Injury Tracking Application (ITA) to allow for direct entry of OSHA Form 300A information. For more information, scan the QR code at the front of the module.

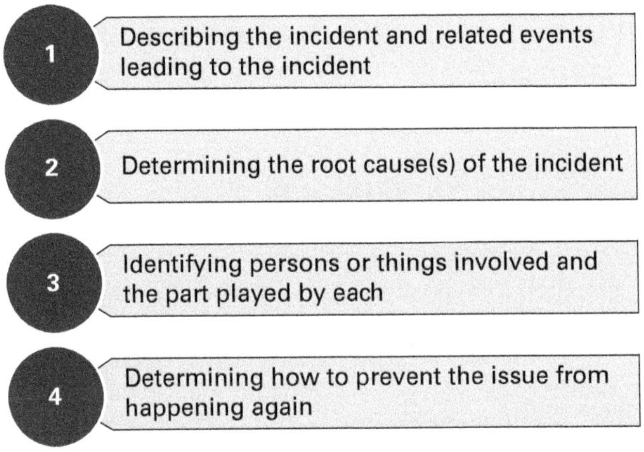

1. Describing the incident and related events leading to the incident

2. Determining the root cause(s) of the incident

3. Identifying persons or things involved and the part played by each

4. Determining how to prevent the issue from happening again

Figure 21 An incident investigation has four major parts.

Root cause: A core issue or fundamental reason for the occurrence of a problem.

Root cause analysis (RCA): A method of problem-solving used for identifying the root causes of faults or problem.

A key step is determining the **root cause** of the incident. A root cause is a core issue or fundamental reason for the occurrence of a problem. Understanding the root cause is critical to avoiding similar incidents in the future. In many cases, the root cause was a flaw in the system that failed to recognize the unsafe condition or the potential for an unsafe act.

Finding a root cause to a problem can be difficult. A **root cause analysis (RCA)** is a method of problem-solving used for identifying the root causes of faults or problems. A *Five Whys Analysis* is one type of root cause analysis that is helpful with incident investigations. The method is simple: When a problem occurs, you start with the broadest explanation for an incident and narrow it down to a more specific problem by asking "why" five times. The technique is even more effective when you include experienced team members who were actively part of the situation or have had similar experiences. It is especially useful when the root cause is not that obvious. *Figure 22* shows an example of how a Five Whys Analysis can apply to an incident investigation.

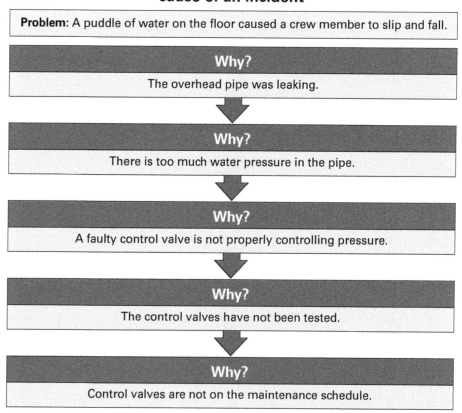

Using a Five Whys Analysis to understand the root cause of an incident

Problem: A puddle of water on the floor caused a crew member to slip and fall.

Why?
The overhead pipe was leaking.

Why?
There is too much water pressure in the pipe.

Why?
A faulty control valve is not properly controlling pressure.

Why?
The control valves have not been tested.

Why?
Control valves are not on the maintenance schedule.

Figure 22 Using a Five Whys Analysis to understand the root cause of an incident.

4.5.0   Supporting Total Health

More than ever, employers must focus not only on their employees' physical well-being, but on their total health. A healthy work environment not only helps to ensure workers return home safely at the end of their shift, but also ensures they can have a healthy life outside of work. Many companies are adopting practices that address not only protection from work-related safety and health hazards, but also overall worker safety, health, and well-being.

Two key aspects of total health for your workforce include addressing **substance abuse** and mental health. As a frontline leader, you will not be

Substance abuse: The inappropriate overuse of drugs and chemicals, whether they are legal or illegal.

able to address the root causes of these issues for your workers. You do, however, have the important responsibility to enforce company policies and procedures in these areas and work with your management or Human Resources Department to help employees in these situations.

4.5.1 Substance Abuse

Substance abuse is a serious safety issue in the workplace. Substance abuse is the inappropriate overuse of drugs and chemicals, whether they are legal or illegal. All substance abuse results in some form of mental, sensory, or physical **impairment**. Impairment is the point where someone's intake of substances affects their ability to perform appropriately. Alcohol can be abused by consuming to the point of intoxication. Legal prescription drugs can be abused by using the wrong dosage, using other people's medications, or self-medicating. Illegal drugs, such as cocaine, crystal meth, or marijuana in some states, can cause impairment.

Impairment: The point where someone's intake of substances affects their ability to perform appropriately.

Crew leaders must enforce company policies and procedures regarding substance abuse. There are legal consequences and safety implications associated with substance abuse. Suppose several crew members go out and smoke marijuana or drink during lunch. Then, they return to work to erect scaffolding for a concrete pour in the afternoon. If you can smell marijuana on the worker's clothing or alcohol on their breath, you must step in and take action. An impaired worker is more likely to cause an incident that could cause serious injury or death to themselves or others (*Figure 23*).

If you observe an employee showing impaired behavior for any reason, immediately contact your supervisor and/or Human Resources Department for assistance. You should work with management to deal with suspected drug and alcohol abuse and should not deal with these situations yourself. Reporting the concern helps you protect the business, and the employee's and other workers' safety.

It is often difficult to detect substance abuse because the effects can be subtle. The best way is to look for physical signs or sudden changes in behavior that are not typical of the employee. *Table 7* lists some examples of potential signs and behavioral changes that might indicate substance abuse.

Figure 23 An impaired worker is a dangerous one.
Source: vejaa/iStock/Getty Images

TABLE 7 Potential Signs of Substance Abuse in a Worker

Physical Signs	Behavioral Changes
• Smelling of alcohol or drugs • Slurring of speech or an inability to communicate effectively • Wearing sunglasses indoors or on overcast days to hide dilated or constricted pupils, conditions which impair vision • Significant changes in personal appearance, cleanliness, or health	• Unscheduled absences • Failure to report to work on time • Significant changes in the quality of work • Unusual energy or tiredness • Sudden and irrational temper flare-ups • Sneaky behaviors, such as disappearing to wooded areas, storage areas, or other private locations • Attempting to borrow money from coworkers • Losing tools or company equipment

4.5.2 Mental Health

Mental health is a critically important, but often overlooked, part of overall construction safety and health. Construction workers have more risk factors than employees in other industries. These risk factors make workers more likely to experience mental health challenges, with a specific set of risks around suicide. Construction workers are six times more likely to die by suicide than they are to die from a fall.

With proper treatment, mental health conditions can be managed and overcome so those experiencing them can have full and productive lives. As with substance abuse, however, these conditions can lead to people being distracted, less productive than normal, and possibly unsafe and unable to perform their normal jobs. Because of this, mental health and suicide prevention need to be safety considerations.

As a leader, you cannot solve these challenges for your workers, but you can create a better environment. Use your toolbox talks to discuss mental health as part of overall total worker safety. In your safety huddles, check in on how your teammates are doing. Encourage your crew to talk to supervisors and colleagues when they feel stressed, anxious, or on the brink of burnout. Help ensure they are aware of any available counseling resources and that they are taking time off when they need it.

Even more so than substance abuse, it can be very difficult to identify a worker struggling with a mental health issue. *Table 8* lists some examples of potential signs and behavioral changes that might indicate someone is struggling with mental health issues.

If you observe an employee showing signs of mental health issues or have concerns for any reason, immediately contact your supervisor and/or Human Resources Department for assistance. You should work with management to deal with suspected mental health issues and should not deal with these situations yourself. Reporting the concern helps you protect the business, and the employee's and other workers' safety.

TABLE 8 Signs That May Indicate Someone Is Struggling with Mental Health Issues

Behavioral Changes	Performance Changes	Verbal Cues	Life Events
• Becoming withdrawn • Acting anxious or reckless • Misusing drugs and alcohol	• Missing work • Showing up late • Being less productive • Being unable to think clearly or solve problems • Having increased safety incidents	• Talking about feeling trapped • Talking about wanting to die • Saying they feel like a burden • Saying they feel like it would be better if they were gone	• Dealing with a divorce or breakup • Handling a death in the family • Working through a major illness or injury • Having financial issues

Source: Construction Industry Alliance for Suicide Prevention (ciasp.org)

4.0.0 Section Review

1. One of a crew leader's most important responsibilities to the employer is to _____.
   a. enforce company safety policies
   b. estimate material costs for a project
   c. make recommendations for setting up a crew
   d. provide input for fixed-price contracts

2. Which of these is *not* a function of OSHA?
   a. Issuing standards and rules for working conditions
   b. Researching occupational incidents and injuries
   c. Conducting workplace inspections
   d. Distributing company safety manuals

3. Which of the following should all employees receive upon being hired?
   a. A copy of the company's safety policies and procedures
   b. A new set of tools
   c. A leadership assessment to help map their career path
   d. First-aid and CPR training

4. A problem-solving method that is useful in investigating incidents and identifies the core issue of a problem is _____.
   a. problem-solving
   b. root cause analysis
   c. incident recordkeeping
   d. event playback

5. A focus on overall worker safety, health, and well-being, including mental health, is _____.
   a. home health
   b. complete worker safety
   c. occupational safety
   d. total health

5.0.0 Project Planning, Scheduling, and Cost and Resource Management

Objective

Demonstrate an understanding of construction project phases, project planning and scheduling, field reporting, and cost and resource management.

a. Identify construction project phases.
b. Summarize crew planning and scheduling responsibilities, including creating lookahead schedules.
c. Describe the frontline leader's role in field reporting.
d. Explain approaches to managing costs and controlling project resources, such as materials, tools, and equipment.

Performance Task

3. Develop and present a lookahead schedule.

This section describes the various phases of a construction project. It examines **estimating**, **planning**, **scheduling**, and **cost control** through managing material, people, tools, and other resources. All workers who participate in a job are responsible at some level for their assigned jobs. For example, the contractor's responsibility begins with obtaining the contract and it doesn't end until the client takes ownership of the project. The project manager is generally the person with overall responsibility for coordinating the project. Finally, the superintendent and frontline leadership, like a foreman, supervisor, or crew leader, are responsible for coordinating the work of one or more workers, one or more crews of workers within the company, or one or more crews of subcontractors.

Estimating: The process of calculating the cost of a project.

Planning: The process of laying out all details of a construction project before construction begins.

Scheduling: Involves defining which activities happen in what order and assigning dates and times to those tasks.

Cost control: The process that helps keep expenses under control by managing labor, material, and other costs to ensure that the project finishes on budget.

5.1.0   Construction Project Phases

A simple way to think about a construction project is in three phases: the *design and development phase,* the *preconstruction phase,* and the *construction phase. Figure 24* shows the flow of a typical project from beginning to end through these phases.

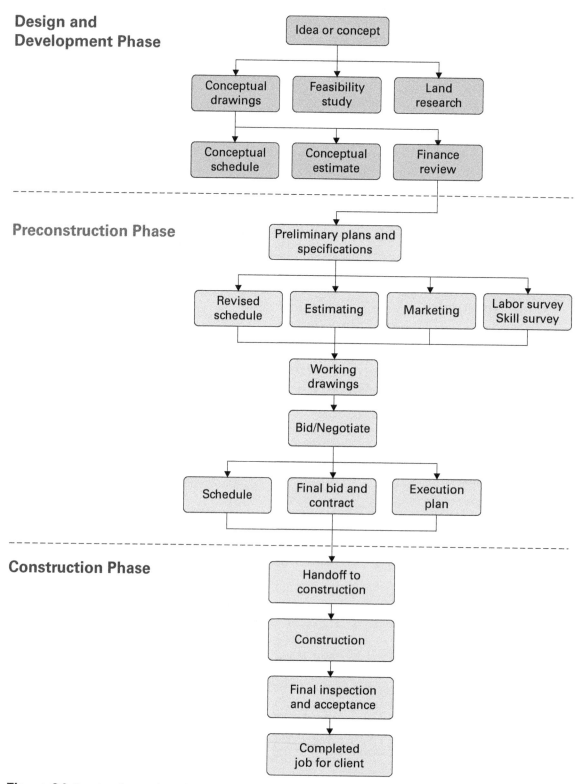

Figure 24 Construction project phases and flow.

5.1.1 Design and Development Phase

The design and development phase is the first stage of a project (*Figure 25*). It starts when an owner decides to build a new facility or add to an existing facility. This process involves land research and feasibility studies to ensure that the project has merit. A **feasibility study** is an analysis that considers all of a project's relevant factors—including budget, technical elements, legal aspects, and scheduling—to determine the likelihood that the project can be completed successfully.

Feasibility study: An analysis that considers all of a project's relevant factors—including budget, technical elements, legal aspects, and scheduling—to determine the likelihood that the project can be completed successfully.

Design and Development Phase

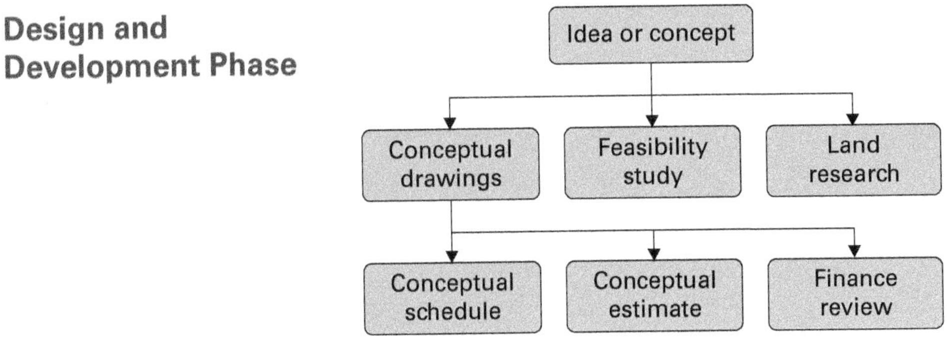

Figure 25 The design and development phase is the first stage of a project.

Architects or engineers develop the conceptual drawings that define the project graphically. They then provide the owner with sketches of room layouts and elevations and make suggestions about what construction materials to use.

During the design and development phase, architects, engineers, and/or the owner develop an estimate for the proposed project and establish a preliminary budget. Once that budget is established, the owner will determine how to pay for, or finance, the project. The owner will also complete a review to determine whether the project costs will exceed its market value and that the project provides a reasonable profit during its existence. If needed, the development team begins preliminary reviews for zoning, building restrictions, landscape requirements, and environmental impact studies.

5.1.2 Preconstruction Phase

If all steps in the design phase are completed successfully, the project moves to the preconstruction phase (*Figure 26*). The preconstruction stage includes planning that occurs before the start of construction. The preconstruction planning process does not always involve you, as a crew leader, but you should understand it to do your job well.

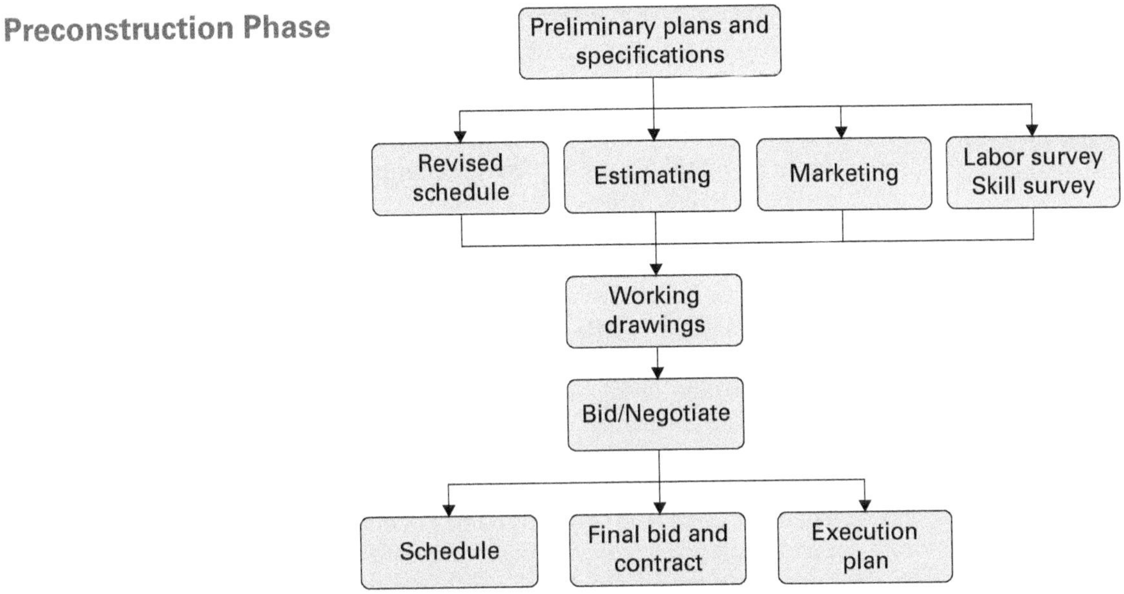

Preconstruction Phase

Figure 26 The preconstruction phase is the second stage of a project.

When the architects and engineers begin to develop the project drawings and specifications, they consult with other design professionals such as structural, mechanical, and electrical engineers. They perform the calculations, make a detailed technical analysis, and check details of the project for accuracy.

The design professionals create preliminary drawings and specifications. They use these drawings and specifications to communicate the necessary information to the contractors, subcontractors, suppliers, and workers who will contribute to the project.

A key next step is developing the proposal, bid, or negotiated price for the job. This is when the estimator, the project manager, and the field superintendent develop a preliminary plan for completing the work. They apply their experience and knowledge from previous projects to develop the plan. The process involves determining what methods, personnel, tools, and equipment the work will require and what level of productivity they can achieve.

Part of this process is determining how much the project will cost. Before building a project, an estimate must be prepared. Estimating is the process of calculating the cost of a project. There are two types of costs to consider, including direct and indirect costs as shown in *Table 9*. Direct costs, also known as *general conditions*, are those that planners can clearly assign to a budget. Indirect costs are overhead costs shared by all projects. Planners generally calculate these costs as an overhead percentage to labor and material costs.

TABLE 9 Examples of Direct and Indirect Costs

Direct Costs	Indirect Costs
• Materials	• Office rent
• Labor	• Utilities
• Tools	• Telecommunications
• Equipment	• Accounting
	• Office supplies and signage

Planners must also consider other factors when estimating project costs, including the following:

- Timing of all phases of work
- Types of materials to be installed and their availability
- Equipment and tools required and their availability
- Personnel requirements and availability
- Site and local conditions, such as soil types, accessibility, or available staging areas
- Climate conditions that should be anticipated during the project
- Relationships with the other contractors and their representatives on the job

On a simple job, crews can handle these items almost automatically. However, larger or more complex jobs require the planner to give these factors more formal consideration and study.

The bid price includes the estimated cost of the project as well as the **profit**. Profit refers to the amount of money that the contractor will make after paying all the direct and indirect costs. If the direct and indirect costs exceed the estimate for the job, the difference between the actual and estimated costs must come out of the company's profit. This reduces what the contractor makes on the job.

Many companies employ professional cost estimators to determine the cost of a project. Estimators will focus on **production** and **productivity rate**. Production is a quantity or amount of construction materials or work put in place. It is the quantity of materials that will be installed on a job, such as 1,000 linear feet of waste pipe installed for a task. On the other hand, the productivity rate depends on the level of efficiency of the work. The productivity rate relates to the quantity of work completed or materials put in place by the crew over a certain period of time. For example, productivity could be the amount of waste pipe installed each day by one worker or a crew.

Did You Know?
Profit Powers Businesses

Profit is the fuel that powers a business. It allows the business to invest in new equipment and facilities, to provide training, and to maintain a reserve fund for times when business is slow. In large companies, profitability attracts investors who provide the capital necessary for the business to grow. For these reasons, contractors can't afford to consistently lose money on projects. If they can't operate profitably, they are forced out of business. As a frontline leader, you can help your company remain profitable by managing budget, schedule, quality, and safety as outlined in the drawings, specifications, and project schedule.

Profit: The amount of money that the contractor will make after paying all the direct and indirect costs.

Production: The amount of construction materials put in place. It is the quantity of materials installed on a job.

Productivity rate: A measure of how much work gets done over a period of time, also called *output per labor hour*.

Production levels are set during the estimating stage. The estimator determines the total amount of materials to be placed based on the drawings, plans, and specifications. After the job is complete, supervision can assess the actual amount of materials installed and can compare the actual production to the estimated production.

During the estimating stage, the estimator uses company records of productivity to determine how much time and labor it will take to place a certain quantity of materials. From this information, the estimator calculates the productivity necessary to complete the job on time.

For example, say you need a crew of two people to paint 8,100 ft^2 in nine days. To calculate the required average productivity, divide 8,100 ft^2 by 9 days. The result is 900 ft^2/day. As the work progresses, you can compare the daily production of any crew of two painters doing similar work with this average (*Figure 27*).

NOTE

Many software programs and mobile apps are available to simplify the cost-estimating process. Estimating programs are typically set up to include a takeoff form and a form for estimating labor by category. Most of these programs include a database that contains current prices for labor and materials, so they automatically price the job and produce a bid. The programs can also generate purchase orders for materials on the job. Many of them are tailored to specific industries, such as construction or manufacturing, as well as specific trades within the industries, such as plumbing, electrical, HVACR, and insulating contractors.

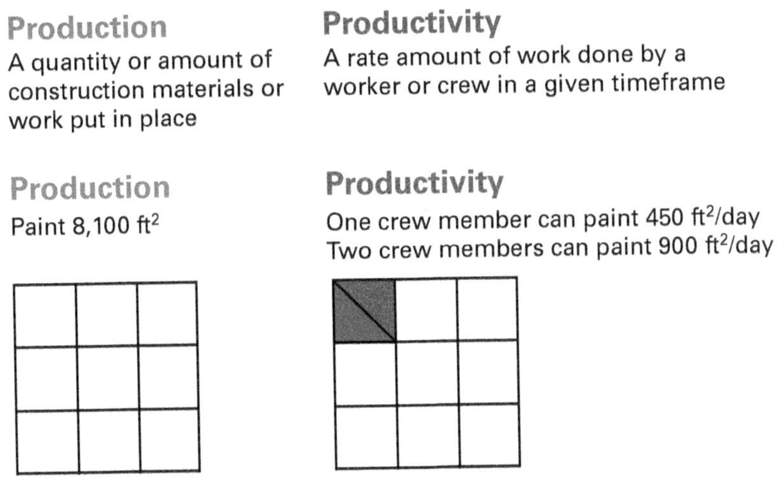

Production
A quantity or amount of construction materials or work put in place

Productivity
A rate amount of work done by a worker or crew in a given timeframe

Production
Paint 8,100 ft^2

Productivity
One crew member can paint 450 ft^2/day
Two crew members can paint 900 ft^2/day

Figure 27 Production and productivity.

During the preconstruction phase, the owners hold many meetings to review and refine estimates, adjust plans to conform to regulations, and secure a construction loan. If the project is one that requires a marketing program, such as a condominium, office building, or shopping center, the marketing may start during preconstruction. In such cases, the selling of the building space often starts before actual construction begins.

Next, the design team produces a complete set of drawings, specifications, and bid documents. After this, the owner will select the method to obtain contractors. The owner may choose to negotiate with several contractors or select one through competitive bidding. Everyone concerned must also consider safety as part of the planning process. A safety crew leader may walk through the site as part of the pre-bid process.

Contracts for construction projects can take many forms. The most common types of contracts are *lump sum* and *time and materials* contracts.

In a *lump sum contract*, also called a *fixed price* or *firm-fixed price contract*, the owner generally provides detailed drawings and specifications, which the contractor uses to calculate the cost of materials and labor. The contractor also adds a percentage to these costs representing company overhead expenses, such as office rent, insurance, and accounting/payroll costs. At the end, the contractor adds a profit factor.

When submitting the bid, the contractor will state very specifically the conditions and assumptions on which the company based the bid. The **scope**, or *scope of work*, is the list of the commitments and activities that all contractors, subcontractors, and suppliers are expected to do during a project.

In a *time and materials contract*, also known as a *T&M contract*, the owner pays back the contractor for labor, materials, and other costs encountered in the performance of the contract. Typically, the contractor and owner agree in advance on hourly or daily labor rates for different categories of worker. These rates include an amount representing the contractor's overhead expense. The owner

Scope: The list of the commitments and activities that all contractors, subcontractors, and suppliers are expected to do during a project. Also called *scope of work*.

also reimburses the contractor for the cost of materials and equipment used on the job. Since the buyer reimburses the contractor for the cost of materials and pays an hourly wage, any unexpected delays, roadblocks, and other changes to the scope of work are covered. T&M contracts thus work best for projects in which the scope of work is not well-defined.

On a time and materials contract project, you play an important role by understanding the scope of the work and making sure that everything is tracked so that your company is reimbursed for the work. You must identify changes or problems that increase the scope or amount of planned labor or materials. You also must ensure you and your team are following your company policies and getting proper authorization before starting any work.

After the client awards the contract, planners create a final schedule and an execution plan that defines the actual work methods and resources needed to perform the work.

5.1.3 Construction Phase

After the contract is signed, the construction phase can start (*Figure 28*). The designated contractor enlists the help of mechanical, electrical, elevator, and other specialty subcontractors to complete the construction phase. The contractor may perform one or more parts of the construction and rely on subcontractors for the remainder of the work. However, the general contractor is responsible for managing all the trades necessary to complete the project.

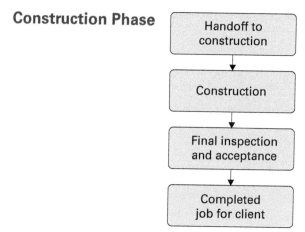

Figure 28 The construction phase is the third stage of a project.

The initial set of drawings for a construction project reflects the completed project as conceived by the architect and engineers. During construction, changes may be necessary because of factors unforeseen during the design phase. For example, when electricians must reroute cabling or conduit, or the installed equipment location is different than shown on the original drawing, such changes must be marked on the drawings. Without this record, technicians called to perform maintenance or modify the equipment later will have trouble locating all the cabling and equipment.

Project supervision must document any changes made during construction or installation on the drawings as the changes occur. Before digital drawings, architects would note changes on hard-copy drawings using a colored pen or pencil. These marked-up versions are commonly called **redline drawings**. With new digital drawings created using building information modeling tools, architects and engineers can revise and share the latest drawing versions almost instantly to mobile phones and tablets. After the drawings have been revised to reflect any redline changes, the final drawings are called **as-built drawings**. As-built drawings are a revised set of drawings submitted by a contractor upon completion of a project or job. These drawings, sometimes called *record drawings*, reflect all changes made during construction and show the exact specifications of the completed work.

NOTE

You may also hear about other types of contracts, such as cost-plus and guaranteed maximum price. Some types of contracts, such as unit-rate and hybrid, combine several aspects of these other contract types. Scan the QR code in the front of this module to access additional resources on construction contract types.

Redline drawings: Marked-up versions of construction drawings documenting any changes made during construction or installation.

As-built drawings: A revised set of drawings that reflect all changes made during the construction process and show the exact specifications of the completed work.

The main part of your role as a frontline leader happens during the construction phase, including daily work planning and scheduling, cost control, and resource management, including material delivery and storage, tool control, and equipment control. All of these tasks drive production and productivity, as well as the overall profitability of the project.

As construction nears completion, the architect/engineer, owner, and government agencies start their final inspections and acceptance of the project. If the general contractor has managed the project, the subcontractors have performed their work, and the architects/engineers have regularly inspected the project to ensure it satisfies the local code, then the inspection process can finish up in a timely manner. This results in a satisfied client and a profitable project for all.

On the other hand, if the inspection reveals faulty workmanship, poor design, incorrect use of materials, or violation of codes, then the inspection and acceptance will take longer. It may result in a dissatisfied client or an unprofitable project.

Construction **project delivery methods** provide different ways to organize the services of development, planning, and construction in order to execute a project.

NOTE

Scan the QR code in the front of this module to access additional resources on construction project delivery methods.

Project delivery methods: Different ways to organize the services of development, planning, and construction in order to execute a project.

5.2.0 Planning and Scheduling

In the previous sections, you learned about the phases of a construction project, including the design and development phase, the preconstruction phase, and the construction phase. As noted, the main part of your role as a frontline leader happens during the construction phase.

During construction, you may be directly involved in daily work planning and scheduling. You also may be involved in cost control and resource management, including material delivery and storage, tool control, and equipment control.

5.2.1 Understanding Overall Schedules

Construction planning and scheduling refers to the process through which you map out what will take place during a project, what resources you need, when key activities will take place, which personnel will be involved, and anything else pertinent to the project.

Construction planning refers to the process of laying out all details of a construction project beforehand. Construction scheduling involves defining which activities happen in what order and assigning dates and times to those tasks. Construction schedules are linked to resources such as materials, labor, and equipment to help create the most efficiency in your project. Construction scheduling is deeply connected to planning.

Planning and scheduling are closely related and are both very important to a successful job. Planning identifies the required activities and how and in what order to complete them. Scheduling involves establishing start and finish times/dates for each activity.

A schedule for a project typically shows the following:

- Operations listed in sequential order
- Units of construction
- Duration, or length of time, of an activity
- Estimated dates to start and complete each activity
- Quantity of materials to be installed

Figure 29 shows an example of a high-level project schedule. As noted, the development of a schedule for a project starts as early as the design and development phase. As the project moves through each phase and the work is better defined, the schedule also becomes more clearly defined.

When a crew is close to completing specific work—say, within three weeks of working on Foundation 1—a **Level 4 schedule** is created. A Level 4 schedule offers additional detail to provide working schedules to the crews doing the

NOTE

Scan the QR code in the front of this module to access additional resources on schedule levels.

Level 4 schedule: A schedule that offers additional detail to provide working schedules to the crews doing the work, usually within three to six weeks of completing specific work.

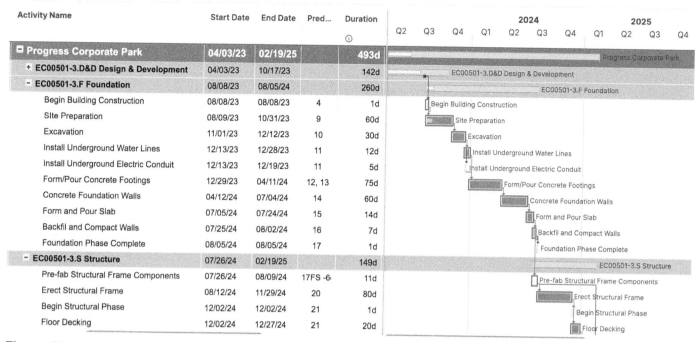

Figure 29 Example of a construction project schedule.

work (*Figure 30*). A Level 4 schedule allows a supervisor to create a plan, schedule handoffs from crew to crew, and estimate all of the resources that must be in place for the work to happen.

In the schedule in *Figure 30*, ironworkers must fabricate and set the rebar and carpenters must build the forms before the handoff is made to place the concrete. Each of these activities is a critical path activity. A **critical path** is the sequence of dependent tasks that form the longest duration. In *Figure 29* and *Figure 30*, the critical path in the schedule is shown in red. In other words, until one activity is complete, workers cannot start the other tasks, and the project will likely be delayed by the amount of time the previous activity was delayed.

Critical path: In manufacturing, construction, and other types of creative processes, the required sequence of tasks that directly controls the ultimate completion date of the project.

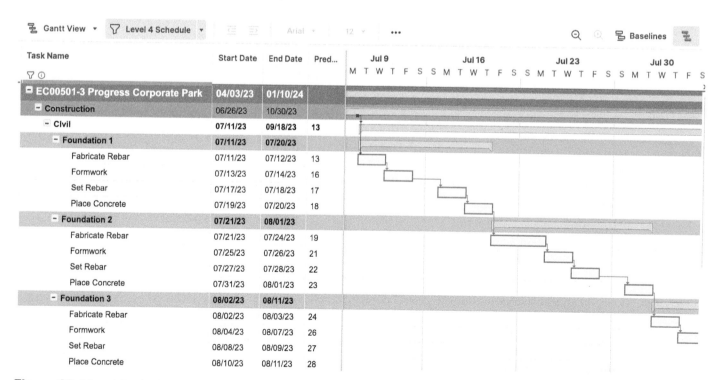

Figure 30 A Level 4 schedule provides enough details for project foremen to forecast handoffs and needed resources.

| Grid View ▾ | ▽ 3 Week Lookahead ▾ | ⇤ ⇥ | Arial ▾ | 10 ▾ | **B** *I* U S̶ ⬧ ▾ |

Task Name ▽	Start Date ▽	End Date	Materials Available	Assigned To ⓘ	Notes	Status ⓘ
− CIvil	**07/11/23**	**09/18/23**				
− Foundation 1	**07/11/23**	**07/20/23**				
Fabricate Rebar	07/11/23	07/12/23	Yes	JKR		Not Started
Formwork	07/13/23	07/14/23	Yes	JKR		Not Started
Set Rebar	07/17/23	07/18/23	N/A	JKR	Quality signoff	Not Started
Place Concrete	07/19/23	07/20/23	Yes	JKR		Not Started
− Foundation 2	**07/21/23**	**08/01/23**				
Fabricate Rebar	07/21/23	07/24/23	Yes	JKR		Not Started
Formwork	07/25/23	07/26/23	Yes	JKR		Not Started
Set Rebar	07/27/23	07/28/23	N/A	JKR	Quality signoff	Not Started
Place Concrete	07/31/23	08/01/23	Yes	JKR		Not Started
− Foundation 3	**08/02/23**	**08/11/23**				
Fabricate Rebar	08/02/23	08/03/23	On Order	JKR		Not Started
Formwork	08/04/23	08/07/23	Yes	JKR		Not Started
Set Rebar	08/08/23	08/09/23	N/A	JKR	Quality signoff	Not Started
Place Concrete	08/10/23	08/11/23	Yes	JKR		Not Started

Figure 31 A step-by-step lookahead in a Level 4 schedule provide details for frontline leaders to plan handoffs.

For work to happen, you will need resources such as materials, available crew, available construction equipment, completed permits, and more. Developing a **lookahead schedule** helps ensure that all resources are available for the project when needed. This is often included in a Level 4 schedule. A lookahead schedule is useful for planning and scheduling work that is happening in the short term, usually within one to three weeks (*Figure 31*).

5.2.2 Creating a Short-Term or Lookahead Schedule

As a frontline leader of a crew, you must be able to create, read and interpret a lookahead schedule. You can use this information to plan work more effectively, set realistic goals, and compare the starts and completions of tasks to those on the schedule. As the leader of the team executing the work, you are a key expert to help produce a plan at this level of detail. *Table 10* provides a checklist for planning and scheduling.

Lookahead schedule: For planning and scheduling work that is happening in the short term, usually within one to three weeks; used for anticipating material, labor, tool, and other resource requirements, as well as identifying potential schedule conflicts or other problems. Also called *short-term schedule*.

TABLE 10 Crew Leader Planning and Scheduling Checklist

Did You...?	Yes/No
Determine the best method for performing the job.	
Identify the responsibilities of each person on the work crew.	
Determine the duration, or length in time, and sequence of each activity.	
Identify the tools and equipment needed to complete the job.	
Ensure that the required materials are at the worksite when needed.	
Make sure that heavy construction and other equipment is available when required.	
Work with other contractors in such a way as to avoid interruptions and delays.	

The following outlines the process you must complete to develop a plan and a schedule and to use the schedule to manage work.

Understand or Establish a Goal

A goal is a specific outcome that one works toward. The goals you must accomplish are often set in the higher-level schedules. For example, the project superintendent of a residential construction project could establish a goal to have a house dried-in or ready for the application of roofing and siding by a certain date. To meet that goal, you and the superintendent would need to agree to a goal to have the framing completed by a given date. You then establish subgoals for your crew to complete each element of the framing (floors, walls, roof) by a set time.

Identify the Required Work Activities and Divide Them into Tasks

To plan effectively, you must be able to break a work activity assignment down into smaller tasks. Large jobs include a greater number of tasks than small ones, but all jobs can be broken down into manageable components.

Make a list of all the activities that the plan requires to complete the job. This includes individual work tasks and special tasks, such as inspections or the delivery of materials. Next, start to divide work activities into tasks. As you do this, remember that almost every work activity has three general parts: preparing for the activity, performing the activity, and cleaning up after the activity.

Figure 32 shows a suitable breakdown for the work activity to install square vinyl floor tiles in a cafeteria. You could create even more detail by breaking down any one of the tasks into subtasks. In this case, however, that much detail is unnecessary and wastes your time and the project's money.

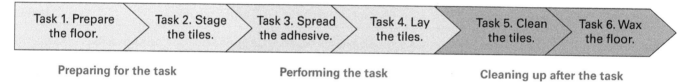

Figure 32 Example task breakdown of the work activity to install square vinyl floor tiles in a cafeteria.

When breaking down an activity into tasks, make each task *identifiable* and *definable*. A task is identifiable when you know exactly the types and amounts of resources it requires. A task is definable if it has a specific duration. The activity breakdown does not need to be overly detailed or complex unless the job has never been done before or must be performed with strictest efficiency.

At this point, you should just be concerned with generating a list, not with determining how to accomplish the activities, who will perform them, how long they will take, or the necessary sequence to complete them.

Plan Resources Needed for Each Task

Once a job has been broken down into activities and tasks, you must identify and account for all of the resources you need to complete them. Resource planning includes considerations for safety, materials, site-specific issues, equipment, tools, and labor (*Table 11*). Each of these is discussed in more detail in the sections that follow.

Organize the Work into a Logical Sequence

The next step in planning is to use the list of activities and resources needed to organize the work activities into a logical sequence. When doing this, keep in mind that certain steps can't happen before the completion of others. For example, concrete placement cannot happen until after rebar is set. Some construction work activities, such as installing 12" footing forms, are done so often that they require little planning. However, other jobs, such as placing a new type of mechanical equipment, require substantial planning.

Assign Responsibility for Each Task

After you have identified the tasks, you must determine who will complete each activity. This requires that you are aware of the experience, skills, and abilities of

TABLE 11 Key Resource Considerations During Planning

Resource Planning Area	Key Responsibilities
Safety Planning	Using the company safety manual as a guide, you must assess the safety issues associated with the job and take necessary measures to minimize any risk to the crew.
Materials Planning	You may be required to estimate quantities of materials during construction. You also have a key role in the receipt, storage, and control of the materials after they reach the jobsite.
Site Planning	Many planning elements are involved in sitework, including access roads, parking, site security, material and equipment staging and storage, sedimentation control, stormwater runoff, and more.
Equipment Planning	You may need to plan the types of equipment needed, the use of the equipment, and the length of time it will be on the site. You must work with the main office to make certain that the equipment reaches the jobsite on time. You must reserve time for equipment maintenance to prevent equipment failure.
Tool Planning	To plan tool usage, you must determine the tools required, tell your crew who will provide the tools, ensure that the crew is qualified to use the tools safely and effectively, and establish proper tool controls.
Labor Planning	In many companies, the project manager or job superintendent determines the size and makeup of the crew. Then, supervision expects you to accomplish the goals and objectives with the crew provided. If you are responsible for planning for labor, you must identify the skills needed to perform the work, determine the number of people having the required skills, and decide who will be on the crew.

everyone in your team. You must put this knowledge to work by matching the crew's skills and abilities to specific tasks. Your job is to draw from their expertise to get the job done well and in a safe and timely manner.

Assign a Duration to Each Task

Next, assign a duration or length of time that it will take to complete each activity and determine the start date and time for each. Then place each activity into a schedule format. This step is important because it helps you compare the task time estimates to the scheduled completion date or time.

Make note of any activities that cannot start until previous work has been completed and ensure those do not start before the other work is complete. You also will need to ensure you have not assigned too much work to any one person or team.

Some of the information to support short-term scheduling comes from the estimate or cost breakdown. *Figure 33* shows how you can use estimate or cost breakdown information to set short-term production goals for a team.

Scenario: A carpentry crew on a retaining wall project is about to form and pour catch basins and put up wall forms. The crew will work on both the basins and the wall forms at the same time. The crew has put in several catch basins, so the crew leader is sure that they can perform the work within the estimate. However, the crew leader is concerned about their production of the wall forms and wants to set a clear target for the amount of work to be done each day to hit the schedule.

The scheduling process in this scenario could be handled as follows:

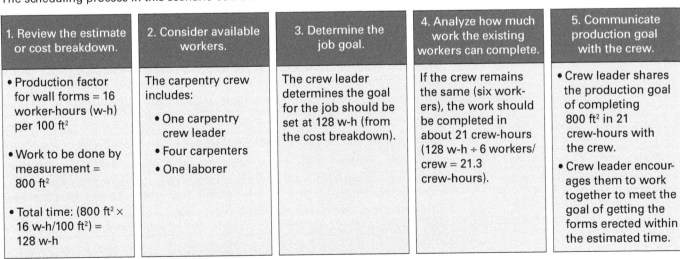

1. Review the estimate or cost breakdown.	2. Consider available workers.	3. Determine the job goal.	4. Analyze how much work the existing workers can complete.	5. Communicate production goal with the crew.
• Production factor for wall forms = 16 worker-hours (w-h) per 100 ft² • Work to be done by measurement = 800 ft² • Total time: (800 ft² × 16 w-h/100 ft²) = 128 w-h	The carpentry crew includes: • One carpentry crew leader • Four carpenters • One laborer	The crew leader determines the goal for the job should be set at 128 w-h (from the cost breakdown).	If the crew remains the same (six workers), the work should be completed in about 21 crew-hours (128 w-h ÷ 6 workers/crew = 21.3 crew-hours).	• Crew leader shares the production goal of completing 800 ft² in 21 crew-hours with the crew. • Crew leader encourages them to work together to meet the goal of getting the forms erected within the estimated time.

Figure 33 Crew leaders can use lookaheads or short-term scheduling to set production goals.

In this example, the crew leader used the short-term schedule to translate production into work hours and to schedule work so that the crew can accomplish it within the estimate. A crew can improve production when they know the amount of work to accomplish and the time available to complete the work, and can provide input when setting goals. In addition, setting production targets provides the motivation to exceed schedule targets.

Communicate the Schedule and Assigned Responsibilities

After the schedule is complete, you must then assign work responsibilities to the crew. Each person in the crew should know what they are responsible for accomplishing on the job.

Use the Schedule to Monitor and Control Work

Schedules only have value if they are actively used as tools to monitor and control work. A lookahead or short-term schedule provides you with the information you need to plan for work and track progress. Before starting a job, actively review the schedule to:

- Determine the materials, tools, equipment, and labor needed to complete the job.
- Determine when the various resources are needed.
- Follow up to ensure that the resources are available on the jobsite when needed.

You should also verify the availability of needed resources three to four working days before the start of the job. This should occur even earlier for larger jobs. Advance preparation will help avoid situations that could potentially delay starting the job or cause it to fall behind schedule. As work continues, use your schedule to monitor whether actual production begins to slip behind estimated production and develop plans to recover the schedule as needed.

Follow Up and Confirm Work Is Completed Well

After you delegate tasks to the appropriate crew members, you must follow up to confirm that the crew completed the tasks correctly and efficiently. To do this effectively, you must be present on the jobsite to make sure all the resources are available to complete the work, to confirm that team members are working on their assigned activities, to answer any questions, and to help to resolve any problems that occur while the work is being done.

5.3.0 Crew Leader Field Reporting Responsibilities

A field-reporting system consists of a series of forms that are completed by you and other leaders on the project to share updates on labor, materials, tools, equipment, and more.

As a frontline leader, you may not be directly involved in updating overall schedules. However, the daily field or progress reports you submit will be used by the company to keep the schedule up to date.

The following sections provide some examples of field reporting. Each company has its own forms and methods for obtaining information. You should always follow your company's policies and procedures for field reporting.

5.3.1 Completing Daily Reports

Job diary: A written record that a supervisor maintains periodically (usually daily) of the events, communications, observations, and decisions made during the course of a project.

You will be involved with many activities on a day-to-day basis. As you work each day, you will want to keep a **job diary** to track events such as job changes, interruptions, and visits. A job diary, also called a *daily report, site diary, construction log,* or *site journal,* is a notebook or digital application in which you record activities or events that take place on the jobsite that may be important later (*Figure 34*).

July 8, 2023

Weather: Hot and Humid

Project: Opal Garage Annex

- The paving contractor crew arrived late (10 am).

- The owner representative inspected the footing foundation at approximately 1 pm.

- The concrete slump test did pass. Two trucks had to be ordered to return to plant, causing a delay.

- David Lopez had an incident on the second floor. I sent him to the on-site medical personnel for medical first aid treatment. The cause of the incident is being investigated.

Figure 34 Sample page from a job diary.

For this comparison process to be of use, the information you and others submit from the field must be accurate. In the event of a legal or contractual conflict with the client over costs, job diaries can be helpful as evidence in legal proceedings or in reaching a settlement. *Table 12* outlines an example scenario where your action of accurate field reporting will help benefit your company.

TABLE 12 Accurate Field Reporting Benefits Your Company

Scenario	Field Reporting Action
You are running a crew of five concrete finishers for a specialty contractor. When you and your crew show up to finish a slab, the general contractor (GC) informs you, "We're a day behind on setting the forms, so I need you and your crew to stand down until tomorrow."	Follow your company's field reporting policy to report the change. For example, you might first call your office to let them know about the delay. Then, you would record it in your job diary.
	A six-man crew for one day represents 48 worker-hours. If your company charges $30 an hour, that's a potential loss of $1,440, which the company would want to recover from the general contractor. If there is a dispute, your entry in the job diary could result in a favorable decision for your employer.

Today, many companies use mobile applications that allow you to complete daily reporting on your mobile phone or tablet while you are on-site. You can quickly enter information throughout the day and attach photos and other documents to add more information. They generally also make it easy to send to your supervisor via email or text message (*Figure 35*). Your supervisor will be notified when the report is complete.

Whether print or digital, when making entries in a job diary, make sure that the information is accurate, factual, complete, consistent, organized, and up to date. Do not include any emotional details; keep the report focused on facts. This is especially true if documenting personnel problems. If the company

Figure 35 Mobile apps make it easier to complete daily reporting on-site.
Source: Fieldwire, https://www.fieldwire.com/

requires maintaining a job diary, follow company policy in determining which events and what details you should record. The information needs to be detailed enough so that, if needed, you could determine what happened on any day years later. If you are unsure what to include, it is better to have more information than too little. Check in with your supervisor to confirm if your daily reporting is including the right information.

5.3.2 Tracking Employee Costs Through Timesheets

When the project budget is set up, many companies create a work breakdown for each project. A **work breakdown structure (WBS)** organizes the entire scope of work in a logical manner based on how the work will be executed. Within the WBS, planners assign each major task a unique charge number or code (*Figure 36*). Anyone working on that task charges that number or code on their timesheet, so that project managers can readily track cost performance.

As a frontline leader, you not only need to ensure everyone is completing their timesheets, but also that they are coding those timesheets to the proper codes in the WBS. Doing so ensures you and your company are tracking and managing labor costs against specific work.

For example, as shown in *Figure 36*, if all timesheets are correctly coded, you and the company can determine the labor cost to install each of the three types of pipe: carbon steel piping, stainless steel piping, and the high-density polyethylene (HDPE) piping. You can also calculate how much it costs to install the piping for the entire project.

Work breakdown structure (WBS): A method of organizing the entire scope of work in a logical manner based on how the work will be executed.

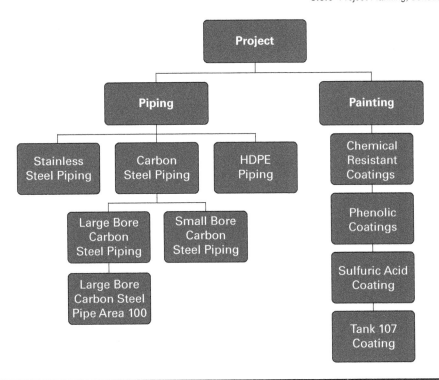

Level	Work	Account Code
Level 0	Piping	B
Level 1	Stainless Steel Piping	B-SS
Level 1	Carbon Steel Piping	B-CS
Level 1	HDPE Piping	B-HP
Level 2	Large Bore Carbon Steel Piping	B-CS-L
Level 2	Small Bore Carbon Steel Piping	B-CS-S
Level 3	Large Bore Carbon Steel Pipe Area 100	B-CS-L-100
Level 0	Painting	L
Level 1	Chemical Resistant Coatings	L-CR
Level 2	Phenolic Coating	L-CR-P
Level 3	Sulfuric Acid Tanks Coating	L-CR-P-ST
Level 4	Tank 107 Coating	L-CR-P-ST-107

Figure 36 An example of a work breakdown structure (WRS) for piping and painting on a project.

5.4.0 Frontline Leader Cost and Resource Management Responsibilities

Being aware of costs and controlling them is the responsibility of every employee on the job. As a frontline leader, it is your job to ensure that employees uphold this responsibility. Cost control is the process by which you keep expenses under control by managing labor, material, and other costs to ensure that the project finishes on budget.

5.4.1 Managing Costs

As you have learned, profit refers to the amount of money that the contractor will make after paying all the direct and indirect costs. If the costs exceed the estimate for the job, the difference between the actual and estimated costs must come out of the company's profit. This reduces what the contractor makes on the job.

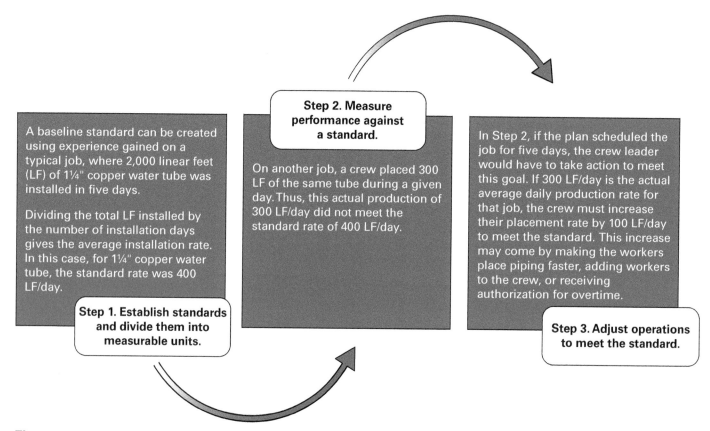

Step 2. Measure performance against a standard.

A baseline standard can be created using experience gained on a typical job, where 2,000 linear feet (LF) of 1¼" copper water tube was installed in five days.

Dividing the total LF installed by the number of installation days gives the average installation rate. In this case, for 1¼" copper water tube, the standard rate was 400 LF/day.

On another job, a crew placed 300 LF of the same tube during a given day. Thus, this actual production of 300 LF/day did not meet the standard rate of 400 LF/day.

In Step 2, if the plan scheduled the job for five days, the crew leader would have to take action to meet this goal. If 300 LF/day is the actual average daily production rate for that job, the crew must increase their placement rate by 100 LF/day to meet the standard. This increase may come by making the workers place piping faster, adding workers to the crew, or receiving authorization for overtime.

Step 1. Establish standards and divide them into measurable units.

Step 3. Adjust operations to meet the standard.

Figure 37 An effective cost control process monitors production and productivity to help manage costs.

NOTE

Before starting any action, you should check with company supervision to see that the action proposed is acceptable and within the company's policies and procedures.

Early identification of cost changes is very important and a key part of your role. As shown in *Figure 37*, an effective cost control process to monitor production and productivity includes several key steps. Identifying changes provides enough notice to correct the causes of overruns and manage those costs. You must learn to anticipate changes from plans and specifications based on experience and take measures to prevent them from occurring. If a situation cannot be fully corrected, early notification allows time for communicating and for managing expectations. No one likes surprises when it comes to increased project costs.

Table 13 lists some of the factors that can make actual costs exceed estimated costs, along with some potential actions you can take to avoid these situations.

A key factor in managing costs is controlling waste. Waste in construction can add up to loss of critical and costly materials and may result in job delays. You need to ensure that every crew member knows how to complete the work assigned and use materials efficiently. One example of waste is a carpenter who saws off a piece of lumber from a full-sized piece, when he could have found the length needed in the lumber scrap pile.

Doing rework is another form of waste (*Figure 38*). For example, less-experienced or less-trained crew members may make mistakes that cause work to have to be done over. You can help avoid these errors by providing on-the-spot training and by checking on newer, less experienced workers more often. Rework also can result from a conflict in craft scheduling that requires removal and reinstallation of finished work. Rework also increases safety risks on a job. Rework often involves demolition, time pressure, and frustration from the team, all of which create a higher likelihood of an incident.

Some issues that cause rework, such as mistakes on drawings, may be outside of your control. However, you can manage this risk by ensuring that you always have the most current set of drawings and examining the drawings carefully before you start work. Finding errors before the work begins increases safety and saves time, labor, materials, and more, which saves costs overall.

TABLE 13 Factors That Drive Cost Changes

Factor	Proactive Action
Client-related changes	Clients may want to change the scope of a project after it has started. You must be able to assess the potential impact of such changes and, if necessary, confer with the employer to determine the course of action. If contractor-related losses are occurring, you and the superintendent will need to work together to get the costs back in line.
Late delivery of materials, tools, and/or equipment	Review and update your lookahead schedules on a regular basis. Plan ahead to ensure that job resources will be available when needed.
Inclement weather	Work with the superintendent and other field leaders to have alternate plans ready.
Higher-than-expected wages or unexpected overtime due to low manpower	Plan ahead to ensure that labor resources will be available when needed. Ensure you are accurately recording daily work using your field reporting systems and job diary. Work with your supervisor or the superintendent to identify other staffing approaches.
Contractor or owner delays due to work availability	Ensure you are accurately recording daily work using your field reporting systems and job diary. Work with the superintendent and other field leaders to have alternate plans ready.
Unmotivated workers	Apply the leadership and team motivation strategies discussed earlier in this course. Remember, different things motivate different people in different ways. This means there is no single best approach to motivating all of your crew members.
Incidents	Continue to lead by example, reinforce the critical nature of safety, and follow the guidelines discussed earlier in the course. Work with the superintendent and other field leaders to help evaluate and improve the safety and health program, if needed.
Owner-related changes	Owners may want to change the scope of a project after it has started. You must be able to assess the potential impact of such changes and, if necessary, confer with the employer to determine the course of action. If contractor-related losses are occurring, you and the superintendent will need to work together to get the costs back in line.
Waste due to rework, material usage, and more	Train less-experienced workers on how to complete work. Review and update your lookahead schedules on a regular basis. Ensure crew members know how to use materials efficiently. Ensure you have the latest drawings and reference them daily.

Figure 38 Rework is another form of waste on a construction site.
Source: Vladimir Cetinski/iStock/Getty Images

5.4.2 Ensuring Crew Productivity

People typically represent more than half the cost of a project and thus have an enormous impact on profitability. For that reason, it is essential for you to manage your crew and their work environment in a way that maximizes their productivity.

One way to manage costs and increase productivity is to make sure you have the right skills on your crew to execute the level of tasks that you have to complete. *Crew mix* is a concept that describes the number of experienced or journey-level workers relative to the number of apprentices or less-experienced workers in a crew. For safety, quality, and productivity reasons, you do not want a worker trying to complete a task for which they are not skilled. However, you also do not want to have higher-wage crew members doing work that can be done by lower-wage crew members. For example, paying a journeyman electrician to pull wire costs more than paying an electrician helper who is skilled enough to do that work under the supervision of a journeyman electrician. As a frontline supervisor, you may not have control of your specific crew mix, but you can work to align their skills levels most appropriate to the specific work. Doing this well can reduce the overall project costs.

Other practices can help to ensure productivity:

- Keeping the crew adequately staffed at all times to avoid job delays
- Ensuring that all workers have the required resources when needed
- Ensuring that everyone knows where to go and what to do after each task is completed
- Making reassignments as needed
- Ensuring that all workers have completed their work properly

The time on the job should be for business, not for taking care of personal problems. Anything not work related should be handled after hours, away from the jobsite, if possible. Even the planning of after-work activities, arranging social functions, or running personal errands should occur after work or during breaks. Limited and necessary exceptions to these rules should be permitted (for example, making or meeting medical appointments), and these should be spelled out in company or crew policies.

Given the nature of construction work, frequent breaks can be needed to ensure worker safety. However, if crew members are often taking unscheduled breaks or goofing off, you will need to confront the behavior and coach these workers. Document any discussions in your job diary. If needed, you should redirect repeated violations to the attention of higher management as guided by company policy.

You can minimize or avoid delays caused by other crews or contractors by carefully tracking the project schedule. In doing so, you can anticipate delays that will affect your work and either take action to prevent the delay or redirect the crew to another task.

The availability and location of materials, equipment, and tools can have a serious effect on a crew's productivity. In the sections that follow, you will learn more about how to manage these resources to minimize such problems.

5.4.3 Handling Material Delivery and Storage

Your specific responsibilities for material delivery and storage will depend on the policies and procedures of the company. In general, you will be responsible for ensuring on-time delivery of materials, verifying the materials were delivered, and storing the materials (*Figure 39*). You also must prevent the waste and theft of materials.

You will also be involved in planning materials for tasks such as job-built formwork and scaffolding or acquiring materials if you run out. If you run out of a specific material, be sure to consult the appropriate supervisor on how to proceed, because most companies have specific purchasing policies and procedures.

It is essential that the materials required for each day's work be on the jobsite when needed. You should confirm in advance the placement of orders for all

NOTE

Scan the QR code in the front of this module to access additional resources on the steps involved in the development of an estimate, including creating a quantity takeoff sheet.

Figure 39 The receipt, storage, and control of materials after they reach the jobsite is an important part of a frontline leader's work.
Source: July Alcantara/E+/Getty Images

Source: metamorworks/Shutterstock

Did You Know?

Just-in-Time (JIT)

Just-in-time (JIT) delivery is a strategy in which materials arrive at the jobsite when needed. This means that a crew may install the materials right off the truck. This method reduces the need for on-site storage and staging areas. It also reduces the risk of loss or damage while moving products about the site. Other modern material management methods include the use of radiofrequency identification (RFID) tags that make it easy to inventory and locate material in crowded staging areas. Using an RFID inventory control system requires special scanners.

materials and that they will arrive on schedule. A week or so before the delivery date, follow up to make sure there will be no delayed deliveries or items on backorder. If other people are responsible for providing the materials for a job, you must follow up to make sure that the materials are available when needed. Otherwise, delays can occur as crew members wait for the delivery of materials.

You may be responsible for the receipt of materials delivered to the worksite. When this happens, you should require a copy of the shipping invoice or similar document and check each item on the invoice against the actual materials to verify delivery of the correct amounts.

You should also check the condition of the materials to verify that nothing is defective before signing the invoice. This can be difficult and time-consuming because it requires the crew to open cartons and examine their contents. However, this step cannot be overlooked, because a signed invoice indicates that the recipient accepted all listed materials and in an undamaged state. If the crew leader signs for the materials without checking them, and then finds damage, no one will be able to prove that the materials came to the site in that condition (*Figure 40*).

Figure 40 When receiving materials, check the condition of the materials to verify that nothing is defective before signing the invoice.
Source: kali9/E+/Getty Images

After checking and signing the shipping invoice, you should give the original or a copy to the superintendent or project manager. The company then will file the invoice for future reference because it serves as the only record the company has to check bills received from the supplier company.

Another essential element of materials control is selecting where to store materials on a jobsite. *Table 14* outlines three factors to consider when selecting an appropriate material storage location.

TABLE 14 Considerations When Selecting a Material Storage Location

Factor	Description
Access	If possible, store materials near where the crew will use them, so the crew can access them easily. Not having to carry the materials long distances will greatly reduce time, effort, and installation costs.
Damage Protection	Store materials in a secure area to avoid damage. Be sure that the storage area suits the materials. For instance, temperature-sensitive materials, such as chemicals or paints, should be stored in climate-controlled areas to prevent waste.
Security	Theft or vandalism of construction materials increases direct and indirect costs. These include replacement materials, lost production time while replacing needed materials, and the costs of additional security precautions. In addition, a contractor's insurance costs may increase due to theft and vandalism.

The best way to avoid theft and vandalism is to have a secure jobsite. At the end of each workday, store unused materials and tools in a secure location, such as a locked construction trailer. If the jobsite is fenced or the building has lockable access, the materials can be stored within. Many sites have security cameras and/or intrusion alarms to help minimize theft and vandalism.

5.4.4 Planning and Controlling Tool Usage

Among companies, various policies govern who provides hand and power tools to employees. Some companies provide all the tools, while others furnish only the larger power tools. You should be familiar with and enforce any company policies related to tools.

As a crew leader, you are responsible for planning tool usage for a job. This task includes the following:

- Determining the tools required
- Informing the workers who will provide the tools (company or worker)
- Making sure the workers are qualified to use the tools safely and effectively
- Determining what controls to establish for tools

Tool control involves controlling the issue, use, and maintenance of all tools provided by the company. It also involves storing the tools in a location that provides easy access to your crew (*Figure 41*).

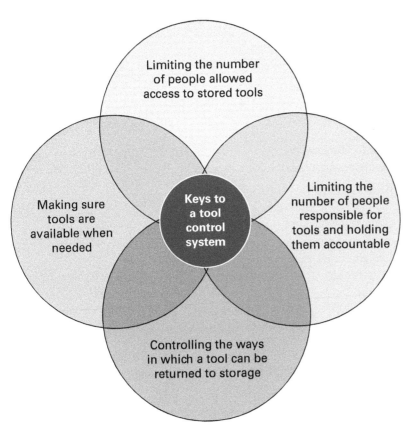

Figure 41 Studies show there are four keys to an effective tool control system.

Using the proper tools correctly saves time and energy. In addition, proper tool use reduces the chance of damage to the tool during use, as well as injury to the user and to nearby workers. This is especially true of edge tools and those with any form of fixed or rotating blade. A dull tool is far more dangerous than a sharp one.

Tools must be adequately maintained and properly stored. Making sure that tools are cleaned, dried, and lubricated prevents rust and ensures that the tools are in the proper working order. This applies to tools supplied by the company as well as tools that belong to the workers.

If a tool is damaged, it is essential to repair or replace it promptly. Otherwise, an incident or injury could result. Users should take care of company-issued tools as if they were their own. Workers should not abuse tools simply because they are not their property.

One of the major sources of low productivity on a job is the time spent searching for a tool. To prevent this from occurring, supervision should establish a storage location for company-issued tools and equipment. You should make sure that crew members return all company-issued tools and equipment to this designated location after use. Similarly, workers should organize their personal toolboxes so that they can readily find the appropriate tools and return their

NOTE

Regardless of whether a worker or the company owns a tool, OSHA holds companies responsible for the consequences of using a tool on a jobsite. The company is accountable if an employee receives an injury from a defective tool. Therefore, the crew leader needs to be aware of any defects in the tools the crew members are using.

Did You Know?

Connected Tools Are Controlled Tools

Tool companies are using technology to help you track tools and equipment on a jobsite. These smart tool systems have a software application that allows you to add tools enabled with Bluetooth® technology to the tracking system. You can also attach powered tags to large equipment or materials and add those to the tracking system. As crew members sign out tools, the assignment is tracked in the system, so you can track which tools are in use, by whom, and where they are on the jobsite at any point. You can also track materials and equipment in the same way. By helping to ensure all of your equipment, tools, and materials are accounted for and easy to find, these systems help companies control costs by saving time and replacement costs.

Source: DEWALT Industrial Tool Co.

tools to their toolboxes when they are finished using them. If needed, coach less-experienced team members on how to master this important professional skill.

5.4.5 Ensuring Equipment Control

Equipment planning includes the types of equipment needed, the use of the equipment, and the length of time it will be on the site. As a frontline leader of a crew, you may not be responsible for long-term equipment control.

However, the equipment required for a specific job is often your responsibility. The first step is to schedule when a crew member must transport the required equipment from the shop or rental yard. You are responsible for informing the shop of the location of its use and confirming a worker returns it to the shop when finished with it. You must work with the main office to make certain that the equipment reaches the jobsite on time.

Coordinating the use of the equipment is also very important. Some equipment operates in combination with other equipment (*Figure 42*). For example,

Figure 42 Coordinating equipment requires knowing which equipment operates in combination with other equipment.
Source: RyanOverman/iStock/Getty Images

dump trucks are generally required when loaders and excavators are used. You should also coordinate equipment with other contractors on the job. Sharing equipment can save time and money and avoid duplication of effort.

It is common for equipment to lie idle at a jobsite because of a lack of proper planning for the job and the equipment arriving early. For example, if wire-pulling equipment arrives at a jobsite before the conduit is in place, this equipment will be out of service while awaiting the conduit installation. In addition to the wasted rental cost, the equipment could also be damaged, lost, or stolen while awaiting use.

You need to control equipment use, ensure that the crew operates the equipment according to its instructions, and make sure they are adhering to time and cost guidelines. You must also assign or coordinate responsibility for maintaining and repairing equipment as indicated by the applicable preventive maintenance schedule. Delaying maintenance and repairs can lead to costly equipment failures.

You must also ensure that crew equipment operators are properly trained. This includes ensuring that the equipment operators have the necessary credentials to operate the equipment, including applicable licenses.

You are also responsible for the proper operation of all other equipment resources, including cars and trucks. Reckless or unsafe operation of vehicles will likely result in damaged equipment, leading to disciplinary action for the workers, fines and repair costs levied by the vehicle vendors, and a loss of confidence in your ability to control the actions of your crew.

You should also arrange to secure all equipment at the close of each day's work to prevent theft. If continued use of the equipment is necessary for the job, you should make sure to lock it in a safe place. If it is not, return it promptly.

5.0.0 Section Review

1. Which of these activities occurs during the design and development phase of a project?
   a. Architect/engineer sketches are prepared, and a preliminary budget is developed.
   b. Final inspections are completed.
   c. Detailed project drawings and specifications are prepared.
   d. Contracts for the project are awarded.

2. Which of these activities is *not* part of creating a lookahead schedule?
   a. Identifying the required work activities and dividing them into tasks
   b. Planning resources needed for each task
   c. Organizing the work into a logical sequence
   d. Assigning cost estimates to each task

3. A job diary should typically record any of the following *except* _____.
   a. job interruptions and visits
   b. personal details and emotions
   c. changes needed to project drawings
   d. the actual time each job task took to complete

4. Which of the following is a *correct* statement regarding project cost?
   a. Cost is handled by the Accounting Department and isn't a concern of the crew leader.
   b. The difference between the estimated cost and the actual cost affects a company's profit.
   c. Wasted material is factored into the estimate and is never a concern.
   d. The contractor's overhead costs aren't included in the cost estimate.

1. Who is generally the on-site field supervisor who is responsible for one or more crew leaders?
   a. Project Manager
   b. Lead Engineer
   c. Superintendent
   d. Pre-Construction Manager

2. Accepting responsibility for job completion and being personally answerable for the results is known as _____.
   a. responsibility
   b. delegation
   c. authority
   d. accountability

3. Which of the following is *not* true of situational leadership?
   a. It involves adjusting the way you lead to the specific situation and needs of the group.
   b. It assumes all situations do not require the same leadership.
   c. It allows you to adapt your leadership style to the situation.
   d. It has four styles, including directing, coaching, supporting, and delegating.

4. When you communicate with your team verbally, be sure to _____.
   a. assess your audience before you speak
   b. use technical language as much as possible
   c. speak quickly to save time
   d. assume everyone has the same basic understanding of the topic

5. The *first* step in problem solving should be _____.
   a. choosing a solution to the problem
   b. evaluating the solution to the problem
   c. defining the problem
   d. testing the solution to the problem

6. Which of these is a way to lower the motivation and morale of your employees and team?
   a. Focusing on task completion only, not professional growth
   b. Giving recognition and rewards
   c. Setting clear expectations and goals
   d. Providing opportunities for training and upskilling

7. During an employee's first week of work on your crew, you should be sure to set aside time to _____.
   a. eat lunch together
   b. do a task verification
   c. quiz the employee on company policies
   d. evaluate social engagement

8. Which of these steps is important when delegating work to an individual or a team?

   a. Letting individuals set their own standards of measurement
   b. Checking in only after the work is complete
   c. Ensuring crew members understand what to do and have the skill to do it
   d. Allowing the team to set their own deadlines and get their own resources

9. Which common jobsite documents can help a crew leader overcome a language barrier with a worker?

   a. OSHA incident records
   b. Contracts and schedules
   c. Plans and drawings
   d. Permits

10. As a leader, you can help to manage performance issues by _____.

    a. ignoring the problem that needs resolving
    b. gathering information and preparing well for the employee conversation
    c. discussing the performance issues publicly with the rest of the team
    d. having the conversation when the employee is unprepared

11. The negative impact of incidents may include all of the following *except* _____.

    a. lower workers' compensation costs
    b. damage to the overall morale of everyone on a project or jobsite
    c. increased project costs to the company
    d. project delays

12. OSHA was established in 1970 to _____.

    a. collect dues for membership
    b. ensure safe and healthful working conditions for workers
    c. manage workers' compensation claims
    d. conduct medical research

13. Which of the following statements regarding safety training sessions is *correct*?

    a. The project manager usually holds them.
    b. They are held only for new employees.
    c. They should be conducted frequently by a frontline leader.
    d. They are required only after an accident has occurred.

14. A key step in an incident investigation is to determine the incident's _____.

    a. root cause
    b. secondary driver
    c. cost impacts
    d. design flaws

15. If you observe an employee showing impaired behavior or signs of mental health issues, you should work with _____.

    a. OSHA
    b. the safety inspector
    c. the owner or client
    d. your supervisor and/or Human Resources

16. What is completed during the design phase to ensure a project can be delivered successfully?

    a. Feasibility study
    b. As-built drawings
    c. Installation work packages
    d. Project schedule

17. A contract in which the client pays the contractor for their labor, materials, equipment, and other costs they incur during the performance of the contract is known as a _____.

    a. fixed-price contract
    b. time-spent contract
    c. time and materials contract
    d. cost-reimbursable contract

18. Which of the following would be tracked in a job diary?

    a. Job changes, interruptions, and visits
    b. The crew's time off
    c. Training needs and career development
    d. Personal information provided by an employee

19. Which of the following is a source of data in the field-reporting system?

    a. The cost-estimate for the project
    b. The daily job diary
    c. The Level 1 schedule
    d. The lookahead schedule

20. To prevent job delays due to late delivery of materials, the crew leader should _____.

    a. demand a discount from the supplier to compensate for the delay
    b. tell a crew member to go look for the delivery truck
    c. refuse the late delivery and re-order the materials from another supplier
    d. check with the supplier in advance of the scheduled delivery

Answers to Section Review Questions

Answer	Section	Objective
Section One		
1. b	1.1.2	1a
2. b	1.2.2	1b
Section Two		
1. d	2.1.2	2a
2. b	2.2.3	2b
3. a	2.3.0	2c
Section Three		
1. b	3.1.1	3a
2. c	3.2.0	3b
3. c	3.3.3	3c
Section Four		
1. a	4.0.0	4a
2. d	4.2.0	4b
3. a	4.3.1	4c
4. b	4.4.2	4d
5. d	4.5.0	4e
Section Five		
1. a	5.1.1	5a
2. d	5.2.2	5b
3. b	5.3.1	5c
4. b	5.4.1	5d

APPENDIX A Odd-Numbered Module Review Answers

MODULE 01 (27213)

Answer	Section Head
1. d	1.0.0
3. b	1.1.0
5. b	1.2.0
7. c	2.1.0
9. d	2.2.0
11. d	2.2.0
13. b	3.1.2
15. c	3.1.3

MODULE 02 (27402)

Answer	Section Head
1. b	1.1.0
3. a	1.2.2
5. d	1.2.3
7. c	1.2.4
9. c	2.1.3
11. a	3.1.2
13. d	4.1.3
15. a	4.2.3

MODULE 03 (27404)

Answer	Section Head
1. a	1.1.0
3. c	1.1.0
5. b	2.1.0
7. a	2.2.0
9. d	2.3.2
11. b	2.3.7
13. b	2.3.5
15. a	2.4.1

MODULE 04 (27202)

Answer	Section Head
1. c	1.1.0
3. c	1.2.2; *Figure 3*
5. c	1.2.4
7. c	2.1.0
9. a	2.3.1
11. c	3.1.0
13. b	3.1.0
15. c	3.3.0
17. a	3.5.0
19. c	3.5.0
21. a	3.7.0
23. a	4.2.2
25. c	4.2.6
27. d	4.3.1
29. b	4.3.3
31. b	4.4.2
33. c	4.4.6
35. b	5.0.0

MODULE 05 (27205)

Answer	Section Head
1. b	1.2.0
3. a	1.2.1
5. c	1.2.3
7. c	1.3.3
9. a	1.3.6
11. d	2.3.0
13. a	2.3.5

MODULE 06 (27203)

Answer	Section Head
1. a	1.0.0
3. c	1.1.1
5. b	1.2.0
7. a	1.2.2
9. c	2.0.0; *Table 5*
11. d	2.1.2
13. a	2.4.1
15. b	2.4.3
17. a	3.0.0
19. c	3.1.1
21. a	3.2.1
23. b	3.3.0
25. b	4.0.0
27. c	4.2.0
29. c	4.2.1; *Table 6*
31. d	5.1.0
33. b	5.3.2
35. a	5.3.2

MODULE 07 (45104)

Answer	Section Head
1. d	1.1.0
3. b	1.1.1
5. d	1.2.1
7. a	1.2.2
9. c	1.3.0
11. b	2.1.0
13. c	2.1.4
15. a	2.2.0
17. d	2.4.0
19. b	2.4.2

MODULE 08 (27215)

Answer	Section Head
1. a	1.1.0
3. b	1.1.3
5. b	2.1.2
7. d	2.1.3
9. d	3.1.1
11. d	3.1.2
13. c	3.2.0
15. c	3.2.1
17. b	3.3.0
19. a	4.1.1
21. c	5.1.2
23. c	6.3.1

MODULE 09 (27208)

Answer	Section Head
1. b	1.1.0
3. c	2.1.0
5. a	2.2.4
7. b	2.3.0
9. c	3.1.5
11. d	3.2.1
13. c	4.2.5

MODULE 10 (27209)

Answer	Section Head
1. b	1.0.0
3. a	1.0.0; *Table 1*
5. b	1.1.3
7. d	1.2.1
9. c	1.2.1
11. a	1.3.2
13. d	2.1.0
15. c	3.2.0
17. c	3.2.2
19. b	3.4.2

MODULE 11 (27405)

Answer	Section Head
1. c	1.1.1
3. a	1.1.2
5. d	1.1.4
7. b	1.2.1
9. b	1.3.1
11. b	2.1.0
13. a	2.1.0
15. c	2.2.0
17. c	3.1.0
19. b	3.1.0

MODULE 12 (27204)

Answer	Section Head
1. b	1.1.0
3. c	1.2.3
5. d	2.3.1
7. c	2.3.3
9. b	3.0.0
11. a	3.2.4
13. b	3.4.0
15. a	3.5.0

MODULE 13 (27214)

Answer	Section Head
1. b	1.0.0
3. c	1.1.0; *Table 1*
5. a	1.2.0
7. a	1.2.0
9. b	1.3.4
11. d	1.3.6, 1.3.7
13. c	2.1.2
15. d	2.2.4
17. b	3.1.2
19. b	3.2.2

MODULE 14 (46100)

Answer	Section Head
1. b	1.1.1
3. d	2.1.3
5. c	2.3.1
7. b	3.1.2
9. c	3.2.5
11. a	4.1.0
13. c	4.3.1
15. d	4.5.2
17. c	5.1.2
19. b	5.3.1

APPENDIX B

Common Terms Used in Cold-Formed Steel Framing Work

AISC: American Institute of Steel Construction.

AISI: American Iron and Steel Institute.

Base steel thickness: The thickness of bare steel exclusive of all coatings.

Bracing: Structural elements that are installed to provide restraint or support (or both) to other framing members so that the complete assembly forms a stable structure.

Ceiling joist: A horizontal structural framing member that supports ceiling components and may be subject to attic loads.

Cold-formed sheet steel: Sheet steel or strip steel that is manufactured by press braking of blanks sheared from sheets or cut lengths of coils or plates, or by continuous roll forming of cold- or hot-rolled coils of sheet steel. Both forming operations are performed at ambient room temperature; that is, without any addition of heat such as would be required for hot forming.

Cold-formed steel: See *cold-formed sheet steel*.

Component assembly: A fabricated assemblage of cold-formed steel structural members that is manufactured by the component manufacturer. The component assembly may also include structural steel framing, sheathing, insulation, or other products.

Component design drawing: The written, graphic, and pictorial definition of the engineering design data of an individual component.

Component designer: The individual or organization responsible for the engineering design of component assemblies.

Component manufacturer: The individual or organization responsible for the manufacturing of component assemblies for the project.

Component placement diagram: The illustration supplied by the component manufacturer identifying the location assumed for each of the component assemblies, which references each individually designated component design drawing.

Cripple stud: A stud that is placed between a header and a window or door head track, a header and a wall top track, or a windowsill and a bottom track to provide a backing to attach finishing and sheathing material.

Design thickness: The steel thickness used in design that is equal to the minimum base steel thickness divided by 0.95.

Edge stiffener: That part of a C-shape framing member that extends perpendicularly from the flange as a stiffening element.

Erection drawings: See *installation drawings*.

Erector: See *installer*.

Flange: That portion of the C-shape framing member or track that is perpendicular to the web.

Floor joist: A horizontal structural framing member that supports floor loads and superimposed vertical loads.

Framing contractor: See *installer*.

Framing material: Steel products, including but not limited to structural members and prefabricated structural assemblies, ordered expressly for the requirements of the project.

General contractor: See *installer*.

Harsh environments: Coastal areas where additional corrosion protection may be necessary.

In-line framing: Framing method where all vertical and horizontal load-bearing members are aligned.

Installation drawings: Field installation drawings that show the location and installation of the cold-formed steel structural framing.

Installer: Party responsible for the installation of cold-formed steel products.

Jack stud: A stud that does not span the full height of the wall and provides bearing for headers. Also called a *trimmer stud*.

King stud: A stud, adjacent to a jack stud, that spans the full height of the wall and supports vertical and lateral loads.

Stiffening Lips: See *edge stiffener*.

Material supplier: An individual or entity responsible for furnishing framing materials for the project.

Nonstructural member: A member in a steel-framed assembly that is limited to a transverse load of not more than 10 lb/ft^2 (480 Pa); a superimposed axial load, exclusive of sheathing materials, of not more than 100 lb/ft (1,460 N/m); or a superimposed axial load of not more than 200 lb (890 N).

Punchout: A hole made during the manufacturing process in the web of a steel framing member.

Shop drawings: Drawings to produce individual component assemblies for the project.

Span: The clear horizontal distance between bearing supports.

Standard cold-formed steel structural shapes: Cold-formed steel structural members that meet the requirements of the *SSMA Product Technical Guide*.

Strap: Flat or coil sheet steel material typically used for bracing and blocking that transfers loads by tension and/or shear.

Structural engineer-of-record: The design professional who is responsible for sealing the contract documents, which indicates that the structural engineer-of-record has performed or supervised the analysis, design, and document preparation for the structure and has knowledge of the requirements for the load-bearing structural system.

Structural member: A floor joist, rim track, structural stud, wall track in a structural wall, ceiling joist, roof rafter, header, or other member that is designed or intended to carry loads.

Structural stud: A stud in an exterior wall or an interior stud that supports superimposed vertical loads and may transfer lateral loads, including full-height wall studs, king studs, jack studs, and cripple studs.

Stud: A vertical framing member in a wall system or assembly.

Trimmer: See *jack stud*.

Truss: A coplanar system of structural members joined together at their ends, usually to construct a series of triangles that form a stable beam-like framework.

Yield strength: A characteristic of the basic strength of the steel material defined as the highest unit stress that the material can endure before permanent deformation occurs, as measured by a tensile test in accordance with *ASTM A370, Standard Test Methods and Definitions for Mechanical Testing of Steel Products*.

GLOSSARY

Accomplishment: An individual's need to set challenging goals and achieve them.

Accountability: In relation to job completion, accepting responsibility and being personally answerable for the results.

Acoustical materials: Types of ceiling panel, plaster, and other materials that have high absorption characteristics for sound waves.

Acoustics: A science involving the production, transmission, reception, and effects of sound. In a room or other location, it refers to those characteristics that control reflections of sound waves and thus the sound reception in the area.

AED (automated external defibrillator): A device used to help those experiencing sudden cardiac arrest.

Air barrier: One or more materials joined together continuously to prevent or restrict the passage of air through a building's thermal envelope and assemblies.

Alidade: The upper part of a transit level or theodolite that contains the telescope and related components.

Annular nails: Nails with rings around the shank, which provide a stronger grip and higher withdrawal resistance than nails with smooth shanks.

Apron: A piece of window trim, sometimes known as *undersill trim*, that is installed under the stool of a finished window frame.

As-built drawings: A revised set of drawings that reflect all changes made during the construction process and show the exact specifications of the completed work.

Asphalt roofing cement: An adhesive that is used to seal down the free tabs of strip shingles. This plastic asphalt cement is mainly used in open valley construction and other flashing areas where necessary for protection against the weather.

Astragal: A piece of molding attached to the edge of an inactive door on a pair of double doors. It serves as a stop for the active door.

Authority: The right to act or make decisions in carrying out an assignment.

A-weighted decibel (dBA): A single number measurement based on the decibel but weighted to approximate the response of the human ear with respect to frequencies.

Axis: An imaginary line or plane.

Backing: A flat, horizontal wood or metal member that provides a supportive attachment surface between stud bays.

Backsight (BS): A reading taken on a leveling rod held on a point of known elevation to determine the height of the leveling instrument.

Backsplash: A protective and decorative wall covering typically installed behind kitchen and bathroom sinks, stoves, and countertops to prevent water damage to the wall surface.

Balustrade system: A collective term that refers to the newels, balusters, and handrail(s) on a stairway.

Base flashing: The protective sealing material placed next to areas vulnerable to leaks, such as chimneys.

Batt: A flat, pre-cut piece of insulation.

Bead: An application of adhesive or other construction material in a sphere or line not less than $3/8$" in diameter.

Behavior: The way a person responds to a specific situation.

Benchmark (BM): A relatively permanent object with a known elevation located near or on a site. It can be iron stakes driven into the ground, a concrete monument with a brass disk in the middle, a chiseled mark at the top of a concrete curb, or similar items.

Blocking: A standard track, brake shape, or flat strap attached to structural members or sheathing panels to transfer shear forces.

Braced wall: A nonbearing exterior wall that is set into, and attached to, the structure of a building. This type of wall is designed to strengthen walls and resist lateral forces.

Breaking the tape: Making measurements using a portion of the full tape's length in a series of steps until the full tape length has been traversed.

Bundle: A package containing a specified number of shingles. The number is related to square-foot coverage and varies with the product.

Calcination: The process of heating gypsum rock enough to evaporate most of the water in its molecular structure, causing a chemical change in the material.

Cap flashing: The protective sealing material that overlaps the base and is embedded in the mortar joints of vulnerable areas of a roof, such as a chimney.

Casing: A type of drywall trim that is used around windows and doors.

Cavity insulation: Insulating materials that are located between framing members.

Ceiling panels: Acoustical ceiling boards that are suspended by a concealed grid mounting system. The edges are often kerfed and cut back.

Ceiling tiles: Any lay-in acoustical boards designed for use with exposed grid mounting systems. Ceiling tiles normally do not have finished edges or precise dimensional tolerances because the exposed grid mounting system provides the trim-out.

Centerline cracking: A crack in a finished drywall joint that can occur as the result of environmental conditions or poor workmanship.

Climate zone: A geographical region based on climate criteria, as determined by the *IECC*®.

Clip angle: An L-shaped piece of steel (normally with a 90-degree bend), typically used for connections.

Cold-formed steel: Sheet steel or strip steel that is manufactured by press braking of blanks sheared from sheets or cut lengths of coils or plates, or by continuous roll forming of cold- or hot-rolled sheet steel coils.

Collated magazine: An attachment for a screw or nail gun that automatically feeds collated fasteners (fasteners arranged side by side on a strip of plastic) into the chamber of the gun for quick application.

Communication: The process of exchanging facts, information, and ideas. Communication comes in many forms, including verbal, written, visual, and nonverbal.

Condensation: The process by which a vapor is converted to a liquid, such as the conversion of the moisture in air to water.

Construction safety plan: A document that outlines the procedures, rules, and regulations that are or will be put in place to protect workers during a construction project.

Continuous insulation: Insulation that runs over a building's structural members seamlessly without breaks or gaps.

Control joints: Deliberate gaps left between long stretches of gypsum panels to allow them to expand, contract, or shift without cracking.

Control points: Specific points created on the site, where they are used as elevation and dimension reference points during site and building layout.

Convection: The movement of heat that either occurs naturally due to temperature differences or is forced by a fan or pump.

Coordinators: Devices used with exit features to hold active doors open until inactive doors are closed.

Coped joint: A joint made by cutting the end of a piece of molding to fit seamlessly to the contoured profile face of adjoining molding.

Corner bead: A metal or plastic angle used to protect and finish outside corners where drywall panels meet.

Cornice: A functional and decorative form of trim or molding located where the exterior walls and roof meet.

Cost control: The process that helps keep expenses under control by managing labor, material, and other costs to ensure that the project finishes on budget.

Countersink: To drive a nail or screw through a gypsum board and into the framing member until the head of the fastener rests just below the surface, without breaking the face paper.

Cove molding: A decorative strip that, when attached to the underside of the tread nosing, covers the joint between the tread and riser.

CPR (cardiopulmonary resuscitation): A lifesaving technique that is used when someone's heartbeat or breathing has stopped.

Crew leader: A person appointed leader of a team or crew, usually an experienced craft professional who has demonstrated leadership qualities. Also known as *front-line leader, frontline supervisor, lead person*, or *foreman*.

Critical path: In manufacturing, construction, and other types of creative processes, the required sequence of tasks that directly controls the ultimate completion date of the project.

Crosshairs: A set of lines, typically horizontal and vertical, placed in a telescope used for sighting purposes.

Cupped-head nails: Nails with a concave head (shaped like a cup) and a thin rim.

Curtain wall: A light, nonbearing exterior wall attached to the concrete or steel structure of the building. Also, a nonbearing exterior wall that is set into, and attached to, the steel or concrete structure of a building.

Decibel (dB): An expression of the relative loudness of sounds in air as perceived by the human ear.

Decision-making: The action or process of making a choice. Decision-making involves starting an action, stopping one, or choosing a different plan or action.

Dew point: The temperature at which air becomes oversaturated with moisture and the moisture condenses.

Diaphragm: A floor, ceiling, or roof assembly designed to resist in-plane forces such as wind or seismic loads.

Differential leveling: A method of leveling used to determine the difference in elevation between two points.

Diffuser: An attachment for duct openings in air distribution systems that distributes the air in wide flow patterns. In lighting systems, it is an attachment used to redirect or scatter the light from a light source.

Diffusion: The movement, often contrary to gravity, of molecules of gas in all directions, causing them to intermingle.

Distractions: Anything that disrupts the communication flow between sender and receiver. Also called *noise*.

Dry lines: A string line suspended from two points and used as a guideline when installing a suspended ceiling.

Duty: A work task an employee is expected to perform.

Duty rating: Load capacity of a ladder.

Earthwork: All construction operations connected with excavating (cutting) or filling earth.

Eave: The projecting edge of a roof, extending horizontally beyond the walls of a structure.

Edge: The paper-bound edge of a gypsum board as manufactured.

Elevations: Vertical distances above a datum point. For leveling purposes, a datum is normally based on the ocean's mean sea level (MSL).

End: The side of a gypsum board perpendicular to the paper-bound edge. The gypsum core is always exposed.

Estimating: The process of calculating the cost of a project.

Ethics: The moral principles that guide an individual's or organization's actions when dealing with others.

Exposure: The distance (in inches) between the exposed edges of overlapping shingles.

Exterior insulation finish system (EIFS): Nonstructural, nonbearing, exterior wall cladding system that consists of an insulation board attached adhesively or mechanically, or both, to the substrate (*IBC*® 2021).

Exterior wall envelope: A system or assembly of exterior wall components, including exterior wall covering materials, that provides protection of the building structural members, including framing and sheathing materials, and conditions interior spaces from the detrimental effects of the environment (*IBC*® 2021, Section 202).

Extruded: The forming of desired shapes by forming or pressing a material through a shaped opening.

Face paper: The paper bonded to the surface of gypsum board during the manufacturing process.

Fascia: A horizontal board that encloses and protects the front face of projecting eaves.

Fastener pops: Protrusions of nails or screws above the surface of a gypsum board, usually caused by shrinkage of wood framing or by incorrect board installation.

Feasibility study: An analysis that considers all of a project's relevant factors—including budget, technical elements, legal aspects, and scheduling—to determine the likelihood that the project can be completed successfully.

Fiberglass insulation: A type of batt, roll, or loose-fill insulation that is made of extremely small pieces (or fibers) of glass.

Fiberglass: A material made of sand and recycled glass.

Field: The inner area of a gypsum panel.

Field notes: A permanent record of field measurement data and related information.

Finger-jointed stock: Paint-grade moldings made in a mill from shorter lengths of wood joined together.

Finishing: The application of joint tape, joint compound, corner bead, and primer or sealer onto gypsum board in preparation for the final decoration of the surface.

Fire blocking: Building materials, or materials approved for use as fire blocking, installed to resist the free passage of flame to other areas of the building through concealed spaces (*IBC*® 2021, Section 202).

Fire-resistance rating (FRR): The period of time a building element, component, or assembly maintains the ability to confine a fire, continues to perform a given structural function, or both, as determined by the tests, or methods based on tests (*IBC*® 2021, Section 202).

Fire-resistance-rated assembly: Construction built with certain materials in a certain configuration that has been shown through testing to restrict the spread of fire.

First aid: Help given to a sick or injured worker on the worksite immediately after an injury occurs.

Fissured: A ceiling-panel or ceiling-tile surface design that has the appearance of splits or cracks.

Flange: The rim of an accessory used to attach it to another object or surface.

Flashing: Thin, water-resistant material that prevents water seepage into a building and directs the flow of moisture in walls.

Flexible insulation: A type of insulation that is made from a flexible material.

Floating angle method: A drywall installation technique used with wood stud framing in which no fasteners are used where ceiling and wall panels intersect in order to allow for structural stresses.

Foresight (FS): A reading taken on a leveling rod held on a point in order to determine a new elevation.

Frequency: Cycles per unit of time, usually expressed in hertz (Hz).

Fur out: To attach furring strips or furring channels to masonry walls, wood framing, or steel framing to create a level surface before applying gypsum board.

Furring channel: A long, narrow piece of metal bent into the shape of a hat (which is why it is also called a *hat channel*), with two flanges (the brim of the hat) on either side of a channel (the crown of the hat), used to create a level surface on uneven masonry or metal framing in preparation for installing gypsum board.

Furring strip: A flat, narrow piece of wood attached to wood framing to create a level surface in preparation for installing gypsum board.

Gable: The triangular section of a wall that is enclosed by the edges of a pitched roof.

Glue block: A block of wood that is attached to the hidden side of a tread and riser joint. This is done for structural soundness and to eliminate squeaking.

Gypsum board: A generic term for paper-covered panels with a gypsum core; also known as *gypsum drywall*.

Hanging stile: The door stile to which the hinges (butts) are fastened.

Harassment: Unwelcome conduct that is based on race, color, religion, sex, national origin, age, disability, or genetic information. Harassment includes verbal or physical conduct that impacts a person's work performance or creates an intimidating, hostile, or offensive work environment.

Hazards: Conditions or activities that, if left uncontrolled, have the potential for harm.

Header: A horizontal structural framing member that supports and transfers weight loads over openings in floors, roofs, or walls. The most common location for headers is above windows and doors.

Heat conduction: The process by which heat is transferred through a material, which is caused by a difference in temperature between two areas.

Height of the Instrument: Elevation of the line of sight of the telescope relative to a known elevation. Determined by adding the backsight elevation to the known elevation.

Hertz (Hz): A unit of frequency equal to one cycle per second.

High-pressure laminate (HPL): Composite material produced by pressing layers of plastic material under intense heat and pressure.

Hypotenuse: The longest side of a right triangle. It is always opposite from the right angle.

Impairment: The point where someone's intake of substances affects their ability to perform appropriately.

Incidents: Unplanned events that result in property damage, fatalities, injuries, illnesses, or near misses.

Inside corners: Locations where two walls meet and face each other.

Invert the telescope: To reverse the direction of an instrument telescope around its horizontal axis.

Job description: A description of the scope and responsibilities of a worker's job so that the individual and others understand what the job entails.

Job diary: A written record that a supervisor maintains periodically (usually daily) of the events, communications, observations, and decisions made during the course of a project.

Job safety analysis (JSA): A technique that focuses on job tasks as a way to identify hazards before they occur. It focuses on the relationship between the worker, the task, the tools, and the work environment. Also known as *job hazard analysis (JHA)* or *activity hazard analysis (AHA)*.

Joint compound: A mixture of gypsum, clay, and resin applied wet during the finishing process to the taped joints between gypsum boards to cover the fasteners, and to the corner bead and accessories to create the illusion of a smooth unbroken surface. Sometimes called *mud* or *taping compound*.

Joint tape: Wide tape applied to the joints between gypsum boards and then covered with joint compound during the finishing process.

Joints: Places where two pieces of wallboard meet.

Kerf: A slot or cut made with a saw.

Knurled: Having a series of small ridges that provide a better gripping surface on metal and plastic.

Kraft-faced insulation: Insulation that has a paper, vinyl, or foil vapor retarder on one side.

Lateral forces: Forces exerted on the sides of a building due to soil loads, wind, or seismic activity.

Lateral: Running side to side; horizontal.

Least count: The finest reading that can be made directly on a vernier of a transit level or micrometer of a theodolite.

Level 4 schedules: A schedule that offers additional detail to provide working schedules to the crews doing the work, usually within three to six weeks of completing specific work.

Lookahead schedule: For planning and scheduling work that is happening in the short term, usually within one to three weeks; used for anticipating material, labor, tool, and other resource requirements, as well as identifying potential schedule conflicts or other problems. Also called a *short-term schedule*.

Lookouts: Joists that extend past the exterior wall of a structure to support roof loads and provide nailing surfaces for fascia.

Loose-fill insulation: Insulation that comes in the form of loose material in bags or bales.

Mineral wool insulation: A type of batt or loose-fill insulation that is made of natural stone fibers or slag.

Moderate contact: The contact between the edges and ends of abutting gypsum boards in a wall or ceiling assembly, which should not be tight or too widely spaced.

Molding: Material used for decorative trim.

Moldings: A defined, decorative element that outlines the edges of a structure and covers gaps at intersections of walls, ceilings, floors, and openings.

Morale: An individual's emotional outlook toward work and the level of satisfaction gained while performing the jobs assigned.

Mortise lockset: A rectangular metal box that houses a lock. It usually has a latch and deadbolt as part of the unit. Options consist of a locking cylinder and/or thumb turn by which it can be secured. It is used on residential entry doors and commercial doors.

Motivate: To make someone feel determined or enthusiastic or to make someone behave in a particular way.

Movement joint: A type of joint that allows a building to move or relieve movement when weather or temperature changes cause any type of structural movement.

Multi-ply construction: A wall or ceiling installation built with more than one layer of gypsum board.

Non-rated assembly: A ceiling or wall assembly that does not exhibit enough fire-resistant properties to qualify for a fire-resistance rating.

Nonverbal communication: Things the speaker and the receiver see or hear while communicating. Examples include facial expressions, body movements, hand gestures, tone of voice, and eye contact.

Occupational Safety and Health Administration (OSHA): A national public health agency that is part of the US Department of Labor. To promote a safe and healthy work environment, OSHA issues standards and rules for working conditions, facilities, equipment, tools, and work processes.

On center (OC): The distance between the center of one framing member or fastener to the center of an adjacent framing member or fastener.

Opening cap: A handrail fitting at the start of a level balustrade system.

Optical plummet: A device incorporated into a transit level, theodolite, or similar instrument that allows the operator to sight a point below that is exactly plumb with the center of the instrument. This enables quick and accurate setup of the instrument over a point.

Organization: A group of people working together.

Organizational chart: A diagram that shows how the various management and operational responsibilities relate to each other within an organization. Named positions are ranked top to bottom, from those functional units with the most and broadest authority to those with the least. Lines connect positions and indicate chains of authority and other relationships.

Outside corners: Locations where two walls meet and face away from each other.

Overhang: The part of a structure that extends beyond the building line. The amount of overhang is always given as a projection from the building line on a horizontal plane.

Over-the-post: A balustrade system that utilizes fittings to go over newel posts for an unbroken, continuous handrail.

Panelization: The process of assembling steel-framed walls, joists, or trusses before they are installed in a structure.

Parallax: The apparent movement of the crosshairs in a surveying instrument caused by movement of the eyes.

Parallel installation: Applying gypsum board so that the edges are parallel to the framing members, meaning the board is oriented vertically.

Peer: Someone who is in a position or level in an organization equal or similar to yours.

Peg test: A procedure used to check for an out-of-adjustment bubble vial on levels and other instruments.

Perimeter relief: A gap left between a ceiling assembly and a wall assembly to keep the two assemblies separate and to allow the gypsum board in the ceiling assembly to move freely.

Perimeter: The outer boundary of a gypsum panel.

Perm: The measure of water vapor permeability. It equals the number of grains of water vapor passing through a 1 ft^2 piece of material per hour, per inch of mercury difference in vapor pressure.

Permeability: The measure of a material's capacity to allow the passage of liquids or gases.

Permeance: The ratio of water vapor flow to the vapor pressure difference between two surfaces.

Perpendicular installation: Applying gypsum board so that the edges are at right angles to the framing members, meaning the board is oriented horizontally.

Personal growth: Improving one's talents and potential, both in and out of the workplace.

Pitch: The number of threads per inch on a screw shank, with more threads producing a finer pitch and fewer threads producing a coarser pitch. Also, the ratio of the rise to the span, indicated as a fraction. For example, a roof with a 6' rise and a 24' span will have a $1/4$ pitch.

Pitch block: A triangular piece of wood used for laying out the rise and run on stringers, and when trimming easement fittings.

Planning: The process of laying out all details of a construction project before construction begins.

Plenum: A chamber or container for moving air under a slight pressure. In commercial construction, the area between the suspended ceiling and the floor or roof above is often used as the HVAC return air plenum. Also, an enclosed space, such as the space between a suspended ceiling and an overhead deck, which is used as a return for heating, ventilating, and air conditioning (HVAC) systems.

Policy: A general statement establishing guidelines for a specific activity.

Post-to-post: Balustrade system where the handrail is not continuous. The handrail is lagged into the face of a square-top newel.

Powder load: A crimped and sealed metal casing that contains a powderized propellant.

Problem-solving: A process used to identify possible solutions to a situation.

Procedures: Formal instructions to carry out and meet policies.

Production: The amount of construction materials put in place. It is the quantity of materials installed on a job.

Productivity rate: A measure of how much work gets done over a period of time, also called *output per labor hour*.

Professional growth: Gaining new skills and work experience that can help one to reach a goal in one's career.

Profit: The amount of money that the contractor will make after paying all the direct and indirect costs.

Project delivery methods: Different ways to organize the services of development, planning, and construction in order to execute a project.

Project manager: Person responsible for managing one or more projects.

Pythagorean theorem: A geometric theorem for right triangles stating that the sum of the squares of the legs of a right triangle is equal to the square of the hypotenuse. It is expressed mathematically as $a^2 + b^2 = c^2$

Racking: Being forced out of plumb by wind or seismic forces.

Radiation: Energy emitted from a source in electromagnetic waves or subatomic particles.

Recognition: Acknowledgement of an employee's accomplishments through verbal and nonverbal communications.

Redline drawings: Marked-up versions of construction drawings documenting any changes made during construction or installation.

Reflective insulation: A type of insulation made of outer layers of aluminum foil bonded to inner layers of various materials.

Resilient furring channel: A furring channel with one flange to attach to another surface, leaving the other side of the channel unattached.

Responsibility: The act of accepting a task and completing it to the best of one's ability.

Returns: A pair of components of a C-shaped stud that extend perpendicularly from the flanges as stiffening elements. These may also be called *stiffening lips*.

Reveal: The distance that the edge of a casing is set back from the edge of a jamb.

Rewards: Tangible items given to an employee as recognition for good work.

Ridge: The horizontal line formed by the two rafters of a sloping roof that have been nailed together. The ridge is the highest point at the top of the roof where the roof slopes meet.

Ridging: A defect in finished gypsum board drywall caused by environmental or workmanship issues.

Rigid or semi-rigid insulation: A type of insulation that comes in formed boards made of mineral fibers.

Rim tracks: Horizontal structural members that connect to the ends of floor joists.

Roof rafter: A horizontal or sloped structural framing member that supports roof loads.

Roof sheathing: Usually 4 × 8 sheets of plywood, but can also be 1 × 8 or 1 × 12 roof boards, or other new products approved by local building codes. Also referred to as *decking*.

Root cause: A core issue or fundamental reason for the occurrence of a problem.

Root cause analysis (RCA): A method of problem-solving used for identifying the root causes of faults or problem.

R-value: A measure of an object or material's amount of thermal resistance.

Saddle: An auxiliary roof deck that is built above the chimney to divert water to either side. It is a structure with a ridge sloping in two directions that is placed between the back side of a chimney and the roof sloping toward it. Also referred to as a *cricket*.

Sanitary stop: A door stop with a 45-degree angle cut at the bottom of the vertical jambs.

Scarf joints: End joints made by overlapping two pieces of molding with angle cuts.

Scheduling: Involves defining which activities happen in what order and assigning dates and times to those tasks.

Scope: The list of the commitments and activities that all contractors, subcontractors, and suppliers are expected to do during a project. Also called *scope of work*.

Scrim: A loosely knit fabric.

Self-drilling screw: A screw with a point shaped like a drill bit to penetrate steel.

Self-piercing screw: A screw with a point sharp enough to penetrate steel.

Self-tapping screw: A screw that bores an internal screw thread in the material into which it is driven, creating a strong hold between the material and the screw.

Shear wall: A wall designed to resist lateral forces such as those caused by earthquakes or wind. Also, a wall that strengthens buildings by resisting lateral forces caused by soil loads, high wind, and/or seismic activity.

Silicosis: A life-threatening lung disease caused by inhaling large amounts of crystalline silica dust.

Single-ply construction: A wall or ceiling installation built with one layer of gypsum board.

Site-specific safety plan (SSSP): A safety plan that includes a focused set of information that accounts for a jobsite's unique hazards.

Situational leadership: A leadership style that involves adjusting to the specific situation or task and the leadership needs of the group.

Slope: A measurement of how much the ground varies from horizontal. Also, the ratio of rise to run. The rise in inches is indicated for every foot of run.

Soffit: A nonloadbearing material beneath eaves that encloses and protects the space between a roofline and exterior wall.

Sound attenuation: The reduction of sound as it passes through a material.

Sound transmission class (STC): A rating that describes how much sound a product or assembly prevents from getting through to the other side. The higher the STC, the more sound at common frequencies has been prevented from traveling through a product or assembly to an adjacent space. Also, a rating that measures the amount of sound that can pass through a door. The greater the rating number, the better the sound reduction.

Sound-rated assembly: Construction built with certain materials in a certain configuration that has been shown through testing to obtain specific acoustical performance.

Square: The quantity of shingles needed to cover 100 ft² of roof surface. For example, square means 10' square, or 10' × 10'.

Station: Instrument-setting locations in differential leveling.

Stool: The bottom horizontal trim piece of a window that lays flat above the apron, commonly known as a *windowsill*.

Striated: A ceiling-panel or ceiling-tile surface design that has the appearance of fine parallel grooves.

Structural insulated panels (SIPs): A type of wall system that consists of an insulated foam core sandwiched between two wood structural panels (WSPs), such as oriented strand board (OSB) or plywood.

Studs: Cold-formed steel structural and nonstructural framing members, consisting of a web, two flanges, and two returns.

Substance abuse: The inappropriate overuse of drugs and chemicals, whether they are legal or illegal.

Superintendent: An on-site supervisor who is responsible for one or more crew leaders or frontline supervisors.

Temporary benchmark: A point of known (reference) elevation determined from benchmarks through leveling that lasts for the duration of the project.

Thermal bridging: When a small area of floor, wall, or roof loses substantially more heat than the surrounding area.

Thermal resistance: A measure of how a material will resist the flow of heat energy.

Thread: The protruding rib of a screw that winds in a helix down its shank.

Three-way communication: A form of communication with an initial message from a sender, which is then repeated by the receiver to confirm accuracy, and then validated by the sender. Also called *repeat-back communication*.

Toolbox talks: Short, informal safety meetings held at the start of each day or shift. Sometimes referred to as *tailgate meetings* or *safety briefings*.

Tooth: A textured surface created mechanically to help a covering material adhere more effectively.

Track: A steel framing member consisting of two flanges and a web. Although similar in shape to studs, track has no returns. Track web depth is measured between the inside edges of the flanges.

Trait: A distinguishing quality of a person.

Transom: A panel above a door that lets light and/or air into a room or is used to fill the space above a door when the ceiling heights on both sides of the door opening allow.

Trim: Millwork placed around wall openings and at wall intersections at floors and ceilings.

Turning point (TP): A temporary point whose elevation is determined by differential leveling. The turning-point elevation is determined by subtracting the foresight reading from the height-of-instrument elevation.

U-factor: A measure of the total heat transmission through a wall, roof, or floor of a structure.

Underlayment: Asphalt-saturated felt protection for sheathing; 15 lb roofer's felt is commonly used. The roll size is 3' × 144', or a little over four squares.

Unfaced insulation: Insulation that does not have a vapor retarder.

Upskilling: A type of training that provides someone with more advanced skills through additional education.

Valley: The internal part of the angle formed by the meeting of two roofs.

Valley flashing: Watertight protection at a roof intersection. Various metals and asphalt products are used; however, materials vary based on local building codes.

Vapor permeable: Permitting the passage of moisture vapor. A vapor permeable material, as defined by the *IBC*®, has a moisture vapor permeance of 5 perms or greater.

Vapor retarder: A material used to retard the flow of vapor and moisture into walls and prevent condensation within them. The vapor retarder must be located on the warm side of the wall.

Vapor retarder class: A measure of a material's ability to limit the amount of moisture that passes through it.

Veneer: A thin layer or sheet of wood intended to be overlaid on a surface to provide strength, stability, and/or an attractive finish. Thicknesses range between $1/16$" and $1/8$" for core plies and between $1/128$" and $1/32$" for decorative faces.

Vent-stack flashing: Flanges that are used to tightly seal pipe projections through the roof. They are usually prefabricated.

Verbal communication: Messages that are sent using voice or other audible sounds. Also called *spoken communication* or *oral communication*.

Vernier: A short auxiliary scale set parallel to a primary scale.

Visual communication: The process of using visual elements, such as photos, graphics, colors, and fonts to convey messages.

Wainscoting: A decorative wall panel covering that protects lower elevations of an interior wall.

Wall flashing: A form of metal shingle that can be shaped into a protective seal interlacing where the roof line joins an exterior wall. Also referred to as *step flashing*.

Wall rail brackets: Metal supports for a wall rail.

Water vapor: Water in a vapor (gas) form, especially when below the boiling point and diffused in the atmosphere.

Water-resistive barrier: One or more materials installed behind exterior wall coverings to prevent water from entering a building.

Weather stripping: Strips of metal or plastic used to keep out air or moisture that would otherwise enter through the spaces between the outer edges of doors and windows and the finish frames.

Web: A portion of a framing member that connects the two flanges.

Wedges: Triangular-shaped wood pieces that are coated with adhesive and used to drive treads and risers tightly into a routed stringer.

Weephole: Channel that allows water that may enter the wall system to drain directly to the exterior.

Withdrawal resistance: The amount of resistance of a nail or screw to being pulled out of a material into which it has been driven.

Work breakdown structure (WBS): A method of organizing the entire scope of work in a logical manner based on how the work will be executed.

Workers' compensation: Insurance that companies have to cover medical expenses, lost wages, and other costs for employees who are injured or become ill while completing work for their job.

Written communication: The process of sending messages or information using words, whether digitally or physically.

Wythe: Single thickness of a masonry wall.

REFERENCES

27213 The Building Process

This module presents thorough resources for task training. The following resource material is suggested for further study.

ASHREA. *https://www.ashrae.org/*.

ASTM International. *https://www.astm.org/*.

Environmental Protection Agency. *https://www.epa.gov*.

Green Building Initiative. *https://thegbi.org/*.

LEED. *https://www.usgbc.org/leed*.

National Electrical Code®. Latest Edition. Quincy, MA: National Fire Protection Association.

International Building Code®. 2021. *https://codes.iccsafe.org/content/IBC2021P1*.

International Residential Code®. 2021. *https://codes.iccsafe.org/content/IRC2021P2*.

UL Solutions. *www.ul.com*.

27402 Site Preparation

This module presents thorough resources for task training. The following resource material is suggested for further study.

Surveying Fundamentals and Practices, Jerry A. Nathanson, Michael T. Lanzafama, Philip Kissam. 2017. Hoboken, NJ: Pearson.

Construction Staking: Step by Step Guide, Jim Curme. 2014, New York, NY: Elite Publishing Services.

EM 1110-1-1005, 1 January 2007. US Army Corps of Engineers. *Engineering and Design Control and Topographic Surveying*.

Setting Out for Construction, A Practical Guide to Site Surveying, Saffron Grant. 2019. London, UK Costello House Publishing.

Surveying and Levelling Volume II, S.S. Bhavikatti. 2016. 2nd Edition. New York, NY: International Publishing House.

27404 Wall Systems and Installations

This module presents thorough resources for task training. The following reference material is suggested for further study.

Building Code Essentials: Based on the 2021 International Building Code. 2021. Washington, DC: International Code Council.

Curtain Wall Systems: A Primer. 2013. Ali M. Memari, ed. ASCE Manuals and Reports on Engineering Practice, No. 126. Reston, VA: American Society of Civil Engineers.

Foundation and Anchor Design Guide for Metal Building Systems. 2014. Alexander Newman, PE, FASCE. New York, NY: McGraw Hill.

The Gypsum Construction Handbook, Seventh Edition. Kingston, MA: R.S. Means Company.

Principles and Practices of Heavy Construction, Tenth Edition. Hoboken, NJ: Pearson.

Residential Code Essentials: Based on the 2021 International Residential Code. 2021. Washington, DC: International Code Council.

UL Solutions. *www.ul.com*

27202 Roofing Applications

This module presents thorough resources for task training. The following resource material is suggested for further study.

Asphalt Roofing Manufacturers Association website. *www.asphaltroofing.org*.

National Roofing Contractors Association website. *www.nrca.net*.

OSHA Safety and Health Standards for the Construction Industry, Part 1926, Subpart M. www.osha.gov.

OSHA Safety and Health Standards for the Construction Industry, Part 1926, Appendices C and D to Subpart M. www.osha.gov.

Roof Coating Manufacturers Association website. *www.roofcoatings.org*.

27205 Steel Framing Systems

This module presents thorough resources for task training. The following resource material is suggested for further study.

American Iron and Steel Institute (AISI). Provides a variety of resources related to the use of cold-formed steel for construction applications, *http://www.steel.org*.

ASTM A370, Standard Test Methods and Definitions for Mechanical Testing of Steel Products. 2012. ASTM International.

National Electrical Code®. 2011. National Fire Protection Association.

National Fire Protection Association (NFPA). Works to reduce the worldwide burden of fire and other hazards by providing and advocating codes and standards, research, training, and education, *http://www.nfpa.org*.

Product Technical Guide. 2012. Steel Stud Manufacturers Association.

Standard for Cold-Formed Steel—General Provisions. 2001. American Iron and Steel Institute.

Steel Framing Alliance (SFA). An advocate of cold-formed steel structures, *https://www.steelframingalliance.com/*.

Steel Stud Manufacturers Association (SSMA). A manufacturers' trade group that promotes the use of steel framing members in structures, *http://www.ssma.com*.

Thermal Design and Code Compliance for Cold-Formed Steel Walls. 2008. Steel Framing Alliance.

27203 Thermal and Moisture Protection

This module presents thorough resources for task training. The following reference material is suggested for further study.

International Building Code®. 2021. International Code Council.

International Energy Conservation Code®. 2021. International Code Council.

International Residential Code®. 2021. International Code Council.

LEED Reference Guide for Building Design and Construction, v4. U.S. Green Building Council.

Standard 90.1-2022—Energy Standard for Sites and Buildings Except Low-Rise Residential Buildings. ANSI/ASHRAE/IES.

Standard 189.1-2020—Standard for the Design of High-Performance Green Buildings. ANSI/ASHRAE/ICC/USGBC/IES.

US Department of Energy website. *www.eere.energy.gov*.

45104 Drywall Installation

This module presents thorough resources for task training. The following reference material is suggested for further study.

The Gypsum Construction Handbook. Seventh Edition, 2014. Chicago: USG.

27209 Suspended and Acoustical Ceilings

This module presents thorough resources for task training. The following reference material is suggested for further study.

Ceiling and Interior Systems Construction Association. *https://www.cisca.org*.

International Building Code®. 2021. International Code Council.

International Residential Code®. 2021. International Code Council.

National Earthquake Hazards Reduction Program. FEMA. *https://www.nehrp.gov/*.

The Gypsum Construction Handbook. 7th edition. 2014. Chicago: USG.

US Geological Survey. *https://www.usgs.gov/*.

27405 Finished Stairs

This module presents thorough resources for task training. The following resource material is suggested for further study.

International Building Code®. 2021. International Code Council.

International Residential Code®. 2021. International Code Council.

27204 Exterior Finish and Trim

This module presents thorough resources for task training. The following resource material is suggested for further study.

Vinyl Siding Institute website. *www.vinylsiding.org*.

Cedar Shake & Shingle Bureau website. *www.cedarbureau.org*.

27214 Mass Timber Construction

This module presents thorough resources for task training. The following reference material is suggested for further study.

American Wood Council. *www.awc.org*.

Forest Service. U.S. Department of Agriculture. *www.fs.usda.gov*.

International Association for Mass Timber Construction. *www.iamtc.org*.

International Building Code®. 2021. International Code Council®.

Naturally:wood®. *www.naturallywood.com*.

Oregon Mass Timber Coalition. *www.masstimbercoalition.org*.

Think Wood®. *www.thinkwood.com*.

Woodworks®. *www.woodworks.org*.

46100 Introduction to Leadership

This module presents thorough resources for task training. The following reference material is suggested for further study.

Architecture, Engineering, and Construction Industry (AEC). *www.aecinfo.com*.

The Center for Leadership Studies. *www.situational.com*.

Conflict Resolution Playbook: Practical Communication Skills for Preventing, Managing, and Resolving Conflict, Jeremy Pollack. 2020. Rockridge Press.

Construction Industry Alliance for Suicide Prevention. *preventconstructionsuicide.com*.

The Constructor Construction Encyclopedia. *theconstructor.org*.

Crucial Conversations: Tools for Talking When Stakes Are High. Kerry Patterson, Joseph Grenny, Ron McMillan, and Al Switzler. 2002. Maidenhead, England: McGraw-Hill Contemporary.

Dare to Lead: Brave Work. Tough Conversations. Whole Hearts. Brene Brown. 2018. New York, NY. Random House Publishing. *www.penguinrandomhouse.com*.

Engineering News-Record. *www.enr.com*.

Equal Employment Opportunity Commission (EEOC). *www.eeoc.gov*.

The Five Dysfunctions of a Team, Lencioni, Patrick M. 2002. London, England: Jossey-Bass.

Mentoring for Craft Professionals, NCCER. 2016. New York, NY: Pearson Education, Inc.

National Association of Women in Construction (NAWIC). *www.nawic.org*.

National Census of Fatal Occupational Injuries (NCFOI). *www.bls.gov*.

National Institute of Occupational Safety and Health (NIOSH). *www.cdc.gov/niosh*.

National Safety Council. *www.nsc.org*.

Occupational Safety and Health Administration (OSHA). *www.osha.gov*.

Smart Choices: A Practical Guide to Making Better Decisions. John S. Hammond, Ralph L. Keeney, Howard Raiffa. 2015. Boston, MA. Harvard Business Publishing Education. *hbsp.harvard.edu*.

Society for Human Resources Management (SHRM). *www.shrm.org*.

Total Worker Health. *www.cdc.gov/niosh*.

United States Census Bureau. *www.census.gov*.

United States Department of Labor. *www.dol.gov*.